Electronic Principles

OTHER BOOKS BY ALBERT PAUL MALVINO

Digital Computer Electronics
Resistive and Reactive Circuits
Transistor Circuit Approximations
Electronic Instrumentation Fundamentals
Digital Principles and Applications (with D. Leach)
Experiments for Electronic Principles (with G. Johnson)

The entire truth about anything is almost impossible to tell within practical limits. No desert is absolutely dry, no liquid absolutely wet. No plain completely flat; no ball-bearing truly round.

Shidle

Electronic Principles

Second Edition

Albert Paul Malvino, Ph.D.

Foothill College
Los Altos Hills, California

West Valley College
Saratoga, California

Gregg Division

McGraw-Hill Book Company

New York St. Louis Dallas San Francisco Auckland
Bogotá Düsseldorf Johannesburg London Madrid
Mexico Montreal New Delhi Panama Paris
São Paulo Singapore Sydney Tokyo Toronto

TO JOANNA,

My brilliant and beautiful wife
without whom I would be nothing.
She always comforts and consoles,
never complains or interferes,
asks nothing and endures all,
and writes my dedications.

Library of Congress Cataloging in Publication Data

Malvino, Albert Paul.
    Electronic principles.

    Includes index.
    1. Electronics. I. Title.
TK7816.M25 1979   621.381   78–11374
ISBN 0–07–039867–4

Electronic Principles, Second Edition

   567890 DO DO 865432

The editors for this book were George J. Horesta and Cynthia Newby, the designer was Eileen Thaxton, the cover designer was Blaise Zito Associates, the art supervisor was George T. Resch, and the production manager was S. Steven Canaris. It was set in Baskerville by Kingsport Press, Inc.

Printed and bound by R. R. Donnelley & Sons Company.

contents

15. Decibels, Miller's Theorem, and Hybrid Parameters   379

15-1. Decibel power gain   15-2. Power gain of cascaded stages
15-3. Bel voltage gain   15-4. Miller's theorem   15-5. Second-approximation formulas
15-6. Hybrid parameters   15-7. Approximate hybrid formulas
15-8. Exact hybrid formulas

16. Frequency Effects   412

16-1. The lag network   16-2. Phase angle versus frequency
16-3. Bel voltage gain versus frequency   16-4. Dc-amplifier response
16-5. Risetime-bandwidth relation   16-6. High-frequency FET analysis
16-7. High-frequency bipolar analysis   16-8. The lead network
16-9. Ac-amplifier response   16-10. The bypass capacitor

17. Integrated Circuits   456

17-1. Making an IC   17-2. The differential amplifier
17-3. Ac analysis of a diff amp   17-4. Cascaded diff amps
17-5. The operational amplifier   17-6. A simplified op-amp circuit
17-7. Op-amp specifications   17-8. Slew rate and power bandwidth
17-9. Other linear ICs

18. Negative Feedback   495

18-1. The four basic feedback connections   18-2. SP closed-loop voltage gain
18-3. SP input and output impedances   18-4. Other benefits of SP negative feedback
18-5. The bandwidth of an SP feedback amplifier   18-6. The gain-bandwidth product
18-7. PP negative feedback   18-8. SS negative feedback   18-9. PS negative feedback
18-10. Summary

19. Positive Feedback   535

19-1. The basic idea   19-2. The phase-shift oscillator
19-3. The Wien-bridge oscillator   19-4. *LC* oscillators   19-5. Quartz crystals
19-6. Unwanted oscillations in amplifiers   19-7. Thyristors
19-8. The unijunction transistor

20. Voltage Regulation   571

20-1. Simple regulators   20-2. SP regulation   20-3. Early IC regulators
20-4. Three-terminal voltage regulators

21. Op-Amp Applications   588

21-1. Comparators   21-2. Amplifiers   21-3. Active diode circuits
21-4. Special amplifiers   21-5. The Miller integrator
21-6. The voltage-controlled oscillator   21-7. Active filters

preface

This second edition reflects suggestions from several reviews, dozens of interviews, and hundreds of questionnaires. Everybody wanted more linear ICs and less discrete material. Also asked for were chapter summaries, more FET applications, thyristors, optoelectronics, etc. To strengthen and modernize the book, therefore, I have added:

1. A chapter on op-amp applications covering comparators, amplifiers, active diode circuits, integrators, VCOs, and active filters.

2. A chapter on voltage regulators, emphasizing the new breed of three-terminal IC regulators.

3. Self-testing reviews at the end of each chapter, designed to get reader participation.

4. More linear IC material throughout the book, emphasizing circuit designs extensively used in linear ICs.

5. Earlier coverage of zener diodes and simple regulators.

6. Material on thyristors and optoelectronic devices.

To make room for all this new material, I shortened almost every chapter in the original edition, rewriting some chapters completely and revising others heavily. Where possible, I simplified the mathematics and emphasized the logical approach. The result is a faster moving book that gets you through discretes and into ICs as soon as possible.

Here are some specific changes. Chapters 4 and 5 now include a discussion of zener diodes and regulators. I simplified the mathematical derivations in Chap. 7 (biasing) and introduced the dc load line as a visual aid. Chapter 9 now has new sections on RC coupling, direct coupling, and other types of interstage coupling. In Chap. 10, I added material on thermal resistances, transistor derating, and heat sinking. Chapter 11 discusses the current mirror, a potent circuit used extensively in linear ICs. Also important, Chaps. 7 through 11 now contain multistage amplifiers showing how the CE, CC, and CB connections are used. Chapter 14 now has all kinds of FET applications. In Chap. 15, I expanded the discussion of h parameters, showing how to convert h values on data sheets into r parameters. Chapter 17 analyzes the internal circuitry of typical op amps and audio ICs; this not only reinforces earlier learning, it allows me to explain fully slew rate and power bandwidth, the two most important large-signal concepts in linear ICs. Chapter 19 now includes direct-coupled positive-feedback circuits like four-layer diodes, SCRs, and UJTs. I made Chaps. 20 and 21 completely new because of the demand for op-amp applications and IC regulators. Finally, Chap. 23 now contains a section on the phase-locked loop. Besides the foregoing, I made many other minor additions and deletions throughout the book.

As before, this is a book for a student taking a first course in linear electronics. The ideal prerequisites are a dc-ac course, algebra, and trigonometry. In some schools, it may be possible to take the ac and trigonometry courses concurrently. As in the first edition, the important equations have a triple asterisk *** attached to them. Other study aids include self-testing reviews and problems. Also note, a correlated laboratory manual *Experiments for Electronic Principles* is available.

If you liked the first edition of this book, I'm sure you will like this new edition even better. It accurately reflects the industrial use of discretes and ICs, retaining enough discrete discussion to lay a solid foundation and introducing enough IC material to get in step with industry. Best of all, this new edition is easier to teach from and easier to read.

I want to acknowledge those who helped me most with this second edition. Thanks to my reviewers, Hank Dinter of Hennepin County Area Vocational-Technical Center, Douglas V. Hall of Hall Electronics Consultants, and Thomas P. Herbert of the University of Akron. And thanks to my colleagues, Raymond P. Dong, F. Hilsenrath, A. Jensen, and William E. Long of Foothill College; Albert Camps and Lorne MacDonald of the College of San Mateo; Kenneth Muchow and Charles Wojslaw of San Jose City College; Joseph T. Livingstone of West Valley College; and Francis S. Horton of Southeastern Community College.

Albert Paul Malvino

1 *introduction*

This chapter is about idealization and approximation, ways to simplify analysis. Included are common approximations for resistors, capacitors, and inductors.

1-1 *idealization*

Most electronic devices are complicated. To get to the main ideas, we often idealize; this means stripping away all unnecessary detail. The device that remains is then ideal or perfect.

need for idealization

You have used idealization many times already. A good example is a piece of wire. In most cases, you treat it as a perfect conductor; but this is far from the truth. Take a piece of copper wire 1 ft long as shown in Fig. 1-1*a*. If it is AWG 22, this piece of wire has a resistance of 0.016 Ω and an inductance of 0.24 μH. If the wire is 1 in above a metallic plane (see Fig. 1-1*b*), it has a capacitance of 3.3 pF. Therefore, the piece of wire acts like the circuit of Fig. 1-1*c*.

Even this is not exact. The truth is you cannot draw a circuit at all because the *R, L,* and *C* are *distributed* over the length of the wire rather than *lumped*

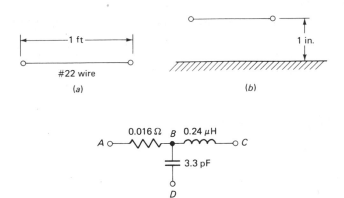

Figure 1-1. A piece of wire and its equivalent circuit.

between nodes *A, B, C,* and *D.* For exact analysis you need to use an approach based on advanced formulas called Maxwell's equations.

Fortunately, you don't need Maxwell's equations below 300 MHz, and you can use lumped-constant circuits like Fig. 1-1c. (Lumped constant means the resistance is between one pair of nodes, the inductance between another pair, and the capacitance between a third pair.) Better yet, for frequencies below 1 MHz the inductive and capacitive reactances are usually negligible, and the wire's resistance is so small compared with other circuit resistance that it too is negligible. In other words, most of the time we *idealize* a piece of wire by neglecting its *R, L,* and *C.*

approximations

Make no mistake about it. At high enough frequencies you do need an equivalent circuit like Fig. 1-1c for a piece of wire. This should give you a clue as to how complicated the exact circuits are for some electronic devices. For most everyday needs, we can approximate; otherwise, we'd be bogged down in unnecessary detail. Throughout this book, we use approximations as much as possible.

The *ideal,* or first approximation, strips away everything but the key ideas behind a device. In this way, we get to the bones of device operation. With ideal devices, circuit analysis is easier. When necessary, we can improve the analysis by using a second approximation, sometimes a third approximation, and once in a while, an exact equivalent circuit. In the chapters to come, look for these levels of approximation:

- Ideal or first approximation: the simplest equivalent circuit. It retains only one or two ideas of how the device works.

- Second approximation: includes extra features to improve analysis. Usually, this is as far as many engineers and technicians go in daily work.

- Third approximation: includes other effects of lesser importance. In some circuits, we will need this approximation.
- Exact circuit: this will be as complete as possible using lumped constants. We almost never use this circuit.

1-2 *resistor approximations*

Every resistor has a small amount of inductance and capacitance. At lower frequencies the unwanted L and C have negligible effect. But as the frequency increases, the resistor no longer acts like a pure resistance.

exact circuit

Figure 1-2*a* shows the exact equivalent circuit of a resistor. The inductance exists because current through the resistor produces a magnetic field. The capacitance exists because voltage across the resistor produces an electric field.

At lower frequencies, the inductive reactance approaches zero and the capacitive reactance approaches infinity. In other words, the inductor appears shorted and the capacitor appears open. In this case, the resistor acts like a pure resistance.

We will refer to the inductance as the *lead inductance* because much of it is produced by the leads going into the resistor. And we will refer to the capacitance as the *stray capacitance* because it represents the lumped capacitance between the ends of the resistor.

second approximation

For most resistors the stray capacitance is more important than the lead inductance. This is why our second approximation is the parallel RC circuit of Fig. 1-2*b*. The stray capacitance of typical resistors (⅛ to 2 W) is in the vicinity of 1 pF, with the exact value determined by the length of the leads, the size of

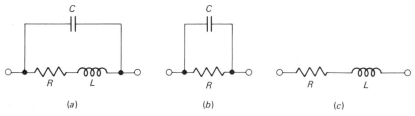

Figure 1-2. Resistor equivalent circuits. (a) *Exact.* (b) *Second approximation.* (c) *Third approximation.*

the resistor body, and other factors. When you need the exact capacitance value, you can measure it on an *RLC* bridge.

When is stray capacitance negligible? Many people use this guide: neglect stray capacitance when reactance is ten times greater than resistance. That is,

$$\frac{X_C}{R} > 10 \tag{1-1}$$

As an example, if a 10-kΩ resistor has 1 pF of stray capacitance, the X_C at 1 MHz equals

$$X_C = \frac{1}{2\pi f C} = \frac{1}{2\pi(10^6)\ 10^{-12}}$$
$$= 159\ \text{k}\Omega$$

The ratio of reactance to resistance is

$$\frac{X_C}{R} = \frac{159\ \text{k}\Omega}{10\ \text{k}\Omega} = 15.9$$

This is greater than 10; therefore, we can neglect the stray capacitance of a 10-kΩ resistor operating at 1 MHz.

third approximation

When the resistance is very low, lead inductance is more important than stray capacitance. In this case, we use the third approximation of a resistor (Fig. 1-2c). This approximation includes lead inductance, approximately 0.02 μH/in for wire sizes commonly used in electronic circuits. (These sizes are from around AWG 18 to 26. AWG 18 has 0.018 μH/in; AWG 26 has 0.024 μH/in.)

When can you neglect lead inductance? The usual rule is this: neglect lead inductance when

$$\frac{R}{X_L} > 10 \tag{1-2}$$

For example, suppose the leads of a 1-kΩ resistor are cut to ½ in on each end. Then, the total lead length is 1 in. Estimating the inductance by 0.02 μH/in, we have a lead inductance of 0.02 μH. At 300 MHz, the reactance is

$$X_L = 2\pi f L = 2\pi(300)10^6(0.02)10^{-6}$$
$$= 37.7\ \Omega$$

and the ratio of resistance to reactance is

$$\frac{R}{X_L} = \frac{1000}{37.7} = 26.5$$

Therefore, even at 300 MHz we can neglect the lead inductance of a 1-kΩ resistor.

selecting an approximation

Which approximation of a resistor should you use? The ideal, the second, or the third? The answer depends on the frequency and the value of resistance.

By setting $X_C/R = 10$ and $R/X_L = 10$, we can plot frequency versus resistance as shown in Fig. 1-3. This graph gives the dividing lines between the ideal, second, and third approximations, assuming a stray capacitance of 1 pF and a lead inductance of 0.02 μH.

Here is how we use the graph. For any point below the two lines, we can idealize the resistor, that is, neglect its capacitance and inductance. On the other hand, if a point falls above either line, we may want to use the second or third approximation. For instance, a 10-kΩ resistor operating at 1 MHz can be idealized because it falls in the ideal region of Fig. 1-3. But if this 10-kΩ resistor operates at 5 MHz, we may include stray capacitance in precise calculations. Similarly, a 20-Ω resistor acts ideally up to 16 MHz, but beyond this frequency, the lead inductance becomes important.

Don't lean too heavily on Fig. 1-3; it is only a guide to help you estimate when to include stray capacitance or lead inductance in your calculations. If working at high frequencies where precise analysis is required, you may have to measure the stray capacitance and the lead inductance on a high-frequency *RLC* bridge.

1-3 *inductor and capacitor approximations*

Figure 1-4a shows the ideal, second, and third approximations of an inductor. For frequencies where only the X_L matters, we use the ideal approximation. But at lower frequencies where X_L is not large compared with R, we must use the second approximation. At higher frequencies, the distributed capacitance between the inductor windings becomes important. In this case, we need the third approximation shown in Fig. 1-4a. We will use these inductor approximations later.

Figure 1-4b shows the approximations for a capacitor. Most of the time, we can idealize a capacitor. For large capacitors, especially the electrolytic type, we may include the leakage resistance (the second approximation). And at the highest frequencies, the lead inductance becomes important; therefore, occasionally, we need the third approximation shown in Fig. 1-4b.

6

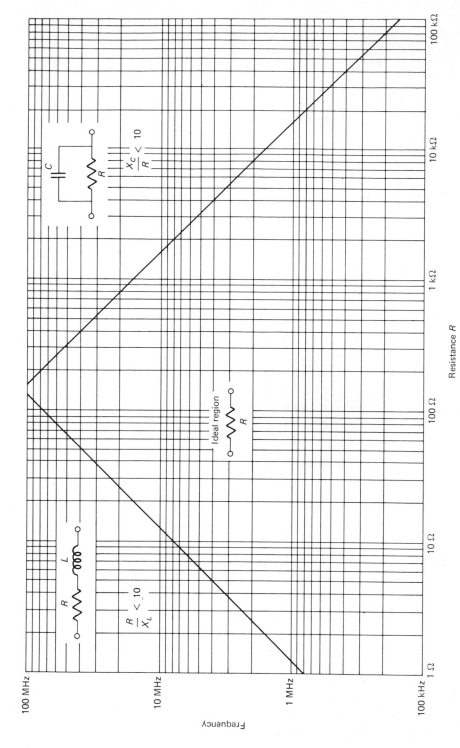

Figure 1-3. Approximation guide for resistors.

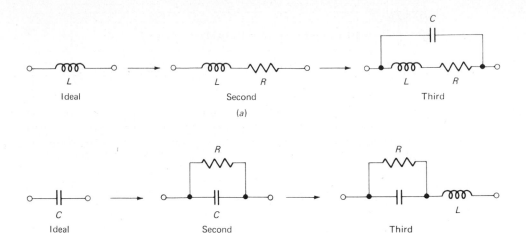

Figure 1-4. (a) *Inductor approximations*. (b) *Capacitor approximations*.

EXAMPLE 1-1
How much inductance does a 5-in piece of wire have? What does its X_L equal at 10 MHz?

SOLUTION
The inductance is approximately 0.02 μH/in; therefore, 5 in of wire has an inductance of approximately 0.1 μH.

At 10 MHz, the inductive reactance equals

$$X_L = 2\pi fL = 2\pi(10)10^6(0.1)10^{-6}$$
$$= 6.28 \ \Omega$$

This inductive reactance may be important; it depends on the circuit. All we can say is this: for the 6.28 Ω to be negligible, the circuit impedance in series with it must be much larger.

Besides having inductance, a piece of wire forms a capacitance with any nearby metal. With increasing frequency, this capacitance becomes important because it provides a path for ac current.

The inductance and capacitance of connecting wires may disturb circuit operation. It depends on the resistances and frequencies involved. As a general rule, keep all connecting wires as short as possible for circuits operating above 100 kHz.

EXAMPLE 1-2
Capacitors have lead inductance. For this reason, they become self-resonant at higher frequencies. Calculate the resonant frequency for a 1000-pF capacitor with ½-in leads on each end.

SOLUTION

Assume a lead inductance of 0.02 μH/in. Then,

$$f = \frac{1}{2\pi\sqrt{LC}} = \frac{1}{2\pi\sqrt{0.02(10^{-6})1000(10^{-12})}}$$
$$= 35.6 \text{ MHz}$$

This is the resonant frequency of the capacitor. Beyond this frequency, the inductive reactance becomes greater than the capacitive reactance. In effect, the capacitor no longer acts like a capacitor above 35.6 MHz; it acts like a small inductor.

1-4 *chassis and ground*

Electronic components are often mounted on a metal base called a *chassis*. For instance, in Fig. 1-5*a* the voltage source and resistors are connected to the chassis. Because of its low resistance, the chassis provides a conducting path, the same as a piece of wire. In fact, because the chassis resistance is so low, the usual practice is to visualize the entire chassis as an equipotential point.

Figure 1-5*b* shows the symbol for the chassis. When you see this symbol, visualize a metal chassis completing the path between components. Often, the power plug has a third prong on it (see Fig. 1-5*c*). When connected to an ac outlet, this third prong is *grounded* (put in contact with the earth). When this kind of plug is used, the third prong places the chassis in contact with the earth; this is why the chassis is often referred to as *ground*.

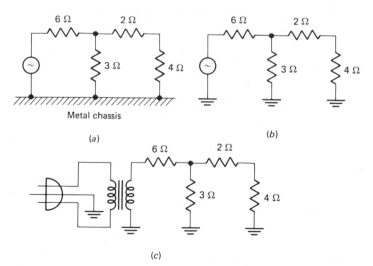

Figure 1-5. (a) *Chassis completes circuit.* (b) *Chassis symbol.* (c) *Power plug grounds chassis.*

Even if the power plug does not have a third prong, it's still customary to refer to the chassis as ground. Because the chassis is an excellent conductor, all ground points represent the same potential.

1-5 *thevenin's theorem*

Briefly, here is a review of the Thevenin theorem for dc circuits. Suppose you have a linear circuit where only dc currents and voltages exist. (Linear means the resistance values do not change with increasing voltage.) Pick any pair of terminals A and B. The Thevenin theorem says these terminals act as if a single battery and resistor are behind them, as shown in Fig. 1-6a. The original circuit may be complicated with many sources and loops. But the Thevenin equivalent has only one source and one loop when you connect R_L (Fig. 1-6b). This one-loop simplicity is the power of the Thevenin circuit; anyone can calculate currents and voltages in a circuit like this.

To find V_{TH}, you calculate or measure the open-circuit voltage in Fig. 1-7a. To get R_{TH}, you reduce all sources to zero, and calculate or measure the resistance from A to B. When measuring the resistance, you may find it inconvenient to reduce all sources to zero. In this case, connect a variable load resistance as shown in Fig. 1-7b; when V_L equals half of V_{TH}, R_L equals R_{TH}. You can then measure R_L with an ohmmeter.

Thevenin's theorem also applies to ac circuits. Combined with the superposition theorem, it is the key to understanding many different electronic circuits.

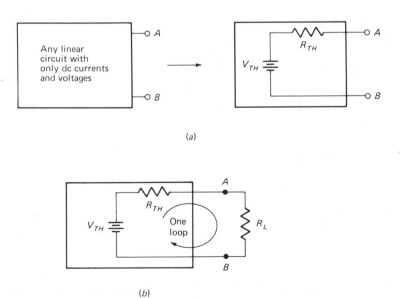

Figure 1-6. Thevenin's theorem.

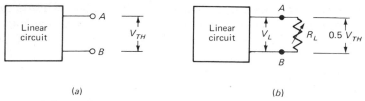

(a)   (b)

Figure 1-7. Finding V_{TH} *and* R_{TH}.

Chapter 9 discusses this combined use of Thevenin's theorem and the superposition theorem.

EXAMPLE 1-3

Figure 1-8*a* shows a circuit with a dc voltage source and a dc current source. *Thevenize* the circuit at the *AB* terminals. (Thevenize means "find the Thevenin equivalent of.")

SOLUTION

The I_{CBO} current source forces a conventional current of I_{CBO} to flow back through the resistor and down through the voltage source as shown in Fig. 1-8*b*. The voltage across the resistor equals $I_{CBO}R_B$. Therefore,

$$V_{TH} = V_{BB} + I_{CBO}R_B$$

To get R_{TH}, reduce all sources to zero. Reducing a voltage source to zero is identical to replacing it with a *short* as shown in Fig. 1-8*c*. Reducing a current

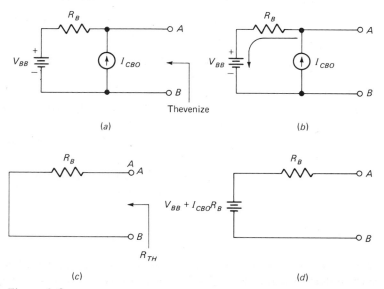

Figure 1-8.

source to zero is the same as replacing it with an *open*. When a current source is open, we can remove it altogether from the circuit. Looking into the *AB* terminals of Fig. 1-8*c*, therefore, we see a Thevenin resistance equal to R_B.

Figure 1-8*d* is the Thevenin equivalent circuit. You will see this circuit again when we analyze the effect of transistor leakage currents.

self-testing review

Read each of the following and provide the missing words. Answers appear at the beginning of the next question.

1. You have used idealization many times already. A good example is a piece of _____. In most cases, you treat it as a perfect _____.

2. *(wire, conductor)* A piece of wire has _____, inductance, and _____. These are distributed over the length of the wire rather than _____.

3. *(resistance, capacitance, lumped)* The _____ or _____ approximation strips away everything but the key ideas behind the device.

4. *(ideal, first)* The stray _____ of a resistor is negligible when the capacitive reactance is ten times larger than the _____.

5. *(capacitance, resistance)* When the resistance is very low, the _____ is more important than the stray _____.

6. *(lead inductance, capacitance)* Besides having inductance, a piece of wire forms a _____ with any nearby metal. As a general rule, keep all connecting wires as _____ as possible for circuits operating above 100 kHz.

7. *(capacitance, short)* Capacitors have lead inductance. For this reason, they become self-resonant at _____ frequencies.

8. *(higher)* Electronic components are often mounted on a metal base called the _____. Because of its low resistance, it provides a conducting path similar to a piece of wire.

9. *(chassis)* The Thevenin theorem says a complicated dc circuit is equivalent to a single _____ and a single resistor. To get V_{TH}, you calculate or measure the open-circuit voltage. To get R_{TH}, you reduce all sources to zero, and calculate or measure the resistance from *A* to *B*.

10. *(battery)* Reducing a voltage source to zero is identical to shorting it, but reducing a current source to zero is identical to opening it.

problems

1-1. Channel 4 on a TV receiver operates at a center frequency of 69 MHz. How much inductive reactance does a 3-in piece of wire have at this frequency? (Use 0.02 μH/in.)

1-2. A 1-MΩ resistor has a stray capacitance of 1 pF. At what frequency does the capacitive reactance equal 1 MΩ? At what frequency will X_C equal 10 MΩ?

1-3. A 1-Ω resistor has a lead inductance of 0.02 μH. What does its X_L equal at 10 MHz? At what frequency does X_L equal 0.1 Ω?

1-4. Using Fig. 1-3 as a guide, estimate the maximum frequency for which the ideal approximation is valid for each of these: 5-Ω resistor, 100-Ω resistor, 1-kΩ resistor, and 56-kΩ resistor.

1-5. You are going to build a circuit that will operate up to 10 MHz. Using Fig. 1-3 as a guide, you can idealize all resistors within what range?

1-6. A 0.01-μF capacitor has a lead inductance of 0.04 μH. What is its resonant frequency?

1-7. What is the Thevenin equivalent circuit for Fig. 1-9a?

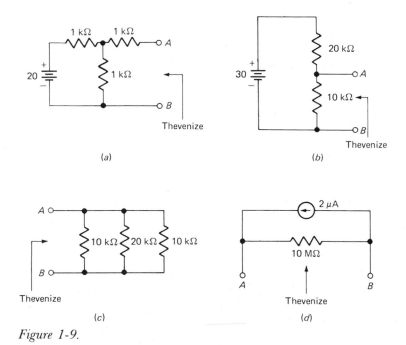

Figure 1-9.

1-8. Thevenize the circuit of Fig. 1-9b.

1-9. What is the Thevenin equivalent circuit of Fig. 1-9c?

1-10. Work out the Thevenin circuit for Fig. 1-9d.

1-11. If the 10-MΩ resistor of Fig. 1-9d opens up, what will the R_{TH} equal? What does the R_{TH} of any ideal current source equal?

2 semiconductor theory

You already know something about atoms, electrons, and protons. This chapter adds to your knowledge. Our treatment will be simple, and sometimes idealized; otherwise, we would be caught up in highly advanced mathematics and physics.

2-1 atomic structure

Bohr idealized the atom. He saw it as a nucleus surrounded by orbiting electrons (Fig. 2-1). The nucleus has a positive charge and attracts the electrons. The electrons would fall into the nucleus without the centrifugal force of their motion. When an electron travels in a *stable* orbit, it has just the right velocity for centrifugal force to balance nuclear attraction. The nearer an electron is to the nucleus, the faster it must travel to offset nuclear attraction.

Three-dimensional drawings like Fig. 2-1 are difficult to show for complicated atoms. Because of this, we often symbolize the atom in two dimensions. An isolated silicon atom (Fig. 2-2a) has 14 protons in its nucleus. Two electrons travel in the first orbit, eight electrons in the second, and four in the outer or *valence* orbit. The 14 revolving electrons neutralize the charge of the nucleus so that from a distance the atom acts electrically neutral.

Figure 2-2b shows an isolated atom of germanium. Notice the 32 protons in the nucleus and the 32 orbiting electrons. Especially important, the outer orbit contains four electrons, the same as silicon. Because of this, silicon and germa-

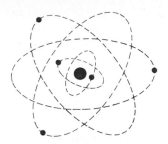

Figure 2-1. Bohr model.

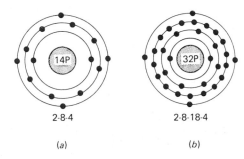

2·8·4

2·8·18·4

(a)

(b)

Figure 2-2. (a) *Silicon atom.* (b) *Germanium atom.*

nium are called *tetravalent* elements (tetravalent means having four valence electrons).

2-2 *orbital radius*

You might think that an electron can travel in an orbit of any radius, provided its velocity has the right value. Modern physics says otherwise; only certain orbit sizes are permitted. In other words, some radius values are *forbidden*.

For example, the smallest orbit in a hydrogen atom has a radius of

$$r_1 = 0.53(10^{-10}) \text{ m}$$

The next permitted orbit has a radius of

$$r_2 = 2.12(10^{-10}) \text{ m}$$

All radii between r_1 and r_2 are forbidden. Regardless of its velocity, an electron cannot remain in a stable orbit if the radius has a value between r_1 and r_2.

Why can an electron travel only in orbits of certain sizes? An electron appears to be two different things at the same time. In some experiments it acts like a *particle,* something with mass lumped in one place; in other experiments it acts like a *wave,* something with periodic vibrations in space.

We cannot go into this topic too far, but we can say this much: the orbit of an electron must satisfy not only equations for its particle nature, but also equations for its wave nature. Because it is a wave, an electron can *fit* only into an orbit whose circumference equals the electron *wavelength* or some multiple of it. (Wavelength is the distance a wave travels in one period.)

2-3 *energy levels*

In Fig. 2-3*a*, it takes energy to move an electron from a smaller to a larger orbit because work has to be done to overcome the attraction of the nucleus. Therefore, the larger the orbit of an electron, the greater its potential energy with respect to the nucleus.

For convenience in drawing, we can visualize the curved orbits as horizontal lines like those of Fig. 2-3*b*. The first orbit represents the first *energy level*, the second orbit is the second energy level, and so on. The higher the energy level, the greater the energy of the electron and the larger its orbit.

If external energy like heat, light, or other radiation bombards an atom, it can lift an electron to a higher energy level (larger orbit). The atom is then described as being in a state of *excitation*. This state does not last long, because the electron soon falls back to its original energy level. As it falls, it gives back the acquired energy in the form of heat, light, or other radiation.

Figure 2-4 summarizes the process of excitation and radiation. The wavy arrow of Fig. 2-4*a* represents incoming energy. As the electron absorbs this energy, its total energy increases; therefore, it can escape the nuclear attraction and move into a larger orbit (see Fig. 2-4*b*). The atom is now in a state of excitation.

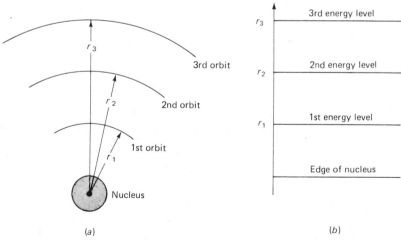

Figure 2-3. (a) *Magnified view of atom.* (b) *Energy levels.*

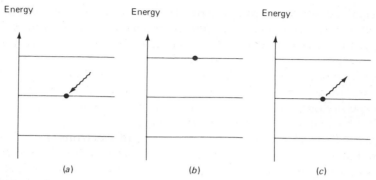

Figure 2-4. (a) *Electron absorbs energy.* (b) *Excitation state.* (c) *Electron radiates energy.*

After a brief period the electron falls back to its original orbit. As it does, it radiates energy, as shown in Fig. 2-4c.

Energy levels are important because they explain how fluorescent displays, light-emitting diodes, photodiodes, and transistors operate.

2-4 *crystals*

When atoms combine to form a solid, they arrange themselves in an orderly pattern called a *crystal*. The forces holding the atoms together are the *covalent bonds*. To describe a covalent bond, we will examine silicon.

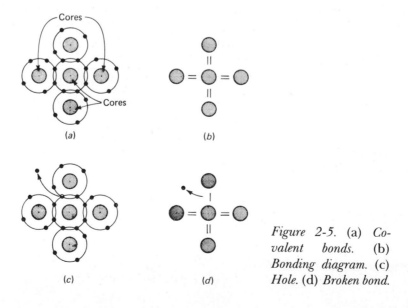

Figure 2-5. (a) *Covalent bonds.* (b) *Bonding diagram.* (c) *Hole.* (d) *Broken bond.*

An isolated silicon atom has four electrons in its valence orbit. For reasons covered by advanced equations, silicon atoms combine so as to have eight electrons in the valence orbit. To manage this, each silicon atom positions itself between four other silicon atoms (see Fig. 2-5a). Each neighbor shares an electron with the central atom. In this way, the central atom has picked up four electrons, making a total of eight in its valence orbit. Actually, the electrons no longer belong to a single atom; they are shared by adjacent atoms. It is this sharing that sets up the covalent bond.

Figure 2-5b symbolizes the mutual sharing of electrons. Each line represents a shared electron. Each shared electron establishes a bond between the central atom and a neighbor. For this reason, we call each line a covalent bond.

When outside energy lifts a valence electron to a higher energy level (larger orbit), the departing electron leaves a vacancy in the outer orbit (see Fig. 2-5c). We call this vacancy a *hole*. The hole is equivalent to a broken covalent bond, symbolized by Fig. 2-5d.

2-5 *energy bands*

When silicon atoms combine into a crystal, the orbit of an electron is influenced not only by charges in its own atom but by the nucleus and electrons of every other atom in the crystal. Since each electron has a different position inside the crystal, no two electrons see exactly the same pattern of surrounding charges. Because of this, the orbit of each electron is different.

Figure 2-6 shows what happens to energy levels. All electrons traveling in first orbits have slightly different energy levels because no two see exactly the same charge environment. Since there are billions of first-orbit electrons, the slightly different energy levels form a cluster or band. Similarly, the billions of second-orbit electrons, all with slightly different energy levels, form the second energy band shown. And all third-orbit electrons form the third band.

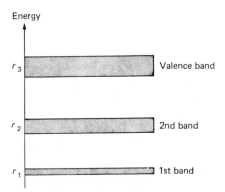

Figure 2-6. Energy bands.

Figure 2-6 shows the energy bands as dark regions. This will be our way of indicating filled or saturated bands; that is, every permitted orbit is occupied by an electron. Figure 2-6 shows the energy bands in a silicon crystal at absolute zero temperature (−273°C).

2-6 *conduction in crystals*

Figure 2-7a shows a bar of silicon with metal end surfaces. An external voltage sets up an electric field between the ends of the crystal. Does current flow? It depends. On what? On whether there are any movable electrons inside the crystal.

absolute zero

At absolute zero temperature, electrons cannot move through the crystal. All electrons are tightly held by the silicon atoms. Inner-orbit electrons are buried deep within atoms; outer-orbit electrons are part of the covalent bonding and cannot break away without receiving outside energy. Therefore, at absolute zero temperature a silicon crystal acts like a perfect insulator.

Figure 2-7b shows the energy-band diagram. The first three bands are filled, and electrons cannot move easily in these bands. But beyond the valence band is a *conduction band.* This band represents the next larger group of radii that satisfy the particle-wave nature of an electron. The orbits in the conduction band are so large that nuclear attraction is almost negligible. In other words, if an electron can be lifted into the conduction band, it is virtually free to move from one atom to the next. This is why electrons in the conduction band are often called *free electrons.*

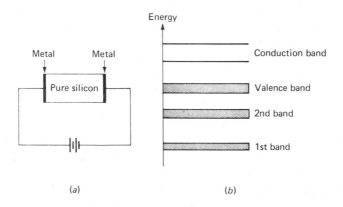

Figure 2-7. (a) *Circuit.* (b) *Energy bands at absolute zero temperature.*

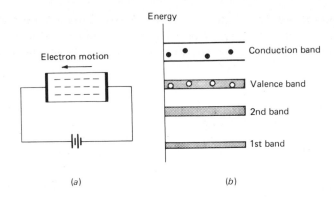

Figure 2-8. (a) *Electron flow.* (b) *Energy bands at room temperature.*

At absolute zero temperature, the conduction band is empty; this means no electron has enough energy to travel in a conduction-band orbit.

above absolute zero

Raise the temperature above absolute zero and things change. The incoming heat energy breaks some covalent bonds, that is, it knocks valence electrons into the conduction band. In this way, we get a limited number of conduction-band electrons symbolized by the minus signs of Fig. 2-8a. Under the influence of the electric field, these free electrons move to the left and set up a current.

Above absolute zero, we visualize the energy bands shown in Fig. 2-8b. Heat energy has lifted some electrons into the conduction band where they travel in orbits of larger radii than before. In these larger conduction-band orbits, the electrons are only loosely held by the atoms and can easily move from one atom to the next.

In Fig. 2-8b, each time an electron is bumped up to the conduction band, a hole is created in the valence band. Therefore, the valence band is no longer saturated or filled: each hole represents an available orbit of rotation.

The higher the temperature, the greater the number of electrons kicked up to the conduction band, and the larger the current in Fig. 2-8a. At room temperature (around 25°C) the current is too small to be useful in most applications. At this temperature, a piece of silicon is neither a good insulator nor a good conductor. For this reason, it is called a *semiconductor*.

silicon versus germanium

A germanium crystal is also a semiconductor at room temperature. But there is crucial difference between silicon and germanium. At room temperature a silicon crystal has *fewer* free electrons than a germanium crystal. This is one of the reasons silicon has become the main semiconductor material in use today.

2-7 *hole current*

Holes also can move and produce a current. In other words, in a semiconductor there are two distinct kinds of current: conduction-band current and hole current.

how holes move

At the extreme right of Fig. 2-9 is a hole. This hole attracts the valence electron at *A*. With only a slight change in energy, the valence electron at *A* can move into the hole. When this happens, the original hole vanishes and a new one appears at position *A*. The new hole at *A* can attract and capture the valence electron at *B*. When the valence electron moves from *B* to *A*, the hole moves from *A* to *B*. The motion of valence electrons can continue along the path shown by the arrows; the holes move in the opposite direction. What it boils down to is this: because holes are present in valence orbits, there is *a second path for electrons to move through a crystal.*

Here is what happens in terms of energy levels. To begin with, *thermal energy* (same as heat energy) bumps an electron from the valence band into the conduction band. This leaves a hole in the valence band as shown in Fig. 2-10. With a slight energy change, the valence electron at *A* can move into the hole. When this happens, the original hole disappears and a new one appears at *A*. Next, the valence electron at *B* can move into the new hole with a slight change in energy. In this way, with minor changes in energy, valence electrons can move along the path shown by the arrows. This is equivalent to the hole moving through the valence band along the *ABCDEF* path.

electron-hole pairs

When we apply an external voltage across a crystal, it forces the electrons to move. In Fig. 2-11*a*, two kinds of movable electrons exist: conduction-band

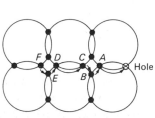

Figure 2-9. Hole current.

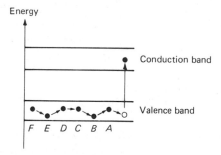

Figure 2-10. Energy diagram of hole current.

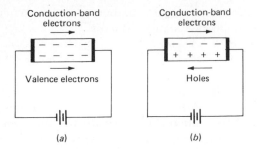

Figure 2-11. Two paths for current.

electrons and valence electrons. The motion of valence electrons to the right means holes are moving to the left.

Most of the time, we prefer to talk about holes rather than valence electrons. In a pure semiconductor, each conduction-band electron means a hole exists in the valence orbit of some atom. In other words, thermal energy produces *electron-hole pairs*. The holes act like positive charges and for this reason are shown as plus signs in Fig. 2-11b. As before, we visualize conduction-band electrons moving to the right. But now, we think of holes (positive charges) moving to the left.

recombination

In Fig. 2-11b, each minus sign is a conduction-band electron in a large orbit, and each plus sign is a hole in a smaller orbit. Occasionally, the conduction-band orbit of one atom may overlap the hole orbit of another. Because of this, a conduction-band electron can fall into a hole every so often. This merging of a conduction-band electron and a hole is called *recombination*. When recombination takes place, the hole does not move elsewhere; it disappears.

Recombination occurs continuously in a semiconductor. Because of this, every hole would eventually be filled except for one thing: Incoming heat energy continuously produces new electron-hole pairs. *Lifetime* is the name given to the average time between the creation and disappearance of an electron-hole pair. Lifetime varies from a few nanoseconds to several microseconds, depending on how perfect the crystal structure is, and other factors.

2-8 *doping*

We will call a *pure* silicon crystal (every atom a silicon atom) an *intrinsic semiconductor*. The only current carriers in an intrinsic semiconductor are the electron-hole pairs. For most applications, not enough of these exist to produce a usable current.

Doping means adding impurity atoms (nontetravalent) to a crystal to increase either the number of free electrons or the number of holes. When a crystal has been doped, it is called an *extrinsic* semiconductor.

n-*type semiconductor*

To get extra conduction-band electrons, we add *pentavalent* atoms; these have *five* electrons in the valence orbit. After adding pentavalent atoms to a pure silicon crystal, we still have mostly silicon atoms. But every now and then, we find a pentavalent atom between four neighbors as shown in Fig. 2-12*a*. The pentavalent atom originally had five electrons in its valence orbit. After forming covalent bonds with four neighbors, this central atom has an extra electron left over. Since the valence orbit can hold no more than eight electrons, the extra electron must travel in a conduction-band orbit.

For a crystal doped by a pentavalent impurity, here is what we find:

1. Many new conduction-band electrons are produced by doping. Since each pentavalent atom contributes one conduction-band electron, we can control the number of conduction-band electrons by the amount of impurity added.
2. Thermal energy still generates a few electron-hole pairs. These are far fewer in number than the conduction-band electrons produced by doping.

Figure 2-12*b* shows a crystal that has been doped by a pentavalent impurity. We have a large number of conduction-band electrons produced mostly by doping. Only a few holes exist, created by thermal energy. For obvious reasons, we call the electrons the *majority carriers* and the holes the *minority carriers*. Doped silicon of this kind is known as *n*-type semiconductor, where *n* stands for negative.

A final point. Pentavalent atoms are often called *donor* atoms because they produce conduction-band electrons. Examples of donor impurities are arsenic, antimony, and phosphorus.

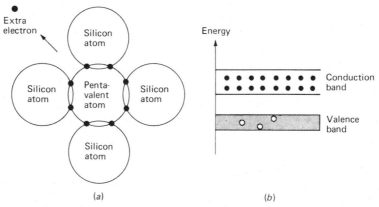

Figure 2-12. Doping with a donor impurity.

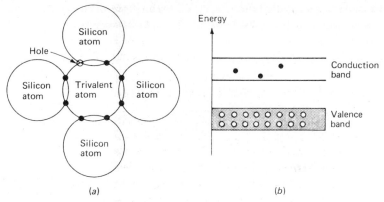

Figure 2-13. Doping with an acceptor impurity.

p-*type semiconductor*

How can we dope a crystal to get extra holes? By using a *trivalent* impurity (one with three electrons in the outer orbit). After adding the impurity, we find each trivalent atom between four neighbors as shown in Fig. 2-13*a*. Since each trivalent atom brought only three valence-orbit electrons with it, only seven electrons will travel in its valence orbit. In other words, one hole appears in each trivalent atom. By controlling the amount of impurity added, we can control the number of holes in the doped crystal.

A semiconductor doped by a trivalent impurity is known as a *p*-type semiconductor; the *p* stands for positive. As shown in Fig. 2-13*b*, the holes of a *p*-type semiconductor far outnumber the conduction-band electrons. For this reason, the holes are the majority carriers in a *p*-type semiconductor, while the conduction-band electrons are the minority carriers.

Trivalent atoms are also known as *acceptor* atoms because each hole they contribute may accept an electron during recombination. Examples of acceptor impurities are aluminum, boron, and gallium.

bulk resistance

A doped semiconductor still has resistance. We call this resistance the *bulk resistance*. A lightly doped semiconductor has a high bulk resistance. As the doping increases, the bulk resistance decreases.

self-testing review

Read each of the following and provide the missing words. The answers appear at the beginning of the next question.

1. The outer orbit is called the _____ orbit. This orbit contains _____ electrons in an isolated silicon atom. For this reason, silicon is classified as a _____ element.

2. *(valence, four, tetravalent)* The smallest orbit has a radius of r_1. The next orbit has a radius of r_2. All radii between r_1 and r_2 are _____. Forbidden orbits exist because an electron is a particle and a _____.

3. *(forbidden, wave)* The larger the orbit, the greater the _____ of the electron. External energy can lift an electron to a _____ energy level. When the electron falls back to its original energy level, it radiates _____.

4. *(energy, higher, energy)* When silicon atoms combine into a solid, they form a _____. The forces holding the atoms together are called _____ bonds.

5. *(crystal, covalent)* When outside energy lifts a valence electron into the conduction band, the departing electron leaves a vacancy known as a _____. At absolute zero temperature a silicon crystal acts like a perfect _____ because there are no _____ electrons and no _____.

6. *(hole, insulator, free, holes)* The merging of a free electron and a hole is called _____. The average time between the creation and disappearance of an electron-hole pair is known as the _____.

7. *(recombination, lifetime)* A pure silicon crystal is an _____ semiconductor. A doped crystal is an _____ semiconductor.

8. *(intrinsic, extrinsic)* To get extra conduction-band electrons, we can add _____ atoms. These are also known as _____ atoms. To get extra holes, we can add _____ atoms, also called _____ atoms.

9. *(pentavalent, donor, trivalent, acceptor)* Free electrons are the _____ carriers in *n*-type semiconductor, and holes are the _____ carriers.

10. *(majority, minority)* A doped semiconductor still has _____ resistance. A lightly doped semiconductor has a _____ bulk resistance.

11. *(bulk, high)* As the doping increases, the bulk resistance decreases.

3 pn *junctions*

The *junction* is where the *p*-type and *n*-type regions meet, and *junction diode* is another name for a *pn* crystal. (The word *diode* is a contraction of *two* electr*ode*, where *di* stands for two.) This chapter describes how junction diodes work and prepares us for the transistor, which combines two junction diodes.

3-1 *the unbiased diode*

Figure 3-1*a* shows a junction diode. The *p* side has many holes and the *n* side many conduction-band electrons. To avoid confusion, no minority carriers are shown—but bear in mind that a few conduction-band electrons are on the *p* side and a few holes on the *n* side.

The diode of Fig. 3-1*a* is *unbiased*, which means no external voltage is applied to it.

the depletion layer

The electrons on the *n* side tend to *diffuse* (spread) in all directions. Some diffuse across the junction. When an electron enters the *p* region, it becomes a minority carrier. With so many holes around it, this minority carrier has a short lifetime; soon after entering the *p* region, the electron will fall into a hole. When this

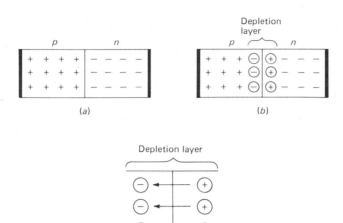

Figure 3-1. (a) *Before diffusion.* (b) *After diffusion.* (c) *Depletion-layer field.*

happens, the hole disappears and the conduction-band electron becomes a valence electron.

Each time an electron diffuses across the junction, it creates a pair of *ions*. Figure 3-1*b* shows these ions on each side of the junction. The circled plus signs are the positive ions, and the circled minus signs are the negative ions. The ions are fixed in the crystal structure because of covalent bonding and cannot move around like conduction-band electrons or holes.

Each pair of positive and negative ions in Fig. 3-1*b* is called a *dipole*. The creation of a dipole means one conduction-band electron and one hole have been taken out of circulation. As the number of dipoles builds up, the region near the junction is emptied of movable charges. We call this charge-empty region the *depletion layer*.

barrier potential

Each dipole has an electric field (Fig. 3-1*c*). The arrows show the direction of force on a positive charge. Therefore, when an electron enters the depletion layer, the field tries to push the electron back into the *n* region. The strength of the field increases with each crossing electron until the field eventually stops diffusion of electrons across the junction.

To a second approximation, we need to include minority carriers. Remember the *p* side has a few thermally produced conduction-band electrons. Those inside the depletion layer are pushed by the field into the *n* region. This slightly reduces the field strength and lets a few majority carriers diffuse from right to left to restore the field to its original strength.

Here is our final picture of equilibrium at the junction:

1. A few minority carriers drift across the junction. They would reduce the field except that
2. A few majority carriers diffuse across the junction and restore the field to its original value.

In Fig. 3-1*b*, having a field between ions is equivalent to a difference of potential called the *barrier potential*. At 25°C, the barrier potential approximately equals 0.3 V for germanium diodes and 0.7 V for silicon diodes.

temperature effects

The barrier potential depends on the temperature at the junction. Higher temperature creates more electron-hole pairs. As a result, the drift of minority carriers across the junction increases. This forces equilibrium to occur at a slightly lower barrier potential.

As a guide, we will use this approximation for changes in barrier potential: for either germanium or silicon diodes, the barrier potential *decreases* 2.5 mV for each Celsius degree rise.[1] In symbols, the change in barrier potential is

$$\Delta V = -0.0025 \, \Delta T \qquad (3\text{-}1)***$$

where Δ stands for "the change in." (As mentioned in the Preface, important equations are marked with a triple asterisk ***.)

EXAMPLE 3-1
Calculate the barrier potential at 75°C for a silicon diode.

SOLUTION
The barrier potential of a silicon diode equals 0.7 V at 25°C. When the temperature rises to 75°C, the barrier potential decreases. The amount of decrease is given by Eq. (3-1):

$$\Delta V = -0.0025(75 - 25) = -0.125 \text{ V}$$

Therefore, the barrier potential at 75°C is

$$V = 0.7 - 0.125 = 0.575 \text{ V}$$

3-2 *the energy hill*

Let's take a close look at the unbiased diode from the energy viewpoint. This will give us more insight into the action at a junction.

[1] This rule is derived in Millman, J., and C. Halkias, *Electronic Fundamentals and Applications for Engineers and Scientists*, McGraw-Hill Book Company, New York, 1976, p. 28.

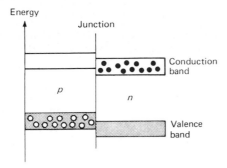

Figure 3-2. Energy
bands before diffusion.

before diffusion

Assuming an abrupt junction (one that suddenly changes from p to n material), what does the energy diagram look like? Figure 3-2 shows the energy bands before electrons have diffused across the junction. The p side has many holes in the valence band, and the n side many electrons in the conduction band. But why are the p bands slightly higher than the n bands?

The p side has trivalent atoms with a core charge of +3. On the other hand, the n side has pentavalent atoms with a core charge of +5. A +3 core attracts an electron less than a +5 core. Therefore, the orbits of a trivalent atom (p side) are slightly larger than those of a pentavalent atom (n side). This is why the p bands of Fig. 3-2 are slightly higher (more energy and larger radius) than the n bands.

A truly abrupt junction is an idealization; the p side does not suddenly end where the n side begins. A better picture allows for a gradual change from one material to the other. To a second approximation, Fig. 3-3a shows the energy diagram for a junction *before* diffusion has occurred.

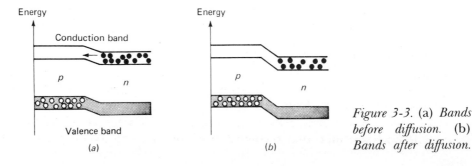

Figure 3-3. (a) Bands
before diffusion. (b)
Bands after diffusion.

at equilibrium

What happens when conduction-band electrons diffuse across the junction? The electrons near the top of the *n* conduction band move across the junction as described earlier. Not only does this create the depletion layer, it also changes the energy levels in the junction area.

Figure 3-3*b* shows the energy diagram after equilibrium is reached. The *p* bands have moved up with respect to the *n* bands. In fact, the bottom of each *p* band is even with the top of the corresponding *n* band. This means electrons on the *n* side no longer have enough energy to get across the junction. What follows is a simplified explanation of the reasons why.

When an electron diffuses across the junction, it fills the hole of a trivalent atom. The extra electron will push the conduction-band orbit farther away from the trivalent atom. Therefore, any other electrons coming into the area will need more energy than before to travel in a conduction-band orbit. This is the same as saying the *p* bands move up with respect to the *n* bands after the depletion layer has built up.

At equilibrium, conduction-band electrons on the *n* side travel in orbits not quite large enough to match the *p*-side orbits. In other words, electrons on the *n* side do not have enough energy to get across the junction. To an electron trying to diffuse across the junction, the path it must travel looks like an *energy hill* (see Fig. 3-3*b*). The electron cannot climb this hill unless it receives energy from an outside source.

3-3 *forward bias*

Figure 3-4*a* shows a dc source across a diode. The negative source terminal connects to the *n*-type material, and the positive terminal to the *p*-type material. We call this connection *forward bias*.

large forward current

Current flows easily in a circuit like Fig. 3-4*a*. Why? To begin with, when those conduction-band electrons move toward the junction, the right end of the crystal becomes slightly positive. This happens because electrons at the right end of the crystal move toward the junction and leave positively charged atoms behind. The positively charged atoms then pull electrons into the crystal from the negative source terminal.

When electrons on the *n* side approach the junction, they recombine with holes. These recombinations occur at varying distances from the junction, depending on how long a conduction-band electron can avoid falling into a hole. The odds are high that recombinations occur close to the junction. To a first

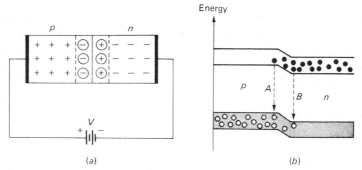

Figure 3-4. Forward bias. (a) Charges. (b) Bands.

approximation, we can visualize all the conduction-band electrons recombining when they reach the junction.

When we look at Fig. 3-4*a*, here is what we see. Electrons are pouring into the right end of the crystal, while the bulk of electrons in the *n* region move toward the junction. The left edge of this moving group disappears as it hits the junction (the electrons fall into holes). In this way, there is a continuous stream of electrons from the negative source terminal toward the junction.

Those conduction-band electrons disappearing at the junction—what happens to them? They become valence electrons. As valence electrons, they can move through holes in the *p* region. We can visualize the valence electrons on the *p* side moving toward the left end of the crystal; this is equivalent to holes moving to the right. When the valence electrons reach the left end of the crystal, they leave the crystal and flow into the positive terminal of the source.

Here are the highlights of what happens to an electron in Fig. 3-4*a*:

1. After leaving the negative source terminal, it enters the right end of the crystal.
2. It travels through the *n* region as a conduction-band electron.
3. Near the junction it recombines and becomes a valence electron.
4. It travels through the *p* region as a valence electron.
5. After leaving the left end of the crystal, it flows into the positive source terminal.

energy bands

Forward bias lowers the energy hill (see Fig. 3-4*b*). Because of this, conduction-band electrons have enough energy to invade the *p* region. Soon after entering the *p* region, each electron falls into a hole (path *A*). As a valence electron, it continues its journey toward the left end of the crystal.

A conduction-band electron may fall into a hole even before it crosses the junction. In Fig. 3-4*b* a valence electron may cross the junction from right to left; this leaves a hole just to the right of the junction. This hole does not live long. A conduction-band electron soon falls into it (path *B*).

No matter where the recombination takes place, the result is the same. A steady stream of conduction-band electrons moves toward the junction and falls into holes near the junction. The captured electrons (now valence electrons) move left in a steady stream through the holes in the *p* region. In this way, we get a continuous flow of electrons through the diode.

3-4 *reverse bias*

Turn the dc source around and you *reverse-bias* the diode as shown in Fig. 3-5*a*. Does the diode still conduct? What happens to the depletion layer? To the energy bands?

depletion layer widens

The externally produced field is in the same direction as the depletion-layer field. Because of this, holes and electrons move toward the ends of the crystal (away from the junction). The fleeing electrons leave positive ions behind, and departing holes leave negative ions; therefore, the depletion layer gets wider. The greater the reverse bias, the wider the depletion layer becomes.

How wide does the depletion layer get? In Fig. 3-5*a*, when the holes and electrons move away from the junction, the newly created ions increase the difference of potential across the depletion layer. The wider the depletion layer, the greater the difference of potential. The depletion layer stops growing when its difference of potential equals the applied reverse voltage.

transient current

How much current is there in a reverse-biased diode? While the depletion layer is adjusting to its new width, holes and electrons move away from the junction.

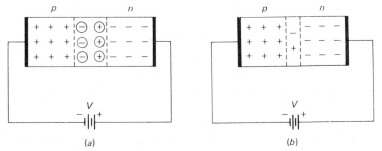

Figure 3-5. Reverse bias. (a) *Charges.* (b) *Minority carriers.*

This means electrons are flowing from the negative source terminal into the left end of the crystal; at the same time, electrons are leaving the right end of the crystal and flowing into the positive source terminal. Therefore, a current flows in the external circuit while the depletion layer is adjusting to its new width. This *transient* current drops to zero after the depletion layer stops growing. Typically, this transient current lasts only a few nanoseconds.

minority-carrier current

Is there any current at all after the depletion layer settles down? Yes. A small current flows. Thermal energy creates electron-hole pairs. In other words, a few minority carriers exist on both sides of the junction. Most of these recombine with the majority carriers. But those inside the depletion layer may live long enough to get across the junction. When this happens, a small current flows in the external circuit.

Figure 3-5b illustrates the idea. When an electron-hole pair is created inside the depletion layer, the field pushes the electron to the right, forcing one electron to leave the right end of the crystal. The hole in the depletion layer is pushed to the left. This extra hole on the *p* side lets one electron enter the left end of the crystal and fall into a hole. Since thermal energy is continuously producing electron-hole pairs near the junction, we get a *small* continuous current in the external circuit.

The reverse current caused by the minority carriers is called the *saturation current*, designated I_S. The name *saturation* reminds us we cannot get more minority-carrier current than is produced by thermal energy. In other words, increasing the reverse voltage will not increase the number of thermally created minority carriers.

Thermal energy produces saturation current; the higher the temperature, the greater the saturation current. A good practical rule to remember is this: I_S approximately doubles for each 10°C rise.[2] For example, if I_S equals 5 nA (nanoamperes) at 25°C, it will approximately equal 10 nA at 35°C, 20 nA at 45°C, 40 nA at 55°C, and so on.

As mentioned earlier, thermal energy produces fewer minority carriers in silicon diodes than in germanium diodes. In other words, a silicon diode has a much smaller I_S than a germanium diode. This immense advantage for silicon is one of the reasons it dominates the semiconductor field.

surface-leakage current

Besides transient current and minority-carrier current, does any other current flow in a reverse-biased diode? Yes. A small current flows on the *surface* of the

[2] This rule is derived in Millman, J., and H. Taub, *Pulse, Digital, and Switching Waveforms*, McGraw-Hill Book Company, New York, 1965, pp. 182–183.

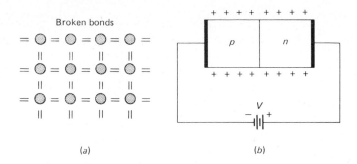

Figure 3-6. Surface-leakage current.

crystal. We call this component the *surface-leakage current*. The exact mechanism for this current is not completely understood, although most people believe it is caused by surface impurities and imperfections.

Here is one way imperfections may produce a surface current. Suppose the atoms at the top of Fig. 3-6*a* are atoms on the surface of the crystal. With no neighbors on top of them, these atoms have broken covalent bonds, that is, holes. Visualize these holes along the surface of the crystal as shown in Fig. 3-6*b*. In effect, the skin of a crystal is like a *p*-type semiconductor. Because of this, electrons can enter the left end of the crystal, travel through the surface holes, and leave the right end of the crystal. In this way, we get a small reverse current along the surface; this current increases when you increase the reverse voltage.

breakdown voltage

Keep increasing the reverse voltage and you eventually reach the *breakdown voltage*. For *rectifier* diodes (those manufactured to conduct better one way than the other), the breakdown voltage is usually greater than 50 V. Once the breakdown voltage is reached, a large number of minority carriers appear in the depletion layer and the diode conducts heavily.

Where do the carriers suddenly come from? Figure 3-7*a* shows a thermally

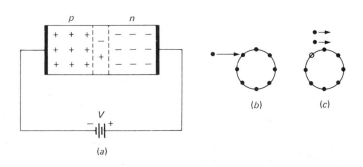

Figure 3-7. Breakdown. (a) *Minority carriers in depletion layer.* (b) *Free electron hits valence electron.* (c) *Two free electrons.*

produced electron-hole pair inside the depletion layer. With reverse bias, the electron is pushed right and the hole left. As it moves, the electron gains speed. The stronger the depletion-layer field, the faster the electron moves. For large reverse voltages, the electron reaches high velocities. This high-speed electron may collide with a valence electron (see Fig. 3-7b). If the high-speed electron has enough energy, it can bump the valence electron into a conduction-band orbit. This results in two conduction-band electrons as shown in Fig. 3-7c. Both of these will now accelerate and may go on to dislodge two more electrons. In this way, the number of minority carriers may become quite large and the diode can conduct heavily.

Most diodes are not allowed to break down. In other words, the reverse voltage across a diode is kept less than the breakdown voltage.

summary

Here's what counts:

1. A forward-biased diode conducts easily.
2. A reverse-biased diode conducts poorly.

As an ideal approximation, a diode acts like a closed switch when forward-biased, and like an open switch when reverse-biased.

3-5 *bipolar and unipolar devices*

The junction diode is a *bipolar device*. (Bipolar is a contraction of *two polar*ities with bi used for two). A bipolar device needs holes *and* conduction-band electrons to work properly.

Some devices are *unipolar;* they need holes *or* electrons to operate normally. A carbon resistor, for instance, is a unipolar device; it uses conduction-band electrons only. If we wanted, we could make a resistor out of *p*-type semiconductor; it would be a unipolar device.

In later chapters, we discuss transistors. Bipolar transistors are older and more widely used; they are also less expensive. Unipolar transistors (called field-effect transistors, or FETs) are important in special applications.

Figure 3-8 shows a *chip*, a small piece of semiconductor material. The dimensions are representative; chips are often smaller than this, occasionally larger. By advanced photographic techniques, a manufacturer can produce circuits on the surface of this chip, circuits containing many diodes, resistors, transistors, etc. The finished network is so small you need a microscope to see the connections. We call a circuit like this an *integrated circuit* (IC).

A final point. A *discrete circuit* is the kind of circuit you build when you connect separate resistors, capacitors, transistors, etc. Each component you add to the

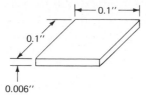

Figure 3-8. Semiconductor chip.

circuit is discrete, that is, distinct or separate from the others. This differs from an integrated circuit where components are atomically part of semiconductor chip.

self-testing review

Read each of the following and provide the missing words. Answers appear at the beginning of the next question.

1. The region near the junction is emptied of movable charges. We call this region the _____ layer. The difference of potential across this region is known as the _____ potential.

2. *(depletion, barrier)* At 25°C the barrier potential equals approximately _____ for germanium diodes and _____ for silicon diodes.

3. *(0.3 V, 0.7 V)* Forward bias lowers the energy hill. Because of this, _____ electrons have enough energy to enter the *p* region, where they fall into _____.

4. *(conduction-band, holes)* The greater the reverse bias, the _____ the depletion layer becomes. The transient current in a _____ diode drops to zero after the depletion layer stops changing.

5. *(wider, reverse-biased)* The reverse current produced by minority carriers is called the _____ current. The higher the temperature, the _____ this current becomes.

6. *(saturation, larger)* A small current flows on the surface of a crystal. We call this the _____ current. Keep increasing the reverse voltage and you eventually reach the _____ voltage.

7. *(surface-leakage, breakdown)* The junction diode is a bipolar device. It needs _____ and _____ electrons to work properly. A carbon resistor is an example of a _____ device.

8. *(holes, conduction-band, unipolar)* A _____ circuit requires separate resistors, capacitors, transistors, etc. This differs from an _____ circuit where the components are atomically part of a semiconductor chip.

9. *(discrete, integrated)* A diode conducts easily when forward-biased; poorly when reverse-biased. Ideally, a forward-biased diode is equivalent to a closed switch, and a reverse-biased diode to an open switch.

problems

3-1. Calculate the approximate barrier potential a silicon diode at 60°C, 90°C, and 150°C.

3-2. What is the approximate barrier potential for a silicon diode at −20°C?

3-3. A silicon diode has a saturation current of 2 nA at 25°C. What is the value of I_S at 75°C? At 125°C?

3-4. At 25°C, a silicon diode has a reverse current of 25 nA. The surface-leakage component equals 20 nA. If the surface current still equals 20 nA at 75°C, what does the total reverse current equal at 75°C?

3-5. When forward-biased in a particular circuit, a signal diode has a current of 50 mA. When reverse-biased, the current drops to 20 nA. What is the ratio of forward to reverse current?

3-6. What is the power dissipation in a forward-biased silicon diode if the diode voltage is 0.7 V and the current is 100 mA?

4 *diodes*

This chapter is about approximations for a diode. With these approximations, you can analyze diode circuits quickly and easily. These approximations are also a big help in transistor-circuit analysis.

4-1 *the rectifier diode*

In Fig. 4-1*a*, the ac source pushes electrons up the resistor during the positive half cycle of input voltage, and down the resistor during the negative half cycle. The up and down currents are equal.

Put a diode in the circuit and things change. In Fig. 4-1*b*, the positive half cycle of input voltage will forward-bias the diode; therefore, the diode conducts during the positive half cycle. But on the negative half cycle, the diode is reverse-biased and only a small reverse current can flow. The arrows symbolize the large upward flow of electrons and the small downward flow. The diode has *rectified* the ac current, that is, changed it from an alternating current to a unidirectional current.

Figure 4-1*c* shows the schematic symbol of a rectifier diode. The *p* side is called the *anode,* and the *n* side the *cathode.* The diode symbol looks like an arrow that points from the *p* side to the *n* side. Because of this, it is a reminder that *conventional* current flows easily from the *p* side to the *n* side.

A forward-biased diode conducts well, and a reverse-biased one poorly. There-

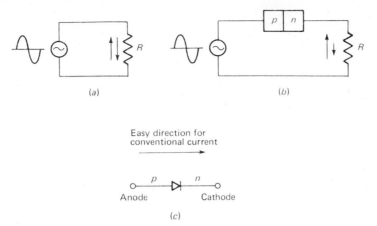

Figure 4-1. (a) *Equal currents.* (b) *Rectified current.* (c) *Schematic symbol of diode.*

fore, when analyzing diode circuits, one of the things to decide is whether the diode is forward- or reverse-biased. This is not always easy to do. But here is something that helps. Ask yourself this question: Is the external circuit trying to push conventional current in the direction of the diode arrow or in the opposite direction? If conventional current is in the same direction as the diode arrow, the diode is forward-biased. On the other hand, if conventional current tries to flow opposite the arrowhead, the diode is reverse-biased.

EXAMPLE 4-1

In each diode circuit of Fig. 4-2, determine whether the diodes are forward- or reverse-biased.

SOLUTION

Fig. 4-2a. Conventional current flows out of the positive source terminal and down through each vertical branch as shown. Since conventional current flows in the direction of the diode arrow, the diode is forward-biased.

Fig. 4-2b. As before, conventional current flows down through each vertical branch. Diode *A* is forward-biased, but diode *B* is reverse-biased. Diode *A* conducts well, diode *B* poorly.

Fig. 4-2c. In this problem, time is important. During the positive half cycle, the source voltage is plus-minus as shown. Therefore, during time interval *A,* conventional current flows in the same direction as the diode arrow and the diode is forward-biased.

During the negative half cycle (time interval *B*), the polarity of the source voltage is opposite that shown. As a result, conventional

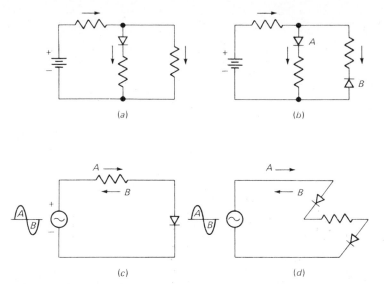

Figure 4-2.

current tries to flow up through the diode. Since this is opposite the diode arrow, the diode is reverse-biased.

Fig. 4-2*d*. During the positive half cycle, conventional current flows in the same direction as the diode arrows; therefore, both diodes are forward-biased. On the other hand, the negative half cycle of input voltage tries to push conventional current opposite the diode arrows. So, both diodes are reverse-biased during the negative half cycle.

4-2 *the forward diode curve*

Figure 4-3*a* shows a circuit you can set up in the laboratory. Since the dc source pushes conventional current in the same direction as the diode arrow, the diode is forward-biased. The greater the applied voltage, the larger the diode current. By varying the applied voltage, you can measure the diode current (use a series ammeter) and the diode voltage (a voltmeter in parallel with the diode). By plotting the corresponding currents and voltages, you get a graph of diode current versus diode voltage.

knee voltage

Figure 4-3*b* shows how the graph looks for a forward-biased silicon diode. It is customary to plot voltage along the horizontal axis because voltage is the

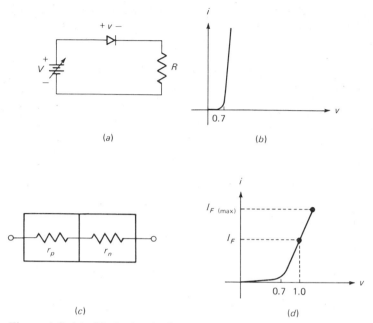

Figure 4-3. (a) *Diode circuit.* (b) *Forward curve.* (c) *Bulk resistance.* (d) *Forward currents.*

independent variable. Each value of diode voltage produces a particular current; the current is a dependent variable and is plotted along the vertical axis.

What does the graph tell us? To begin with, the diode doesn't conduct well until we overcome the barrier potential. This is why the current is small for the first few-tenths volt. As we approach 0.7 V, conduction-band electrons and holes start crossing the junction in larger numbers. This is the reason the current starts increasing rapidly. Above 0.7 V, each 0.1-V increase produces a sharp increase in current.

The voltage where the current starts to increase rapidly is called the *knee voltage* of the diode. For a silicon diode, the knee voltage equals the barrier potential, approximately 0.7 V. A germanium diode, on the other hand, has a knee voltage of about 0.3 V.

bulk resistance

Above the knee voltage, the diode current increases rapidly; small increases in diode voltage cause large increases in diode current. The reason is this: after overcoming the barrier potential, all that impedes current is the resistance of the p and n regions, symbolized by the r_p and r_n in Fig. 4-3c. Since any conductor has some resistance, both the p region and the n region also have some resistance. The sum of these resistances is called the *bulk resistance* of the diode. In symbols,

$$r_B = r_p + r_n$$

The value of bulk resistance r_B depends on the doping and size of the p and n regions; typically, r_B is from 1 to 25 Ω.

how to calculate bulk resistance

It helps to have the approximate value of r_B when analyzing a diode circuit. Here is a way to estimate r_B. A manufacturer's data sheet often gives you the amount of forward current I_F at 1 V (see Fig. 4-3*d*). For a silicon diode, the first 0.7 V is dropped across the depletion layer; the final 0.3 V is dropped across the r_B of the diode. Therefore,

$$r_B \cong \frac{0.3}{I_F} \qquad \text{for silicon diodes} \qquad \text{(4-1)}\text{***}$$

where I_F is the dc forward current at 1 V. (The symbol $\cong$ means "approximately equal to.")

As an example, a 1N456 is a silicon diode with an I_F of 40 mA at 1 V. With Eq. (4-1),

$$r_B \cong \frac{0.3}{40(10^{-3})} = 7.5 \ \Omega$$

In a circuit using a 1N456, the first 0.7 V appears across the depletion layer. Any additional diode voltage is dropped across the 7.5 Ω of bulk resistance.

maximum dc forward current

If the current in a diode is too large, excessive heat will destroy it. Even approaching the burnout value without reaching it can shorten diode life and degrade other properties. For this reason, a manufacturer's data sheet specifies the maximum current a diode can safely handle without shortening its life or degrading its characteristics.

The maximum dc forward current is one of the maximum ratings usually given on a data sheet. In Fig. 4-3*d*, this current is designated $I_{F(\text{max})}$. It represents the largest dc forward current the diode can safely handle. For instance, the 1N456 has an $I_{F(\text{max})}$ rating of 135 mA. This means it can safely handle a continuous forward current of 135 mA.

current-limiting resistor

Which brings us to why a resistor is almost always used in series with a diode. In Fig. 4-3*a*, R is called a *current-limiting resistor*. The larger we make R, the smaller the diode current. The exact size of R will depend on what we are trying to do with the circuit. We will give examples later. For now it is enough to know the diode depends on R to keep the current less than the maximum rated current.

EXAMPLE 4-2

Estimate the bulk resistance of each of these silicon diodes:

1. A 1N662 with an I_F of 10 mA at 1 V.
2. A 1N3070 with an I_F of 100 mA at 1 V.

SOLUTION

1. The 1N662 has a bulk resistance of approximately

$$r_B = \frac{0.3}{0.010} = 30 \ \Omega$$

2. A 1N3070 has an approximate bulk resistance of

$$r_B = \frac{0.3}{0.100} = 3 \ \Omega$$

4-3 *the diode curve*

When you reverse-bias a diode (Fig. 4-4*a*), you get only a small current. By measuring diode current and voltage, you can plot the reverse curve; it will look something like Fig. 4-4*b*. There are no surprises here; diode current is very small for all reverse voltages less than the breakdown voltage *BV;* at breakdown the current increases rapidly for small increases in voltage.

By using positive values for forward current and voltage, and negative values for reverse current and voltage, we can plot the forward and reverse curves in

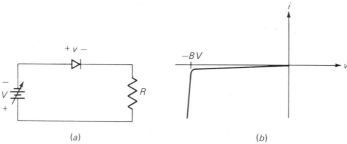

Figure 4-4. (a) *Reverse bias.* (b) *Reverse curve.*

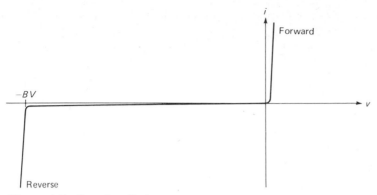

Figure 4-5. Complete diode curve.

a single graph (see Fig. 4-5). This graph summarizes the action of a diode; it tells us how much diode current flows for each value of diode voltage.

4-4 *the ideal diode*

The *ideal-diode* approximation strips away all but the bones of diode operation. What does a diode do? It conducts well in the forward direction and poorly in the reverse direction. Boil this down to its essence, and this is what you get: ideally, a diode acts like a perfect conductor (zero voltage) when forward-biased and like a perfect insulator (zero current) when reverse-biased, as shown in Fig. 4-6a.

In circuit terms, an ideal diode acts like an *automatic switch.* When conventional current tries to flow in the direction of the diode arrow, the switch is closed (see Fig. 4-6b). If conventional current tries flowing the other way, the switch is open. This is rock bottom; we cannot simplify beyond this point without losing the main idea of a diode.

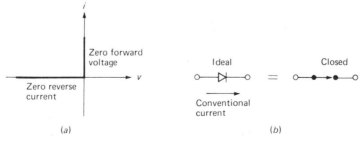

Figure 4-6. (a) *Ideal-diode curve.* (b) *Closed-switch analogy.*

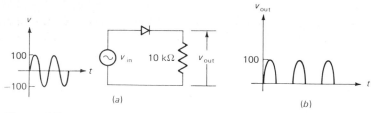

Figure 4-7.

Extreme as the ideal-diode approximation seems at first, it gives good answers for most diode circuits. There will be times when the approximation breaks down; for this reason, we need a second and third approximation. But for all preliminary analysis of diode circuits, the ideal diode is an excellent approximation.

EXAMPLE 4-3
Use the ideal-diode approximation to find the output waveform in Fig. 4-7a.

SOLUTION

The input waveform is a sine wave with a positive peak of 100 V and a negative peak of -100 V. During the positive half cycle, the diode is forward-biased because conventional current tries to flow in the direction of the diode arrow. During the negative half cycle, the source tries pushing conventional current against the diode arrow. For this reason, the diode looks like an open switch during the negative half cycle.

We can summarize the circuit action like this: each positive half cycle of input voltage appears across the output; each negative half cycle is blocked from the output. Figure 4-7b shows the output waveform.

EXAMPLE 4-4
Calculate the peak current through the diode in Fig. 4-7a. Also, what is the maximum voltage across the diode?

SOLUTION

The largest current we can get through the diode occurs when the input voltage reaches its positive peak. The diode current equals the current through the 10-kΩ resistor:

$$i_{\text{peak}} = \frac{100}{10,000} = 10 \text{ mA}$$

During forward bias, the ideal-diode voltage drop equals zero. But during reverse bias, the diode voltage gets quite large. How large? Look at Fig. 4-7a.

Kirchhoff's voltage law implies that 100 V appears across the open switch (the reverse-biased diode). Since this occurs at the negative peak, it is the maximum voltage that appears across the diode.

4-5 *the second approximation*

We need about 0.7 V before a silicon diode really conducts well. When we have a large input voltage, this 0.7 V is too small to matter. But when the input voltage is not large, we may want to take the knee voltage into account.

Figure 4-8a shows the graph for the *second approximation*. The graph says no current flows until 0.7 V appears across the diode. At this point the diode turns on. No matter what forward current flows, we allow only 0.7 V drop across a silicon diode. (Use 0.3 V for germanium diodes.)

Figure 4-8b shows the equivalent circuit for the second approximation. We think of the diode as a switch in series with a 0.7-V battery. If the external circuit can force conventional current in the direction of the diode arrow, the switch is closed and the diode voltage equals 0.7 V; otherwise, the switch is open.

EXAMPLE 4-5
Use the second approximation in Fig. 4-9a to find the output waveform. Also, calculate the peak forward current and maximum reverse voltage across the diode.

SOLUTION
During the positive half cycle of input voltage, the first 0.7 V is wasted in overcoming the barrier potential; thereafter, the diode can conduct. By Kirchhoff's voltage law, the voltage across the resistor equals 9.3 V. Therefore, the current through the resistor equals

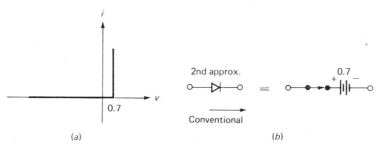

Figure 4-8.(a) *Second approximation.* (b) *Switch-battery equivalent circuit.*

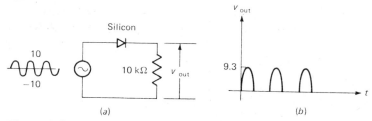

Figure 4-9.

$$i_{\text{peak}} = \frac{9.3}{10,000} = 0.93 \text{ mA}$$

Because the diode drops 0.7 V when conducting, the peak output voltage equals 9.3 V rather than the full 10 V. Figure 4-9b shows the output waveform.

What about the maximum reverse voltage? When the diode is reverse-biased, it is open. No current flows through the resistor; therefore, to satisfy Kirchhoff's voltage law, all of the applied voltage must appear across the diode. At the peak, this voltage equals −10 V.

4-6 *the third approximation*

In the third approximation of a diode, we include bulk resistance r_B. Figure 4-10a shows the effects of r_B. After the silicon diode turns on, the current produces a voltage across r_B. The greater the current, the larger the voltage.

The equivalent circuit for the third approximation is a switch in series with a 0.7-V battery and a resistance of r_B (see Fig. 4-10b). After the external circuit has overcome the barrier potential, it forces conventional current in the direction of the diode arrow. Therefore, the total voltage across the silicon diode equals

$$V_F = 0.7 + I_F r_B \qquad (4\text{-}2)$$

(For a germanium diode, use 0.3 V instead of 0.7 V.)

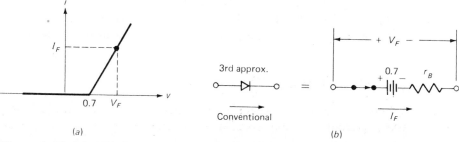

Figure 4-10. (a) *Third approximation.* (b) *Equivalent circuit.*

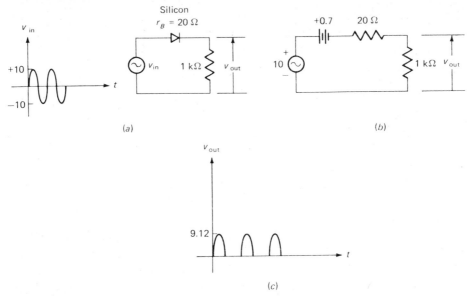

Figure 4-11.

EXAMPLE 4-6
Use the third approximation with an r_B of 20 Ω to calculate the peak current in Fig. 4-11a. Also, show the output waveform.

SOLUTION
Figure 4-11b shows the circuit at the instant the input voltage reaches its positive peak of 10 V. Therefore, the peak current equals

$$i_{\text{peak}} = \frac{10 - 0.7}{1020}$$

$$= 9.12 \text{ mA}$$

When this flows through the 1-kΩ resistor, it produces an output voltage of

$$v_{\text{out(peak)}} = 9.12(10^{-3})10^3$$
$$= 9.12 \text{ V}$$

Figure 4-11c shows the output waveform. (If we had used an ideal diode, the output waveform would be a rectified signal with a peak of 10 V.)

4-7 *reverse resistance*

When reverse-biased, a diode has a small reverse current. One way to estimate the importance of this current is with the *reverse resistance* of a diode, defined as

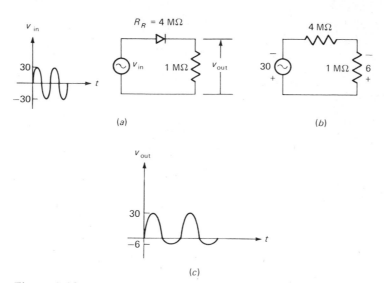

Figure 4-12.

$$R_R = \frac{V_R}{I_R} \qquad\qquad (4\text{-}3)***$$

where I_R is the reverse current at a reverse voltage V_R. As an example at 25°C the 1N456 has an I_R of 25 nA for a V_R of 25 V. Therefore, the reverse resistance equals

$$R_R = \frac{25}{25(10^{-9})} = 1000 \text{ M}\Omega$$

At 150°C, the data sheet of the 1N456 gives an I_R of 5 μA for a V_R of 25 V. So, the reverse resistance at 150°C is

$$R_R = \frac{25}{5(10^{-6})} = 5 \text{ M}\Omega$$

EXAMPLE 4-7
Sketch the output waveform for the circuit of Fig. 4-12a, taking reverse resistance into account. Neglect knee voltage and bulk resistance.

SOLUTION
The reverse resistance is given as 4 MΩ, and the resistance in series with the diode is 1 MΩ. Figure 4-12b shows the equivalent circuit at the negative peak of the input voltage. The voltage divider delivers 6 V to the output. Therefore, the output waveform is the poorly rectified signal of Fig. 4-12c. To improve the circuit, either reduce the 1-MΩ resistor or use a diode with a higher reverse resistance.

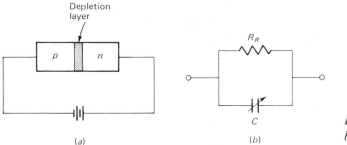

Figure 4-13. Reverse-biased diode.

4-8 *diode capacitance*

Like any component with leads, a diode has stray capacitance that may affect high-frequency operation; this external capacitance is usually less than 1 pF. More important than this external capacitance, however, is the internal capacitance built into the junction of a diode. We call this internal capacitance the *transition capacitance,* designated C_T. The word "transition" refers to the transition from p-type to n-type material. Transition capacitance is also known as depletion-layer capacitance, barrier capacitance, and junction capacitance.

What is transition capacitance? Figure 4-13a shows a reverse-biased diode. As discussed earlier, the depletion layer widens until its difference of potential equals the applied reverse voltage. The greater the reverse voltage, the wider the depletion layer. Because the depletion layer has almost no charge carriers, it acts like an insulator or dielectric. The doped p and n regions, on the other hand, act like fairly good conductors. With a little imagination, we can visualize the p and n regions separated by the depletion layer as a parallel-plate capacitor. This parallel-plate capacitance is the same as the transition capacitance.

When you increase reverse voltage, you make the depletion layer wider. It's as though you have moved the parallel plates farther apart. In effect, the transition capacitance decreases when the reverse voltage increases. Figure 4-13b shows how diode capacitance appears during reverse bias; C includes transition capacitance and stray capacitance. With increasing frequency, X_C gets smaller.

4-9 *charge storage*

Transition capacitance is the largest capacitive effect in a reverse-biased diode. But when a diode is forward-biased, a new capacitive effect takes place.

basic idea

Figure 4-14a shows a forward-biased diode, and Fig. 4-14b illustrates the energy bands. As you can see, conduction-band electrons have diffused across the junction and traveled into the p region before recombining (path A). Similarly, holes

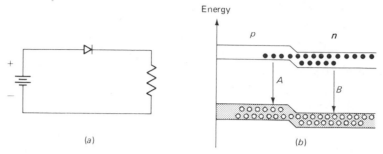

Figure 4-14. (a) *Forward biased-diode.* (b) *Stored charges in energy bands.*

cross the junction and travel into the *n* region before recombination occurs (path *B*). If the lifetime equals 1 μs, conduction-band electrons and holes exist for an average of 1 μs before recombination takes place.

Because of the lifetime of minority carriers, we have a new capacitive effect: charges in a forward-biased diode are temporarily *stored* in different energy bands near the junction. The greater the forward current, the larger the number of stored charges. This capacitive effect is referred to as *charge storage.*

reverse recovery time

Charge storage is important when you try to switch a diode from *on* to *off.* Why? Because if you suddenly reverse-bias a diode, the stored charges can flow in the reverse direction for a while. The greater the lifetime, the longer these charges can contribute to reverse current.

For example, suppose a forward-biased diode is suddenly reverse-biased as shown in Fig. 4-15*a*. Then, a large reverse current can exist for a while because of the stored charges shown in Fig. 4-15*b*. Until the stored charges either cross the junction or recombine, the reverse current can remain large.

The time it takes to turn off a forward-biased diode is called the *reverse recovery*

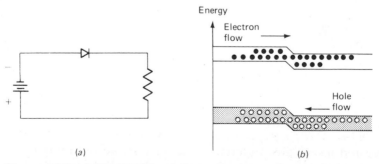

Figure 4-15. (a) *Reverse-biased diode.* (b) *Stored charges flow in reverse direction.*

time t_{rr}. The conditions for measuring t_{rr} vary from one manufacturer to the next. As a guide, t_{rr} is the time it takes for reverse current to drop to 10 percent of forward current. For instance, the 1N4148 has a t_{rr} of 4 ns. If this diode has a forward current of 10 mA and it is suddenly reverse-biased, it will take approximately 4 ns for the reverse current to decrease to 1 mA.

diffusion capacitance

Charge storage is equivalent to a capacitance called the *diffusion capacitance,* designated C_D. This capacitance is directly proportional to lifetime and forward current. Furthermore, the diffusion capacitance is much larger than the transition capacitance of a reverse-biased diode. Diffusion capacitance limits the high-frequency response of transistor circuits (Chap. 16).

4-10 other types of diodes

We have been discussing the rectifier diode, the kind optimized to conduct better forward than reverse. Here are a few other types.

light-emitting diodes

In a forward-biased diode, conduction-band electrons cross the junction and fall into holes (see Fig. 3-4*b*). As these electrons fall from the conduction band to the valence band, they radiate energy. In a rectifier diode, this energy goes off as heat. But in a *light-emitting* diode (LED), the energy radiates as light.

By using elements like gallium, arsenic, and phosphorus, a manufacturer can produce LEDs that radiate red, green, yellow, and infrared (invisible). LEDs that produce visible radiation are useful in instrument displays, calculators, digital watches, etc. The infrared LED finds applications in burglar-alarm systems and other areas requiring invisible radiation.

The advantages of a LED over an incandescent lamp are long life (more than 20 years), low voltage (1 to 2 V), and fast on-off switching (nanoseconds). Figure 4-16*a* is the schematic symbol of a LED.

photodiodes

Thermal energy produces minority carriers in a diode. The higher the temperature, the greater the current in a reverse-biased diode.

Light energy also can produce minority carriers. By using a small window to expose the junction, a manufacturer can build a *photodiode*. When external light falls upon the junction of a reverse-biased photodiode, electron-hole pairs are created inside the depletion layer. The stronger the light, the greater the number of light-produced carriers and the larger the reverse current. Because

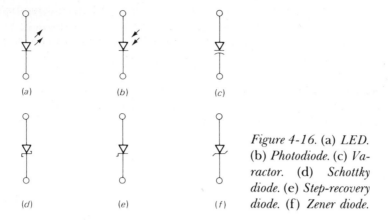

Figure 4-16. (a) LED. (b) Photodiode. (c) Varactor. (d) Schottky diode. (e) Step-recovery diode. (f) Zener diode.

of this, photodiodes make excellent light detectors. Figure 4-16*b* shows the schematic symbol of a photodiode.

varactors

As discussed in Sec. 4-8, the transition capacitance of a diode decreases when the reverse voltage increases. Silicon diodes optimized for this variable-capacitance effect are called *varactors*. Figure 4-16*c* is the schematic symbol of a varactor.

Varactors are replacing mechanically tuned capacitors in many applications. In other words, a varactor in parallel with an inductor gives us a resonant tank circuit; by varying the reverse voltage across the varactor, we can vary the resonant frequency. This electronic control of resonant frequency is useful in remote tuning, sweep oscillators, and other applications to be discussed later.

schottky diodes

The *Schottky diode* (Fig. 4-16 *d*) uses a metal like gold, silver, or platinum on one side of the junction and doped silicon (usually *n* type) on the other side. This kind of diode is a unipolar device because free electrons are the majority carriers on both sides of the junction. Furthermore, the Schottky diode has no depletion layer or charge storage. As a result, it can switch on and off much faster than a bipolar diode. The net effect is a device that can rectify frequencies above 300 MHz, far beyond the capability of the bipolar diode with its reverse-recovery-time limitations.

step-recovery diode

By reducing the doping level near the junction, a manufacturer can produce a *step-recovery diode,* a device that takes advantage of charge storage. During forward conduction, the diode acts like an ordinary diode. When reverse-biased, the step-recovery diode conducts while the depletion layer is adjusting; then all of a sudden, the reverse current drops to zero. It's as though the diode suddenly snaps open like a switch. This is why the step-recovery diode is often called a *snap diode*.

Step-recovery diodes are used in pulse and digital circuits for generating very fast pulses. The sudden snap-off can produce on-off switching of less than 1 ns. These special diodes are also used in *frequency multipliers*. Figure 4-16e is the schematic symbol of a step-recovery diode.

zener diodes

The zener diode is widely used, second only to the rectifier diode. This silicon diode is optimized for operation in the breakdown region. Sometimes called a breakdown diode, the zener diode is the backbone of *voltage regulators*. Figure 4-16f shows the schematic symbol of a zener diode and the next section continues the discussion of this important diode.

4-11 *the zener diode*

The zener diode is made for operation in the breakdown region. By varying the doping level, a manufacturer can produce zener diodes with breakdown voltages from about 2 to 200 V. By applying reverse voltages that exceed the zener breakdown voltage, we have a device that acts like a *constant-voltage* source.

avalanche and zener breakdown

When the applied reverse voltage reaches the breakdown value, minority carries in the depletion layer are accelerated and reach high enough velocities to dislodge valence electrons from outer orbits. The newly liberated electrons can then gain high enough velocities to free other valence electrons. In this way, we get an *avalanche* of free electrons. Avalanche occurs for reverse voltages greater than 6 V or so.

The *zener effect* is different. When a diode is heavily doped, the depletion layer is very narrow. Because of this, the electric field across the depletion layer is very intense. When the field strength reaches approximately 300,000 V per centimeter, the field is intense enough to pull electrons out of valence orbits. The creation of free electrons in this way is called zener breakdown (also known as high-field emission).

The zener effect is predominant for breakdown voltages less than 4 V, the avalanche effect is predominant for breakdown voltages greater than 6 V, and both effects are present between 4 and 6 V. Originally, people thought the zener effect was the only breakdown mechanism in diodes. For this reason, the name "zener diode" came into widespread use before the avalanche effect was discovered. All diodes optimized for operation in the breakdown region are therefore still called zener diodes.

breakdown voltage and power rating

Figure 4-17 shows the current-voltage curve of a zener diode. Negligible reverse current flows until we reach the breakdown voltage V_Z. In a zener diode, the breakdown has a very sharp knee, followed by an almost vertical increase in current. Note that the voltage is approximately constant, equal to V_Z over most of the breakdown region. Data sheets usually specify the value of V_Z at a particular *test current* I_{ZT}, which is beyond the knee (see Fig. 4-17).

The power dissipation of a zener diode equals the product of its voltage and current. In symbols,

$$P_Z = V_Z I_Z \qquad\qquad (4\text{-}4)***$$

For instance, if $V_Z = 12$ V and $I_Z = 10$ mA,

$$P_Z = 12 \times 0.01 = 0.12 \text{ W}$$

As long as P_Z is less than the power rating $P_{Z(max)}$, the zener diode will not be destroyed. Commercially available zener diodes have power ratings from $\frac{1}{4}$ W to more than 50 W.

Data sheets often specify the maximum current a zener diode can handle without exceeding its power rating. This maximum current is designated I_{ZM} (see Fig. 4-17). The relation between I_{ZM} and power rating is given by

$$I_{ZM} = \frac{P_{Z(max)}}{V_Z} \qquad\qquad (4\text{-}4a)$$

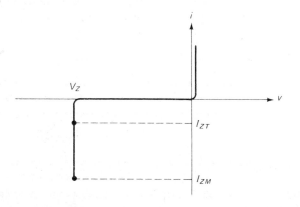

Figure 4-17. Zener-diode curve.

zener impedance

When a zener diode is operating in the breakdown region, a small increase in voltage produces a large increase in current. This implies that a zener diode has a small impedance. We can calculate this impedance with

$$Z_Z = \frac{\Delta v}{\Delta i}$$

For example, if a curve tracer shows changes of 80 mV and 20 mA, the zener impedance is

$$Z_Z = \frac{0.08}{0.02} = 4 \ \Omega$$

Data sheets specify zener impedance at the same test current used for V_Z. The zener impedance at this test current is designated Z_{ZT}. For instance, a 1N3020 has a V_Z of 10 V and a Z_{ZT} of 7 Ω for an I_{ZT} of 25 mA.

temperature coefficient

The *temperature coefficient* T_c is the percent change in zener voltage per degree Celsius. If $V_Z = 10$ V at 25°C and if $T_c = 0.1$ percent,

$$
\begin{aligned}
V_Z &= 10 \text{ V} &&(25°C) \\
V_Z &= 10.01 \text{ V} &&(26°C) \\
V_Z &= 10.02 \text{ V} &&(27°C) \\
V_Z &= 10.03 \text{ V} &&(28°C)
\end{aligned}
$$

and so on.

As a formula, the change in zener voltage is given by

$$\Delta V_Z = T_c \times \Delta T \times V_Z \tag{4-5}$$

Given $T_c = 0.004$ percent and $V_Z = 15$ V at 25°C, the change in zener voltage from 25°C to 100°C is

$$\Delta V_Z = 0.004 \ (10^{-2}) \ (100 - 25) \ 15 = 0.045 \text{ V}$$

Therefore, at 100°C, $V_Z = 15.045$ V.

zener approximations

For all preliminary analysis, we can approximate the breakdown region as vertical. This means the voltage is constant even though the current changes. Figure 4-18a shows the *ideal approximation* of a zener diode. To a first approximation, a zener diode operating in the breakdown region is equivalent to a battery of V_Z volts.

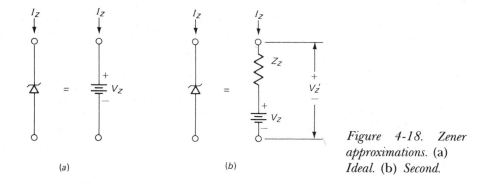

Figure 4-18. Zener approximations. (a) Ideal. (b) Second.

To improve the analysis, we take the slope of the breakdown region into account. The breakdown region is not quite vertical, implying a small zener impedance. Figure 4-18b shows the *second approximation* of a zener diode. Because of the zener impedance, the total zener voltage V_Z' is given by

$$V_Z' = V_Z + I_Z Z_Z \qquad (4\text{-}6)***$$

EXAMPLE 4-8
The zener diode of Fig. 4-19a has $V_Z = 10$ V and $Z_{ZT} = 7$ Ω. Find the value of V_{OUT} with the ideal approximation. Also, calculate the minimum and maximum zener current.

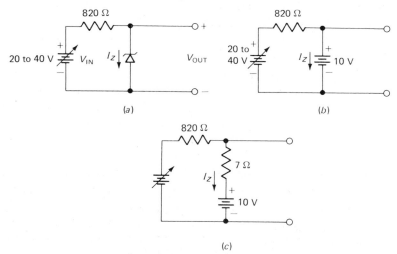

Figure 4-19. An example of voltage regulation.

SOLUTION

The applied voltage (20 to 40 V) is always greater than the breakdown voltage of the zener diode. Therefore, we can visualize the zener diode like the battery shown in Fig. 4-19*b*. The output voltage is

$$V_{OUT} = V_Z = 10 \text{ V}$$

No matter what value the source voltage has between 20 and 40 V, the output voltage always equals 10 V. When the source voltage is 20 V, the series-limiting resistor has 10 V across it; when the source voltage is 40 V, the series-limiting resistor has 30 V across it. Any changes in source voltage, therefore, appear across the series-limiting resistor. The output voltage is ideally constant.

The minimum zener current $I_{Z(min)}$ occurs for minimum source voltage. With Ohm's law,

$$I_{Z(min)} = \frac{V_{IN(min)} - V_Z}{R} = \frac{20 - 10}{820} = 12.2 \text{ mA}$$

The maximum zener current exists when the source voltage is maximum:

$$I_{Z(max)} = \frac{V_{IN(max)} - V_Z}{R} = \frac{40 - 10}{820} = 36.6 \text{ mA}$$

EXAMPLE 4-9

Use the second approximation to calculate the minimum and maximum output voltages in Fig. 4-19*a*.

SOLUTION

Example 4-8 gave $Z_{ZT} = 7 \text{ }\Omega$. Although exact only at the specified test current, Z_{ZT} is a good approximation for Z_Z anywhere in the breakdown region.

We found $I_{Z(min)} = 12.2$ mA and $I_{Z(max)} = 36.6$ mA. When these currents flow through the zener diode of Fig. 4-19*c*, the minimum and maximum output voltages are

$$\begin{aligned} V_{OUT(min)} &\cong V_Z + I_{Z(min)}Z_Z \\ &= 10 + 0.0122(7) = 10.09 \text{ V} \end{aligned}$$

and

$$\begin{aligned} V_{OUT(max)} &\cong V_Z + I_{Z(max)}Z_Z \\ &= 10 + 0.0366(7) = 10.26 \text{ V} \end{aligned}$$

The point of these examples is to illustrate *voltage regulation* (maintaining constant output voltage). Here we have a source that varies from 20 to 40 V, a 100 percent change. The output voltage varies from 10.09 to 10.26 V, a 1.7 percent change. The zener diode has reduced an input change of 100 percent to an output change of only 1.7 percent. Voltage regulation is the main use of zener diodes.

self-testing review

Read each of the following and provide the missing words. Answers appear at the beginning of the next question.

1. The diode symbol points from the *p* side to the *n* side. Because of this, _____ current flows easily from the *p* side to the *n* side.

2. *(conventional)* For a silicon diode, the knee voltage equals the barrier potential, approximately _____. A germanium diode, on the other hand, has a knee voltage of about _____.

3. *(0.7 V, 0.3 V)* Since any conductor has some resistance, both the *p* region and the *n* region also have some resistance. The sum of these resistances is called the _____ resistance.

4. *(bulk)* In circuit terms, the ideal diode acts like an automatic switch. When conventional current tries to flow in the direction of the diode arrow, the switch is _____. If conventional current tries to flow the other way, the switch is _____.

5. *(closed, open)* In the second approximation of a silicon diode, approximately _____ is dropped across a silicon diode when it is forward-biased.

6. *(0.7 V)* In the third approximation of a diode, we include _____ resistance r_B. After a silicon diode turns on, the current produces a voltage across r_B. The greater the current, the _____ the voltage.

7. *(bulk, larger)* When reverse-biased, a diode has a small reverse _____. One way to estimate the importance of this _____ is with the reverse resistance, defined as the ratio of reverse voltage to _____ _____.

8. *(current, current, reverse current)* If you suddenly reverse-bias a diode, the _____ charges can flow in the reverse direction for a while. The greater the _____, the longer these charges can contribute to reverse current.

9. *(stored, lifetime)* The time it takes to turn off a forward-biased diode is called the _____ recovery time t_{rr}. As a rough guide, t_{rr} is approximately the time it takes for _____ current to drop to 10 percent of forward current when a diode is suddenly switched from forward to reverse bias.

10. *(reverse, reverse)* The LED is a light emitter, and the photodiode is a light detector. A varactor has a variable _____. The Schottky diode has no charge storage, which allows it to switch on and off much faster than a bipolar diode. The _____ diode is the second most widely used diode.

11. *(capacitance, zener)* Data sheets specify the value of V_Z at a particular test current I_{ZT}. The zener impedance at this test current is designated _____.

12. *(Z_{ZT})* To a first approximation, a zener diode operating in the breakdown region is equivalent to a battery of V_Z volts. Therefore, even though the zener current varies, the zener voltage remains constant.

problems

4-1. In Fig. 4-20*a*, are the diodes forward- or reverse-biased?

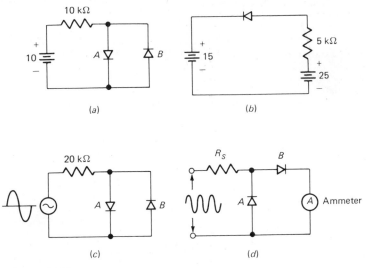

Figure 4-20.

4-2. Is the diode of Fig. 4-20*b* forward- or reverse-biased?

4-3. At the peak of the positive half cycle in Fig. 4-20*c*, which diode is forward-biased? Which one is reverse-biased at the negative input peak?

4-4. Figure 4-20*d* shows a circuit often used in a VOM to measure ac voltages. The ammeter has current through it when diode *B* is forward-biased. During which half cycle of input voltage does this happen? During the negative half cycle of input voltage, is diode *A* forward- or reverse-biased?

4-5. The circuit of Fig. 4-21*a* has a sine wave induced across the secondary winding. When the upper end of the secondary is positive with respect to the lower end, which diode is forward-biased and which reverse-biased?

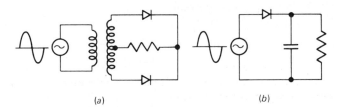

(a)   (b)   *Figure 4-21.*

4-6. In Fig. 4-21*b*, the diode is forward-biased at either the positive or negative input peak. Which of these is it?

4-7. The resistance of the *p* region of a diode equals 5 Ω and the resistance of the *n* region is 3 Ω. What does the bulk resistance of the diode equal?

4-8. A data sheet for a silicon diode says the forward current equals 50 mA at 1 V. Estimate the bulk resistance.

4-9. A silicon diode has a forward current of 30 mA at 1 V. What is the approximate bulk resistance of this diode?

4-10. Figure 4-22*a* shows the forward curve of a silicon diode. Estimate the bulk resistance.

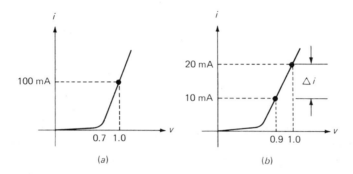

Figure 4-22.

4-11. Figure 4-22*b* gives two values of diode current for corresponding diode voltages. An alternative way to calculate bulk resistance is

$$r_B = \frac{\Delta v}{\Delta i}$$

where the symbol Δ means "the change in." As shown, Δ*i* is the difference in the two values of current. The quantity Δ*v* is the difference in the two values of corresponding voltages. Calculate r_B for Fig. 4-22*b*.

4-12. The diodes of Fig. 4-20*a* are ideal. How much current is in each diode?

4-13. Treat the diode of Fig. 4-20*b* as ideal. How much current is in the 5-kΩ resistor?

4-14. The sine wave of Fig. 4-20*c* has a positive peak voltage of 40 V and a negative peak of −40 V. What does the current in diode A equal at the positive peak? At the negative peak? (Use ideal diodes.)

4-15. Assuming an ideal diode in Fig. 4-23*a*, what is the peak forward current in the circuit? Sketch the output voltage waveform.

4-16. Use the ideal-diode approximation and Thevenin's theorem to find the peak forward current through the diode of Fig. 4-23*b*.

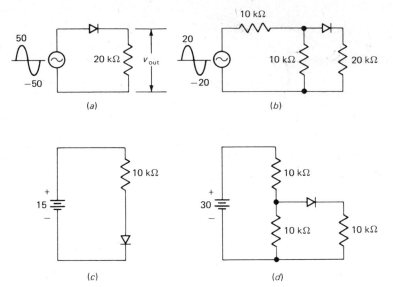

Figure 4-23.

4-17. If the diode of Fig. 4-23c is ideal, how much current is there in the 10-kΩ resistor?

4-18. Use Thevenin's and the ideal-diode approximation to find the current in Fig. 4-23d.

4-19. In Fig. 4-23b, sketch the voltage waveform across the 20-kΩ resistor.

4-20. When the diode of Fig. 4-23a is reverse-biased, what is the maximum voltage across it?

4-21. What is the maximum reverse voltage across the diode of Fig. 4-23b?

4-22. Use the second approximation in Fig. 4-24a to calculate the peak forward current. Sketch the output waveform.

4-23. Calculate the peak current in the 12.5-kΩ resistor using the second approximation for the diode (Fig. 4-24b).

4-24. How much current is there in Fig. 4-24c? (Use the second approximation.) And in Fig. 4-24d?

4-25. Use the third approximation in Fig. 4-25a to calculate the peak forward current through the diode. What is the peak voltage across the 300-Ω resistor?

4-26. The 1N3062 of Fig. 4-25b has these specifications: $I_{F(max)} = 115$ mA and $I_F = 20$ mA at 1 V. Use the third approximation of a silicon diode to calculate the current through the diode.

4-27. A 1N5429 has a reverse current of 50 nA for a reverse voltage of 125 V (at a

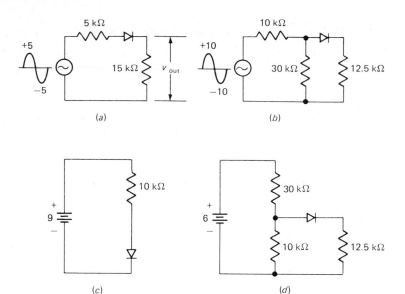

Figure 4-24.

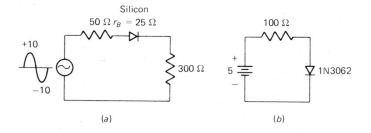

Figure 4-25.

temperature of 25°C). Calculate the reverse resistance. At 150°C, $I_R = 50$ μA for a $V_R = 125$ V. What is the reverse resistance at this elevated temperature?

4-28. The 1N3595 is a low-leakage silicon diode. At 25°C, it has an I_R of 1 nA for a V_R of 125 V. What does its reverse resistance equal? Even at 125°C, $I_R = 500$ nA for a $V_R = 125$ V. Calculate the reverse resistance at 125°C.

4-29 A zener diode has 15 V across it and 20 mA through it. What is the power dissipation?

4-30. If a zener diode has a power rating of 5 W and a zener voltage of 20 V, what does its I_{ZM} equal?

4-31. In the breakdown region of a zener diode, a change of 15 mV produces a change of 2 mA. What is the zener impedance?

4-32. A zener diode has a temperature coefficient of 0.01 percent/°C. If the zener voltage is 8.2 V at 25°C, what does V_Z equal at 75°C?

4-33. $V_Z = 18$ V and $Z_{ZT} = 12$ Ω. If the zener current equals 10 mA, what does V_Z' equal?

4-34. The zener diode of Fig. 4-26a has $V_Z = 15$ V and $P_{Z(max)} = 0.5$ W. If $V_{IN} = 40$ V, what is the minimum value of R that prevents the diode from being destroyed?

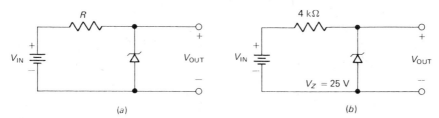

(a)   (b)

Figure 4-26.

4-35. Same data as the preceding problem, except $R = 2$ kΩ. What does the zener current equal? The power dissipation P_Z?

4-36. In Fig. 4-26a, $V_Z = 18$ V, $Z_{ZT} = 2$ Ω, $R = 68$ Ω, and $V_{IN} = 27$ V. What does I_Z equal? And V_Z'? If V_{IN} changes to 40 V, what does V_Z' equal?

4-37. In Fig. 4-26b, what is the minimum value of V_{IN} that keeps the zener current greater than zero?

4-38. $V_{IN} = 50$ V in Fig. 4-26b. Ideally, what does V_{OUT} equal? If $Z_{ZT} = 10$ Ω, what does V_{OUT} equal?

4-39. How much current is there in the LED of Fig. 4-27a if the LED voltage equals 1.7 V?

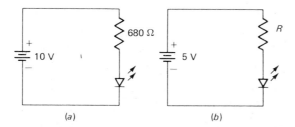

(a)   (b)   *Figure 4-27.*

4-40. The LED of Fig. 4-27b has a voltage drop of 2.2 V. What value should R have to get a current of 40 mA?

5
diode circuits

Power companies in the United States supply a nominal line voltage of 120 V rms at 60 Hz. In Europe it is 240 V at 50 Hz. To get dc voltage, electronics equipment includes a *power supply,* a circuit that converts ac voltage to dc voltage.

5-1 *the half-wave rectifier*

When connected to a power outlet, the three-wire plug of Fig. 5-1*a* grounds the chassis (middle prong) and delivers 120 V rms to the circuit (two outside prongs). Figure 5-1*b* shows how this line voltage looks. As explained in basic courses, the relation between rms and peak values is

$$V_{RMS} = \frac{V_P}{\sqrt{2}} \qquad\qquad (5\text{-}1)\text{★★★}$$

Figure 5-1*a* is a *half-wave rectifier,* a circuit that converts ac voltage into pulsating dc voltage. On the positive half cycle of line voltage, the diode is forward-biased. On the negative half cycle, it's reverse-biased. This is why the voltage across R_L is the half-wave signal shown in Fig. 5-1*c.*

average value and output frequency

The *average* value, also known as the dc value, of a half-wave signal is

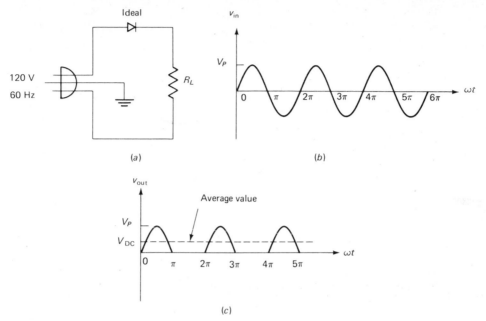

Figure 5-1. Half-wave rectifier. (a) Circuit. (b) Input signal. (c) Output signal.

$$V_{DC} = \frac{V_P}{\pi} \qquad (5\text{-}2)\text{***}$$

where $\pi \cong 3.14$. For example, if the peak value of the half-wave signal is 170 V, the average value is

$$V_{DC} = \frac{170}{\pi} = 54.1 \text{ V}$$

In Fig. 5-1c, the period of the output signal is the same as the period of the input signal. Each cycle of input produces one cycle of output. This is why the output frequency of a half-wave rectifier equals the input frequency:

$$f_{out} = f_{in} \qquad (5\text{-}3)$$

transformer and peak inverse voltage

Most electronics equipment is transformer-coupled at the input, as shown in Fig. 5-2a. The transformer allows us to step the voltage up or down. Another advantage of the transformer is isolation from the power line; this reduces the risk of electrical shock.

Figure 5-2b shows the half-wave rectifier at the peak of the negative half cycle

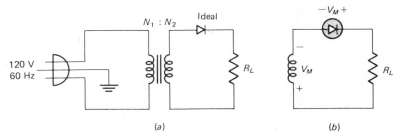

Figure 5-2. (a) *Transformer input.* (b) *Peak inverse voltage.*

of secondary voltage. The diode is shaded or dark to indicate it's off. Since there is no current through the diode, the maximum secondary voltage appears across the diode. This maximum voltage is known as the *peak inverse voltage* (PIV). It represents the maximum voltage the diode must withstand during the reverse part of the cycle. In symbols,

$$\text{PIV} = V_M \qquad (5\text{-}4)$$

EXAMPLE 5-1
The turns ratio of Fig. 5-2a is $N_1 : N_2 = 4 : 1$. What is the dc load voltage? The PIV?

SOLUTION

$$V_P = \sqrt{2}\ V_{RMS} = \sqrt{2} \times 120\ \text{V} = 170\ \text{V}$$

This is the maximum primary voltage. The maximum secondary voltage is

$$V_M = \frac{N_2}{N_1}\ V_P = \frac{1}{4} \times 170 = 42.5\ \text{V}$$

Ignoring the small diode drop, the dc load voltage is

$$V_{\text{DC}} = \frac{V_M}{\pi} = \frac{42.5\ \text{V}}{\pi} = 13.5\ \text{V}$$

and

$$\text{PIV} = V_M = 42.5\ \text{V}$$

5-2 *the center-tap rectifier*

Figure 5-3a shows a *center-tap* rectifier. During the positive half cycle of secondary voltage, the upper diode is forward-biased and the lower diode is reverse-biased;

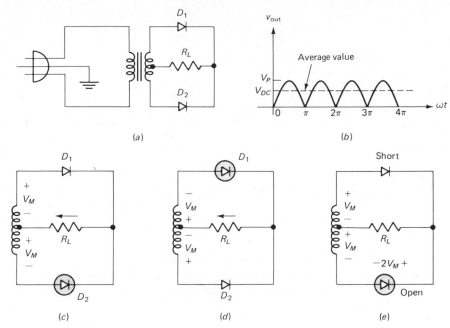

Figure 5-3. Center-tap rectifier.

therefore, current is through the upper diode, the load resistor, and the upper half winding (Fig. 5-3c). During the negative half cycle, current is through the lower diode, the load resistor, and the lower half winding (Fig. 5-3 d). In Figs. 5-3 c and d, the load current is *in the same direction.* This is why the load voltage is the full-wave signal shown in Fig. 5-3 b.

average value and output frequency

The average or dc value of a full-wave signal is

$$V_{DC} = \frac{2 V_P}{\pi} \qquad (5\text{-}5)***$$

For instance, if the peak value of the full-wave signal is 170 V, the average value is

$$V_{DC} = \frac{2 \times 170}{\pi} = 108 \text{ V}$$

In Fig. 5-3b, the period of the output signal is half the period of the input signal. In other words, each cycle of input produces two cycles of output. This is why the output frequency of a center-tap rectifier is twice the input frequency:

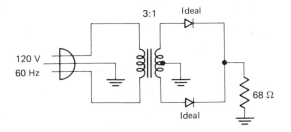

Figure 5-4.

$$f_{out} = 2 f_{in} \qquad (5\text{-}6a)\text{***}$$

If the input frequency is 60 Hz, the output frequency is 120 Hz.

peak inverse voltage

Figure 5-3e shows the circuit at the instant the secondary voltage reaches its maximum value. V_M is the voltage across half the secondary winding; therefore, the reverse voltage across the nonconducting diode is $2 V_M$. In other words,

$$\text{PIV} = 2 V_M \qquad (5\text{-}6b)$$

EXAMPLE 5-2
In Fig. 5-4, the maximum voltage across half the secondary winding is 28.3 V. Ignore diode drop and calculate the average load voltage. Also, what are the output frequency and peak inverse voltage?

SOLUTION
The average load voltage is

$$V_{DC} = \frac{2 V_M}{\pi} = \frac{2 \times 28.3}{\pi} = 18 \text{ V}$$

The output frequency is

$$f_{out} = 2 f_{in} = 2 \times 60 \text{ Hz} = 120 \text{ Hz}$$

The peak inverse voltage is

$$\text{PIV} = 2 V_M = 2 \times 28.3 \text{ V} = 56.6 \text{ V}$$

5-3 *the bridge rectifier*

Figure 5-5a shows the *bridge* rectifier, the most widely used of all. During the positive half cycle of secondary voltage, diodes D_2 and D_3 are forward-biased;

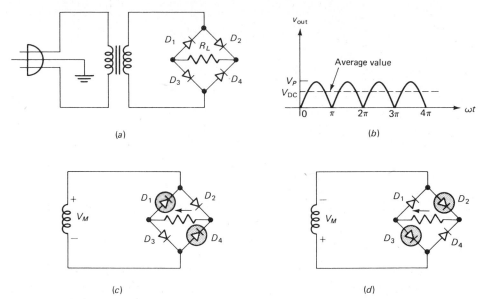

Figure 5-5. Bridge rectifier.

therefore, the load current is to the left (see Fig. 5-5c). During the negative half cycle, diodes D_1 and D_4 are forward-biased, and load current is to the left (Fig. 5-5 d). In Figs. 5-5c and d, the load current is *in the same direction*. This is why the load voltage is the full-wave signal shown in Fig. 5-5b.

The average load voltage is given by

$$V_{\text{DC}} = \frac{2\,V_P}{\pi} \qquad (5\text{-}7)***$$

Also, the output frequency is twice the input:

$$f_{\text{out}} = 2\,f_{\text{in}}$$

And the peak inverse voltage is

$$\text{PIV} = V_M \qquad (5\text{-}8)$$

where V_M is the maximum secondary voltage, shown in Fig. 5-5c.

Table 5-1 compares the rectifiers discussed so far. These rectifiers are called *average* rectifiers because their dc output equals the average value of a rectified sine wave. Besides power supplies, average rectifiers are used in ac voltmeters where they convert ac inputs to direct current suitable for driving dc meters.

As mentioned earlier, the bridge rectifier is the most widely used. Its main disadvantage is having four diodes, two of which conduct on alternate half cycles.

Table 5-1 Ideal Average Rectifiers

	Half-wave	Center-tap	Bridge
Number of diodes	1	2	4
Transformer necessary	No	Yes	No
Peak rectified output	V_M	V_M	V_M
Dc output (unfiltered)	V_M/π	$2V_M/\pi$	$2V_M/\pi$
Peak inverse voltage	V_M*	$2V_M$	V_M
Output frequency	f_{in}	$2f_{in}$	$2f_{in}$

* *Note:* With a capacitor-input filter (Sec. 5-5), the PIV of a half-wave circuit becomes $2V_M$.

This creates a problem when the secondary voltage is low because the two diode drops (1.4 V) become significant. For this reason, the center-tap rectifier is preferred for low-voltage applications because it has only one diode drop (0.7 V). In some low-voltage applications, a center-tap rectifier with germanium diodes may be used, since this results in a diode drop of only 0.3 V.

EXAMPLE 5-3
In Fig. 5-6, the maximum secondary voltage is 68 V. What are the dc load voltage, output frequency, and peak inverse voltage?

SOLUTION
The average load voltage is

$$V_{DC} = \frac{2 V_M}{\pi} = \frac{2 \times 68}{\pi} = 43.3 \text{ V}$$

The output frequency is

$$f_{out} = 2 f_{in} = 2 \times 60 = 120 \text{ Hz}$$

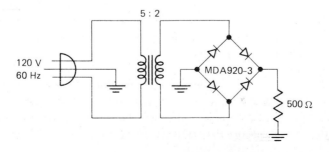

Figure 5-6.

and the peak inverse voltage across each diode is

$$PIV = V_M = 68 \text{ V}$$

The MDA920–3 of Fig. 5-6 is a commercially available bridge-rectifier assembly. It consists of hermetically sealed diodes interconnected and encapsulated in plastic to provide a single rugged package.

A final point. The barrier or *offset* potential of a diode is designated ϕ. When necessary, you can improve the accuracy of your answers by subtracting diode drop as follows:

$$V_P = V_M - \phi \qquad \text{(half-wave or center-tap)}$$
$$V_P = V_M - 2\phi \qquad \text{(bridge)}$$

where $\phi = 0.3$ V for germanium diodes and 0.7 V for silicon diodes. In our example, this means

$$V_P = V_M - 2\phi = 68 - 1.4 = 66.6 \text{ V}$$

and

$$V_{DC} = \frac{2 V_P}{\pi} = \frac{2 \times 66.6}{\pi} = 42.4 \text{ V}$$

5-4 *the choke-input filter*

The uses for a pulsating direct voltage are limited to charging batteries, running dc motors, and a few other applications. What we really need is a dc voltage that is constant in value, similar to the voltage from a battery. To convert half-wave and full-wave signals into constant dc voltage, we must *filter* or smooth out the ac variations. This section is about the *choke-input* filter.

basic idea

Figure 5-7a shows a full-wave rectifier driving a *choke* (iron-core inductor), capacitor, and load resistor. The full-wave signal out of the rectifier has a dc component (we want this) and an ac component (unwanted). The choke allows the dc component to pass through easily because X_L is zero for dc or constant current. Since the capacitor appears open at zero frequency, all the dc current out of the choke flows through load resistance R_L.

The ac component out of the rectifier has a frequency of 120 Hz. The choke blocks this ac component because X_L is high at this frequency. Furthermore, any ac current that does manage to get through the choke passes through the capacitor (very low X_C) rather than through R_L. In other words, the choke and capacitor act like an ac voltage divider that *attenuates* (reduces) the ac component.

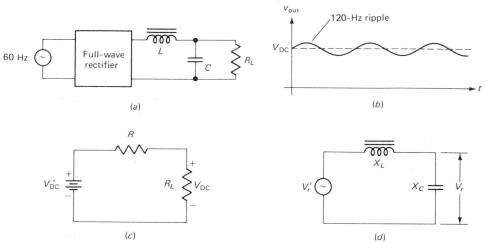

Figure 5-7. Choke-input filter. (a) Circuit. (b) Output. (c) Dc equivalent circuit. (d) Ac equivalent circuit.

dc output

Figure 5-7b shows the filtered output, a large dc component and a small ac component. The dc component is given by

$$V_{DC} = \frac{R_L}{R + R_L} V'_{DC}$$ (5-9)

where V_{DC} = dc voltage across load resistance
R = dc resistance of choke
R_L = load resistance
V'_{DC} = dc voltage from full-wave rectifier

In other words, at zero frequency the choke's resistance in series with the load resistance forms a *voltage divider* as shown in Fig. 5-7c. Usually, R is much smaller than R_L; therefore, almost all of the dc voltage reaches the load.

output ripple

The full-wave signal has a frequency of 120 Hz. After being filtered, this 120 Hz appears greatly attenuated as shown in Fig. 5-7 b. This undesired ac component is called the *ripple;* it's a fluctuation superimposed on the dc component. The ripple is small because X_L is much greater than X_C, and X_C is much smaller than R_L. For these conditions, the circuit acts like the ac voltage divider of Fig. 5-7d, and the output ripple is given by

$$V_r \cong \frac{X_C}{X_L} V_r' \qquad\qquad (5\text{-}10)$$

where V_r = rms output ripple
V_r' = rms input ripple

Typically, X_C/X_L is less than 0.01, which means the ripple is reduced by a factor of more than 100.

An exact analysis of filtering requires *harmonics* (Chap. 22). With harmonic analysis and Eq. (5-10), we can derive the following. For a ripple frequency of 120 Hz,

$$V_r = 5.28(10^{-7}) \frac{V_P}{LC} \qquad \text{(full-wave)} \qquad\qquad (5\text{-}11)$$

Incidentally, a full-wave signal is preferred to a half-wave signal because the ripple frequency is 120 Hz instead of 60 Hz; this means a full-wave rectifier can use smaller L and C. Unless otherwise indicated, all rectifiers discussed from now on are *full-wave* rectifiers (either center-tap or bridge).

ripple factor

The *ripple factor* r is a figure of merit (number used for comparison) for power supplies. In percent, it's defined as

$$r = \frac{V_r}{V_{\mathrm{DC}}} \times 100\% \qquad\qquad (5\text{-}12)\text{***}$$

As an example, if a power supply delivers 10 V dc with a ripple of 0.5 V rms, the ripple factor is

$$r = \frac{0.5}{10} \times 100\% = 5\%$$

If another power supply delivers 25 V dc with a ripple of 1 mV rms, its ripple factor is

$$r = \frac{0.001}{25} \times 100\% = 0.004\%$$

In general, the lower r is, the better.

critical inductance

To work properly, the choke-input filter requires choke current throughout the cycle. If the choke is too small, this condition is not satisfied. The *critical inductance* is the minimum inductance that gives good filtering. The critical inductance for a full-wave rectifier at a line frequency of 60 Hz is given by

$$L_{\text{critical}} \cong \frac{R_L}{1000} \tag{5-13}$$

As long as the inductance is greater than this value, we get normal filtering.

EXAMPLE 5-4

In Fig. 5-8, the full-wave signal at the input to the choke has a peak of 25.7 V and an average value of 16.4 V. If the choke has a dc resistance of 25 Ω, what is the dc output voltage? The output ripple? The ripple factor?

SOLUTION
With Eq. (5-9),

$$V_{\text{DC}} = \frac{R_L}{R + R_L} V'_{\text{DC}} = \frac{750}{25 + 750} \, 16.4 = 15.9 \text{ V}$$

Equation (5-11) gives

$$V_r = 5.28(10^{-7}) \frac{V_P}{LC} = 5.28(10^{-7}) \frac{25.7}{10(500)10^{-6}} = 2.71 \text{ mV}$$

The ripple factor is

$$r = \frac{2.71(10^{-3})}{15.9} \times 100\% = 0.017\% \tag{5-14}$$

An oscilloscope across the load resistance of Fig. 5-8 would display an average voltage of 15.9 V, plus an extremely small ripple of 2.71 mV. This means the output is essentially a constant dc voltage, similar to that from a battery.

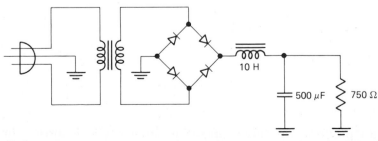

Figure 5-8.

5-5 *the capacitor-input filter*

The choke-input filter is excellent for attenuating ripple, but chokes are bulky and expensive. This has led to the *capacitor-input* filter. Instead of depending on average detection, the capacitor-input filter relies on *peak detection.*

basic idea

Figure 5-9*a* shows a capacitor-input filter. Using a capacitor instead of a choke changes the operation from average detection to peak detection. During the first quarter cycle of input voltage, the diode is forward-biased. Ideally, it looks like a closed switch (see Fig. 5-9*b*). Since the diode connects the source *directly* across the capacitor, the capacitor charges to the peak voltage V_P.

Just past the positive peak, the diode stops conducting, which means the switch opens as shown in Fig. 5-9*c*. Why? Because the capacitor has $+V_P$ volts across it. With the source voltage slightly less than $+V_P$, the capacitor will try to force current back through the diode. This reverse-biases the diode.

With the diode off, the capacitor starts to discharge through the load resistance

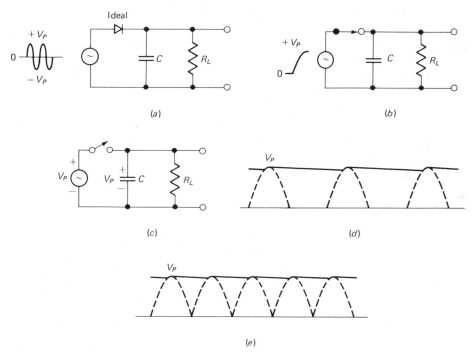

Figure 5-9. Capacitor-input filter. (a) *Circuit.* (b) *During first quarter cycle.* (c) *Just past the positive peak.* (d) *Dc output voltage with ripple.* (e) *Output from full-wave circuit.*

R_L. But here is the key idea behind a *peak rectifier* (also called a peak detector): the $R_L C$ time constant is *much greater* than the period T of the input signal. Because of this, the capacitor will lose only a small part of its charge. Near the next positive input peak, the diode turns on briefly and recharges the capacitor.

output

Figure 5-9*d* shows the kind of output waveform we get with a capacitor-input filter. The maximum voltage equals V_P. When the diode is off, the capacitor discharges through the load resistance. With a long RC time constant, the output voltage drops only slightly as shown; we see only the earliest part of an exponential discharge. Near the next positive peak, the diode turns on briefly. This replaces the lost capacitor charge, and the output voltage rises to V_P.

The signal of Fig. 5-9*d* is almost a constant voltage. The only deviation from a pure dc voltage is the small ripple caused by charging and discharging the capacitor. The smaller the ripple is, the better.

A center-tap or bridge rectifier working into a capacitor produces even better peak rectification because the capacitor is charged twice as often (see Fig. 5-9*e*). As a result, the ripple is smaller and the dc output voltage more closely approaches the peak voltage. Full-wave peak rectifiers are used far more often than half-wave. From now on, our analysis emphasizes the full-wave capacitor-input filter (either center-tap or bridge).

long time constant

In full-wave circuits driven by a line frequency of 60 Hz, the output period is

$$T = \frac{1}{f} = \frac{1}{120} = 8.33 \text{ ms} \tag{5-15}$$

To have a long $R_L C$ time constant, $R_L C$ must be much greater than 8.33 ms. How much greater? At least 10 times, that is,

$$R_L C \geqslant 83.3 \text{ ms} \tag{5-16}$$

is considered a long time constant. When this condition is satisfied, the following approximations can be used for full-wave peak rectifiers:

$$V_{DC} = \left(1 - \frac{0.00417}{R_L C}\right) V_P \tag{5-17}\text{***}$$

and

$$V_r = \frac{0.0024 \, V_P}{R_L C} \tag{5-18}\text{***}$$

where V_r is the rms ripple.

Also useful is the formula for minimum capacitance:

$$C_{MIN} = \frac{0.24}{rR_L} \qquad (5\text{-}19)***$$

Given the ripple factor in percent and the load resistance, you can use this formula to calculate the minimum capacitance needed for filtering.

EXAMPLE 5-5
Given a peak secondary voltage of 30 V in Fig. 5-10, what are the dc output voltage and ripple? (Ignore diode drop.)

SOLUTION
First, check for a long time constant as follows:

$$R_L C = 220 \times 470(10^{-6}) = 103 \text{ ms}$$

Since this is greater than 83.3 ms, Eqs. (5-17) through (5-19) are valid approximations.
 Equation (5-17) gives

$$V_{DC} = \left(1 - \frac{0.00417}{0.103}\right) 30 = 28.8 \text{ V}$$

and Eq. (5-18) gives

$$V_r = \frac{0.0024 \times 30}{0.103} = 0.699 \text{ V}$$

EXAMPLE 5-6
A full-wave peak rectifier has to meet these specifications: $r = 2$ percent and $R_L = 10$ kΩ. What minimum filter capacitance is needed?

SOLUTION
Use Eq. (5-19):

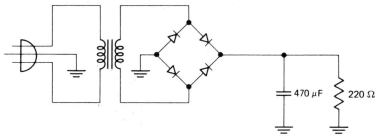

$470 \, \mu F$ $220 \, \Omega$

Figure 5-10.

$$C_{MIN} = \frac{0.24}{rR_L} = \frac{0.24}{2 \times 10,000} = 12 \ \mu F$$

This means the filter capacitor has to be at least 12 μF to ensure a ripple factor of 2 percent or less.

5-6 *nonideal peak rectification*

When the load current is heavy, it may be impossible to get a long time constant. In this case, the discharge between peaks is big, as shown in Fig. 5-11a.

thevenin circuit

Figure 5-11b shows a bridge peak rectifier, and Fig. 5-11c is the equivalent circuit. Here is what each quantity represents:

$$\left(\frac{N_2}{N_1}\right)^2 R_1 = \text{reflected primary resistance}$$
$$R_2 = \text{secondary resistance}$$
$$2\phi = \text{two diode drops}$$
$$2r_B = \text{two bulk resistances}$$

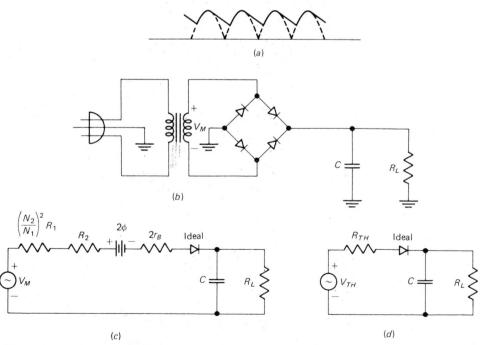

Figure 5-11. Nonideal peak rectification. (a) *Large ripple.* (b) *Circuit.* (c) *Including all nonideal effects.* (d) *Thevenin equivalent circuit.*

By combining resistances and voltages, we get the Thevenin equivalent circuit of Fig. 5-11*d*, where

$$V_{TH} = V_M - 2\phi \tag{5-20}$$

and

$$R_{TH} = \left(\frac{N_2}{N_1}\right)^2 R_1 + R_2 + 2r_B \tag{5-21}$$

output graph

With calculus, it's possible to analyze Fig. 5-11*d*. Figure 5-12 summarizes the results. The top curve is for zero Thevenin resistance; this applies when the primary and secondary winding resistances are negligible, and when the diode

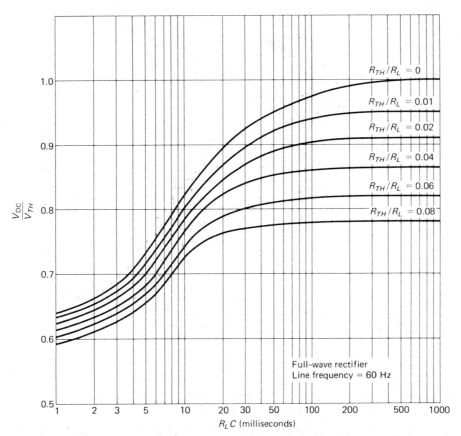

Figure 5-12. Normalized dc output of full-wave rectifier for line frequency of 60 Hz.

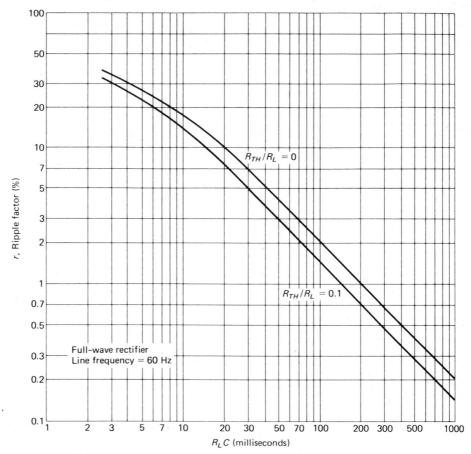

Figure 5-13. Ripple factor of full-wave rectifier for line frequency of 60 Hz.

bulk resistances are small enough to ignore. As you see, a long time constant results in V_{DC}/V_{TH} approaching unity. But when the time constant is short, the dc voltage falls off.

The second curve is for $R_{TH}/R_L = 0.01$. This means R_{TH} is 1 percent of R_L. With a long time constant, V_{DC}/V_{TH} is no more than 0.95. The remaining curves are for larger Thevenin resistances.

Figure 5-13 shows another useful graph. Here we have ripple factor as a function of time constant. The top curve is for zero Thevenin resistance, and the bottom curve is for a Thevenin resistance equal to 10 percent of load resistance. You can interpolate for intermediate values of Thevenin resistance. Notice how the ripple factor decreases as the time constant increases.

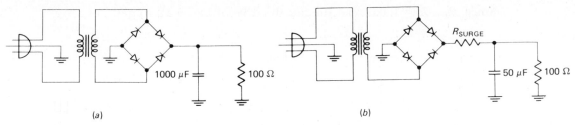

Figure 5-14. Surge current.

surge current

Before power is turned on in Fig. 5-14a, the filter capacitor is uncharged. At the instant the circuit is energized, the capacitor looks like a short; therefore, the initial charging current may be quite large. This sudden gush of current is called the *surge current.*

In the worst case, we can energize the circuit at the instant line voltage is at its maximum. This means V_M is across the secondary and the capacitor is still uncharged; only R_{TH} can impede current at this instant. Therefore, the worst-case surge current is

$$I_{\text{SURGE}} = \frac{V_M}{R_{TH}} \qquad (5\text{-}22)***$$

As an example, if $V_M = 25$ V and $R_{TH} = 1$ Ω,

$$I_{\text{SURGE}} = \frac{25 \text{ V}}{1 \text{ }\Omega} = 25 \text{ A}$$

In some power supplies the surge current is so high that it would destroy the diodes. In this case, it is necessary to add a surge resistor as shown in Fig. 5-14b. For instance, if we want to limit the surge current to approximately 5 A in the preceding example, we can add a 3.9-Ω surge resistor to get

$$I_{\text{SURGE}} = \frac{V_M}{R_{TH}} = \frac{25}{1 + 3.9} = 5.1 \text{ A}$$

EXAMPLE 5-7
R_{TH} is 4 Ω in Fig. 5-14a. Ignoring the offset voltage of the diodes, what are the dc output voltage and ripple factor for a V_M of 25 V?

SOLUTION
The time constant of this circuit is

$$R_L C = 100(1000)10^{-6} = 100 \text{ ms}$$

Since $R_{TH}/R_L = 0.04$, read the fourth curve of Fig. 5-12 to get

$$\frac{V_{DC}}{V_{TH}} \cong 0.86$$

Because we are ignoring offset voltage, $V_{TH} = V_M = 25$ V. Therefore,

$$V_{DC} = 0.86\, V_{TH} = 0.86(25) = 21.5 \text{ V}$$

Next, interpolate Fig. 5-14 to get

$$r \cong 1.9\%$$

EXAMPLE 5-8

Use the data of Example 5-7, but this time include the effect of offset voltage for silicon diodes.

SOLUTION
The Thevenin voltage is

$$V_{TH} = V_M - 2\phi = 25 - 1.4 = 23.6 \text{ V}$$

and

$$V_{DC} = 0.86\, V_{TH} = 0.86 \times 23.6 = 20.3 \text{ V}$$

The ripple factor is still 1.9 percent.

EXAMPLE 5-9

In Fig. 5-14*b*, R_{TH} (includes R_{SURGE}) equals 2 Ω. What are the dc output voltage and ripple factor for $V_M = 25$ V? Include silicon offset voltage.

SOLUTION
This time, $R_{TH}/R_L = 0.02$. The time constant is

$$R_L C = 100(50)10^{-6} = 5 \text{ ms}$$

The third curve of Fig. 5-12 gives

$$\frac{V_{DC}}{V_{TH}} \cong 0.7$$

or

$$V_{DC} = 0.7\, V_{TH} = 0.7(25 - 1.4) = 16.5 \text{ V}$$

Interpolating Fig. 5-13 gives

$$r \cong 26\%$$

5-7 *rc and lc filters*

When the discharging time constant is long, the ripple out of a peak rectifier may be small enough to ignore. But when the time constant is short, the ripple is large and we have to use additional filtering to reduce it. This section is about RC and LC filters, one way to attenuate the ripple.

rc filter

Figure 5-15*a* shows an RC filter between the input capacitor and load resistor. By deliberate design, R is much greater than X_C. Therefore, the circuit acts like an ac voltage divider (Fig. 5-15*b*). Because R is much greater than X_C, the output ripple is much smaller than the input ripple. Typically, R is at least 10 times X_C; this means the output ripple is attenuated or reduced by a factor of at least 10.

We can use more than one section, as shown in Fig. 5-15*c*. Since each section acts like an ac voltage divider, the overall attenuation equals the product of the individual attenuations; if each section reduces the ripple by a factor of 15, the overall attenuation is 225.

Figure 5-16 shows the *attenuation* α for one, two, or three sections. For instance, if $RC = 20$ ms,

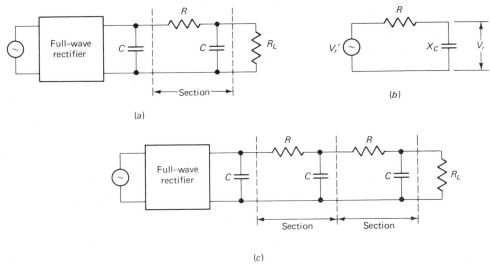

Figure 5-15. (a) *One-section* RC *filter.* (b) *Ac equivalent circuit.* (c) *Two-section* RC *filter.*

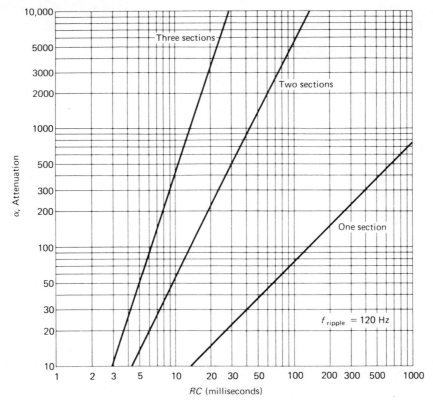

Figure 5-16. Attentuation of RC *filter versus filter time constant.*

$$\alpha = 15 \qquad \text{(one section)}$$
$$\alpha = 225 \qquad \text{(two sections)}$$
$$\alpha = 3300 \qquad \text{(three sections)}$$

This means one section reduces the ripple by a factor of 15, two sections by 225, and three sections by 3300.

The main disadvantage of an RC filter is the loss of dc voltage across resistance R. Since R is in series with R_L, we get voltage-divider action. On the one hand, we need a large R for good filtering action. On the other, we need a small R to prevent excessive loss of dc voltage. These conflicting requirements mean the RC filter is practical only for small load currents (large R_L).

lc filter

When the load current is heavy, an LC filter is preferred. Figure 5-17a shows a one-section LC filter. As discussed earlier, the X_L is much greater than the

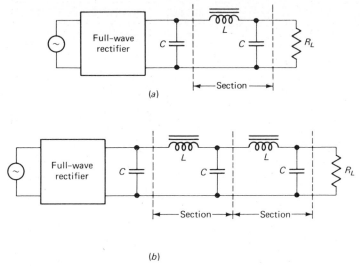

Figure 5-17. LC *filter.* (a) *One section.* (b) *Two sections.*

X_C. Because of this, the ripple is attenuated. Typically, X_L is at least 10 times greater than X_C; therefore, the ripple is attenuated by at least a factor of 10.

We can use more than one section, as shown in Fig. 5-17*b*. Since each section acts like an ac voltage divider, the overall attenuation equals the product of the individual attenuations. For example, if each section reduces the ripple by a factor of 20, then the overall attenuation is 400.

Figure 5-18 shows the attenuation for one, two, or three *LC* sections.

EXAMPLE 5-10

A two-section *LC* filter has $L = 1$ H and $C = 68$ μF. What is the attenuation for a ripple frequency of 120 Hz? The dc output voltage if each choke's resistance is 1 Ω, V_{DC} is 50 V, and R_L is 1 kΩ?

SOLUTION

The *LC* value (*L* in henries, *C* in microfarads) is

$$LC = 1 \times 68 = 68$$

Reading the second curve of Fig. 5-18 gives

$$\alpha = 1400$$

The dc output voltage is

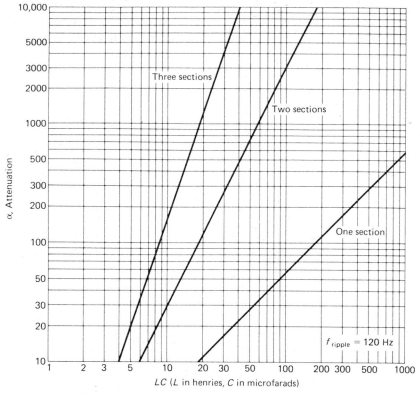

Figure 5-18. Attenuation of LC *filter.*

$$V_{\text{DC}} = \frac{R_L}{2R + R_L} \, V_{\text{DC}}' = \frac{1000}{2 + 1000} \, 50 = 49.9 \text{ V}$$

5-8 *voltage multipliers*

A *voltage multiplier* is two or more peak rectifiers that produce a dc voltage equal to a multiple of the peak input voltage ($2V_P$, $3V_P$, $4V_P$, and so on). These power supplies are used for high voltage/low current applications like supplying cathode-ray tubes (the picture tubes in TV receivers, oscilloscopes, and computer displays).

voltage doubler

Figure 5-19a is a *voltage doubler,* a connection of two peak rectifiers. At the peak of the negative half cycle, D_1 is forward-biased and D_2 is reverse-biased. This

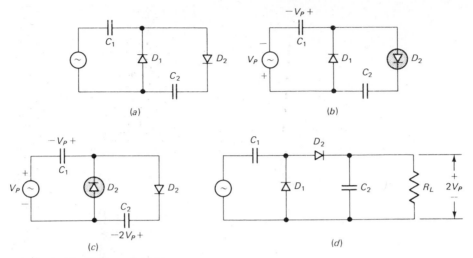

Figure 5-19. Voltage doubler.

charges C_1 to the peak voltage V_P with the polarity shown in Fig. 5-19b. At the peak of the positive half cycle, D_1 is reverse-biased and D_2 is forward-biased. Because the source and C_1 are in series, C_2 will try to charge toward $2V_P$. After several cycles, the voltage across C_2 will equal $2V_P$ as shown in Fig. 5-19c.

By redrawing the circuit and connecting a load resistance, we get Fig. 5-19d. As long as R_L is large, the output voltage approximately equals $2V_P$. That is, provided the load is light (long time constant), the output voltage is double the peak input voltage.

voltage tripler

By connecting another section, we get the *voltage tripler* of Fig. 5-20a. The first two peak rectifiers act like a doubler. At the peak of the negative half cycle,

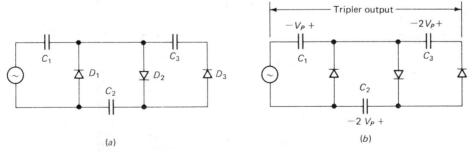

Figure 5-20. Voltage tripler.

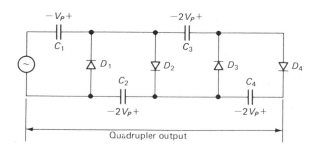

Figure 5-21. Voltage quadrupler.

D_3 is forward-biased. This charges C_3 to $2V_P$ with the polarity shown in Fig. 5-20*b*. The tripler output appears across C_1 and C_3.

The load resistance is connected across the tripler output. As long as the time constant is long, the output approximately equals $3V_P$.

voltage quadrupler

Figure 5-21 is a *voltage quadrupler,* four peak rectifiers in cascade (one after another). The first three are a tripler, and the fourth makes the overall circuit a quadrupler. As shown in Fig. 5-21, the first capacitor charges to V_P; all others charge to $2V_P$. The quadrupler output is across the series connection of C_2 and C_4. As usual, a large load resistance (long time constant) is needed to have an output of approximately $4V_P$.

Theoretically, we can add sections indefinitely; however, *voltage regulation* becomes extremely poor. This is why you rarely see n greater than four.

5-9 *voltage regulation*

The load resistance connected to a power supply may be a single resistor or it may be the equivalent resistance of several circuits in parallel. In either case, this load resistance is usually variable, able to change from a low to a high value. In Fig. 5-22*a*, load current I_{DC} passes through a variable R_L. When R_L varies from infinity to lower values, I_{DC} increases from zero to higher values. Figure 5-22*b* shows the typical decrease in V_{DC} for an increase in I_{DC}. The dropoff in V_{DC} is caused by filter resistance, decreasing R_LC time constant, source resistance, etc.

One way to compare the performance of power supplies is with a figure of merit called the *voltage regulation:*

$$VR = \frac{V_{NL} - V_{FL}}{V_{FL}} \times 100 \ \%$$

(5-23)***

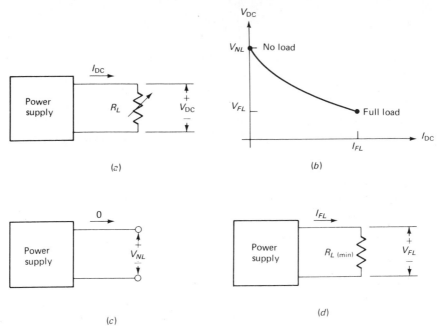

Figure 5-22. Voltage regulation. (a) *Circuit.* (b) *Regulation curve.* (c) *No-load condition.*
(d) *Full-load condition.*

where VR = voltage regulation in percent
V_{NL} = dc output voltage with no load
V_{FL} = dc output voltage with full load

No load means zero load current, equivalent to infinite R_L (Fig. 5-22c). *Full load* means maximum load current, the same as minimum R_L (Fig. 5-22d).

·In a well designed power supply, the full-load voltage is only slightly less than the no-load voltage. In Eq. (5-23), this means VR approaches zero. In other words, the lower VR is, the better the power supply. As an example, the VR of commercially available power supplies is less than 1 percent. This means the full-load voltage is within 1 percent of the no-load voltage.

When using a commercial power supply, it's useful to know the minimum load resistance allowed. By applying Ohm's law to Fig. 5-22d,

$$R_{L \text{ (min)}} = \frac{V_{FL}}{I_{FL}} \qquad (5\text{-}24)^{***}$$

As an example, suppose a data sheet specifies an output voltage of 50 V at a maximum rated current of 0.2 A. Then,

$$R_{L \text{ (min)}} = \frac{V_{FL}}{I_{FL}} = \frac{50}{0.2} = 250 \ \Omega$$

This is the smallest load resistance you can connect to the supply.

A final point. The choke-input and capacitor-input rectifiers are examples of *unregulated* power supplies, where output voltage decreases significantly when load current increases. The *VR* of unregulated supplies is usually greater than 10 percent. Chapter 21 will introduce *regulated* power supplies; they use sophisticated electronic circuits to keep the output voltage constant. A regulated power supply typically has a *VR* of less than 1 percent.

EXAMPLE 5-11

Figure 5-23*a* shows the *regulation curve* (graph of V_{DC} versus I_{DC}) for a power supply that can deliver a maximum rated load current of 0.5 A. Calculate the voltage regulation and the minimum load resistance.

SOLUTION
With Eq. (5-23),

$$VR = \frac{30 - 20}{20} \times 100 \ \% = 50 \ \%$$

The minimum load resistance you can connect to the supply is

$$R_{L \text{ (min)}} = \frac{V_{FL}}{I_{FL}} = \frac{20}{0.5} = 40 \ \Omega$$

EXAMPLE 5-12

Figure 5-23*b* is the regulation curve for another power supply. Calculate *VR* and $R_{L \text{ (min)}}$.

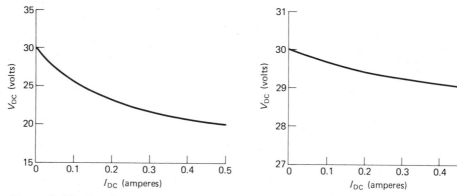

Figure 5-23. Regulation curves.

SOLUTION
With Eq. (5-23),

$$VR = \frac{30 - 29}{29} \times 100 \ \% = 3.45 \ \%$$

The minimum allowable load resistance is

$$R_{L \ (min)} = \frac{V_{FL}}{I_{FL}} = \frac{29}{0.5} = 58 \ \Omega$$

This power supply is much better than the preceding power supply. A regulation of 50 percent is terrible, useless in all but a few applications. On the other hand, a regulation of 3.45 percent means the full-load voltage is only slightly less than the no-load voltage; a power supply like this is very useful.

5-10 *the zener regulator*

A simple way to improve voltage regulation is with a *zener regulator,* shown in Fig. 5-24. The voltage from an unregulated power supply is used as the input voltage V_{IN} to the zener regulator. As long as V_{IN} is greater than V_Z, the zener diode operates in the breakdown region. A series-limiting resistor R_S prevents the zener current from exceeding its rated maximum of I_{ZM}.

Ideally, the zener diode acts like a battery; therefore, the load voltage is constant. For instance, suppose the voltage out of the unregulated power supply changes. As long as this voltage is greater than the zener's breakdown voltage, the zener operates in the breakdown region and the load voltage stays constant.

fundamental relations

In Fig. 5-24, the current through the series-limiting resistor is given by Ohm's law:

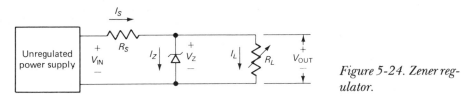

Figure 5-24. Zener regulator.

$$I_S = \frac{V_{\text{IN}} - V_{\text{OUT}}}{R_S} \tag{5-25}$$

This current splits at the junction of the zener diode and the load resistor. With Kirchhoff's current law,

$$I_Z = I_S - I_L \tag{5-26}$$

Ignoring the small zener impedance,

$$V_{\text{OUT}} \cong V_Z \tag{5-27}***$$

and

$$I_L = \frac{V_{\text{OUT}}}{R_L} \tag{5-28}$$

Equations (5-25) through (5-28) are adequate for a preliminary analysis of zener-diode circuits. To improve accuracy, we can modify Eq. (5-27) to include the effect of zener impedance:

$$V_{\text{OUT}} = V_Z + I_Z Z_Z \tag{5-29}***$$

maximum limiting resistance

For a zener regulator to hold the output voltage constant, the zener diode must remain in the breakdown region under all operating conditions. In other words, there must be zener current for all source voltages and load currents. The worst case occurs for minimum source voltage and maximum load current because the zener current drops to a minimum.

By solving Eqs. (5-25) through (5-28) simultaneously for R_S, we get the maximum allowable series-limiting resistance:

$$R_{S\text{ (max)}} = \frac{V_{\text{IN (min)}} - V_{\text{OUT}}}{I_{L\text{ (max)}}} \tag{5-30}***$$

where $R_{S\text{ (max)}}$ = largest permitted limiting resistor
$V_{\text{IN (min)}}$ = smallest possible source voltage
V_{OUT} = zener breakdown voltage (approximately)
$I_{L\text{ (max)}}$ = largest possible load current

If you try to use a resistance larger than the value given by this equation, the zener regulator will stop regulating for low source voltages and high load currents.

EXAMPLE 5-13
Calculate the current through the series-limiting resistor of Fig. 5-25a, the minimum and maximum load current, and the minimum and maximum zener current.

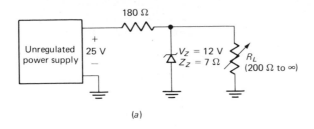

(a)

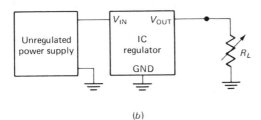

(b)

Figure 5-25.

SOLUTION
With Eq. (5-25),

$$I_S = \frac{V_{IN} - V_{OUT}}{R_S} = \frac{25 - 12}{180} = 72 \text{ mA}$$

The minimum load current occurs for infinite R_L:

$$I_{L \text{ (min)}} = 0$$

The maximum load current occurs when R_L is 200 Ω. With Eq. (5-28),

$$I_{L \text{ (max)}} = \frac{V_{OUT}}{R_{L \text{ (min)}}} = \frac{12}{200} = 60 \text{ mA}$$

Because of Eq. (5-26),

$$I_{Z \text{ (min)}} = I_S - I_{L \text{ (max)}} = 72 \text{ mA} - 60 \text{ mA} = 12 \text{ mA}$$

Conversely,

$$I_{Z \text{ (max)}} = I_S - I_{L \text{ (min)}} = 72 \text{ mA} - 0 = 72 \text{ mA}$$

Here's the point. The current I_S through the series-limiting resistor is constant, equal to 72 mA. When the load current increases from 0 to 60 mA, the zener current decreases from 72 to 12 mA, maintaining I_S current constant in value. This is normal operation for a zener regulator: I_S and V_{OUT} remain fixed in spite of changes in load current or source voltage.

EXAMPLE 5-14

Use Eq. (5-29) to calculate the minimum and maximum load voltage in Fig. 5-25a. Also, work out the VR.

SOLUTION

In the preceding example, $I_{Z(min)} = 12$ mA and $I_{Z(max)} = 72$ mA. With Eq. (5-29), the minimum zener current gives

$$V_{OUT} = V_Z + I_Z Z_Z = 12 + 0.012(7) = 12.1 \text{ V}$$

The maximum zener current results in

$$V_{OUT} = 12 + 0.072(7) = 12.5 \text{ V}$$

The voltage regulation is

$$VR = \frac{12.5 - 12.1}{12.1} \times 100\% = 3.3\%$$

A final point. By combining zener diodes and transistors on a chip, manufacturers are producing three-terminal *IC* voltage regulators (Fig. 5-25b). The input to an *IC* regulator is the voltage from an unregulated supply, such as a bridge peak rectifier. The output of the regulator goes to the load resistance. The new *IC*s are easy to use, offer high reliability, and regulate to better than 1 percent.

EXAMPLE 5-15

A zener regulator has an input voltage from 15 to 20 V and a load current from 20 to 100 mA. To hold load voltage constant under all conditions, what value should the series-limiting resistor have if $V_Z = 10$ V?

SOLUTION

The worst case occurs for minimum source voltage and maximum load current. With Eq. (5-30),

$$R_{S(max)} = \frac{15 - 10}{0.1} = 50 \text{ } \Omega$$

If R_S is greater than 50 Ω, the zener regulator will stop regulating for low source voltages and high load currents. In other words, the zener diode will not be operating in the breakdown region when $V_{IN} = 15$ V and $I_L = 100$ mA.

5-11 *the clipper*

In radar, digital computers, and other electronic systems, we sometimes want to remove signal voltages above or below a specified voltage level. Diode *clippers* are one way to do this.

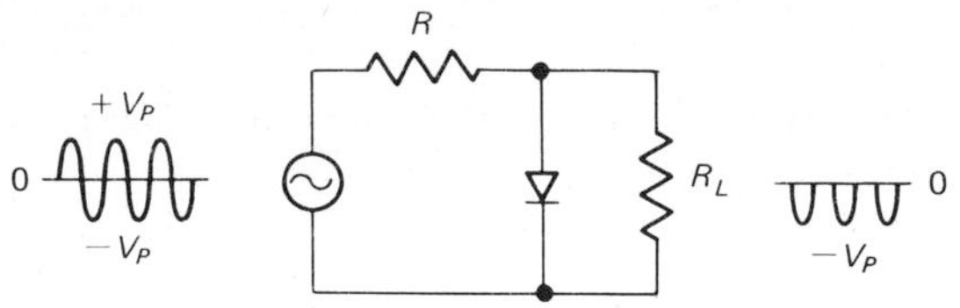

Figure 5-26. Positive clipper.

the positive clipper

Figure 5-26 shows a *positive clipper*, a circuit that removes positive parts of the signal. As shown, the output voltage has all positive half cycles clipped off. The circuit works as follows: During the positive half cycle of input voltage, the diode conducts heavily. To a first approximation, we visualize the diode as a closed switch. The voltage across a short must equal zero; therefore, the output voltage equals zero during each positive half cycle. (All the voltage is dropped across R.)

During the negative half cycle, the diode is reverse-biased and looks open. In effect, the circuit is a voltage divider with an output of

$$V_{\text{out}} = \frac{R_L}{R + R_L} V_P$$

Normally, R_L is much greater than R, so that

$$V_{\text{out}} \cong - V_P$$

So, during each positive half cycle, the diode conducts heavily and most of the voltage is dropped across R; almost none of the voltage appears across R_L. During each negative half cycle, the diode is off. Because R_L is much greater than R, most of the negative half cycle appears across R_L.

Figure 5-26 shows the output waveform. All signal above the 0-V level has been clipped off. The positive clipper is also called a *positive limiter* because the output voltage is limited to a maximum of 0 V.

The clipping is not perfect. To a second approximation, a conducting silicon diode drops 0.7 V. Because the first 0.7 V is used to overcome the barrier potential, the output signal is clipped near +0.7 V rather than 0 V.

In Fig. 5-26, what happens if we turn the diode around? Right, it conducts better on negative half cycles. Stated simply, if you reverse the polarity of the diode in Fig. 5-26, you get a negative clipper that removes all signal below 0 V.

the biased clipper

In some applications, you may want clipping levels different from 0 V. With a *biased clipper* you can move the clipping level to a desired positive or negative level.

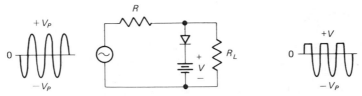

Figure 5-27. Biased positive clipper.

Figure 5-27 shows a biased clipper. For the diode to turn on, the input voltage must be greater than $+V$. When v_{in} is greater than $+V$, the diode acts like a closed switch (ideally) and the voltage across the output equals $+V$. This output voltage stays at $+V$ as long as the input voltage exceeds $+V$.

When the input voltage is less than $+V$, the diode opens and the circuit reverts to a voltage divider. As usual, R_L should be much greater than R; in this way, most of the input voltage appears across the output.

The output waveform of Fig. 5-27 summarizes the circuit action. The biased clipper removes all signal above the $+V$ level.

a combination clipper

You can combine biased positive and negative clippers as shown in Fig. 5-28. Diode D_1 turns on when the input voltage is greater than $+V_1$. Therefore, the output voltage equals $+V_1$ when v_{in} is greater than $+V_1$.

On the other hand, when v_{in} is more negative than $-V_2$, diode D_2 turns on. With D_2 ideally shorted, the output voltage equals $-V_2$ as long as the input voltage is more negative than $-V_2$.

When v_{in} lies between $-V_2$ and $+V_1$, neither diode is on. With R_L much greater than R, most of the input voltage appears across the output.

When the input signal is large, that is, when V_P is much greater than the clipping levels, the output signal resembles a square wave like that of Fig. 5-28.

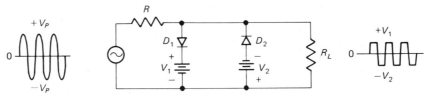

Figure 5-28. Combination clipper.

5-12 *the clamper*

All a clamper does is add a dc component to the signal. In Fig. 5-29*a* the input signal is a sine wave with a peak-to-peak value of 20 V. The clamper pushes the signal upward, so that the negative peaks fall on the 0-V level. As you can see, the shape of the original signal is preserved; all that happens is a vertical shift of the signal. We describe an output signal like that of Fig. 5-29*a* as *positively clamped.*

the key idea

Before getting to a particular clamper circuit, realize this: a clamper has to add a dc voltage to the incoming signal. For instance, in Fig. 5-29*b* the lower source represents the incoming signal. The clamper adds 10 V dc to the signal.

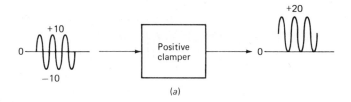

(a)

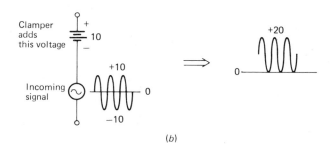

(b)

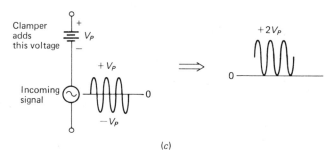

(c)

Figure 5-29. Positive clamping.

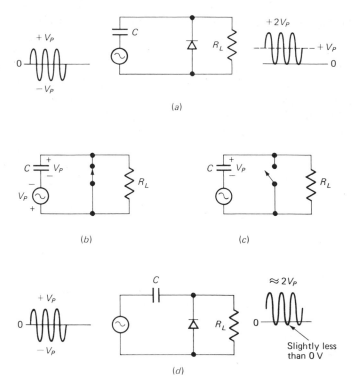

Figure 5-30. Action of a positive clamper.

Therefore, the sum of the incoming signal and the 10 V dc is the positively clamped sine wave shown in Fig. 5-29b.

Given an incoming signal like the one in Fig. 5-29c, the clamper will add a dc voltage of V_P. As a result, the output signal is positively clamped as shown.

the positive clamper

Figure 5-30a shows a positive clamper. Ideally, here is how it works. On the first *negative* half cycle of input voltage, the diode turns on as shown in Fig. 5-30b. At the negative peak, the capacitor must charge to V_P with the polarity shown.

Slightly beyond the negative peak, the diode shuts off as shown in Fig. 5-30c. The $R_L C$ time constant is deliberately made much greater than the period T of the incoming signal. For this reason, the capacitor remains almost fully charged during the off time of the diode. To a first approximation, the capacitor acts like the battery of Fig. 5-29c. This is why the output voltage in Fig. 5-30a is a positively clamped signal.

Figure 5-30d shows the circuit as it is usually drawn. Since the diode drops a few-tenths volt when conducting, the capacitor voltage does not quite reach V_P. For this reason, the clamping is not perfect, and the output generally dips slightly below the 0-V level as shown.

What happens if we turn the diode in Fig. 5-30*d* around? The polarity of capacitor voltage reverses, and the circuit becomes a *negative clamper*. Both positive and negative clampers are widely used. Television receivers, for instance, use a clamper to add a dc voltage to the video signal. In television work, the clamper is usually called a *dc restorer*.

5-13 *the peak-to-peak detector*

If you cascade a positive clamper and a peak detector, you get a peak-to-peak detector (see Fig. 5-31*a*). The input sine wave is positively clamped; therefore,

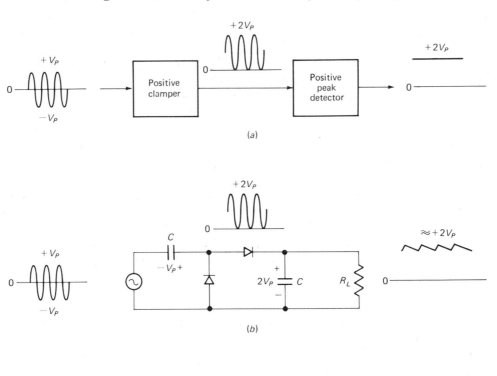

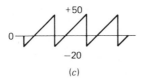

Figure 5-31. Peak-to-peak detector.

the input to the peak detector has a peak value of $2V_P$. This is why the output of the peak detector is a dc voltage equal to $2V_P$.

Figure 5-31b shows a peak-to-peak detector. As usual, the discharge time constant R_LC must be much greater than the period of the incoming signal. By satisfying this condition, you get good clamping action and good peak detection. The output ripple will therefore be small.

Where are peak-to-peak detectors used? Sometimes, the output of a peak-to-peak detector is applied to a dc voltmeter. In this way, the combination acts like a peak-to-peak ac voltmeter. For instance, if the sawtooth wave of Fig. 5-31c is measured with such a voltmeter, the reading would be 70 V pp. Another use for the peak-to-peak detector is in power supplies to get twice as much dc output voltage as with an ordinary peak rectifier. When used in this way, the peak-to-peak detector is the voltage doubler discussed earlier.

5-14 *the dc return*

One of the most baffling things that may happen in the laboratory is this: you connect a signal source to a circuit; for some reason, the circuit will not work; yet nothing is defective in the circuit or in the signal source. As a concrete example, Fig. 5-32a shows a sine-wave source driving a half-wave rectifier. When you look at the output with an oscilloscope, you see no signal at all; the rectifier refuses to work. To add to the confusion, you may try another sine-wave source and find a normal half-wave signal across the load (Fig. 5-32b).

The phenomenon just described is a classic in electronics; it occurs again and again in practice. It may happen with diode circuits, transistor circuits, integrated circuits, etc. Unless you understand why one kind of source works

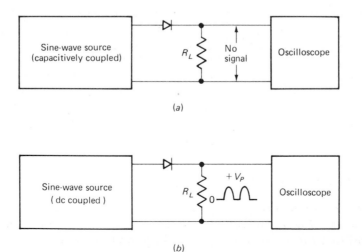

Figure 5-32. The dc-return problem.

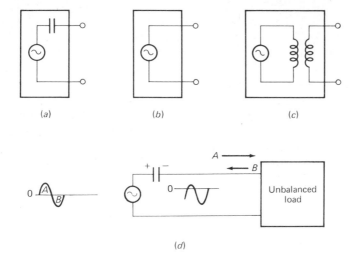

Figure 5-33. (a) *Capacitively coupled source.* (b) *Direct-coupled source.* (c) *Transformer-coupled source.* (d) *Unbalanced load causes unequal currents.*

and another doesn't, you will be confused and possibly discouraged every time the problem arises.

types of coupling

The signal source of Fig. 5-33*a* is capacitively coupled; this means it has a capacitor in the signal path. Many commercial signal generators use a capacitor to *dc-isolate* the source from the load, that is, to prevent dc current between the source and load. The idea of a capacitively coupled source is to let only the ac signal pass from source to load.

The dc-coupled source of Fig. 5-33*b* is different. It has no capacitor; therefore, it provides a path for both ac and dc currents. When you connect this kind of source to a load, it is possible for the load to force a dc current through the source. As long as this dc current is not too large, no damage occurs to the source. Many commercial signal generators are dc-coupled like this.

Sometimes, a signal source is transformer-coupled like Fig. 5-33*c*. The advantage is that it passes the ac signal from source to load, and at the same time provides a dc path through the secondary winding.

All circuits discussed earlier in this chapter work with dc-coupled and transformer-coupled sources. It is only with the *capacitively coupled sources* that trouble may arise.

unbalanced diode circuits

A capacitively coupled source will cause unwanted clamping action when the load is *unbalanced.* Figure 5-33*d* shows what we mean by an unbalanced load; it is any load that causes *unequal* currents during alternate half cycles. Because

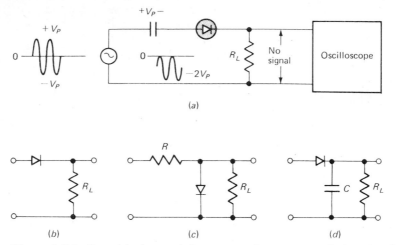

Figure 5-34. Capacitively coupled source produces unwanted clamping in some circuits.

of this, the amount of charge deposited on the capacitor plates is different on one half cycle from the next. When current A is greater than current B, the capacitor voltage increases during each cycle. As we saw with clamper circuits, a charged capacitor like this will push the ac signal up or down depending on the polarity of the capacitor voltage. In other words, whether we like it or not, *an unbalanced load will cause clamping of the signal out of a capacitively coupled source.*

Now we know why a half-wave rectifier won't work if connected to a capacitively coupled source. In Fig. 5-34*a*, the capacitor charges to V_P during the first few cycles. Because of this, the signal coming from the source is negatively clamped and the diode cannot turn on after the first few cycles. This is why we see no signal on the oscilloscope.

Among the diode circuits discussed earlier, the following are unbalanced loads: half-wave rectifier, clipper, peak detector, clamper, and peak-to-peak detector. The last two are supposed to clamp the signal; therefore, they work fine with a capacitively coupled source. But the half-wave rectifier, the clipper, and the peak detector of Fig. 5-34*b*, *c,* and *d* will not work with a capacitively coupled source because of unwanted clamping action.

dc-return resistor

Is there a remedy for unwanted clamping action? Yes. You can add a *dc-return resistor* across the input to the unbalanced circuit (see Fig. 5-35*a*). This resistor R_D allows the capacitor to discharge during the *off* time of the diode. In other words, any charge deposited on the capacitor plates is removed during the alternate half cycle.

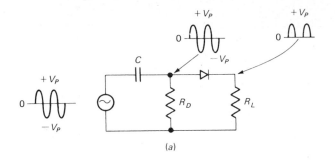

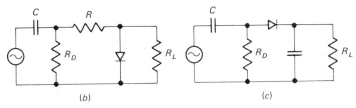

Figure 5-35. Dc return eliminates unwanted clamping.

The size of R_D is not critical. The main idea in preventing clamping action is to keep the discharging resistance R_D less than or equal to the charging resistance in series with the diode. In Fig. 5-35a, this means

$$R_D \leqslant R_L$$

When this condition is satisfied, the capacitor voltage cannot build up significantly and only a small amount of clamping takes place. When possible, make R_D less than one-tenth of R_L.

The dc return for a peak detector (Fig. 5-35c) also requires keeping R_D less than R_L and, if possible, making R_D less than one-tenth R_L.

The clipper of Fig. 5-35b is slightly different. When the diode is conducting, the charging resistance in series with the diode equals R instead of R_L. Therefore, in Fig. 5-35b, the rule is

$$R_D \leqslant R$$

and when possible, R_D should be less than one-tenth of R.

balanced diode circuits

Some diode circuits are *balanced loads*. Examples are the center-tap rectifier, the bridge rectifier, the full-wave peak rectifier, etc. These circuits work fine with a capacitively coupled source. In other words, no dc return is needed because the equal and opposite half-cycle currents produce an average capacitor voltage

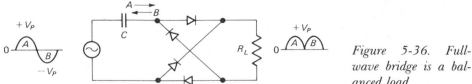

Figure 5-36. Full-wave bridge is a balanced load.

of zero. For instance, Fig. 5-36 shows a capacitively coupled source driving a bridge rectifier. Assuming identical diodes, currents *A* and *B* are equal and opposite, so capacitor voltage does not build up and the bridge rectifier works normally.

In summary, you may get clamping action with a capacitively coupled source. This unwanted clamping may occur in diode circuits, transistor circuits, integrated circuits, etc. In general, whenever a capacitor drives a device that conducts for only *part of the ac cycle,* you may get unwanted clamping action. If this happens, you usually can eliminate it by adding a dc return.

self-testing review

Read each of the following and provide the missing words. Answers appear at the beginning of the next question.

1. The half-wave rectifier is a one-diode circuit that converts ac voltage to pulsating direct voltage. During one half of the cycle, the diode is _____ (ideally shorted); on the other half cycle the diode is _____.

2. *(forward-biased, reverse-biased)* The peak _____ voltage PIV is the maximum voltage across a diode during the reverse part of the cycle.

3. *(inverse)* The _____ rectifier is a two-diode circuit always used with a _____ secondary winding. The diodes conduct on alternate half cycles and produce a _____ load voltage.

4. *(center-tap, center-tapped, full-wave)* The output frequency of a full-wave rectifier is _____ the input frequency.

5. *(twice)* The most widely used of all rectifiers is the _____ rectifier. This four-diode circuit produces a _____ load voltage.

6. *(bridge, full-wave)* A choke-input filter allows the dc component to pass through because X_L is _____ at zero frequency. The choke blocks the ac component because X_L is _____. The small amount of ac component reaching the load resistance is called the _____.

7. *(zero, high, ripple)* Because chokes are bulky and expensive, capacitor-input filters have been developed. Instead of depending on average detection, the capacitor-input filter relies on _____ detection.

8. *(peak)* For a long time constant, equivalent to light load current, the output of a capacitor-input filter approximately equals the _____ input voltage. The output ripple is very _____.

9. *(peak, small)* For a short time constant, or heavy load current, the output of a capacitor-input filter is significantly less than the _____ input voltage, and the ripple is _____.

10. *(peak, large)* When power is first applied to a capacitor-input filter, the initial gush of current is called the _____ current. In some cases, it's necessary to include a _____ resistor to protect the diodes.

11. *(surge, surge)* When working with *RC* and *LC* filters, it's convenient to use _____, defined as the input ripple divided by the output ripple. In practice, _____ sections are the best compromise when high attenuation is needed.

12. *(attenuation, two)* A voltage multiplier is two or more peak rectifiers that produce a no-load dc voltage equal to a _____ of the peak input voltage.

13. *(multiple)* A simple way to improve voltage regulation is with a _____ regulator. As long as V_{IN} is greater than V_Z, the diode operates in the _____ region.

14. *(zener, breakdown)* In a zener regulator, changes in load current produce changes in zener current but not in zener voltage. For the zener regulator to hold the output voltage constant, there must be _____ current for all source voltages and load currents.

15. *(zener)* The positive clipper removes _____ parts of the incoming signal. A combination clipper removes part of the positive and _____ half cycles.

16. *(positive, negative)* The positive clamper shifts the signal vertically, so that the negative peaks fall ideally at _____. A peak-to-peak detector is a cascade of a positive _____ and a positive peak detector.

17. *(0 V, clamper)* A capacitively coupled source will cause unwanted clamping when the load is unbalanced. To prevent unwanted clamping, we can add a dc-return resistor.

problems

5-1. Line voltage typically is 120 V rms but may be as low as 105 V rms or as high as 125 V rms. Calculate the peak value for each of these extremes.

5-2. The transformer of Fig. 5-37a has a turns ratio of $N_1 : N_2 = 4 : 1$. What is the peak voltage across the load resistance? The average voltage? The average current through the load resistance?

5-3. If the transformer of Fig. 5-37a has a turns ratio of $2 : 1$, what is the peak load current? The dc load current?

5-4. Diode rating I_0 is the average rectified current a diode can withstand without dam-

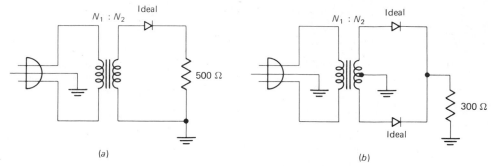

Figure 5-37.

age. For instance, the 1N4002 has an I_0 rating of 1 A, meaning the rectified current can have a dc value up to 1 A. Here are some diodes and their I_0 ratings:

(a) 1N914: $I_0 = 50$ mA
(b) 1N3070: $I_0 = 100$ mA
(c) 1N4002: $I_0 = 1$ A
(d) 1N1183: $I_0 = 35$ A

If the turns ratio is 1:1 in Fig. 5-37a, which of the foregoing diode types can be used?

5-5. Diode rating V_{RWM} is the PIV rating of a diode; the subscripts RWM stand for reverse working maximum. As an example, a 1N4002 has a V_{RWM} rating of 100 V; therefore, it can withstand a PIV of 100 V. Here are some diodes and their V_{RWM} ratings:

(a) 1N914: $V_{RWM} = 20$ V
(b) 1N1183: $V_{RWM} = 50$ V
(c) 1N4002: $V_{RWM} = 100$ V
(d) 1N3070: $V_{RWM} = 175$ V

Given a turns ratio of 2:1 in Fig. 5-37a, what does the PIV across the diode equal? Which of the foregoing diode types can be used?

5-6. In Fig. 5-37b, the turns ratio is 3:1. What is the peak load voltage? The dc load voltage? The dc load current?

5-7. If the turns ratio is 1:1 in Fig. 5-37b, which of the diodes listed in Probs. 5-4 and 5-5 has (have) sufficient I_0 and V_{RWM} ratings to be used?

5-8. The turns ratio is 2:1 in Fig. 5-37b. Which of the diodes in Probs. 5-4 and 5-5 has (have) adequate ratings to be used?

5-9. Given a turns ratio of 3:1 in Fig. 5-37b, calculate the dc load current and the PIV across each diode. What is the average rectified current through each diode?

5-10. If the turns ratio is 4:1 in Fig. 5-38a, what is the dc load voltage? The dc load current? The PIV across each diode?

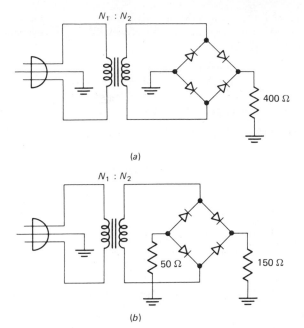

(a)

(b)

Figure 5-38.

5-11. The turns ratio is $2:1$ in Fig. 5-38a. What is the dc load current? The dc current through each diode? The PIV across each diode?

5-12. In Fig. 5-38b, the turns ratio is $3:1$. What are the dc load voltages? The dc load currents? The dc current through each diode? The PIV across each diode?

5-13. The diodes of Fig. 5-38a have an I_O rating of 150 mA and a V_{RWM} rating of 75 V. Will these diodes be adequate if the turns ratio is $3:1$?

5-14. If the diodes of Fig. 5-38b have an I_O rating of 0.5 A and a V_{RWM} rating of 50 V, will they be adequate with a turns ratio of $2:1$?

5-15. The choke of Fig. 5-39 has a dc resistance of 40 Ω. What is the dc load voltage if the full-wave signal into the choke has a peak value of 25 V?

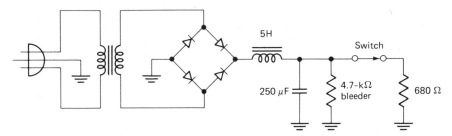

Figure 5-39.

5-16. The full-wave signal has a peak value of 35 V at the choke input in Fig. 5-39. What does the output ripple equal? If the choke has a dc resistance of 40 Ω, what is the dc output voltage?

5-17. Calculate the critical inductance for Fig. 5-39. If the switch is opened, what is the value of critical inductance? *(Note:* the bleeder resistor is always in the circuit whether the final load resistance is switched in or out. The bleeder resistor improves voltage regulation and discharges the capacitor when power is turned off.)

5-18. The maximum secondary voltage is 20 V in Fig. 5-38. What is the ripple factor at the output?

5-19. A bridge peak rectifier has a dc output voltage of 80 V with a ripple factor of 5%. What is the output ripple?

5-20. The dc output from a bridge peak rectifier is 30 V, and the ripple factor is 2 percent. What is the ripple voltage?

5-21. A bridge peak rectifier has a peak output voltage of 25 V. If the $R_L C$ time constant is 120 ms, what is the ripple voltage?

5-22. The secondary voltage has a maximum value of 30 V in Fig. 5-40a. What is the dc load voltage if $C = 220$ μF? The rms ripple? The ripple factor?

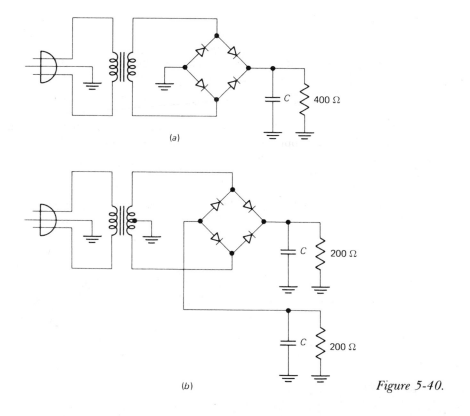

(a)

(b)

Figure 5-40.

5-23. Figure 5-40b is a *split-supply*. Because of the grounded center tap, the output voltages are equal and opposite in polarity. What are the dc output voltages for $V_{2\text{ (peak)}} = 25$ V and $C = 500$ μF? The ripple voltage? The ripple factor?

5-24. Select a capacitor that meets the following specifications in Fig. 5-40a: $V_{DC} = 40$ V, $V_r = 1$ V, and $R_L = 2$ kΩ.

5-25. We want a ripple factor of 2 percent in Fig. 5-40b. What size should C be?

5-26. A center-tap peak rectifier has the following data: $V_M = 15$ V, $\phi = 0.7$ V, $N_2/N_1 = 1/3$, $R_1 = 6$ Ω, $R_2 = 1.5$ Ω, and $r_B = 1$ Ω. Calculate V_{TH} and R_{TH}.

5-27. In Fig. 5-40a, $R_{TH} = 8$ Ω, $C = 100$ μF, $V_M = 20$ V, and $\phi = 0.7$ V. What is the dc output voltage? The ripple factor? The rms ripple?

5-28. In Fig. 5-40b, $R_{TH} = 16$ Ω, $C = 100$ μF, $V_{2\text{ (peak)}} = 30$ V, and $\phi = 0.7$ V. What is the dc output voltage? The ripple factor? *(Hint:* this *split supply* acts like two center-tap rectifiers of opposite polarity.)

5-29. C equals 50 μF in Fig. 5-40a. What does V_{DC}/V_{TH} equal for the following source resistances:
    (a) $R_{TH} = 0$
    (b) $R_{TH} = 4$ Ω
    (c) $R_{TH} = 8$ Ω
    (d) $R_{TH} = 16$ Ω
    (e) $R_{TH} = 24$ Ω
    (f) $R_{TH} = 32$ Ω

5-30. R_{TH} is 8 Ω in Fig. 5-40a. To get $V_{DC}/V_{TH} = 0.85$, what value of C do we need?

5-31. If R_{TH} is 4 Ω in Fig. 5-40b, what size should C be to get a ripple factor of 10%? *(Hint:* this split-supply acts like two center-tap rectifiers.)

5-32. The surge resistance in Fig. 5-41 is 10 Ω. If $V_M = 35$ V, what is the maximum possible value of peak surge current?

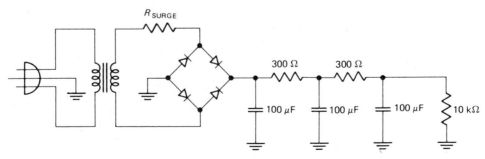

Figure 5-41.

5-33. The input to the two-section RC filter of Fig. 5-41 is $V_{DC} = 20$ V and $V_r = 0.04$ V. What are the dc output voltage and rms ripple?

5-34. To get an attenuation of 400 with an RC filter, what is the RC time constant for a one-section filter? A two-section filter? A three-section filter?

5-35. In Fig. 5-41, $V_M = 12$ V and $\phi = 0.7$ V. If $R_{TH} = 0$, what is the dc output voltage? The rms output ripple?

5-36. To get an attenuation of 1000 with an LC filter, what is the value of LC (C in microfarads) for a two-section filter? A three-section filter?

5-37. The input to the LC filter of Fig. 5-42 is $V_{DC} = 15$ V and $V_r = 0.54$ V. Each choke has a dc resistance of 20 Ω. What is the dc output voltage? The rms output ripple?

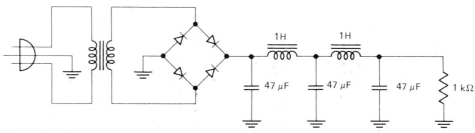

Figure 5-42.

5-38. A peak rectifier has an unloaded dc output voltage of 45 V. What is the voltage regulation if the dc output voltage is 35 V at full load?

5-39. A power supply has the regulation curve shown in Fig. 5-43a. What is the voltage regulation of this supply? The $R_{L(min)}$?

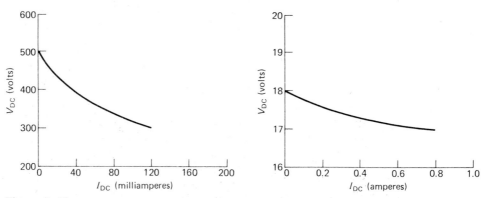

Figure 5-43.

5-40. Figure 5-43*b* is the regulation curve of a power supply. What is the *VR* for this supply? The $R_{L(min)}$?

5-41. A power supply has a *VR* of 1 percent. If the no-load voltage is 20 V, what is the full-load voltage?

5-42. In Fig. 5-44*a*, what is the no-load voltage when R_L is infinite? The PIV across each diode?

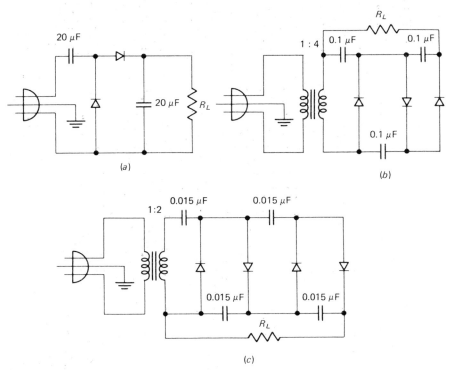

Figure 5-44.

5-43. R_L is infinite in Fig. 5-44*b*. What is the no-load voltage? The PIV across each diode?

5-44. Given an infinite R_L in Fig. 5-44*c*, what is the load voltage? The PIV across each diode? The voltage across each capacitor?

5-45. A voltage quadrupler has a no-load voltage of 3000 V. If the full-load voltage is 2200 V, what is the *VR*?

5-46. In Fig. 5-45*a*, what is the approximate value of zener current for each of these load resistances:

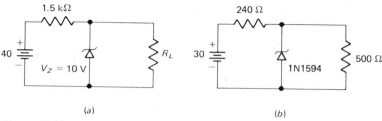

Figure 5-45.

(a) $R_L = 100$ kΩ
(b) $R_L = 10$ kΩ
(c) $R_L = 1$ kΩ

5-47. To keep the zener current greater than zero in Fig. 5-45a, what is the minimum allowable load resistance?

5-48. R_L can vary from infinity to 1 kΩ in Fig. 5-45a. If $Z_Z = 10$ Ω, what are the minimum and maximum values of V_{OUT}? The VR?

5-49. The 1N1594 of Fig. 5-45b has $V_Z = 12$ V and $Z_Z = 1.4$ Ω. Ideally, what is the load voltage? To a second approximation, what is it?

5-50. Same data as the preceding problem, except V_{IN} changes from 30 to 40 V. To a second approximation, what is the load voltage?

5-51. Same data as Prob. 5-49, except the load resistance may vary from infinity to 500 Ω. What is the VR ideally? To a second approximation?

5-52. A zener regulator has $V_Z = 15$ V. V_{IN} may vary from 22 to 40 V. R_L may vary from 1 to 50 kΩ. To ensure regulation under all conditions, what is the maximum value of the series-limiting resistor?

6 *bipolar transistors*

You can dope a semiconductor to get an *npn* crystal or a *pnp* crystal. A crystal like this is called a *junction transistor*. The *n* regions have mostly conduction-band electrons, and the *p* regions mostly holes. For this reason, the junction transistor is often called a *bipolar transistor*.

Shockley worked out the theory of the junction transistor in 1949, and the first one was produced in 1951. The transistor's impact on electronics has been enormous. Besides starting the multibillion-dollar semiconductor industry, the transistor has led to all kinds of related inventions like integrated circuits, opto-electronic devices, and microprocessors.

6-1 *the three doped regions*

Figure 6-1*a* shows an *npn* crystal. The *emitter* is heavily doped; its job is to emit or inject electrons into the *base*. The base is lightly doped and very thin; it passes most of the emitter-injected electrons on to the *collector*. The doping of the collector is between the heavy doping of the emitter and the light doping of the base. The collector is so named because it collects or gathers electrons from the base. The collector is the largest of the three regions; it must dissipate more heat than the emitter or base.

The transistor of Fig. 6-1*a* has two junctions, one between the emitter and the base, and another between the base and the collector. Because of this, the

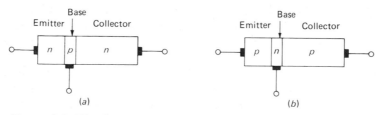

Figure 6-1. *The three transistor regions.* (a) npn *transistor.* (b) pnp *transistor.*

transistor is like two diodes. We call the diode on the left the *emitter-base diode* or simply the *emitter diode*. The diode on the right is the *collector-base diode* or the *collector diode*.

Figure 6-1*b* shows the other possibility: a *pnp* transistor. The *pnp* transistor is the *complement* of the *npn* transistor; this means opposite currents and voltages are involved in the action of a *pnp* transistor. To avoid confusion, we will concentrate on the *npn* transistor during our early discussions.

6-2 *the unbiased transistor*

Figure 6-2*a* shows the majority carriers before any have moved across the junctions. The free electrons diffuse across the junctions, which results in two depletion layers (Fig. 6-2*b*). For each of these depletion layers, the barrier potential

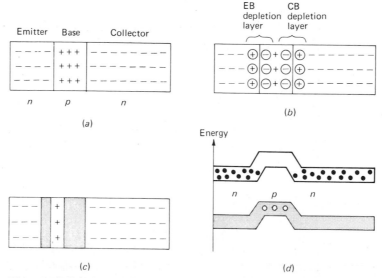

Figure 6-2. (a) *Before diffusion.* (b) *After diffusion.* (c) *Depletion layers.* (d) *Energy bands.*

approximately equals 0.7 V at 25°C for a silicon transistor (0.3 V for a germanium transistor). As with diodes, we emphasize silicon transistors because of their greater importance. In all discussions, the transistors are silicon unless otherwise indicated.

Because the three regions have different doping levels, the depletion layers do not have the same width. The more heavily doped a region is, the greater the concentration of ions near the junction. This means the depletion layer penetrates only slightly into the emitter region (it's heavily doped) but deeply into the base, which is lightly doped. The other depletion layer also extends well into the base, and penetrates the collector region to a lesser amount. Figure 6-2c shows what it boils down to. From now on, we will shade depletion layers to indicate they have no majority carriers.

To complete the picture of the unbiased transistor, Fig. 6-2d shows the energy diagram. Because we have two depletion layers, we have two energy hills. Especially important, the conduction-band electrons in the emitter do not have enough energy to enter the base region. Or in terms of radius, these electrons are traveling in conduction-band orbits smaller than the smallest permitted in the base. Unless we forward-bias the emitter diode and lower the hill, the emitter electrons cannot enter the base.

6-3 *ff and rr bias*

There are several ways to bias a transistor. This section discusses two methods we need for later discussions.

forward-forward bias

Figure 6-3a illustrates *forward-forward* (FF) bias, so named because the emitter diode and the collector diode are forward-biased. The circuits driving the emitter

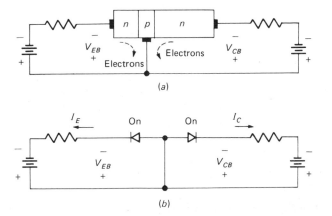

Figure 6-3. FF bias. (a) *Actual circuit.* (b) *Equivalent circuit.*

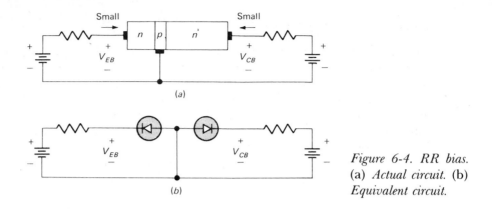

Figure 6-4. RR bias.
(a) *Actual circuit.* (b)
Equivalent circuit.

and collector diodes may be lumped as shown or may represent Thevenin equivalent circuits. Either way, carriers cross the junctions and flow down through the base into the external base lead as shown.

Figure 6-3b shows the equivalent circuit for FF bias. Emitter-base voltage V_{EB} forward-biases the emitter diode and produces a conventional current I_E. Likewise, collector-base voltage V_{CB} forward-biases the collector diode, causing conventional current I_C to flow.

reverse-reverse bias

Another possibility is to *reverse-reverse* (RR) bias the transistor as shown in Fig. 6-4a. Now both diodes are reversed-biased. For this condition, only small currents flow consisting of thermally produced saturation current and surface-leakage current. The thermally produced component is temperature-dependent, and approximately doubles for every 10° rise. The surface-leakage component, on the other hand, increases with voltage. These reverse currents are usually negligible.

Figure 6-4b is the equivalent circuit for RR bias; both diodes are open unless V_{EB} or V_{CB} exceed the breakdown voltages of the diodes.

6-4 *forward-reverse bias*

Forward-bias the emitter diode, reverse-bias the collector diode, and the unexpected happens. Figure 6-5a shows *forward-reverse* (FR) bias. We expect a large emitter current because the emitter diode is forward-biased. But we do not expect a large collector current because the collector diode is reverse-biased. Nevertheless, this is exactly what we get; this is precisely why the transistor is the great invention it is.

preliminary explanation

Here is a brief explanation of why we get that large collector current in Fig. 6-5*a*. At the instant the forward bias is applied to the emitter diode, electrons in the emitter have not yet entered the base region (see Fig. 6-5*b*). If V_{EB} is greater than the barrier potential, many emitter electrons enter the base region as shown in Fig. 6-5*c*. These electrons in the base can flow in either of two directions: down the thin base into the external base lead, or across the collector junction into the collector region.

Which way will they go? For the electrons to flow down through the base region, they must first fall into holes, that is, recombine with base holes. Then, as valence electrons they can flow down through adjacent base holes and into the external base lead. This downward component of base current is called *recombination current*. It is small because the base is lightly doped, that is, has only a few holes.

A second crucial idea in transistor action is that the base is very thin. In Fig. 6-5*c*, the base is teeming with injected conduction-band electrons, causing diffusion into the collector depletion layer. Once inside this layer, they are pushed by the depletion-layer field into the collector region (see Fig. 6-5*d*). These collector electrons can then flow into the external collector lead as shown.

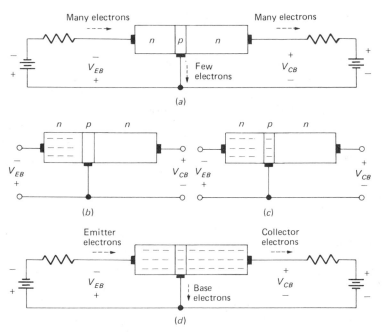

Figure 6-5. FR bias. (a) *Actual circuit.* (b) *Free electrons in emitter.* (c) *Free electrons enter base.* (d) *Free electrons diffuse into collector.*

Our picture of FR bias is this. In Fig. 6-5*d* we visualize a steady stream of electrons leaving the negative source terminal and entering the emitter region. The forward bias forces these emitter electrons to enter the base region. The thin, lightly doped base gives almost all these electrons enough lifetime to diffuse into the collector depletion layer. The depletion-layer field then pushes a steady stream of electrons into the collector region. These electrons leave the collector, enter the external collector lead, and flow into the positive terminal of the collector voltage source. In most transistors, more than 95 percent of the emitter-injected electrons flow to the collector; less than 5 percent fall into base holes and flow out the external base lead.

the energy viewpoint

An energy diagram is the next step to a deeper understanding of the transistor. Forward-biasing the emitter diode lowers its energy hill (see Fig. 6-6). Therefore, conduction-band electrons in the emitter now have enough energy to move into the base conduction band. In other words, the orbits of some emitter electrons are now large enough to match some of the available base orbits. Because of this, emitter electrons can diffuse from the emitter conduction band to the base conduction band.

On entering the base conduction band, the electrons become minority carriers because they are inside a *p* region. In almost any transistor, more than 95 percent of these minority carriers have a long enough lifetime to diffuse into the collector depletion layer and fall down the collector energy hill. As they fall, they give up energy, mostly in the form of heat. The collector must be able to dissipate this heat, and for this reason, it is usually the largest of the three doped regions. Less than 5 percent of the emitter-injected electrons fall along the recombination path shown in Fig. 6-6; those that become valence electrons flow through base holes into the external base lead.

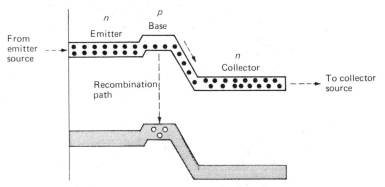

Figure 6-6. Transistor energy bands for FR bias.

We can summarize as follows:

1. The forward bias on the emitter diode controls the number of electrons injected into the base. The larger V_{EB}, the larger the number of injected electrons.
2. The reverse bias on the collector diode has little influence on the number of electrons that enter the collector. Increasing V_{CB} steepens the collector hill but cannot significantly change the number of electrons arriving at the collector depletion layer.

dc alpha

Saying that more than 95 percent of injected electrons reach the collector is the same as saying collector current almost equals emitter current. The *dc alpha* of a transistor indicates how close in value the two currents are; it is defined as

$$\alpha_{dc} = \frac{I_C}{I_E} \qquad (6\text{-}1)***$$

For instance, if we measure an I_C of 4.9 mA and an I_E of 5 mA,

$$\alpha_{dc} = \frac{4.9}{5} = 0.98$$

The thinner and more lightly doped the base is, the higher the α_{dc}. Ideally, if all injected electrons went on to the collector, α_{dc} would equal unity. Many transistors have α_{dc} greater than 0.99, and almost all have α_{dc} greater than 0.95. Because of this, we can approximate α_{dc} as 1 in preliminary analysis.

base-spreading resistance

With two depletion layers penetrating the base, the base holes are confined to a thin channel of *p*-type semiconductor as shown in Fig. 6-7. Increasing the reverse bias on the collector diode (equivalent to increasing V_{CB}) widens the

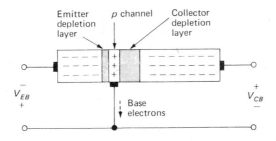

Figure 6-7. Base-spreading resistance r'_b.

collector depletion layer; this reduces the width of the channel containing base holes. In other words, there will be fewer base holes available for recombination current. The resistance of the p channel in the base is called the *base-spreading resistance r_b'*.

Recombination current in the base must flow down through r_b'. When it does, it produces a voltage. We discuss the importance of this voltage later. For now, realize r_b' exists and depends on the width of the p channel in Fig. 6-7 as well as on the doping of the base. In rare cases, r_b' may be as high as 1000 Ω. Typically, it is in the range of 50 to 150 Ω. In preliminary analysis, we will neglect r_b'.

breakdown voltages

Since the two halves of the transistor are diodes, too much reverse voltage on either diode can cause breakdown. With FR bias, we worry only about the collec-

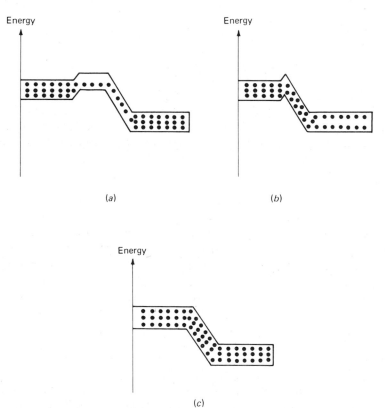

Figure 6-8. Reach-through effect.

tor diode. When V_{CB} is too large, the collector diode breaks down because of avalanche (previously described) or because of *reach-through effect* (also known as *punch-through*).

Reach-through means the collector depletion layer becomes so wide it reaches the emitter depletion layer. When this happens, emitter electrons are injected directly into the collector depletion layer. With even the slightest overlap of depletion layers, the collector current can become large enough to destroy the transistor.

Figure 6-8*a* gives us more insight; reach-through has not yet occurred because the two depletion layers do not yet overlap. With an increase in collector voltage, the collector depletion layer widens. Figure 6-8*b* shows the condition when the two depletion layers overlap. Now, more electrons can flow from emitter to collector. Increasing the collector voltage further may wipe out the emitter hill altogether, as shown in Fig. 6-8*c*. In this case, the enormous current would destroy the transistor.

Avalanche and reach-through are undesirable in the ordinary transistor. We can avoid both by keeping the collector voltage less than the breakdown voltage specified on a manufacturer's data sheet.

6-5 *the ce connection*

In Fig. 6-9*a*, the emitter and the two voltage sources connect to the common point shown. Because of this, we call the circuit a *common-emitter* (CE) connection. The circuit first discussed (Fig. 6-5*a*) is a *common-base* (CB) connection because the base and two voltage sources are connected to a common point.

electron flow

Just because you change from a CB connection (Fig. 6-5*a*) to a CE connection (Fig. 6-9*a*), you do not change the way a transistor operates. Majority carriers move exactly the same as before. That is, the emitter is teeming with conduction-band electrons (Fig. 6-9*b*). When V_{BE} is greater than 0.7 V or so, the emitter injects these electrons into the base (Fig. 6-9*c*). As before, the thin, lightly doped base gives almost all these electrons enough lifetime to diffuse into the collector depletion layer. With a reverse-biased collector diode, the depletion-layer field pushes the electrons into the collector region where they flow out to the external voltage source (Fig. 6-9*d*).

dc beta

We have related the collector current to the emitter current by using α_{dc}. We can also relate collector current to base current by defining the *dc beta* of a transistor as

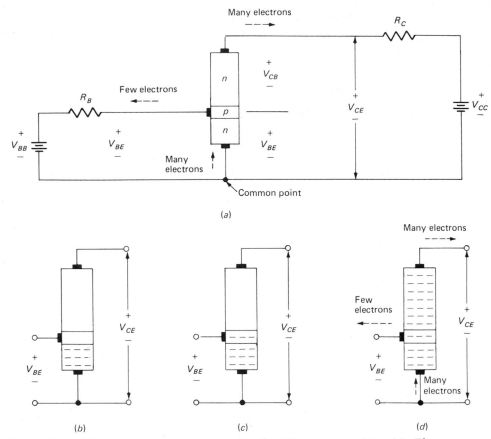

Figure 6-9. CE connection. (a) *Actual circuit.* (b) *Electrons in emitter.* (c) *Electrons enter base.* (d) *Electrons diffuse into collector.*

$$\beta_{\text{dc}} = \frac{I_C}{I_B} \qquad\qquad (6\text{-}2)\text{***}$$

For instance, if we measure a collector current of 5 mA and a base current of 0.05 mA, the transistor has a β_{dc} of

$$\beta_{\text{dc}} = \frac{5 \text{ mA}}{0.05 \text{ mA}} = 100$$

In almost any transistor, less than 5 percent of the emitter-injected electrons recombine with base holes to produce I_B; therefore, β_{dc} is almost always greater than 20. Usually, it is from 50 to 200. And some transistors have β_{dc} as high as 1000. In another system of analysis called h parameters (Sec. 15-6), β_{dc} is called the dc current gain and is designated by h_{FE} (note the FE subscripts are capitalized). In other words, $\beta_{dc} = h_{FE}$. This is important to remember because data sheets usually give the value of h_{FE}.

relation between α_{dc} and β_{dc}

Kirchhoff's current law tells us

$$I_E = I_C + I_B \qquad\qquad (6\text{-}3)\text{***}$$

Dividing by I_C gives

$$\frac{I_E}{I_C} = 1 + \frac{I_B}{I_C}$$

or

$$\frac{1}{\alpha_{dc}} = 1 + \frac{1}{\beta_{dc}}$$

With algebra, we can rearrange to get

$$\beta_{dc} = \frac{\alpha_{dc}}{1 - \alpha_{dc}} \qquad\qquad (6\text{-}4)$$

As an example, if $\alpha_{dc} = 0.98$, the value of β_{dc} is

$$\beta_{dc} = \frac{0.98}{1 - 0.98} = 49$$

Occasionally we need a formula for α_{dc} in terms of β_{dc}. With algebra we can rearrange Eq. (6-4) to get

$$\alpha_{dc} = \frac{\beta_{dc}}{\beta_{dc} + 1} \qquad\qquad (6\text{-}5)$$

For instance, for a β_{dc} of 100,

$$\alpha_{dc} = \frac{100}{100 + 1} = 0.99$$

two equivalent circuits

To crystallize the main ideas of transistor action, we can use the equivalent circuit of Fig. 6-10a. The voltage V'_{BE} is the voltage across the emitter depletion layer. When this voltage is greater than 0.7 V or thereabouts, the emitter injects electrons into the base. As mentioned earlier, the current in the emitter diode controls the collector current. For this reason, the collector-current source forces a current of $\alpha_{dc}I_E$ to flow in the collector circuit. All this assumes V_{CE} is greater

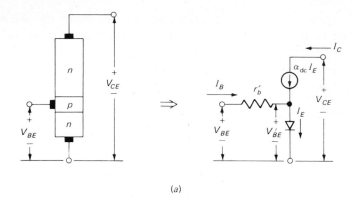

(a)

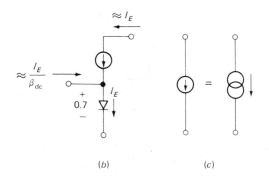

(b)   (c)

Figure 6-10. DC equivalent circuits. (a) Accurate model. (b) Bias model. (c) Alternative symbol.

than a volt or so; otherwise, the collector diode is not reverse-biased and the transistor cannot work normally.

The internal voltage V'_{BE} differs from the applied voltage V_{BE} by the drop across r'_b. That is,

$$V_{BE} = V'_{BE} + I_B r'_b$$

When the $I_B r'_b$ drop is small enough to neglect, $V_{BE} \cong V'_{BE}$.

For most practical work, the equivalent circuit of Fig. 6-10*b* is accurate enough. First, we assume the voltage across the emitter diode is 0.7 V. Second, we disregard the $I_B r'_b$ voltage, equivalent to treating r'_b as negligibly small. Third, emitter current I_E sets up a current of approximately I_E in the collector circuit. Fourth, the base current approximately equals I_E/β_{dc}. This follows from

$$\beta_{dc} = \frac{I_C}{I_B}$$

or

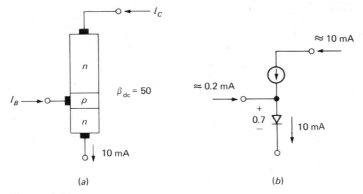

Figure 6-11.

$$I_B = \frac{I_C}{\beta_{dc}} = \frac{\alpha_{dc} I_E}{\beta_{dc}} \cong \frac{I_E}{\beta_{dc}}$$

because α_{dc} is very close to unity.

Figure 6-10*c* shows an alternative schematic symbol for the collector current source.

EXAMPLE 6-1

Figure 6-11*a* shows a silicon transistor with a β_{dc} of 50 and an emitter current of 10 mA. The external circuitry producing this emitter current is not shown, but you may assume the transistor is FR-biased. Show as many currents as possible.

SOLUTION

With an emitter current of 10 mA, the collector current is approximately 10 mA, and the base current equals

$$I_B \cong \frac{10 \ \text{mA}}{50} = 0.2 \ \text{mA}$$

The voltage across the emitter diode is approximately 0.7 V. Figure 6-11*b* shows the currents.

6-6 *transistor curves*

One way to see as many details as possible is with graphs that relate transistor currents and voltages.

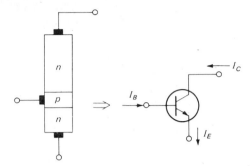

Figure 6-12. Transistor schematic symbol.

schematic symbol

Figure 6-12 shows the schematic symbol for an *npn* transistor. The emitter has an arrowhead, but not the collector. Especially important, the arrowhead points in the direction of conventional emitter current. In other words, electrons flow into the emitter and out the base and collector. Conventional currents flow in opposite directions as shown in Fig. 6-12. Conventional emitter current flows out of the emitter; conventional base and collector current flow into the *npn* transistor.

collector curves

You can get data for CE collector curves by setting up a circuit like Fig. 6-13*a* or by using a transistor curve tracer. Either way, the idea is to vary the V_{BB} and V_{CC} supplies to set up different transistor voltages and currents.

To keep things orderly, the usual procedure is to set a value of I_B and hold it fixed while you vary V_{CC}. By measuring I_C and V_{CE}, you can get data for graphing I_C versus V_{CE}. For instance, suppose $I_B = 10$ μA in Fig. 6-13*a*. Next, we vary V_{CC} and measure the resulting I_C and V_{CE}. If we plot the data, we get Fig. 6-13*b*. Notice we label this curve $I_B = 10$ μA because we got this curve by keeping I_B fixed during all measurements.

There is nothing mysterious about Fig. 6-13*b*. It echoes our explanation of transistor action. When V_{CE} is zero, the collector diode is not reverse-biased; therefore, the collector current is negligibly small. For V_{CE} between 0 and 1 V or thereabouts, the collector current rises sharply and then becomes almost constant. This is tied in with the idea of reverse-biasing the collector diode. It takes approximately 0.7 V or so to reverse-bias the collector diode; once you reach this level, the collector is gathering all electrons that reach its depletion layer.

Above the knee, the exact value of V_{CE} is not too important because making the collector hill steeper cannot appreciably increase the collector current. The

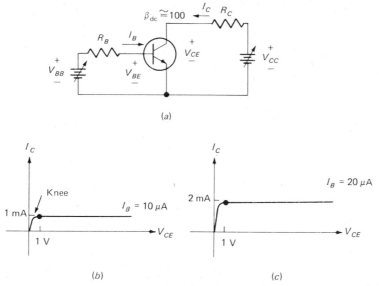

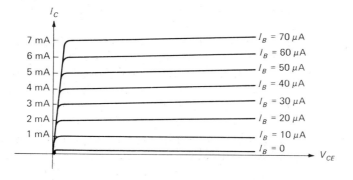

Figure 6-13. Getting collector curves.

small increase in collector current with increasing V_{CE} is caused by the collector depletion layer getting wider and capturing a few more base electrons before they fall into holes.

By repeating the measurements of I_C and V_{CE} for $I_B = 20$ μA, we can plot the graph of Fig. 6-13c. The curve is similar, except that above the knee the collector current approximately equals 2 mA. Again, an increase in V_{CE} produces a small increase in collector current because the widening depletion layer captures a few additional base electrons.

If several curves for different I_B are shown on the same graph, we get the collector curves of Fig. 6-14. Since we used a transistor with a β_{dc} of approxi-

Figure 6-14. Collector curves.

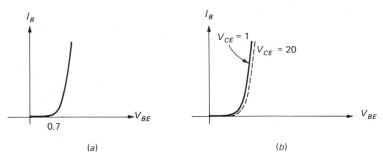

Figure 6-15. Base curves.

mately 100, the collector current is approximately 100 times greater than the base current for any point above the knee of a given curve. Because collector current increases slightly with an increase in V_{CE}, β_{dc} increases slightly with an increase in V_{CE}.

Collector curves are available from manufacturers. You can also use a *curve tracer* to display the collector curves of a transistor.

the base curve

In the circuit of Fig. 6-13a, we can get data for graphing I_B versus V_{BE}. Figure 6-15a shows the kind of graph we get. There is nothing surprising here. Since the base-emitter section of a transistor is a diode, we expect to see something resembling a diode curve. There is a fine point worth mentioning, however. As the collector depletion layer widens with increasing collector voltage, the base current decreases slightly because the collector depletion layer captures a few more base electrons. For this reason, if we plot another base curve for a different collector voltage, the new curve will look slightly different from the old.

Figure 6-15b shows what we mean. For the same value of V_{BE}, the curve with higher V_{CE} has a smaller base current because the collector depletion layer captures a few more base electrons before they fall into holes; as a result, the recombination base current I_B is reduced. The gap between the curves is small, and we take it into account when seeking highly accurate answers. For all preliminary analysis, we disregard the effect a changing collector voltage has on base current.

the current-gain curve

The β_{dc} of a transistor is an important quantity; we should be aware of how it changes with collector current and temperature. Figure 6-16 shows a typical

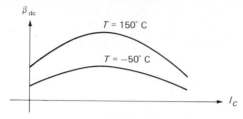

Figure 6-16. Variation
in dc beta.

variation in β_{dc}. At a fixed temperature, β_{dc} increases to a maximum when collector current increases. For further increases in I_C, the β_{dc} drops off. The variation in β_{dc} may be as much as $2:1$ over the useful current range of the transistor; it depends on the type of transistor.

Increasing the temperature will increase β_{dc} at a given collector current. Over a large temperature range, the variations in β_{dc} may be greater than $3:1$, depending on the transistor type.

cutoff current and breakdown voltage

In the collector curves of Fig. 6-14, the lowest curve is for zero base current. The condition $I_B = 0$ is equivalent to having an *open base lead* (see Fig. 6-17a). The collector current with an open base lead is designated I_{CEO}, where subscripts *CEO* stand for collector to emitter with open base. I_{CEO} is caused partly by thermally produced carriers and partly by surface-leakage current.

Figure 6-17b shows the $I_B = 0$ curve. With a large enough collector voltage, we reach a breakdown voltage labeled BV_{CEO}, where the subscripts again stand for collector to emitter with open base. For normal transistor operation, we must keep V_{CE} less than BV_{CEO}. Most transistor data sheets list the value of BV_{CEO} among the maximum ratings. This breakdown voltage may be from less than 20 to over 200 V, depending on the transistor type.

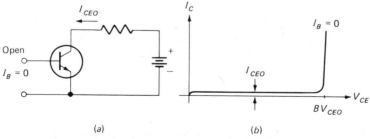

Figure 6-17. Cutoff current and breakdown voltage.

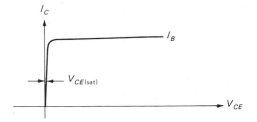

Figure 6-18. Satura-tion region.

collector saturation voltage

For normal transistor operation, the collector diode must be reverse-biased; this requires a V_{CE} greater than a volt or so, depending on how much collector current flows. As a guide, many data sheets list the $V_{CE(sat)}$ of a transistor.

Figure 6-18 shows what we mean by $V_{CE(sat)}$. It is the value of V_{CE} at some point below the knee with the exact position specified on the data sheet. Typically, $V_{CE(sat)}$ is only a few-tenths volt, although at very large collector currents it may exceed a volt. Incidentally, the part of the curve below the knee of Fig. 6-18 is known as the *saturation region*.

summary

Here are the main ideas of this section:

1. V_{CE} should be greater than $V_{CE(sat)}$ but less than BV_{CEO}. This ensures a reverse-biased collector diode.
2. β_{dc} varies with I_C and temperature. These variations are undesirable; Chap. 7 shows you how to get around them.
3. There is a small collector current when $I_B = 0$. The importance of this will be handled in Chap. 7.

self-testing review

Read each of the following and provide the missing words. Answers appear at the beginning of the next question.

1. The emitter is heavily _____, but the base is _____ doped. When the emitter of an *npn* transistor is forward-biased, conduction-band electrons have enough lifetime to diffuse into the _____.

2. *(doped, lightly, base)* Because the base is very thin and lightly doped, most of the conduction-band electrons have enough lifetime to diffuse into the _____. Only a small percent of the electrons injected into the base have enough time to _____ and flow out the external base lead.

3. *(base, recombine)* The forward bias on the emitter diode controls the _____ of electrons injected into the base. The larger V_{BE}, the _____ the number of injected electrons. The reverse bias on the collector diode has _____ influence on the number of electrons that enter the collector.

4. *(number, greater, little)* The dc alpha of a transistor equals the ratio of _____ current to _____ current.

5. *(collector, emitter)* With two depletion layers penetrating the base, holes are confined to a thin channel of semiconductor. The resistance of this narrow channel is called the _____ resistance r'_b.

6. *(base-spreading)* As an approximation, V_{CE} must be greater than 1 V to make sure the collector diode stays _____-biased.

7. *(reverse)* The dc beta equals the ratio of _____ current to _____ current.

8. *(collector, base)* The collector current with an _____ base lead is designated I_{CEO}. With a large enough collector voltage, we reach a _____ voltage labeled BV_{CEO}, where the subscripts stand for collector to _____ with an _____ base.

9. *(open, breakdown, emitter, open)* The part of the collector curve below the knee is known as the _____ region. V_{CE} should be _____ than $V_{CE(\text{sat})}$ but _____ than BV_{CEO}.

10. *(saturation, greater, less)* The dc equivalent circuit of a transistor has a diode for the emitter and a current source for the collector. Emitter and collector currents are approximately equal.

problems

6-1. The barrier potential of the emitter depletion layer is approximately 0.7 V for a temperature of 25°C. Estimate the barrier potential at 0 and 50°C.

6-2. In Fig. 6-4*a*, a small reverse current flows in the collector circuit. The thermally produced component of this current equals 0.1 nA at 25°C. Estimate the values of this component at 75°C and at −15°C.

6-3. For the transistor circuit of Fig. 6-5*d*, suppose that only 2 percent of the electrons injected into the base recombine with base holes. If 10 million electrons enter the emitter in 1 μs, how many electrons come out the base lead in this period? How many come out the collector lead during this time?

6-4. When the electrons in Fig. 6-6 fall down the collector hill, they give up energy, mostly in the form of heat. You can calculate the power by the product of VI, where V is voltage across the collector depletion layer and I is the current flowing through this depletion layer. If $V = 20$ V and $I = 10$ mA, how much power is being dissipated? If the voltage remains the same, but the current increases to 500 mA, how much power is there?

6-5. In a particular transistor, the collector current equals 5.6 mA and the emitter current equals 5.75 mA. What is the value of α_{dc}?

6-6. The base current in a transistor equals 0.01 mA and the emitter current equals 1 mA. What is the value of α_{dc}?

6-7. If $V_{CB} = 4$ V and $V_{BE} = 0.7$ V, what does V_{CE} equal?

6-8. In a transistor, we measure an I_C of 100 mA and an I_B of 0.5 mA. What does β_{dc} equal?

6-9. A transistor has a β_{dc} of 150. If the collector current equals 45 mA, what does the base current equal?

6-10. Suppose the β_{dc} of a particular transistor can vary from 20 to 100. Over what ranges does α_{dc} vary?

6-11. A transistor has an α_{dc} of 0.995. What does its β_{dc} equal? When the temperature is 100°C, the α_{dc} increases to 0.998. What value does β_{dc} have at this higher temperature?

6-12. In the transistor equivalent circuit of Fig. 6-19a, the α_{dc} equals 0.99. Calculate I_C, I_B, and V_{BE}.

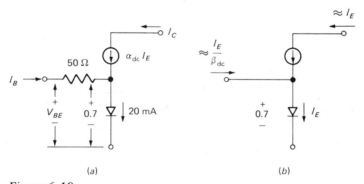

(a)   (b)

Figure 6-19.

6-13. In Fig. 6-19a, suppose we increase the emitter current from 20 mA to a very high value. If $\alpha_{dc} = 0.99$ and we measure a V_{BE} of 1 V, how much emitter current is there?

6-14. A 2N5067 is a power transistor (one that can dissipate more than approximately ½ W of power). Power transistors normally have small r'_b. The typical 2N5067 has an r'_b of 10 Ω. How much $I_B r'_b$ drop is there when $I_B = 1$ mA? When $I_B = 10$ mA? And when $I_B = 100$ mA?

6-15. A 2N3298 has a typical β_{dc} of 90. If Fig. 6-19b represents the 2N3298, calculate the collector and base currents for an emitter current of 10 mA.

6-16. In Fig. 6-19b, suppose the base current equals 1 mA and the emitter current equals 200 mA. What is the value of β_{dc}?

6-17. A transistor has a β_{dc} of 400. In Fig. 6-19b, what does the base current equal when the emitter current equals 50 mA?

6-18. Figure 6-20a shows one of the collector curves for a particular transistor. Calculate the β_{dc} at point A and at point B. If the collector current changes at the same rate with increasing V_{CE}, what will β_{dc} equal at $V_{CE} = 15$ V?

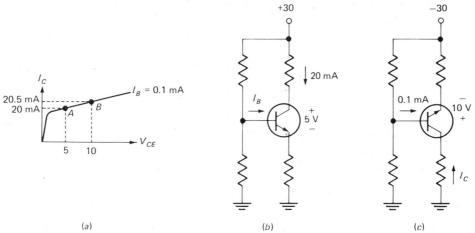

Figure 6-20.

6-19. Figure 6-20b shows a circuit to be analyzed in Chap. 7. Right now, you know enough to answer this question. If the transistor has the collector curve of Fig. 6-20a, what value does I_B have? Figure 9-20c shows the same circuit modified slightly. What does I_C equal?

6-20. A 2N5346 has the base curve shown in Fig. 6-21a. The voltage drop across r'_b

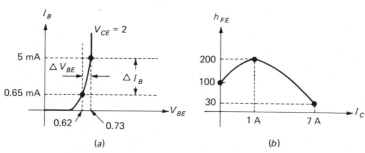

Figure 6-21.

causes V_{BE} to increase when the base current increases. Because of this, here is a way to estimate the value of r'_b:

$$r'_b = \frac{\Delta V_{BE}}{\Delta I_B} \qquad \text{well above the knee}$$

Apply this formula to Fig. 6-21a to get an estimate for the r'_b of a 2N5346.

6-21. A 2N5346 has the β_{dc} variations shown in Fig. 6-21b. If the transistor is operating at collector currents much smaller than 1 A, what is the approximate value of β_{dc}? How much base current is there when $I_C = 1$ A? And when $I_C = 7$ A?

6-22. Figure 6-22a shows a transistor circuit with an open base lead. If we measure a V_{CE} of 9 V, what values does I_{CEO} have? If we change the collector resistor from 10 MΩ to 10 kΩ as shown in Fig. 6-22b, what is the new value of V_{CE}? (Assume I_{CEO} remains the same.)

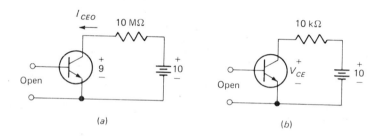

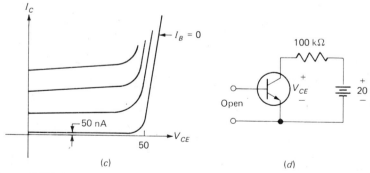

Figure 6-22.

6-23. A transistor has the collector curves of Fig. 6-22c. If this transistor is used in the circuit of Fig. 6-22d, what will V_{CE} equal? What is the value of BV_{CEO}? Is the transistor of Fig. 6-22d in danger of breakdown?

7 *transistor biasing circuits*

Linear transistor circuits operate with the emitter diode forward-biased and the collector diode reverse-biased. This chapter discusses common ways to bias a transistor for linear operation.

7-1 *base bias*

Figure 7-1*a* is an example of *base bias*. A voltage source V_{BB} forward-biases the emitter diode through a current-limiting resistor R_B. Kirchhoff's voltage law tells us the voltage across R_B is $V_{BB} - V_{BE}$. Ohm's law gives the base current:

$$I_B = \frac{V_{BB} - V_{BE}}{R_B} \qquad (7\text{-}1)***$$

where $V_{BE} = 0.7$ V for silicon transistors (0.3 V for germanium).

dc load line

In the collector circuit, voltage source V_{CC} reverse-biases the collector diode through R_C. With Kirchhoff's voltage law,

$$V_{CE} = V_{CC} - I_C R_C \qquad (7\text{-}2)***$$

In a given circuit, V_{CC} and R_C are constants; V_{CE} and I_C are variables.

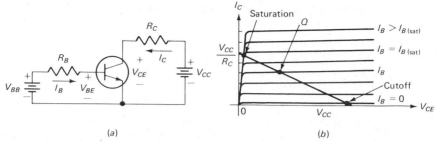

Figure 7-1. (a) *Base bias.* (b) *Dc load line.*

We can rearrange Eq. (7-2) to get

$$I_C = -\frac{V_{CE}}{R_C} + \frac{V_{CC}}{R_C} \tag{7-3}$$

This is a *linear* equation, similar to

$$y = mx + b$$

As proved in basic mathematics, the graph of a linear equation is always a straight line with a slope of m and a vertical intercept of b.

Figure 7-1*b* shows the graph of Eq. (7-3) superimposed on collector curves. The vertical intercept is V_{CC}/R_C, the horizontal intercept is V_{CC}, and the slope is $-1/R_C$. This line is called the *dc load line* because it represents all possible operating points. The intersection of the dc load line with the base current is the operating point of the transistor.

cutoff and saturation

The point where the load line intersects the $I_B = 0$ curve is known as *cutoff*. At this point, base current is zero and collector current is negligibly small (only leakage current I_{CEO} exists). At cutoff, the emitter diode comes out of forward bias and normal transistor action is lost. To a close approximation, the collector-emitter voltage is

$$V_{CE\,(\text{cutoff})} = V_{CC} \tag{7-4}$$

The intersection of the load line and the $I_B = I_{B\,(\text{sat})}$ curve is called *saturation*. At this point, base current equals $I_{B\,(\text{sat})}$ and collector current is *maximum*. At saturation, the collector diode comes out of reverse bias and normal transistor action is lost. To a close approximation, the collector current at saturation is

$$I_{C\,(\text{sat})} \cong \frac{V_{CC}}{R_C} \tag{7-5}***$$

and the base current that just produces saturation is

$$I_{B\,(\text{sat})} = \frac{I_{C\,(\text{sat})}}{\beta_{\text{dc}}} \qquad\qquad (7\text{-}6)\,\text{***}$$

The collector-emitter voltage at saturation is

$$V_{CE} = V_{CE\,(\text{sat})}$$

where $V_{CE\,(\text{sat})}$ is specified on data sheets, typically a few-tenths volt.

If base current is greater than $I_{B\,(\text{sat})}$, collector current cannot increase because the collector diode is no longer reverse-biased. In other words, the intersection of the load line and a higher base curve still results in the same saturation point in Fig. 7-1 b.

active region

All operating points between cutoff and saturation are the *active region* of a transistor. In the active region, the emitter diode is forward-biased and the collector diode is reverse-biased. With Eq. (7-1), we can find the base current in any base-bias circuit. The intersection of this base current and the load line is the *quiescent (Q)* point of Fig. 7-1 b.

EXAMPLE 7-1
The 2N3904 of Fig. 7-2a is a silicon transistor with a β_{dc} of 100. What would a dc voltmeter read across the collector-emitter terminals?

SOLUTION
First, get the base current. It equals the drop across the base resistor divided by the resistance:

$$I_B = \frac{V_{BB} - V_{BE}}{R_B} = \frac{10 - 0.7}{10^6} = 9.3 \ \mu\text{A}$$

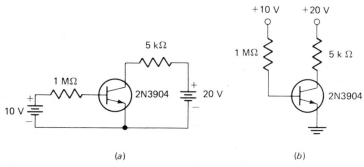

(a)   *(b)*

Figure 7-2.

The collector current is

$$I_C = \beta_{dc} I_B = 100 \times 9.3 \ \mu A = 0.93 \ mA$$

The collector-emitter voltage equals the collector supply voltage minus the drop across the collector resistor:

$$V_{CE} = V_{CC} - I_C R_C = 20 - 0.93(10^{-3})5(10^3)$$
$$= 15.4 \ V$$

Figure 7-2*b* shows the same circuit in a negative-ground system. For simplicity, we need show only the supply voltages, +10 and +20 V. When you see a simplified schematic like this, remember it means the negative terminals of the power supplies are grounded to get a complete path for current.

EXAMPLE 7-2

The 2N4401 of Fig. 7-3*a* is a silicon transistor with β_{dc} of 80. Draw the dc load line. Where is the Q point if $R_B = 390 \ k\Omega$?

SOLUTION
With Eq. (7-5),

$$I_{C(sat)} = \frac{V_{CC}}{R_C} = \frac{30 \ V}{1.5 \ k\Omega} = 20 \ mA$$

With Eq. (7-4),

$$V_{CE(cutoff)} = V_{CC} = 30 \ V$$

Figure 7-3*b* shows the dc load line.
  Get the Q point as follows. With Eq. (7-1),

$$I_B = \frac{V_{BB} - V_{BE}}{R_B} = \frac{30 - 0.7}{390(10^3)} = 75.1 \ \mu A$$

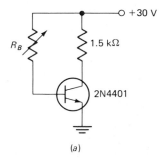

(a)

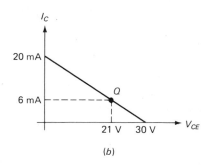

(b)

Figure 7-3.

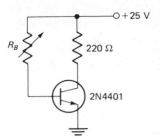

Figure 7-4.

The collector current is

$$I_C = \beta_{dc}I_B = 80 \times 75.1 \ \mu A = 6 \ mA$$

With Eq. (7-2),

$$V_{CE} = V_{CC} - I_C R_C = 30 - 6(10^{-3})1.5(10^3)$$
$$= 21 \ V$$

Figure 7-3*b* shows the Q point: its coordinates are $I_C = 6$ mA and $V_{CE} = 21$ V. Notice that the Q point lies on the dc load line because the load line represents all possible operating points. If we were to change the value of R_B, the Q point would shift to another point on the load line.

EXAMPLE 7-3

The transistor of Fig. 7-4 has a β_{dc} of 80 and $V_{CE(sat)} = 0.1$ V. R_B is adjusted to get transistor saturation. What is the value of $I_{C(sat)}$? The corresponding value of R_B?

SOLUTION

As we decrease R_B, base current increases, collector current increases, and the voltage drop across R_C increases; this decreases the collector-emitter voltage. Eventually, V_{CE} decreases to 0.1 V. At this point, the collector diode just loses reverse bias, preventing a further increase in collector current. The transistor is saturated, and its collector current is

$$I_{C(sat)} \cong \frac{V_{CC}}{R_C} = \frac{25}{220} = 114 \ mA$$

This is the maximum collector current you can get in Fig. 7-4.

The base current is

$$I_{B(sat)} = \frac{I_{C(sat)}}{\beta_{dc}} = \frac{114 \ mA}{80} = 1.43 \ mA$$

and the base resistance is

$$R_B = \frac{V_{BB} - V_{BE}}{I_{B\,(\text{sat})}} = \frac{25 - 0.7}{1.43(10^{-3})} = 17 \text{ k}\Omega$$

If you continue to decrease R_B, base current will increase, but collector current will remain at 114 mA.

Incidentally, the exact value of collector saturation current is

$$I_{C\,(\text{sat})} = \frac{V_{CC} - V_{CE\,(\text{sat})}}{R_C}$$

In low-power transistors, $V_{CE\,(\text{sat})}$ is only a few-tenths volt, small enough to ignore. As an approximation, many people visualize the collector-emitter terminals *shorted*, equivalent to $V_{CE} = 0$. When the transistor of Fig. 7-4 is saturated, therefore, its collector is ideally shorted to ground.

The collector current through the 220-Ω resistor flows downward, disappearing into the collector, similar to water disappearing down a drain. This is why a transistor with a grounded emitter is called a *current sink;* collector current flows down through the current sink into ground.

EXAMPLE 7-4
Draw the dc load line of Fig. 7-5a for $R_C = 20$ kΩ.

SOLUTION
When the transistor operates in the saturation region, its collector is ideally grounded. In this case, all the supply voltage appears across the collector resistor and

$$I_{C\,(\text{sat})} \cong \frac{V_{CC}}{R_C} = \frac{20 \text{ V}}{20 \text{ k}\Omega} = 1 \text{ mA}$$

Therefore, at saturation the transistor sinks 1 mA to ground.

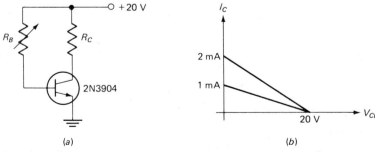

(a)   (b)

Figure 7-5.

When the transistor operates in the cutoff region, its collector-emitter terminals appear open; therefore, all the supply voltage appears across these open terminals, and

$$V_{CE(\text{cutoff})} = 20 \text{ V}$$

Figure 7-5*b* shows the dc load line (the lower one).

EXAMPLE 7-5
Repeat Example 7-4 for $R_C = 10 \text{ k}\Omega$.

SOLUTION
When the transistor is saturated, it will sink

$$I_{C(\text{sat})} = \frac{20 \text{ V}}{10 \text{ k}\Omega} = 2 \text{ mA}$$

The cutoff voltage is still 20 V. Figure 7-5*b* shows the dc load line (the upper one).

EXAMPLE 7-6
Capacitive reactance is inversely proportional to frequency. For this reason, capacitors have infinite reactance to zero frequency, equivalent to being *open* to dc.

What are the collector-emitter voltages in Fig. 7-6? (Use dc betas of 100 for the 2N3904 and 80 for the 2N4401.)

SOLUTION
First, visualize all capacitors as open to dc. Then, the dc currents in one transistor have no effect upon the other. The base current in the 2N3904 is

$$I_B = \frac{V_{BB} - V_{BE}}{R_B} = \frac{20 - 0.7}{2(10^6)} = 9.65 \ \mu\text{A}$$

and the collector current is

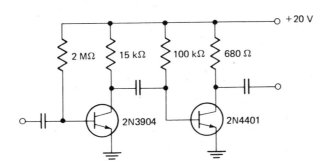

Figure 7-6. Two-stage circuit.

$$I_C = \beta_{dc}I_B = 100 \times 9.65 \ \mu A = 0.965 \ mA$$

The collector-emitter voltage of the first transistor is

$$V_{CE} = V_{CC} - I_C R_C = 20 - 0.965(10^{-3})15(10^3)$$
$$= 5.53 \ V$$

The base current of the 2N4401 is

$$I_B = \frac{V_{BB} - V_{BE}}{R_B} = \frac{20 - 0.7}{100(10^3)} = 0.193 \ mA$$

and the collector current is

$$I_C = \beta_{dc}I_B = 80 \times 0.193 \ mA = 15.5 \ mA$$

The collector-emitter voltage of the second transistor is

$$V_{CE} = V_{CC} - I_C R_C = 20 - 15.5(10^{-3})680$$
$$= 9.46 \ V$$

EXAMPLE 7-7

The TIL222 of Fig. 7-7 is a green LED with a forward voltage of 2.3 V when conducting. If the 2N3904 has a β_{dc} of 150, what is the LED current when the transistor is saturated? The value of V_{IN} that just saturates the transistor?

SOLUTION

When V_{IN} increases, more base current flows, producing more collector current. A large enough V_{IN} will saturate the transistor. When the transistor is saturated, its collector is ideally grounded. The voltage across the 1-kΩ resistor equals the supply voltage minus the drop across the LED; therefore,

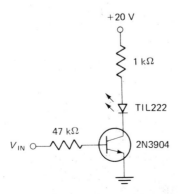

Figure 7-7.

$$I_{LED} = \frac{V_{CC} - V_{LED}}{R_C} = \frac{20 - 2.3}{1000}$$
$$= 17.7 \text{ mA}$$

The base current that just saturates the transistor is

$$I_{B\,(\text{sat})} = \frac{I_{C\,(\text{sat})}}{\beta_{\text{dc}}} = \frac{17.7 \text{ mA}}{150} = 0.118 \text{ mA}$$

Applying Kirchhoff's voltage law to the base circuit gives

$$V_{\text{IN}} = I_B R_B + V_{BE} = 0.118(10^{-3})47(10^3) + 0.7$$
$$= 6.25 \text{ V}$$

As long as V_{IN} is greater than or equal to 6.25 V, the base current is large enough to saturate the transistor. When this happens, the transistor sinks 17.7 mA of LED current to ground.

7-2 *voltage-divider bias*

Figure 7-8 shows *voltage-divider* bias, the most widely used bias in linear discrete circuits. The name "voltage divider" comes from the voltage divider formed by R_1 and R_2. The voltage across R_2 forward-biases the emitter diode. As usual, the V_{CC} supply reverse-biases the collector diode.

emitter current

A typical voltage-divider bias circuit works like this. The base current in Fig. 7-8 is very small compared to the current in R_1 and R_2. As a result, we can apply the voltage-divider theorem to find the voltage across R_2:

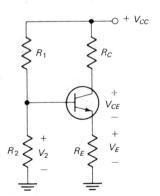

Figure 7–8. Voltage-divider bias.

$$V_2 \cong \frac{R_2}{R_1 + R_2} V_{CC} \qquad (7\text{-}7)\texttt{***}$$

Kirchhoff's voltage law gives

$$V_E = V_2 - V_{BE} \qquad (7\text{-}8)$$

This says the voltage across the emitter resistor equals the voltage across R_2 minus the V_{BE} drop. Therefore, the emitter current is

$$I_E = \frac{V_2 - V_{BE}}{R_E} \qquad (7\text{-}9)\texttt{***}$$

collector-emitter voltage

The collector-to-ground voltage V_C equals the supply voltage minus the drop across the collector resistor:

$$V_C = V_{CC} - I_C R_C \qquad (7\text{-}10)$$

The emitter-to-ground voltage is

$$V_E = I_E R_E \qquad (7\text{-}11)$$

The collector-to-emitter voltage is

$$V_{CE} = V_C - V_E = V_{CC} - I_C R_C - I_E R_E$$

or $\qquad V_{CE} \cong V_{CC} - I_C(R_C + R_E) \qquad (7\text{-}12)\texttt{***}$

because I_C and I_E are approximately equal.

If too much collector current flows in Fig. 7-8, the transistor goes into saturation. Ideally, this means a short between the collector-emitter terminals, with a saturation current of

$$I_{C(\text{sat})} \cong \frac{V_{CC}}{R_C + R_E} \qquad (7\text{-}13)\texttt{***}$$

On the other hand, if the transistor operates in the cutoff region, no collector current flows and all the supply voltage appears across the collector-emitter terminals:

$$V_{CE(\text{cutoff})} = V_{CC} \qquad (7\text{-}14)$$

The dc load line therefore passes through a vertical intercept of $V_{CC}/(R_C + R_E)$ and a horizontal intercept of V_{CC}. The Q point will lie on this load line, its position specified by Eqs. (7-9) and (7-12).

EXAMPLE 7-8
Draw the dc load line for Fig. 7-9a. Where is the Q point?

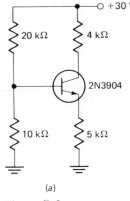

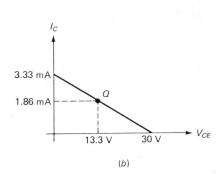

(a)　　　　　　　　　　　　(b)

Figure 7-9.

SOLUTION

When the transistor operates in the cutoff region, all the supply voltage appears across the collector-emitter terminals, giving

$$V_{CE\,(\text{cutoff})} = V_{CC} = 30 \text{ V}$$

When the transistor operates in the saturation region, it appears shorted and all the supply voltage appears across the series connection of R_C and R_E. Therefore,

$$I_{C\,(\text{sat})} = \frac{V_{CC}}{R_C + R_E} = \frac{30 \text{ V}}{9 \text{ k}\Omega} = 3.33 \text{ mA}$$

Figure 7-9*b* shows the dc load line.

The voltage across the 10-kΩ base resistor is 10 V (apply the voltage-divider theorem). The emitter diode drops 0.7 V, leaving 9.3 V across the emitter resistor R_E. So,

$$I_E = \frac{V_2 - V_{BE}}{R_E} = \frac{9.3 \text{ V}}{5 \text{ k}\Omega} = 1.86 \text{ mA}$$

Because α_{dc} is close to unity,

$$I_C \cong I_E = 1.86 \text{ mA}$$

The collector-emitter voltage is

$$V_{CE} = V_{CC} - I_C(R_C + R_E) = 30 - 1.86(10^{-3})9000$$
$$= 13.3 \text{ V}$$

Figure 7-9*b* shows the *Q* point.

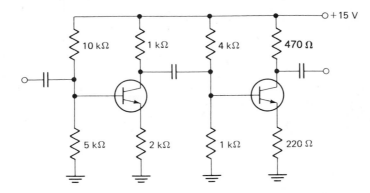

Figure 7-10.

EXAMPLE 7-9

Calculate I_C and V_{CE} for each *stage* in Fig. 7-10. (A stage is each transistor with its biasing resistors including R_C and R_E.)

SOLUTION

The capacitors are open to dc; therefore, we can analyze each stage separately because the dc currents and voltages do not interact. The voltage across the 5-kΩ resistor of the first stage is 5 V (use the voltage-divider theorem). Subtract 0.7 V for the drop across the emitter diode. This leaves 4.3 V across the emitter resistor of the first stage. Therefore, the emitter current is

$$I_E = \frac{4.3 \text{ V}}{2 \text{ k}\Omega} = 2.15 \text{ mA}$$

To a close approximation, $I_C = 2.15$ mA and

$$V_{CE} = V_{CC} - I_C(R_C + R_E) = 15 - 0.00215(3000)$$
$$= 8.55 \text{ V}$$

In the second stage, the voltage across the R_2 resistor is 3 V (one-fifth of the supply voltage). Subtract 0.7 V to get 2.3 V across the 220-Ω resistor. The collector current is

$$I_C \cong I_E = \frac{2.3 \text{ V}}{220 \text{ }\Omega} = 10.5 \text{ mA}$$

This 10.5 mA flows through the 470- and 220-Ω resistors. The drop across these resistors subtracted from the supply voltage gives the voltage across the transistor:

$$V_{CE} = V_{CC} - I_C(R_C + R_E) = 15 - 0.0105(690)$$
$$= 7.76 \text{ V}$$

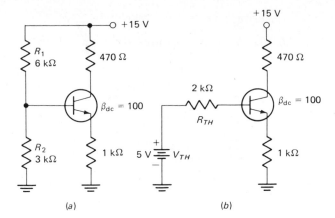

(a)   (b)   *Figure 7-11.*

In summary, the Q point of the first stage has $I_C = 2.15$ mA and $V_{CE} = 8.55$ V; the Q point of the second stage has $I_C = 10.5$ mA and $V_{CE} = 7.76$ V.

EXAMPLE 7-10
Work out an accurate value of emitter current in Fig. 7-11a using a V_{BE} drop of 0.7 V.

SOLUTION
If we open the base lead, the voltage divider is unloaded. The Thevenin voltage of the voltage divider is

$$V_{TH} = \frac{R_2}{R_1 + R_2} V_{CC} \qquad (7\text{-}15)$$

and the Thevenin resistance is

$$R_{TH} = R_1 \| R_2 \qquad (7\text{-}16)$$

where the vertical lines mean "in parallel with." We use this convenient notation in all our formulas involving two parallel resistances. Given R_1 and R_2, we can find the equivalent parallel resistance with

$$R_1 \| R_2 = \frac{R_1 R_2}{R_1 + R_2}$$

Substituting the values of Fig. 7-11a gives

$$V_{TH} = 5 \text{ V}$$
$$R_{TH} = 2 \text{ k}\Omega$$

Figure 7-11b is the equivalent circuit.

By applying Kirchhoff's voltage law to the base-emitter loop of Fig. 7-11b,

$$I_B R_{TH} + V_{BE} + I_E R_E - V_{TH} = 0$$

Since $I_C \cong I_E$ and $I_B \cong I_E / \beta_{dc}$, we can rearrange the equation to

$$I_E = \frac{V_{TH} - V_{BE}}{R_E + R_{TH} / \beta_{dc}} \tag{7-17}$$

Substituting the values of Fig. 7-11b gives

$$I_E = \frac{5 - 0.7}{1000 + 2000/100} = \frac{4.3}{1020} = 4.22 \text{ mA}$$

Notice how the first denominator term is much greater than the second denominator term (1000 versus 20). This is typical of a well-designed voltage-divider bias circuit.

In fact, the definition of a well-designed circuit is one that satisfies this condition:

$$R_E \gg \frac{R_{TH}}{\beta_{dc}} \tag{7-18}$$

This forces Eq. (7-17) to reduce to

$$I_E = \frac{V_{TH} - V_{BE}}{R_E} = \frac{V_2 - V_{BE}}{R_E}$$

because $V_2 = V_{TH}$ when condition (7-18) is satisfied.

Here's why voltage-divider bias has become the most widely used bias for linear discrete circuits. In mass production of transistor circuits, one of the main problems is the variation in β_{dc}. It varies from transistor to transistor— of the same type. A 2N3904, for instance, has a minimum β_{dc} of 100 and a maximum β_{dc} of 300 for an I_C of 10 mA and a temperature of 25°C. This means that when working with thousands of 2N3904s, we can get as much as a 3:1 variation in β_{dc}. Therefore, we cannot get a stable value of I_E unless I_E is virtually independent of β_{dc}. By satisfying condition (7-18), we almost eliminate the effect of β_{dc} on the value of emitter current. This is why β_{dc} does not appear in the approximate formula for emitter current.

7-3 *collector-feedback bias*

Figure 7-12 illustrates *collector-feedback* bias; it offers simplicity (only two resistors) and good low-frequency response (discussed later). Note carefully how the base resistor is returned to the collector rather than to the power supply. This is what distinguishes collector-feedback bias from base bias.

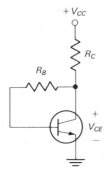

Figure 7-12. Collector-feedback bias.

the feedback concept

Instead of having a fixed supply voltage applied to the base resistor, we use the collector voltage to drive the base resistor. This introduces *feedback,* which helps to reduce the effect of β_{dc} upon the Q point.

Here's how feedback works. Suppose the temperature increases, causing β_{dc} to increase in Fig. 7-12. This will result in more collector current. But as soon as the collector current increases, the collector-emitter voltage decreases (there's a greater drop across R_C). This means less voltage drives the base resistor, causing a decrease in base current. The lower base current therefore reduces the original increase in collector current. Without feedback, collector current is directly proportional to β_{dc}; but with feedback, collector current does not increase as rapidly as β_{dc}. The next section nails down the effect of changes in β_{dc}.

analysis and design formulas

As will be proved in Example 7-13, the collector current is closely approximated by

$$I_C \cong \frac{V_{CC} - V_{BE}}{R_C + R_B / \beta_{dc}} \qquad (7\text{-}19)\text{***}$$

This is the current in the active region. As usual, the collector-emitter voltage is the supply voltage minus the drop across the collector resistor. Since base current is much smaller than collector current,

$$V_{CE} \cong V_{CC} - I_C R_C \qquad (7\text{-}20)\text{***}$$

Also proved in Example 7-13, the base resistance that sets up *midpoint bias* is given by

$$R_B = \beta_{dc} R_C \qquad (7\text{-}21)\text{***}$$

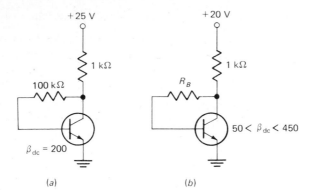

(a) (b) *Figure 7-13.*

Here is the meaning. When R_B equals $\beta_{dc} R_C$, the collector-emitter voltage equals approximately half the supply voltage. Midpoint bias $(V_{CE} = V_{CC}/2)$ means the Q point is at the center of the dc load line.

geometric average of β_{dc}

As mentioned earlier, the β_{dc} of a 2N3904 varies over a 3∶1 range when thousands of 2N3904s are used. On top of this, we can get an additional 3∶1 variation caused by temperature changes. In other words, up to a 9∶1 variation is possible for the β_{dc} of 2N3904s. (The 2N3904 is widely used. It's a low-power silicon transistor that can handle up to 200 mA of collector current.)

What β_{dc} should we use in Eq. (7-21)? The minimum, the maximum, or the average? Whenever large variations may occur in β_{dc}, designers use the *geometric* average given by

$$\beta_{dc} = \sqrt{\beta_{dc(min)}\ \beta_{dc(max)}} \qquad (7\text{-}22)\,\text{***}$$

This geometric average does a much better job of centering the Q point than a simple arithmetic average.

EXAMPLE 7-11
Where are the values of I_C and V_{CE} in Fig. 7-13a?

SOLUTION
With Eq. (7-19),

$$I_C = \frac{V_{CC} - V_{BE}}{R_C + R_B/\beta_{dc}} = \frac{25 - 0.7}{10^3 + 10^5/200}$$
$$= 16.2 \text{ mA}$$

The collector-emitter voltage is

$$V_{CE} = V_{CC} - I_C R_C = 25 - 0.0162(1000)$$
$$= 8.8 \text{ V}$$

EXAMPLE 7-12

The β_{dc} of Fig. 7-13b varies over the 9:1 range shown. Select a value of R_B to set up midpoint bias.

SOLUTION

The geometric average is

$$\beta_{dc} = \sqrt{50 \times 450} = 150$$

With Eq. (7-21),

$$R_B = \beta_{dc} R_C = 150 \times 1 \text{ k}\Omega$$
$$= 150 \text{ k}\Omega$$

EXAMPLE 7-13

Prove Eqs. (7-19) and (7-21).

SOLUTION

In Fig. 7-12, the collector-emitter voltage is

$$V_{CE} = V_{CC} - (I_C + I_B)R_C = V_{CC} - I_E R_C$$

or

$$V_{CE} = V_{CC} - \frac{I_C R_C}{\alpha_{dc}} \tag{7-23}$$

We can also sum voltages around the base circuit to get

$$V_{CE} = I_B R_B + V_{BE}$$

or

$$V_{CE} = \frac{I_C R_B}{\beta_{dc}} + V_{BE} \tag{7-24}$$

Equating the right-hand members of Eqs. (7-23) and (7-24) results in

$$\frac{I_C R_B}{\beta_{dc}} + V_{BE} = V_{CC} - \frac{I_C R_C}{\alpha_{dc}}$$

Solving for I_C, we get

$$I_C = \frac{V_{CC} - V_{BE}}{R_C/\alpha_{dc} + R_B/\beta_{dc}} \tag{7-25}$$

Because α_{dc} is greater than 0.98 for most transistors, the foregoing is closely approximated by

$$I_C \cong \frac{V_{CC} - V_{BE}}{R_C + R_B/\beta_{\text{dc}}} \qquad (7\text{-}26)$$

Next, we can prove Eq. (7-21) as follows. The dc load line of Fig. 7-12 has a saturation current of V_{CC}/R_C. Half of this is $V_{CC}/2R_C$, the current at midpoint bias. When we substitute the right-hand member of Eq. (7-21) into Eq. (7-26), we get

$$I_C = \frac{V_{CC} - V_{BE}}{2R_C} \cong \frac{V_{CC}}{2R_C}$$

when V_{BE} is negligible. In other words, Eq. (7-21) is a close approximation for setting up midpoint bias, provided you can ignore V_{BE}.

7-4 *emitter bias*

Figure 7-14*a* shows *emitter bias,* popular when a split supply is available. (A split supply means positive and negative voltages. See Prob. 5-23.) The name "emitter bias" is used because a negative supply V_{EE} forward-biases the emitter diode through resistor R_E. As usual, the V_{CC} supply reverse-biases the collector diode. Figure 7-14*b* is a simplified way to draw the circuit.

emitter current

The key to analyzing the typical emitter-bias circuit is this. The voltage from emitter to ground is less than 1 V. Because V_{EE} is much greater than 1 V, we can treat the upper end of R_E as an *approximate ground point.* Figure 7-14*c* emphasizes this crucial idea. Because of the ground, virtually all the V_{EE} supply voltage appears across R_E. Therefore,

$$I_E \cong \frac{V_{EE}}{R_E} \qquad (7\text{-}27)***$$

other quantities

As usual, I_C equals I_E to a close approximation. Since the emitter acts like an approximate ground point, the collector-emitter voltage equals the V_{CC} supply voltage minus the drop across the collector resistor:

$$V_{CE} \cong V_{CC} - I_C R_C \qquad (7\text{-}28)***$$

If the transistor is saturated, the collector-emitter terminals act ideally shorted and the sum of the supply voltages appears across R_C and R_E. Because of this, the maximum possible current is

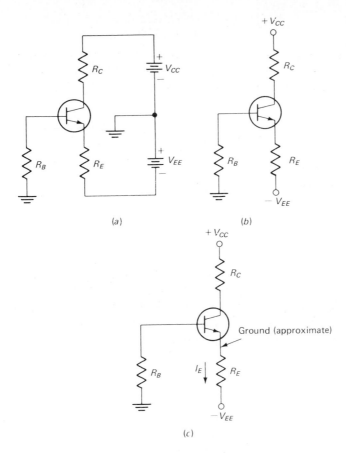

$$I_{C(\text{sat})} = \frac{V_{CC} + V_{EE}}{R_C + R_E} \qquad (7\text{-}29)$$

EXAMPLE 7-14
Calculate I_C and V_{CE} for Fig. 7-15. Also, work out the collector saturation current.

SOLUTION
Visualize the emitter as grounded. Then all the V_{EE} supply voltage (20 V) appears across the emitter resistor (10 KΩ). So, the emitter current is

$$I_E \cong \frac{V_{EE}}{R_E} = \frac{20 \text{ V}}{10 \text{ k}\Omega} = 2 \text{ mA}$$

This means 2 mA of collector current flows through the 5-kΩ resistor. The drop across R_C subtracted from V_{CC} gives V_{CE}:

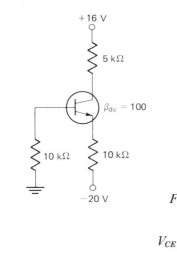

Figure 7-15.

$$V_{CE} = V_{CC} - I_C R_C = 16 - 0.002(5000)$$
$$= 6 \text{ V}$$

If the transistor were saturated, the sum of the supply voltages would appear across R_C and R_E. This is why the saturation current is

$$I_{C(\text{sat})} = \frac{16 + 20}{5000 + 10,000} = 2.4 \text{ mA}$$

EXAMPLE 7-15

Work out an accurate value of emitter current in Fig. 7-15.

SOLUTION

By applying Kirchhoff's voltage law to the emitter-base loop, we get

$$I_B R_B + V_{BE} + I_E R_E - V_{EE} = 0$$

Since $I_C \cong I_E$ and $I_B \cong I_E/\beta_{\text{dc}}$, we can rearrange the equation:

$$I_E = \frac{V_{EE} - V_{BE}}{R_E + R_B/\beta_{\text{dc}}} \tag{7-30}$$

Substituting the values of Fig. 7-15 gives

$$I_E = \frac{20 - 0.7}{10,000 + 10,000/100} = \frac{19.3}{10,100}$$
$$= 1.91 \text{ mA}$$

(*Note:* this is within 5 percent of the approximate answer found in example 7-14.)

In a well-designed emitter-bias circuit,

$$V_{EE} \gg V_{BE} \tag{7-30a}$$

and

$$R_E \gg \frac{R_B}{\beta_{\mathrm{dc}}}$$ (7-30b)

When these two conditions are satisfied, Eq. (7-30) reduces to

$$I_E \cong \frac{V_{EE}}{R_E}$$

which means the emitter acts like an approximate ground point in a well-designed emitter-bias circuit.

7-5 pnp *biasing circuits*

Figure 7-16 shows a *pnp* transistor. Since the emitter and collector diodes point in directions opposite to those of an *npn* transistor, all currents and voltages are reversed when the *pnp* transistor is FR-biased. In other words, to forward-bias the emitter diode of a *pnp* transistor, V_{BE} has the minus-plus polarity shown in Fig. 7-16. To reverse-bias the collector diode, V_{CB} must have the minus-plus polarity indicated. From this, it follows that V_{CE} is minus-plus as shown. Since the emitter diode points *in,* conventional *emitter current* flows *into* the *pnp* transistor; base and collector currents flow out.

complementary circuits

The *pnp* transistor is called the *complement* of the *npn* transistor. The word *complement* signifies all voltages and currents are opposite those of the *npn* transistor.

Every *npn* circuit has a complementary *pnp* circuit. To find the complementary *pnp* circuit, all you do is

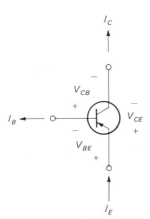

Figure 7-16. pnp *transistor.*

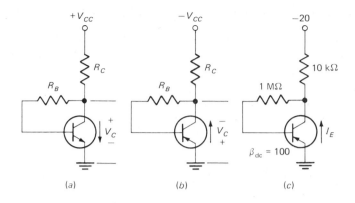

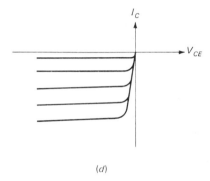

(d)

Figure 7-17. Complementary circuits.

1. Replace the *npn* transistor by a *pnp* transistor.
2. Complement or reverse all voltages and currents.

As an example, Fig. 7-17*a* shows collector-feedback bias using an *npn* transistor. Emitter current flows down and collector voltage is positive with respect to ground. Figure 7-17*b* shows the complementary *pnp* transistor circuit; all we have done is complement voltages and currents, and replace the *npn* by a *pnp* transistor. Now, emitter current flows up and the collector voltage is negative.

analyzing pnp *circuits*

If you use *magnitudes*, all formulas derived for *npn* circuits apply to *pnp* circuits. For instance, suppose we want the emitter current in Fig. 7-17*c*. With Eq. (7-19),

$$I_C = \frac{V_{CC} - V_{BE}}{R_C + R_B/\beta_{dc}} = \frac{20 - 0.7}{10^4 + 10^6/100} = 0.965 \text{ mA}$$

pnp *quantities are negative compared to* npn

We can define the *positive* directions of voltage and current to be those of an FR-biased *npn* transistor; therefore, the voltages and currents in an FR-biased *pnp* transistor are *negative* with respect to the *npn* directions. This is the reason some data sheets give negative values of current and voltage for *pnp* transistors. Also, a transistor curve tracer uses this convention because it simplifies the design of the curve tracer. This is why a curve tracer shows the collector curves of a *pnp* transistor in the third quadrant (Fig. 7-17*d*) rather than the first quadrant.

7-6 *moving ground around*

Don't get the idea the circuits we discuss work only with the ground locations shown. Ground is a reference point you can move around as you please. For

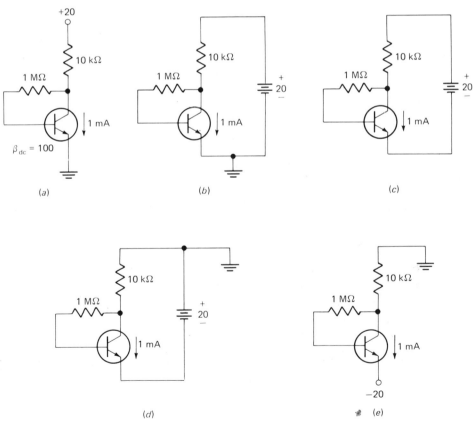

Figure 7-18. Moving ground around.

instance, Fig. 7-18a shows collector-feedback bias. Figure 7-18b shows the same circuit where we do not rely on ground to conduct current. In a circuit like this, we can remove the ground to get the *floating circuit* of Fig. 7-18c. In this floating circuit, all transistor currents and voltages have the same values as before; the same emitter current is flowing and the same collector-emitter voltage is set up.

Following this line of thinking, we can ground the positive terminal of the supply as indicated in Fig. 7-18d. And finally, we can draw the circuit as shown in Fig. 7-18e. The voltages with respect to ground differ in Figs. 7-18a through e; $V_E = 0$ in Fig. 7-18a, but $V_E = -20$ V in Fig. 7-18e.

What is important is that the three transistor currents (I_E, I_C, I_B) and the three transistor voltages (V_{BE}, V_{CE}, V_{CB}) have the same values in Figs. 7-18a through e.

7-7 *the upside-down convention for* pnp *transistors*

More and more schematic diagrams are showing *pnp* transistors drawn upside down. Once you get used to this convention, it simplifies analysis and design because *all emitter currents flow down.*

For instance, Fig. 7-19a shows a *pnp* collector-feedback biasing circuit. If we float the supply, we get the circuit of Fig. 7-19b. And, grounding the negative terminal of the supply, we get Fig. 7-19c. It does not matter how a transistor is oriented in space; therefore, it works just as well upside down as shown in Fig. 7-19d. The emitter current in this circuit has exactly the same value as the emitter current of Fig. 7-19a, b, and c.

Besides unifying analysis and design, drawing *pnp* transistors upside down has another advantage. Often, both *npn* and *pnp* transistors are used in the same circuit. By drawing the *pnp* transistors upside down, the drafter can produce a simpler schematic diagram because fewer lines have to be crossed.

EXAMPLE 7-16
Figure 7-20 shows a three-stage circuit with a V_{CC} supply of +20 V. GND stands for ground. If all transistors have a β_{dc} of 100, what are the I_C and V_{CE} of each stage?

SOLUTION
The first stage is base-biased with

$$I_B = \frac{V_{BB} - V_{BE}}{R_B} = \frac{20 - 0.7}{10^5} = 193 \text{ } \mu A$$
$$I_C = \beta_{dc} I_B = 100 \times 193 \text{ } \mu A = 19.3 \text{ mA}$$
$$V_{CE} = V_{CC} - I_C R_C = 20 - 0.0193(560) = 9.19 \text{ V}$$

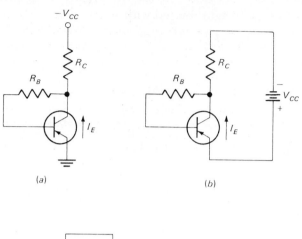

(a)

(b)

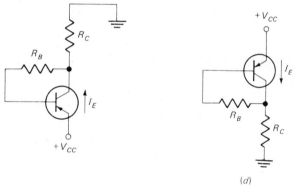

Figure 7-19. Drawing pnp *transistors upside down.*

(d)

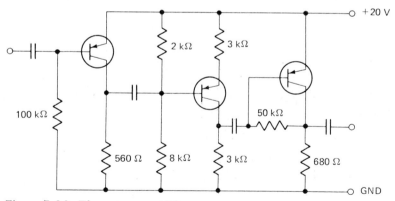

Figure 7-20. Three-stage amplifier.

The second stage is voltage-divider biased with

$$I_C \cong I_E = \frac{V_2 - V_{BE}}{R_E} = \frac{4 - 0.7}{3000} = 1.1 \text{ mA}$$
$$V_{CE} = V_{CC} - I_C(R_C + R_E) = 20 - 0.0011(6000) = 13.4 \text{ V}$$

The third stage is collector-feedback biased with

$$I_C = \frac{V_{CC} - V_{BE}}{R_C + R_B/\beta_{dc}} = \frac{20 - 0.7}{680 + 50,000/100} = 16.4 \text{ mA}$$
$$V_{CE} = V_{CC} - I_C R_C = 20 - 0.0164(680) = 8.85 \text{ V}$$

In summary, here are the Q points:

$$\text{1st stage: } I_C = 19.3 \text{ mA}, \ V_{CE} = \ 9.19 \text{ V}$$
$$\text{2nd stage: } I_C = \ 1.1 \text{ mA}, \ V_{CE} = 13.4 \text{ V}$$
$$\text{3rd stage: } I_C = 16.4 \text{ mA}, \ V_{CE} = \ 8.85 \text{ V}$$

7-8 *collector cutoff current*

All our biasing formulas neglect thermally produced current and surface-leakage current. This is reasonable, *provided you satisfy one condition.* This section tells you what the condition is.

Almost any data sheet lists the value of I_{CBO} at 25°C; this is the current from

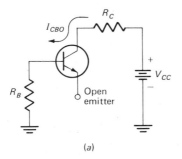

(a)

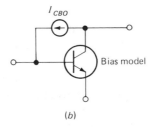

(b)

Figure 7-21. Meaning of I_{CBO}.

collector to base with an open emitter (see Fig. 7-21a). I_{CBO} is the reverse current in the collector diode, partly caused by thermally produced carriers and partly by leakage along the surface. Often, data sheets give values of I_{CBO} at several temperatures besides 25°C. If not, you can estimate I_{CBO} at other temperatures as follows: I_{CBO} roughly doubles for every 10°C rise.

effect of I_{CBO} on base-biased circuits

Even though I_{CBO} is measured with the emitter open (Fig. 7-21a), it still has an effect when the emitter is not open, that is, when the transistor is FR-biased. To find the effect it has on an FR-biased transistor, we first need a model or equivalent circuit that accounts for I_{CBO}. Figure 7-21b shows this model.[1] All we do is visualize an I_{CBO} current source in shunt with the bias model of the transistor. The bias model produces the desired component of collector current. The I_{CBO} current source, on the other hand, produces an undesired component of collector current.

When we FR-bias a transistor, the I_{CBO} source is still in shunt with the bias model (see Fig. 7-22a). To determine the effect of I_{CBO}, it helps to draw the equivalent circuit of Fig. 7-22b. The two I_{CBO} sources produce exactly the same effect as the single source of Fig. 7-22a. Why? Because a current of I_{CBO} still flows into node A, and a current of I_{CBO} still flows out of node B. As a result, the voltages and currents of Fig. 7-22a and b are the same.

We can now put the current source on the left into the base circuit and the current source on the right into the collector circuit, as shown in Fig. 7-22c. The reason for doing this is to allow us to Thevenize the base circuit. Visualize the base lead broken; this gives a Thevenin base voltage of

$$V_{TH} = V_{BB} + I_{CBO}R_B$$

Next, we reduce all sources to zero; this means shorting the voltage source V_{BB} and opening the current source I_{CBO}. Under this condition, we get a Thevenin resistance of

$$R_{TH} = R_B$$

Figure 7-23 shows the Thevenized base circuit. Now we can see exactly what effect I_{CBO} will have on any base-biased circuit. The I_{CBO} of a transistor makes it appear as though the transistor is driven by a $V_{BB} + I_{CBO}R_B$ voltage source instead of a V_{BB} source.

Data sheets list the worst-case I_{CBO} (the maximum value) for the transistor type. Since I_{CBO} varies from transistor to transistor of the same type, the only

---

[1] This is an approximation because I_{CBO} consists of surface-leakage current as well as thermally produced current. The former is voltage-dependent; because of this, using an I_{CBO} current source is not exact.

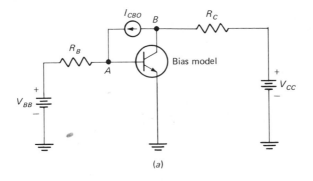

(a)

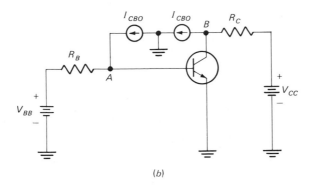

(b)

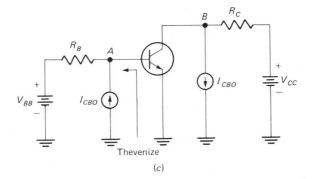

(c)

Figure 7-22. Deriving the effect of I_{CBO} on bias.

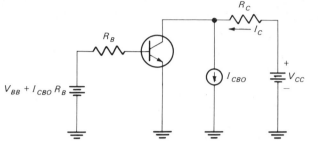

Figure 7-23.

satisfactory way to handle it is to make it negligible. We can do this by satisfying the condition:

$$V_{BB} \gg I_{CBO}R_B \tag{7-31}$$

other biasing circuits

By applying a similar derivation to voltage-divider bias, the condition to satisfy is

$$V_2 \gg I_{CBO}R_{TH} \tag{7-32}$$

For collector-feedback bias, make

$$V_{CC} \gg I_{CBO}R_B \tag{7-33}$$

And for emitter-bias,

$$V_{EE} \gg I_{CBO}R_B \tag{7-34}$$

where I_{CBO} is at the highest expected temperature.

7-9 *summary*

In linear transistor circuits the emitter diode must remain forward-biased and the collector diode reverse-biased. We intend to drive the transistor with an ac voltage to produce changes in the transistor currents and voltages. To prevent this ac voltage from reverse-biasing the emitter diode or forward-biasing the collector diode, we must first set up a quiescent operating point. Then, if the ac signal is not too large, the transistor remains in the active region throughout the cycle.

This chapter has discussed four basic types of biasing a transistor, that is, setting up a Q point in the active region. Table 7-1 lists some advantages and disadvantages for each type of bias. As shown, base bias is used mainly in digital

Table 7-1 Types of Bias

Type	Advantage	Disadvantage	Where used
Base bias	Simplicity	Sensitive to changes in β_{dc}	Digital and switching circuits
Voltage-divider	Stable Q point	Needs four resistors	General-purpose amplifiers
Collector-feedback	Low-frequency response	Partially depends on β_{dc}	Small-signal amplifiers
Emitter bias	Stable Q point	Needs a split power supply	General-purpose amplifiers

and switching circuits. Voltage-divider bias has a stable Q point and is widely used in general-purpose *amplifiers* (circuits that increase the amplitude of an ac signal). Collector-feedback bias is partially dependent upon the value of β_{dc}; therefore, it's used only with small-signal amplifiers. Finally, emitter bias is all right for general-purpose amplifiers, provided a split power supply is available.

self-testing review

Read each of the following and provide the missing words. Answers appear at the beginning of the next question.

1. Linear transistor circuits operate with the _____ diode forward-biased and the collector diode _____-biased.

2. *(emitter, reverse)* In a base-biased circuit, a voltage source V_{BB} forward-biases the emitter diode through a _____ resistor R_B. The base current equals the _____ across this resistor divided by the resistance.

3. *(current-limiting, voltage)* The dc _____ line represents all possible operating points. The intersection of this line with the correct base current gives the actual _____ point of the transistor.

4. *(load, operating)* The upper end of the dc load line is called the _____ point. At this point, the collector current is _____, and the collector diode comes out of _____. If the base current is greater than $I_{B(sat)}$, the collector current cannot increase because the collector diode has lost its reverse bias.

5. *(saturation, maximum, reverse bias)* All operating points between cutoff and saturation are the _____ region of a transistor. The intersection of the base current and the dc load line is the _____ (Q) point.

6. *(active, quiescent)* Many people visualize the collector-emitter terminals of a saturated

transistor as _____, equivalent to $V_{CE} = 0$. In a current sink, the collector current flows downward through the sink into ground.

7. *(shorted)* The most widely used bias in linear discrete circuits is _____ bias. The voltage out of a voltage divider forward-biases the emitter diode. The emitter current equals V_2 minus the V_{BE} drop divided by the _____ resistance.

8. *(voltage-divider, emitter)* Collector-feedback bias uses _____, which helps reduce the effect of β_{dc} upon the Q point. To set up midpoint bias, R_B should equal _____ times _____. Whenever large variations in β_{dc} occur, we use the geometric average.

9. *(feedback, β_{dc}, R_C)* Emitter bias is often used when a _____ power supply is available. Because the emitter is an approximate ground point, almost all the V_{EE} supply voltage appears across the _____ resistor. This is why the emitter current equals V_{EE} divided by R_E.

10. *(split, emitter)* When *pnp* transistors are drawn upside-down, all emitter and collector currents flow downward. Often, both _____ and _____ transistors are used in the same circuit. By drawing *pnp* transistors upside-down, simpler schematic diagrams result.

11. *(npn, pnp)* For I_{CBO} to have negligible effect on the Q point, the product of _____ times _____ must be small compared with V_{BB} in a base-biased circuit, V_2 in a voltage-divider-biased circuit, V_{CC} in a collector-feedback-biased circuit, and V_{EE} in an emitter-biased circuit.

12. *(I_{CBO}, R_B)* Base bias is used mainly in digital and switching circuits. Voltage-divider bias has a stable Q point and is widely used in general-purpose amplifiers. Collector-feedback bias is partially dependent upon the value of β_{dc}; therefore, it's used only with small-signal amplifiers. Emitter bias is all right for general-purpose amplifiers, provided a split supply is available.

problems

7-1. GND stands for ground in Fig. 7-24. What is the V_{CE} of the first stage if $V_{CC} = 25$ V?

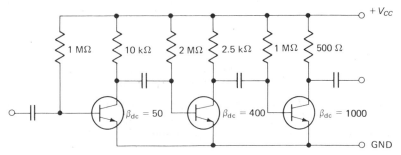

Figure 7-24.

7-2. If $V_{CC} = 15$ V in Fig. 7-24, what would a voltmeter read between the collector of the second stage and ground (GND)?

7-3. What are the values of I_C and V_{CE} in the third stage of Fig. 7-24 if $V_{CC} = 20$ V?

7-4. What is the saturation current for each stage of Fig. 7-24 if $V_{CC} = 10$ V?

7-5. Given $V_{CC} = 30$ V in Fig. 7-24, draw the dc load line for each stage.

7-6. What are the coordinates of the Q point for each stage of Fig. 7-24 if $V_{CC} = 10$ V?

7-7. If $V_{CC} = 20$ V and the base resistors of Fig. 7-24 have a tolerance of ± 10 percent, what is the minimum V_{CE} for each stage? The maximum V_{CE}? (Assume collector resistors are exact.)

7-8. Base bias is not used much in linear circuits because the Q point is unstable, varying with temperature, transistor replacement, aging, etc. As an example of what can happen to the Q point, suppose β_{dc} changes from 50 to 100 when the temperature increases from 25°C to 75°C. In Fig. 7-24, what is V_{CE} in the first stage at 25°C and 75°C if $V_{CC} = 20$ V? What region is the transistor operating in at 25°C and 75°C?

7-9. If the transistor of Fig. 7-25 has $\beta_{dc} = 125$ and the LED has a voltage drop of 1.7 V, how much current does the transistor sink when it's saturated? What is the value of V_{IN} that drives the transistor into the saturation region?

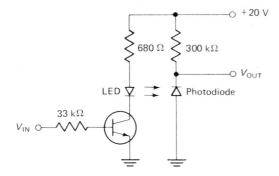

Figure 7-25.

7-10. When the transistor of Fig. 7-25 is saturated, the light from the LED falls upon the photodiode and produces 50 μA of reverse current through the photodiode. When the transistor is cut off, no reverse current flows through the photodiode. What does V_{OUT} equal when the transistor is saturated? When it's cut off? (Incidentally, an *optocoupler* combines a LED and a photodetector in a single package. The key advantage of an optocoupler is electrical isolation between the LED and the photodetector. This isolation is typically greater than 10^{10} Ω because the only contact between the circuits is the light traveling from the LED to the photodetector.)

7-11. Draw the dc load line for the first stage of Fig. 7-26. Where is the Q point? (Use $V_{CC} = 10$ V.)

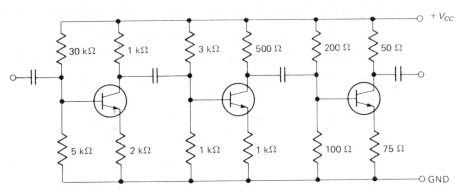

Figure 7-26.

7-12. What does the saturation current of the second stage equal in Fig. 7-26 if $V_{CC} = 15$ V? What is the cutoff voltage for this stage?

7-13. Draw the dc load line and Q point for the third stage of Fig. 7-26 with $V_{CC} = 20$ V.

7-14. Calculate I_C and V_{CE} for each stage of Fig. 7-26 with $V_{CC} = 25$ V.

7-15. $V_{CC} = 30$ V in Fig. 7-26. What would a dc voltmeter read from each collector to ground? From each emitter to ground?

7-16. The β_{dc} of the second stage in Fig. 7-26 varies from 100 to 300 with transistor replacement, temperature change, etc. Use Eq. (7-17) to calculate the minimum and maximum values of I_E in the second stage for a V_{CC} of 15 V. What does this prove about the relation between I_E and β_{dc}?

7-17. What are the values of I_C and V_{CE} for each stage of Fig. 7-27 if $V_{CC} = 10$ V? (Use $\beta_{dc} = 100$ for the second stage.)

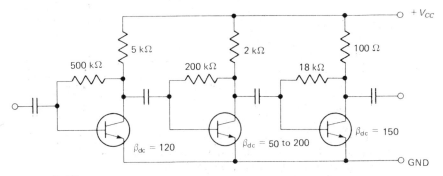

Figure 7-27.

7-18. What does V_{CE} equal in the second stage of Fig. 7-27 when $\beta_{dc} = 50$ and $V_{CC} = 15$ V? What does V_{CE} equal if β_{dc} changes to 200?

7-19. A collector-feedback bias circuit has $R_C = 750\ \Omega$ and $\beta_{dc} = 175$. What should R_B equal for midpoint bias?

7-20. A transistor whose β_{dc} varies from 75 to 250 is used in a collector-feedback bias circuit with an R_C of 1.5 kΩ. What value should R_B have to get midpoint bias in a mass-produced circuit?

7-21. What are the coordinates of the Q point for each stage of Fig. 7-27 if $V_{CC} = 20$ V? (Use $\beta_{dc} = 100$ for the second stage.)

7-22. What are the values of I_C and V_{CE} in the first stage of Fig. 7-28 if $V_{CC} = 15$ V and $V_{EE} = 10$ V?

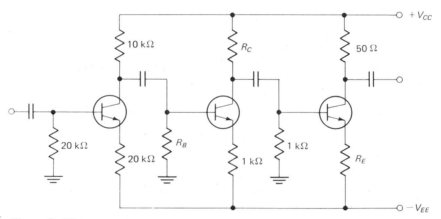

Figure 7-28.

7-23. In Fig. 7-28, $V_{CC} = 30$ V and $V_{EE} = 25$ V. What value should R_C have to get a V_{CE} of approximately 15 V in the second stage?

7-24. A rule of thumb for satisfying condition (7-30b) is to make R_B from 3 to 10 times the value of R_E; 3 times for low betas like 50, and 10 for high betas like 200. What size should R_B be in Fig. 7-28 if the β_{dc} of the second stage is 50? If it is 200?

7-25. To get 100 mA of emitter current in the third stage of Fig. 7-28, what value should R_E have if $V_{CC} = 25$ V and $V_{EE} = 15$ V?

7-26. $V_{EE} = V_{CC} = 20$ V in Fig. 7-28. In the second stage, $R_B = 5$ kΩ and $R_C = 470\ \Omega$. In the third stage, $R_E = 120\ \Omega$. What are the coordinates of the Q point for each stage?

7-27. What is the voltage from each collector to ground in Fig. 7-29 if $V_{CC} = 15$ V?

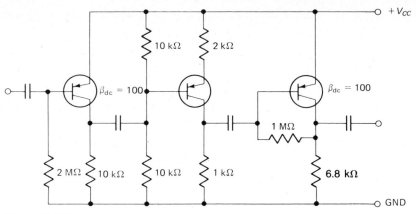

Figure 7-29.

7-28. What are the values of I_C and V_{CE} for each stage in Fig. 7-29 when $V_{CC} = 20$ V?

7-29. What are the coordinates of the Q point in each stage of Fig. 7-29 if $V_{CC} = 25$ V?

8
ac equivalent circuits

We can apply an ac voltage across the emitter diode; this forces transistor currents and voltages to have ac variations of the same frequency. By proper design, we can *amplify* the ac signal, that is, increase its peak-to-peak value. To analyze amplifier circuits, we first must discuss *ac equivalent circuits*.

8-1 *coupling and bypass capacitors*

Most capacitors in transistor circuits are either *coupling* or *bypass* capacitors. A coupling capacitor passes an ac signal from one ungrounded point to another ungrounded point. For instance, in Fig. 8-1a the ac voltage at point A also appears at point B. For this to happen, the capacitive reactance X_C must be very small compared with the resistances.

In Fig. 8-1a the circuit to the left of point A may be a single ac source and resistor, or it may be the Thevenin equivalent circuit of something more complicated. Likewise, resistance R_L may be a single load resistor, or it may be the equivalent resistance of a more complex network. It doesn't matter what the actual circuits are on either side of the capacitor; as long as we can reduce the circuit to a single loop as shown, the ac current flows through a total resistance of $R_{TH} + R_L$.

As you recall from basic circuit theory, the ac current in a one-loop RC circuit equals

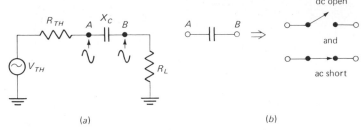

Figure 8-1. Coupling capacitor.

$$I = \frac{V}{\sqrt{R^2 + X_C^2}} \qquad (8\text{-}1)$$

where R is the total resistance of the loop. In Fig. 8-1a, $R = R_{TH} + R_L$. The maximum ac current flows when X_C is much smaller than R. In other words, the capacitor couples the signal properly from A to B when $X_C \ll R$.

capacitor size

The size of the coupling capacitor depends on the lowest frequency you are trying to couple because X_C increases when frequency decreases. We will use this rule:

$$T = RC \quad \text{(at lowest frequency)} \qquad (8\text{-}2)\text{***}$$

where T is the period, that is, $1/f$. In other words, we will use a capacitor large enough to satisfy Eq. (8-2) at the lowest frequency to be coupled; then, all higher frequencies will be well coupled.

For instance, suppose we are trying to couple frequencies from 20 Hz to 50 kHz into an amplifier. Then, the lowest frequency 20 Hz has a period of $T = 1/f = 1/20 = 0.05$ s. If the total resistance in the one-loop circuit (Fig. 8-1a) is 10 kΩ, the coupling capacitor must satisfy

$$T = RC$$
$$0.05 = 10^4 C$$

or
$$C = \frac{0.05}{10^4} = 5 \ \mu\text{F}$$

This size capacitance will do an excellent job of coupling all frequencies above 20 Hz. (Another widely used rule is to keep X_C less than one-tenth R; we prefer $T = RC$ because it is easier to work with.)

Ideally, a capacitor looks open to dc current. For this reason, we can think of a coupling capacitor as shown in Fig. 8-1b. It acts like a switch that is open

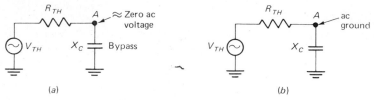

Figure 8-2. Bypass capacitor.

to dc current but shorted to ac current. This discriminating action allows us to get an ac signal from one stage into another without disturbing the dc biasing of each stage.

ac ground

A bypass capacitor is similar to a coupling capacitor except it couples an ungrounded point to a *grounded point* as shown in Fig. 8-2a. Again, V_{TH} and R_{TH} may be a single source and resistor as shown, or may be a Thevenin circuit. To the capacitor it makes no difference. It sees a total resistance of R_{TH}. Equations (8-1) and (8-2) still apply because we have a one-loop RC circuit. The only difference is $R = R_{TH}$.

Bypass capacitors bring a new idea with them. In Fig. 8-2b, the capacitor ideally looks like a short to an ac signal. Because of this, point A is shorted to ground as far as the ac signal is concerned. This is why we have labeled point A as *ac ground*. A bypass capacitor will not disturb the dc voltage at point A because it looks open to dc current. However, a bypass capacitor automatically makes point A an ac ground point.

In the normal frequency range of an amplifier, all coupling and bypass capacitors look like ac shorts. For this reason, we will approximate all capacitors as dc opens and ac shorts, unless otherwise indicated.

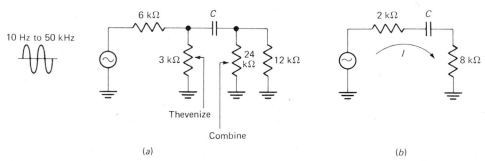

Figure 8-3.

EXAMPLE 8-1

The input signal of Fig. 8-3*a* can have a frequency between 10 Hz and 50 kHz. For the coupling capacitor to work properly, what size should it be?

SOLUTION

The first step is to reduce the circuit to a single loop. When you Thevenize the circuit to the left of capacitor, you get an R_{TH} of 2 kΩ. When you combine the two load resistors, you get an equivalent resistance of 8 kΩ.

Figure 8-3*b* shows the original circuit reduced to a single loop. The total resistance in the loop is 10 kΩ. The lowest frequency, 10 Hz, has a period of $T = 1/f = 1/10 = 0.1$ s. Therefore, the coupling capacitor must satisfy

$$T = RC$$
$$0.1 = 10^4 C$$

or
$$C = \frac{0.1}{10^4} = 10 \ \mu F$$

EXAMPLE 8-2

We want an ac ground on point *A* in Fig. 8-4*a*. What size should the bypass capacitor be?

SOLUTION

Again, we must reduce the circuit to a single loop. Looking back into point *A*, we see a Thevenin resistance of

$$R_{TH} = 125 \,\|\, 10{,}000 \cong 125 \ \Omega$$

Figure 8-4*b* shows the circuit in one-loop form. The lowest frequency is 10 Hz. So,

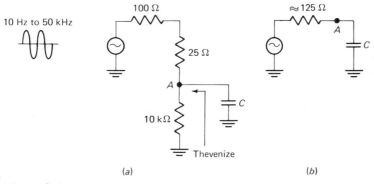

(a)   (b)

Figure 8-4.

$$T = RC$$
$$0.1 = 125C$$

or
$$C = \frac{0.1}{125} = 800 \ \mu\text{F}$$

Therefore, to bypass point A to ground we need at least 800 μF.

8-2 *the superposition theorem for ac-dc circuits*

In a transistor amplifier, the dc sources set up dc currents and voltages. The ac source produces fluctuations in the transistor currents and voltages. The simplest way to analyze the action of transistor circuits is to split the analysis into two parts: a dc analysis and an ac analysis. In other words, we can analyze transistor circuits by applying the superposition theorem in a special way. Instead of taking one source at a time, we take all dc sources at the same time and work out the dc currents and voltages using the methods of Chap. 7. Next, we take all ac sources at the same time and calculate the ac currents and voltages. By adding the dc and ac currents and voltages, we get the total currents and voltages.

ac and dc equivalent circuits

Here are the steps in applying superposition to transistor circuits:

1. Reduce all ac sources to zero; open all capacitors. The circuit that remains is all that matters as far as dc currents and voltages are concerned. Because of this, we call this circuit the *dc equivalent circuit*. Using this circuit, we calculate whatever dc currents and voltages we are interested in.
2. Reduce all dc sources to zero; short all coupling and bypass capacitors. The circuit that remains is all that matters as far as ac currents and voltages are concerned. We call this circuit the *ac equivalent circuit*. This is the circuit to use in calculating ac currents and voltages.
3. The total current in any branch is the sum of the dc current and ac current through that branch. And, the total voltage across any branch is the sum of the dc voltage and ac voltage across that branch.

Here is how to apply the superposition theorem to the transistor amplifier of Fig. 8-5a. First, reduce all ac sources to zero and open all capacitors. All that remains is the circuit of Fig. 8-5b; this is the dc equivalent circuit. This is all that matters as far as dc currents and voltages are concerned. With this circuit, we can find any dc current or voltage of interest.

Second, reduce all dc sources to zero and short all coupling and bypass capacitors. What remains is the ac equivalent circuit shown in Fig. 8-5c. Especially

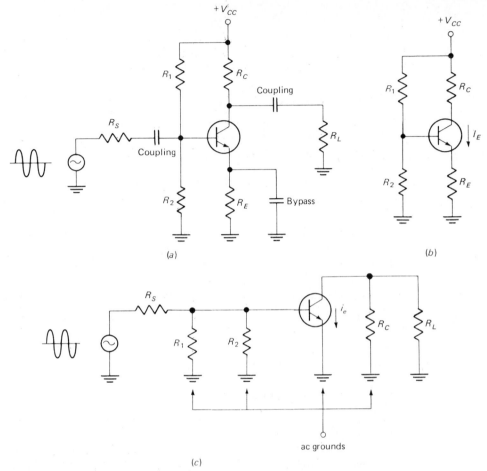

Figure 8-5. Superposition theorem. (a) *Original circuit.* (b) *Dc equivalent circuit.* (c) *Ac equivalent circuit.*

important, reducing a voltage source to zero is the same as shorting it. This is why R_B and R_C are shorted to ac ground through the V_{CC} source. Also, the bypass capacitor places the emitter at ac ground. With the ac equivalent circuit of Fig. 8-5c, we will be able to calculate any ac currents and voltages of interest.

notation

To avoid confusing ac and dc currents and voltages, we will use capital letters and subscripts for dc quantities. That is, we will use

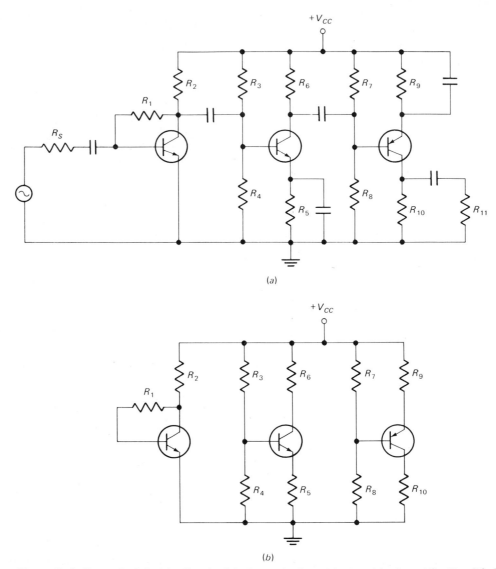

Figure 8-6. Example 8-3. (a) Circuit. (b) Dc equivalent. (c) Ac equivalent. (d) Simplified ac equivalent.

I_E, I_C, I_B for the dc currents
V_E, V_C, V_B for the dc voltages to ground
V_{BE}, V_{CE}, V_{CB} for the transistor dc voltages

For the ac currents and voltages we will use lowercase letters and subscripts as follows:

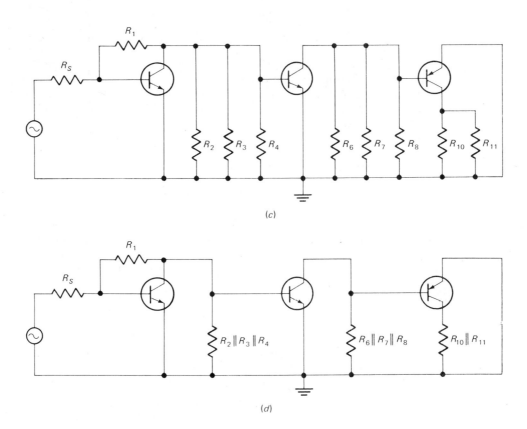

(c)

(d)

i_e, i_c, i_b for the ac currents

v_e, v_c, v_b for the ac voltages to ground

v_{be}, v_{ce}, v_{cb} for the transistor ac voltages

These are standard notations used by most people; you should become familiar with them because we use them extensively in the dc and ac analysis of transistor circuits.

EXAMPLE 8-3

Show the dc and ac equivalent circuits for the three-transistor amplifier of Fig. 8-6a.

SOLUTION

Working from left to right, we open each capacitor. All that remains when we are finished is the dc equivalent circuit of Fig. 8-6*b*. Here we recognize the first *stage* (a transistor and its associated circuitry) as an *npn* collector-feedback bias circuit. The second stage is an *npn* voltage-divider circuit. And the third stage is an upside-down *pnp* voltage-divider circuit. With the methods of Chap. 7, we can calculate any dc current or voltage of interest.

Next, reduce V_{CC} to zero in Fig. 8-6*a* (equivalent to grounding the V_{CC} line) and short all coupling and bypass capacitors. What remains is the ac equivalent circuit of Fig. 8-6*c*. By combining parallel resistances, we get the simple circuit of Fig. 8-6*d;* later chapters show you how to analyze this circuit.

8-3 *transistor ac equivalent circuits*

Remember what happened to a piece of wire when we took everything into account (Chap. 1)?

Figure 8-7 shows the three doped regions and the two depletion layers of a transistor. Each affects ac current differently from dc current. Why? For one thing, there are diode capacitances in the transistor; these look open to dc currents but may affect the ac currents.

the exact ac model

By taking as many effects as possible into account, we get the ac equivalent circuit of Fig. 8-8. Starting at the left, the ac emitter current i_e must flow through the inductance of the emitter lead. After this current enters the emitter region, it must flow through the bulk resistance $r_{e(\text{bulk})}$ of the emitter. When the current reaches the emitter depletion layer, part of it may flow through capacitance C'_e and the rest of it through r'_e. Next, the ac emitter current i_e flows through a small voltage source $\mu v'_c$.

On reaching the central node, i_e splits into i_c and i_b. The much smaller i_b flows down through r'_b shunted by C'_b and out the transistor through the inductance of the base lead. The ac collector current i_c flows through the parallel

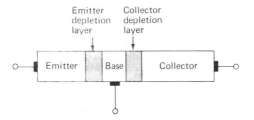

Figure 8-7. Regions of a transistor.

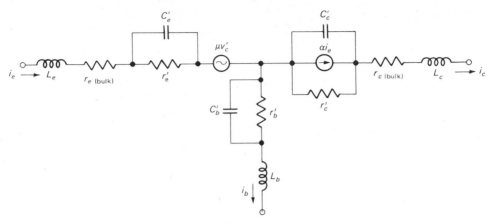

Figure 8-8. Exact ac equivalent circuit of a transistor.

network of C'_c, r'_c, and the αi_e current source. Then, i_c flows through the bulk resistance of the collector and finally through the inductance of the collector lead.

A circuit like Fig. 8-8 is fit only for a computer to work with. We have no intention of using it except as a reminder of some of the smaller effects in a transistor. What we have to do is eliminate as much as possible from Fig. 8-8 until we have reasonable equivalent circuits.

the low-frequency model

To begin with, we can usually neglect the lead inductances when the frequency is less than 100 MHz. Also, because the emitter and collector are larger and more heavily doped than the base, $r_{e(bulk)}$ and $r_{c(bulk)}$ are small enough to neglect. Capacitance C'_b is so small that it drops out. What remains is the equivalent circuit of Fig. 8-9.

But even Fig. 8-9 is too complicated for everyday work. The internal capacitances C'_e and C'_c shunt ac currents away from desirable paths. These capacitances

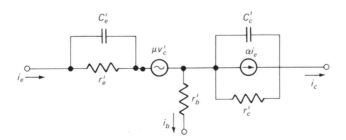

Figure 8-9. Approximation that neglects lead inductances, emitter and collector bulk resistance, and base capacitance.

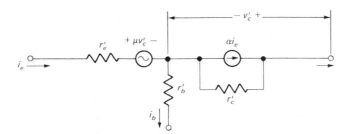

Figure 8-10. Third approximation.

are important only at higher frequencies called the *cutoff frequencies* (Chap. 16). As long as we operate a transistor *below* its cutoff frequencies, we can neglect C'_e and C'_c.

Figure 8-10 shows the low-frequency model of a transistor. Voltage v'_c is the difference of potential across the collector depletion layer. When the collector voltage changes, the width of the depletion layer changes. As you recall, this affects the base current slightly (Fig. 6-15*b*). To account for this effect, the $\mu v'_c$ generator is included in the emitter circuit. We will call Fig. 8-10 the *third approximation.* This or something like it is what we will use for accuracies better than 1 percent.

Figure 8-10 is not good enough. If we are trying to analyze and design new circuits, an equivalent circuit like Fig. 8-10 will so confuse and cloud our thinking that we will not see important relations. What we need is an equivalent circuit that gets to the bones of transistor action.

8-4 *the ideal-transistor approximation*

Accuracy may be important near the end of an analysis or design, but during the early stages it loses its value. If you try for high accuracy from the start, you often waste time. The reason is simple enough: when you try to analyze or design a new circuit, you make many false starts until you find the right approach. If you are trying to find the right approach with an equivalent circuit like Fig. 8-10, you are sure to get lost.

We can prune Fig. 8-10 further. The $\mu v'_c$ generator has values in the millivolt range and can be neglected in a first approximation. The quantity α is the ac alpha, the ratio of ac collector current to ac emitter current, i_c/i_e. For almost any transistor, α is greater than 0.95; therefore, a reasonable approximation is $\alpha \cong 1$. Resistance r'_c accounts for the slight increase in collector current when the collector depletion layer penetrates deeper into the base. For most transistors, r'_c is in megohms and is high enough to neglect in a first approximation.

Finally, r'_b may be negligible. It depends on how much dc collector current is flowing. For I_C less than about 10 mA, the ac voltage across r'_b is almost always

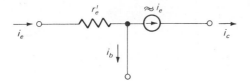

Figure 8-11. Ideal-transistor approximation.

small. In other words, for I_C less than 10 mA or thereabouts, the $i_b r'_b$ voltage drop is usually small enough to neglect in a first approximation (Chap. 10 deals with I_C greater than 10 mA).

With these approximations, Fig. 8-10 reduces to Fig. 8-11. Now we have arrived. This is the *ideal-transistor* approximation. This model, simple as it is, hangs onto the main ideas of transistor action. It is satisfactory for much transistor analysis and design. Using Fig. 8-11, a creative mind can sail through all those false starts that are part of preliminary analysis and design. Then, if necessary, the ideal answers can be improved by using second and third approximations to be discussed later.

Figure 8-11 gets to the point. As far as the ac signal is concerned, the emitter diode acts like a *resistance* of r'_e and the collector diode like a current source of approximately i_e. We will use the ideal transistor as the foundation for understanding ac action of transistors. The ideal transistor is not far from the truth; actual transistors would not be as useful as they are unless the higher-order effects were minor.

What we are doing with the transistor is similar to what everybody does with a piece of wire, a resistor, a capacitor, or an inductor. We are neglecting higher-order effects with the understanding that they are important in special cases. For many people, the ideal-transistor approximation is adequate for more than 90 percent of the analysis and design they do. And for the remaining 10 percent, the second and third approximations can be used.

8-5 *emitter-diode ac resistance*

To use the ideal-transistor approximation, we need to know more about r'_e. Figure 8-12 shows the typical diode curve relating I_E and V_{BE}. Point Q is the *quiescent point;* the coordinates of this point are the dc emitter current and dc base-emitter voltage. In the ideal-transistor approximation, we neglect the effects of r'_b; this means that V_{BE} is the total voltage appearing across the emitter diode.

small signal required

When an ac signal drives an FR-biased transistor, it forces the emitter current and voltage to change. If the signal is small, the operating point swings from

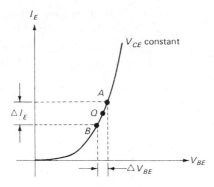

Figure 8-12. Graphical meaning of r'_e.

Q to A, back to Q, down to B, and back to Q. This action repeats for the next cycle. Since A and B are close to Q, only a small arc of the diode curve is used. Whenever a small piece of a curve is used like this, the operation is approximately *linear*. All this means is that the arc from A to B is essentially a straight line. Because of this, the changes in voltage and current are directly proportional. In symbols,

$$\Delta V_{BE} = K \Delta I_E$$

where Δ stands for "the change in" and K is a constant of proportionality.

If we reduce the ac signal, the changes in voltage and current are still directly proportional with the same value of K as before. In fact, changing the peak value of any small ac signal does not change the value of K.

This is what r'_e is all about. It is nothing more than the constant of proportionality K, that is,

$$r'_e = K = \frac{\Delta V_{BE}}{\Delta I_E} \tag{8-3}$$

Figure 8-12 illustrates these changes between points A and B. Since the arc between A and B is essentially linear, you can use any two points between A and B to measure the changes. For instance, if desired, you could measure the change in voltage and current between Q and A. The ratio of these changes still gives the same value of r'_e.

Getting the same value of r'_e for all points between A and B requires a small ac signal. How small? Theoretically, it should be infinitesimally small. As a practical guide, however, we will accept a signal as small when the peak-to-peak change in emitter current is less than 10 percent of the quiescent current. For example, if the dc current is 10 mA, we will consider an ac signal small if the peak-to-peak change in emitter current is less than 1 mA. This is arbitrary, but we need some basic rule to work with until we discuss *harmonic distortion*.

ac signal identical to changes

Since the ac signal causes the operating point to swing from Q to A on one half cycle and from Q to B on the other half cycle, the changes in current and voltage are the same as the ac current and voltage. In symbols,

$$i_e = \Delta I_E$$

and

$$v_{be} = \Delta V_{BE}$$

If the changes are from points B to A, i_e and v_{be} are peak-to-peak values. On the other hand, if the changes are from Q to A, i_e and v_{be} are peak values. Unless otherwise indicated, we will use peak-to-peak values. That is, i_e and v_{be} represent the peak-to-peak ac emitter current and voltage.

Because ac current and voltage are identical to the changes in total current and voltage, we can rewrite Eq. (8-3) as

$$r_e' = \frac{v_{be}}{i_e} \tag{8-4}$$

For instance, if $v_{be} = 25$ mV and $i_e = 1$ mA, r_e' will equal 25 Ω. Once we have the value of r_e', we can use it for other voltages and currents within the small arc.

the formula for r_e'

Look at Fig. 8-12. Since r_e' is the ratio of the change in V_{BE} to the change in I_E, its value depends on the location of Q. The higher Q is up the curve, the smaller r_e' becomes because the same change in voltage produces a larger change in current. In other words, the slope of the diode curve at point Q determines the value of r_e'.

By using calculus to find this slope, we can prove that

$$r_e' = \frac{25 \text{ mV}}{I_E} \tag{8-5}***$$

For instance, if the Q point has an I_E of 0.1 mA

$$r_e' = \frac{25 \text{ mV}}{0.1 \text{ mA}} = 250 \ \Omega$$

Or, for a higher Q point with $I_E = 1$ mA, r_e' decreases to

$$r_e' = \frac{25 \text{ mV}}{1 \text{ mA}} = 25 \ \Omega$$

We already know how to calculate I_E using the biasing formulas of Chap. 7. Therefore, once we have I_E, we can find the corresponding value of r_e'.

Eq. (8-5) is a room-temperature formula; it applies to temperatures near 65°F or 18°C. Also, Eq. (8-5) is based on a rectangular pn junction like the one shown in Fig. 8-7. Because the shape of a diode curve changes slightly with nonrectangular junctions, the value of r_e' may differ slightly from Eq. (8-5). Nevertheless, 25 mV/I_E is an excellent approximation for any transistor, germanium or silicon, no matter what the actual shape of its base-emitter junction. A final condition worth mentioning is that I_E must be greater than zero, always satisfied when a transistor is FR-biased.[1]

at any temperature

Occasionally, we need the value of r_e' for temperatures above and below room temperature. The formula is

$$r_e' = \frac{C + 273}{291} \frac{25 \text{ mV}}{I_E} \tag{8-6}$$

where C is the junction temperature in Celsius degrees. For instance, at 100°C,

$$r_e' = \frac{100 + 273}{291} \frac{25 \text{ mV}}{I_E}$$

$$= \frac{32 \text{ mV}}{I_E}$$

Therefore, at 100°C an I_E of 1 mA means r_e' equals 32 Ω.

8-6 ac beta

Figure 8-13 shows a typical graph of I_C versus I_B. β_{dc} is the ratio of total collector current I_C to total base current I_B. Since the graph is nonlinear, β_{dc} depends on where the Q point is located. This is why data sheets specify β_{dc} for a particular value of I_C.

The ac beta (designated β_{ac} or simply β) is a small-signal quantity and depends on the location of point Q. In terms of Fig. 8-13, ac beta is defined as

$$\beta = \frac{\Delta I_C}{\Delta I_B} \tag{8-7}$$

or since ac currents are the same as the changes in total currents,

[1] For a derivation of r_e', see Millman, J., and C. Halkias, *Electronic Fundamentals and Applications for Engineers and Scientists*, McGraw-Hill Book Company, New York, 1976, p. 29.

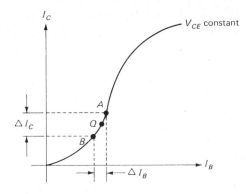

Figure 8-13. Graphical meaning of β.

$$\beta = \frac{i_c}{i_b} \qquad\qquad (8\text{-}8)\,\text{*}\,\text{*}\,\text{*}$$

Graphically, ac beta is the slope of the curve at point Q. For this reason, it has different values for different Q locations.

Data sheets sometimes list the value of h_{fe}, which is the same as β. Especially note the subscripts on h_{fe} are lowercase letters, whereas the subscripts on h_{FE} are capital letters. Therefore, when reading data sheets, do not confuse these current gains. Quantity h_{FE} is the dc current gain and is equivalent to β_{dc}. Quantity h_{fe} is the small-signal current gain, equivalent to β. You will find h_{FE} is almost always listed on data sheets, but h_{fe} is not shown as often. If a data sheet does not give the value of h_{fe}, use the dc current gain h_{FE} as an approximation for h_{fe}. The two quantities are usually close enough in value to justify this as a first approximation.

Table 8-1 summarizes the relations between r' parameters and h parameters. Memorize these relations so you can convert data sheet information into the quantities we need for simplified analysis.

Table 8-1 Relations

r' parameter	h parameter
β_{dc}	h_{FE}
β	h_{fe}
r'_e	h_{ib}

8-7 *the ideal model*

For almost any transistor, α is greater than 0.95 and β is greater than 20. Therefore, we can list the following approximations for ac analysis:

$$\alpha > 0.95 \Rightarrow \alpha \cong 1 \Rightarrow i_c \cong i_e$$

$$i_c \cong i_e \Rightarrow \frac{i_c}{i_b} \cong \frac{i_e}{i_b} \cong \beta$$

$$\beta > 20 \Rightarrow \beta \gg 1 \Rightarrow \beta + 1 \cong \beta$$

$$\beta \gg 1 \Rightarrow i_c + i_b \cong i_c$$

To keep the ac analysis similar to the dc analysis of the preceding chapter, we can redraw the ideal transistor as shown in Fig. 8-14. During the positive half cycle of the ac signal, a voltage of v_{be} appears across the emitter diode with the plus-minus polarity shown. This sets up an ac emitter current of i_e. Because of this, the ac collector current approximately equals i_e and the ac base current approximately equals i_e/β. Since Fig. 8-14 is important in our analysis, we will call it the *ideal model.*

EXAMPLE 8-4
What is the value of r_e' in Fig. 8-15a?

SOLUTION
We recognize voltage-divider bias. The dc voltage across the 10-kΩ resistor is 10 V. Therefore,

$$I_E = \frac{10 - 0.7}{5000} = 1.86 \text{ mA}$$

With Eq. (8-5),

$$r_e' = \frac{25 \text{ mV}}{I_E} = \frac{25 \text{ mV}}{1.86 \text{ mA}} = 13.4 \ \Omega$$

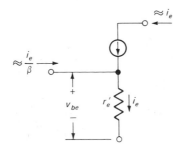

Figure 8-14. Ideal model of transistor.

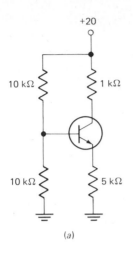

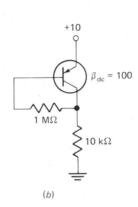

Figure 8-15.

(a)   (b)

EXAMPLE 8-5

What value does r_e' have in Fig. 8-15b?

SOLUTION

This is an upside-down *pnp* collector-feedback bias circuit. Therefore, the dc emitter current equals

$$I_E \cong \frac{V_{CC} - V_{BE}}{R_C + R_B/\beta_{dc}} = \frac{10 - 0.7}{10^4 + 10^6/100} = 0.465 \text{ mA}$$

and the ac emitter resistance is

$$r_e' = \frac{25 \text{ mV}}{I_E} = \frac{25 \text{ mV}}{0.465 \text{ mA}} = 53.8 \text{ } \Omega$$

self-testing review

Read each of the following and provide the missing words. Answers appear at the beginning of the next question.

1. Most capacitors in transistor circuits are either _____ or _____ capacitors.

2. *(coupling, bypass)* A coupling capacitor passes an ac signal from one _____ point to another. Ideally, a capacitor appears _____ to dc current. Therefore, a coupling capacitor appears open to dc current but _____ to ac current.

3. *(ungrounded, open, shorted)* This discriminating action allows the coupling capacitor to pass an ac signal from one circuit to another without disturbing the dc _____ of each circuit.

4. *(biasing)* A bypass capacitor is similar to a coupling capacitor, except that it couples an ungrounded point to a _____ point. A bypass capacitor produces an ac _____.

5. *(grounded, ground)* In the normal frequency range of an amplifier, we usually approximate all coupling and bypass capacitors as dc _____ and ac _____.

6. *(opens, shorts)* If you reduce all ac sources to zero and open all capacitors, the circuit that remains is called the _____ equivalent circuit. If you reduce all dc sources to zero and short all coupling and bypass capacitors, the circuit that remains is the _____ equivalent circuit.

7. *(dc, ac)* In the ideal-transistor approximation the emitter diode acts like a resistance of _____, and the collector diode like a current source of approximately _____.

8. *(r'_e, i_e)* At room temperature, _____ approximately equals 25 mV divided by _____. This resistance _____ with an increase in temperature.

9. *(r'_e, I_E, increases)* The ac beta equals the ratio of ac _____ current to ac _____ current. The value of h_{fe} is the same as the value of _____.

10. *(collector, base, β)* The quantity _____ is the dc _____ gain and is equivalent to the dc _____.

11. *(h_{FE}, current, beta)* An ac voltage v_{be} appears across the emitter diode. This sets up an ac emitter current of _____. Because of this, the ac collector current approximately equals i_e, and the ac base current approximately equals _____.

12. *(i_e, i_e/β)* For most transistors, α is greater than 0.95 and β is greater than 20.

problems

8-1. The ac source of Fig. 8-16a can have a frequency between 100 Hz and 200 kHz. To couple the signal properly over this range, what size should the coupling capacitor be?

8-2. We want the coupling capacitor of Fig. 8-16b to couple all frequencies from 500 Hz to 1 MHz. What size should it have?

8-3. To bypass point A to ground in Fig. 8-16c for all frequencies from 20 Hz up, what size should the bypass capacitor have?

8-4. We want point A in Fig. 8-16d to look like ac ground from 10 Hz to 200 kHz. What size should the bypass capacitor be?

8-5. In Fig. 8-3a, if the lowest frequency is changed from 10 to 1 Hz, what size coupling capacitor do we need?

8-6. Suppose we change the lowest frequency in Fig. 8-4a to 2 Hz. What size does the bypass capacitor need to be? Is this large size commercially available?

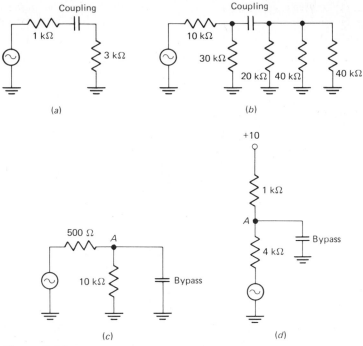

Figure 8-16.

8-7. Draw the dc equivalent circuit for the amplifier of Fig. 8-17*a*; label the three currents with standard dc notation. Next, draw the ac equivalent circuit.

8-8. Draw the dc and ac equivalent circuits for Fig. 8-17*b*. What is the approximate value of dc emitter current?

8-9. Show the dc and ac equivalent circuits for the upside-down *pnp* amplifier of Fig. 8-17*c*.

8.10. Approximately how much dc emitter current is there in the amplifier of Fig. 8-17*d*? Draw the ac equivalent circuit.

8-11. Draw the dc and ac equivalent circuits for the *pnp* amplifier shown in Fig. 8-18*a*.

8-12. For the collector-feedback-biased amplifier of Fig. 8-18*b*, show the ac equivalent circuit.

8-13. Draw the dc and ac equivalent circuits for the three-transistor amplifier of Fig. 8-18*c*.

8-14. In the voltage-divider-biased amplifier of Fig. 8-17*b*, to have small-signal operation what is the largest permitted peak-to-peak change in total emitter current?

8-15. To have small-signal operation in the emitter-biased amplifier of Fig. 8-17*d*, what is the largest permitted peak-to-peak emitter current?

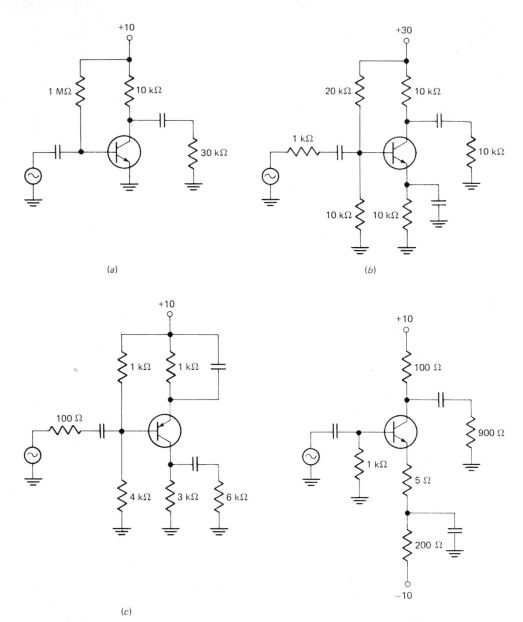

Figure 8-17.

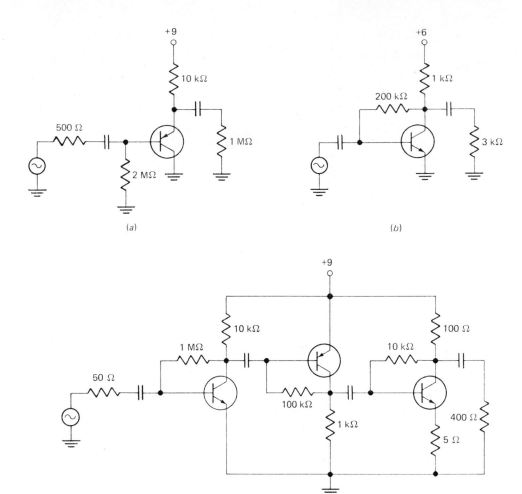

Figure 8-18.

8-16. In measuring ac currents and voltages, suppose you find that $v_{be} = 10$ μV and $i_e = 1$ μA. What is the value of r_e' ?

8-17. If the peak-to-peak change in base-emitter voltage is 5 mV and the peak-to-peak change in emitter current is 0.01 mA, what is the value of r_e' ?

8-18. With Eq. (8-5), calculate the value of r_e' for each of these dc emitter currents: 0.01 mA, 0.05 mA, 0.1 mA, 0.5 mA, 1 mA, 5 mA, and 10 mA.

8-19. What is the value of r_e' in the amplifier of Fig. 8-17*b*?

8-20. In the emitter-biased amplifier of Fig. 8-17*d*, what value does r_e' have?

8-21. If the transistor of Fig. 8-18b has a β_{dc} of 100, what value does r'_e have?

8-22. A transistor has a dc emitter current of 1 mA. What is the value of r'_e at $-50°C$? At $0°C$? At $150°C$?

8-23. The transistor amplifier of Fig. 8-17b operates in outer space and has a junction temperature of $-30°C$. What is the value of r'_e?

8-24. In Fig. 8-13, suppose the changes between points B and A are as follows: $\Delta I_C = 0.1$ mA and $\Delta I_B = 0.002$ mA. What is the value of β? And h_{fe}?

8-25. Suppose the Q point of Fig. 8-13 has coordinates of $I_C = 10$ mA and $I_B = 0.05$ mA. What value does β_{dc} have? If point A has coordinates of $I_C = 10.3$ mA and $I_B = 0.051$ mA, what is the value of β?

8-26. A transistor amplifier has an $I_E = 1$ mA. Figure 8-19 shows the ideal model of the transistor. If $i_e = 0.05$ mA, what value does v_{be} have? For $\beta = 100$, what are the values of i_b and i_c?

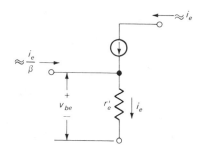

Figure 8-19.

8-27. Suppose the r'_e in Fig. 8-19 equals 10 Ω. If v_{be} equals 1 mV, what is the value of i_e? If you increase v_{be} to 5 mV, what is the new value of i_e?

9
small-signal amplifiers

After biasing a transistor in the active region, we can apply an ac voltage across the emitter diode to produce fluctuations in collector current. When this ac collector current flows through an external resistor, it produces an output signal that is larger than the input signal. This enlargement of the signal is called *amplification*.

In this chapter, we discuss *small-signal* amplifiers, the kind where the changes in collector current are small compared to the quiescent collector current. Small-signal amplifiers are used near the front end of receivers, stereo amplifiers, and measuring instruments.

9-1 *base drive and emitter drive*

As discussed in Chap. 8, we can apply the superposition theorem to calculate the dc and ac currents in a transistor circuit. After finding the Q point, we can reduce all dc sources to zero and short all coupling and bypass capacitors to get the *ac equivalent circuit*. Despite many different amplifier designs, most amplifier stages will reduce to one of two basic forms: *base-driven* or *emitter-driven*.

Figure 9-1a shows a base-driven circuit, so called because source v_{bb} drives the base through resistance r_B. Source v_{bb} and resistance r_B may be a single source and resistor as shown, or they may represent the Thevenin equivalent circuit for whatever is driving the base. Resistor r_E may be a single resistor or

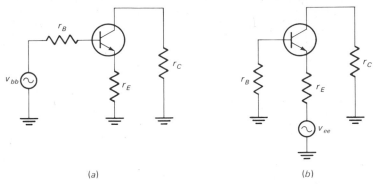

Figure 9-1. (a) *Base-driven.* (b) *Emitter-driven.*

may represent the combined resistance of several resistors. Similarly, r_C may be a single resistor as shown or may be a combined resistance. Regardless of what the actual ac equivalent circuit is, if you can reduce it to the form of Fig. 9-1a you will have the circuit in base-driven form.

Sometimes, Thevenin's and other reducing theorems result in a circuit like Fig. 9-1b. This is emitter-driven because the source v_{ee} drives the emitter through resistance r_E.

9-2 *base-driven formulas*

Emitter current plays the *central role* in transistor action. Once we have found the value of i_e, we can get i_c (approximately equal to i_e) or we can get i_b (divide by β).

formula for ac emitter current

Figure 9-2a shows the three ac currents that flow in a transistor. It is immaterial whether we analyze the circuit for the positive half cycle of input voltage or the negative half cycle. The results for either half cycle apply to the other half when we complement all currents and voltages. For this reason, we can analyze Fig. 9-2a for the positive half cycle of input voltage; this is why v_{bb} has the plus-minus polarity shown. During this positive half cycle of input voltage, base and collector currents flow into the transistor and emitter current flows out.

Visualize the transistor as the ideal model discussed in Chap. 8. Figure 9-2b shows the resulting circuit. Starting with the v_{bb} source, we can sum voltages in a clockwise direction to get

$$-v_{bb} + \frac{i_e}{\beta} r_B + i_e(r_e' + r_E) \cong 0$$

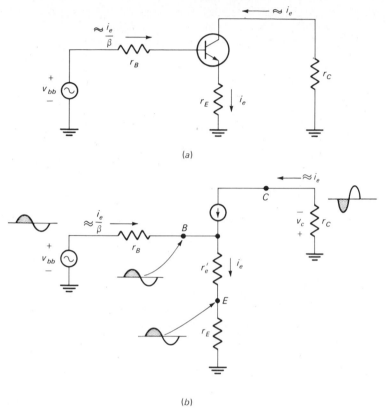

Figure 9-2. (a) *Base-driven form.* (b) *Ideal equivalent circuit.*

After factoring and solving for i_e, we have

$$i_e \cong \frac{v_{bb}}{r_E + r'_e + r_B/\beta} \qquad (9\text{-}1)$$

Because many amplifiers can be reduced to base-driven form, we can use Eq. (9-1) to calculate the ac emitter current in these amplifiers. Once we have i_e, all other ac calculations are easy.

voltage formulas

When you build a transistor amplifier, you often measure transistor voltages to ground with an oscilloscope or voltmeter. For this reason, you should know approximately what ac voltages exist in an amplifier. The three basic ac voltages in any transistor amplifier are the collector-to-ground voltage v_c, the emitter-to-ground voltage v_e, and the base-to-ground voltage v_b. By referring to Fig. 9-2b, we can work out a formula for each of these in terms of i_e.

Start with v_c. The ac voltage from collector to ground must equal the voltage across r_C. Since the current flows up through r_C during the positive half cycle of input voltage, v_c has the minus-plus polarity shown in Fig. 9-2b. The magnitude of v_c is

$$v_c \cong i_e r_C \qquad (9\text{-}1a)$$

When the source voltage reverses polarity, that is, during the negative half cycle of source voltage, the voltage across r_C reverses polarity, making the collector positive with respect to ac ground. Therefore, the ac collector voltage is 180° out of phase with the base voltage. This *phase inversion* between the base and collector happens in all base-driven amplifiers.

Next, by looking at Fig. 9-2b we can see that i_e flows down through r_E during the positive half cycle of input voltage. Therefore, the ac emitter-to-ground voltage is

$$v_e = i_e r_E \qquad (9\text{-}1b)$$

Since the voltage is plus-minus across r_E, the phase of the emitter voltage is the same as the phase of the base voltage.

The third voltage of interest is the base-to-ground voltage v_b. In Fig. 9-2b, v_b is identical to the ac voltage across $r'_e + r_E$. Therefore,

$$v_b = i_e(r'_e + r_E) \qquad (9\text{-}1c)$$

In summary, if we can reduce the ac equivalent circuit of an amplifier to base-driven form, we can use Eq. (9-1) to calculate the ac emitter current. Once we have i_e, the rest is easy. We can find the three transistor voltages by using Eqs. (9-1a) through (9-1c). The phase relations are always the same: v_c is *inverted* with respect to v_b, and v_e *follows* or is in phase with v_b.

voltage gain

The whole point of an amplifier is to increase the peak-to-peak signal. In this chapter, we are interested in *voltage amplifiers,* those that increase the signal voltage. *Voltage gain* from point x to point y is defined as

$$A = \frac{v_y}{v_x} \qquad (9\text{-}2)$$

where v_y is the ac voltage from point y to ground and v_x is the ac voltage from point x to ground. For instance, if $v_y = 1$ V and $v_x = 0.1$ V, we would have a voltage gain of

$$A = \frac{1 \text{ V}}{0.1 \text{ V}} = 10$$

This tells us the ac voltage at point y is 10 times greater than the ac voltage at point x.

One of the most important voltage gains in an amplifier is the voltage gain from base to collector, that is, the ratio of ac collector voltage to ac base voltage. In symbols, the voltage gain from base to collector equals

$$A = \frac{v_c}{v_b}$$

When you build transistor amplifiers and measure v_c and v_b, you find that v_c is normally much greater than v_b. For example, you may find $v_c = 10$ V and $v_b = 0.05$ V. In this case, the amplifier has a voltage gain from base to collector of

$$A = \frac{10 \text{ V}}{0.05 \text{ V}} = 200$$

Because the voltage gain from base to collector is so important, we need a formula for it. With Eqs. (9-1a) and (9-1c),

$$A = \frac{v_c}{v_b} \cong \frac{i_e r_C}{i_e(r_e' + r_E)}$$

or
$$A \cong \frac{r_C}{r_E + r_e'} \qquad (9\text{-}3)\text{***}$$

EXAMPLE 9-1

Work out the value of r_e' in Fig. 9-3a. Then reduce the amplifier to base-driven form.

SOLUTION

To get r_e', we need the dc emitter current I_E. Visualize all coupling and bypass capacitors as open. What remains is the dc equivalent circuit of Fig. 9-3b. At this point, you should be able to see that the dc voltage from base to ground is 10 V; therefore, the dc voltage from emitter to ground is almost 10 V (ignore V_{BE}). And finally, the dc emitter current I_E equals approximately 1 mA. As a result,

$$r_e' = \frac{25 \text{ mV}}{I_E} = \frac{25 \text{ mV}}{1 \text{ mA}} = 25 \ \Omega$$

Return to Fig. 9-3a. The first step in reducing this amplifier to its base-driven form is this: visualize all capacitors shorted, and reduce the V_{CC} supply voltage

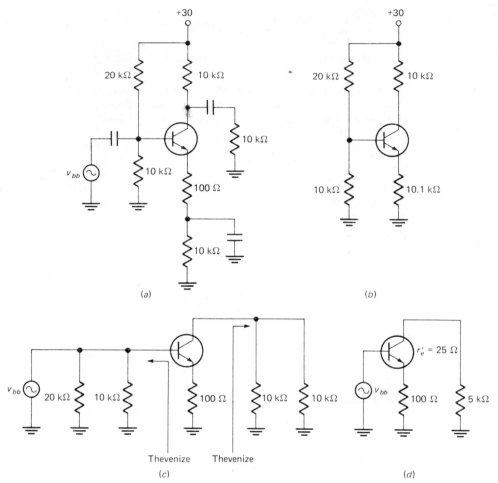

Figure 9-3.

to zero. What remains is the ac equivalent circuit of Fig. 9-3c. To reduce this ac equivalent circuit as much as possible, we Thevenize the base and collector circuits. In the base circuit, the Thevenin voltage is v_{bb} and the Thevenin resistance is zero. In the collector circuit, the Thevenin voltage is zero, and the Thevenin resistance is 5 kΩ.

Figure 9-3d shows the amplifier after Thevenizing the base and collector circuits. We recognize this as base-driven with $r_B = 0$, $r_E = 100$ Ω, and $r_C = 5$ kΩ.

EXAMPLE 9-2
Calculate ac emitter current i_e for the amplifier of Fig. 9-3d. After you have it, work out the three ac voltages v_c, v_e, and v_b. Use a v_{bb} of 1-mV peak.

SOLUTION

We start by using Eq. (9-1).

$$i_e = \frac{v_{bb}}{r_E + r_e' + r_B/\beta} = \frac{0.001}{100 + 25 + 0}$$
$$= 8 \ \mu A$$

With this emitter current, we can work out each ac voltage as follows:

$$v_c \cong i_e r_C = 8(10^{-6})5(10^3) = 40 \ \text{mV}$$
$$v_e = i_e r_E \cong 8(10^{-6})100 = 0.8 \ \text{mV}$$
$$v_b = i_e(r_e' + r_E) = 8(10^{-6})(25 + 100) = 1 \ \text{mV}$$

The last calculation (for v_b) is unnecessary in this particular example because $r_B = 0$. Whenever $r_B = 0$, the base voltage must equal the source voltage, that is, $v_b = v_{bb}$.

Also note that all answers are peak values because we used the peak value of v_{bb} when we calculated i_e. If v_{bb} were a sine wave, you could convert all answers to rms values by multiplying each peak value by 0.707.

EXAMPLE 9-3

Calculate the voltage gain from base to collector for the amplifier of Fig. 9-3*a*.

SOLUTION

We will do this in two different ways. First, working directly with the v_c and v_b found in the preceding example, we get

$$A = \frac{v_c}{v_b} = \frac{40 \ \text{mV}}{1 \ \text{mV}} = 40$$

The second approach is the formula approach using Eq. (9-3).

$$A = \frac{r_C}{r_E + r_e'} = \frac{5000}{100 + 25} = 40$$

9-3 *the common-emitter amplifier*

A special case of the base-driven circuit is the case of $r_E = 0$. When $r_E = 0$, there is no ac resistance between the emitter and ground. In other words, the emitter is at ac ground. This special case is called the *grounded-emitter* or *common-emitter* (CE) amplifier.

Figure 9-4*a* shows the base-driven form of a CE amplifier. During the positive half cycle of source voltage, the ac base and collector currents flow into the transistor, and the ac emitter current flows out. As in any base-driven amplifier, the ac collector voltage is 180° out of phase with ac base voltage.

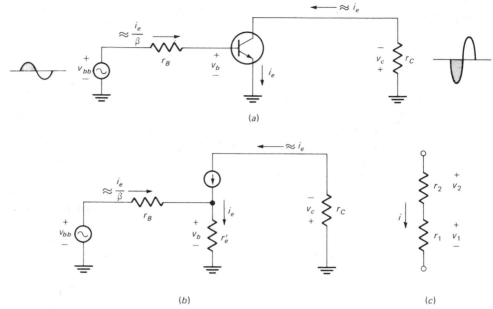

Figure 9-4. Common-emitter amplifier.

What is the voltage gain from base to collector in a CE amplifier? The easy way to see this is to replace the transistor by its ideal model shown in Fig. 9-4b. The magnitude of the collector voltage is

$$v_c \cong i_e r_C$$

and the base voltage is

$$v_b = i_e r_e'$$

because v_b appears directly across r_e' in Fig. 9-4b. By taking the ratio of v_c to v_b, we get the voltage gain from base to collector.

$$A = \frac{v_c}{v_b} \cong \frac{i_e r_C}{i_e r_e'}$$

or

$$A \cong \frac{r_C}{r_e'} \qquad \text{(9-3a)}\text{***}$$

You can also get the same result by substituting $r_E = 0$ into the gain formula (9-3).

As an example, suppose we short the 100-Ω resistor in the emitter circuit of Fig. 9-3a. Then, in the final prototype form of Fig. 9-3d the emitter is at ac ground, that is, $r_E = 0$. In this case, the voltage gain from base to collector equals

$$A \cong \frac{r_C}{r_e'} = \frac{5000}{25} = 200$$

With this kind of gain, a 1-mV base voltage will produce a 200-mV collector voltage.

Equation (9-3a) makes sense. With the same current (approximately) flowing through r_C and r_e', the *ratio of voltages* has to equal the *ratio of resistances*. In other words, look at Fig. 9-4c. The same current i flows through r_1 and r_2; this sets up voltages v_1 and v_2. Since voltage is directly proportional to resistance, the ratio v_2/v_1 equals the ratio r_2/r_1. In any CE amplifier, i_c equals i_e to a close approximation. Therefore, essentially the same current flows through r_C and r_e'. Because of this, the voltage ratio v_c/v_b must equal the resistance ratio r_C/r_e'.

Since the CE amplifier is used so often, you should remember that its approximate voltage gain from base to collector equals r_C/r_e'.

EXAMPLE 9-4

What is the voltage gain from base to collector for each CE amplifier shown in Fig. 9-5?

SOLUTION

You can calculate the dc emitter current I_E in Figs. 9-5a through c. If you do this, you will find $I_E \cong 1$ mA in each amplifier. Therefore, $r_e' = 25 \; \Omega$ for each amplifier.

If you now short all capacitors and ground the dc supplies, you get the ac equivalent circuit of each amplifier. After Thevenizing the base and collector circuits (similar to Example 9-1), you get the same form for each amplifier, shown in Fig. 9-5d.

The voltage gain from base to collector equals

$$A \cong \frac{r_C}{r_e'} = \frac{5000}{25} = 200$$

EXAMPLE 9-5

Figure 9-6a shows a *pnp* CE amplifier. What is the voltage gain from base to collector?

SOLUTION

As far as the ac equivalent circuit is concerned, a *pnp* transistor acts exactly the same as an *npn* transistor, provided it remains in the active region through the ac cycle. During the positive half cycle of ac source voltage, ac base and

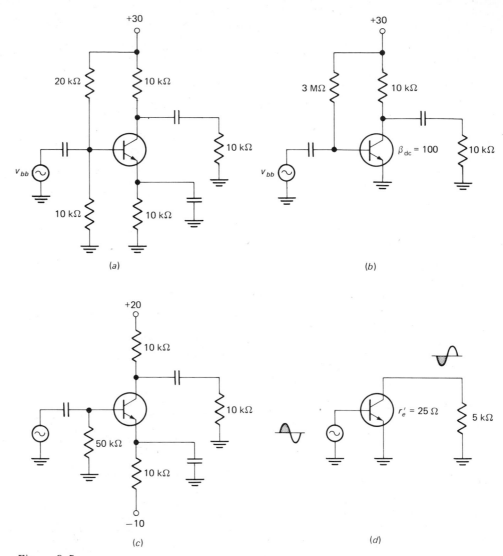

Figure 9-5.

collector currents flow into the transistor, and ac emitter current flows out of the transistor whether the transistor is *npn* or *pnp*. In other words, the ideal model is the same for either an *npn* or a *pnp* transistor; for this reason, all ac formulas we derive for *npn* circuits apply to *pnp* circuits.

Visualize all capacitors open in Fig. 9-6*a*. Then, 10 V appears across the

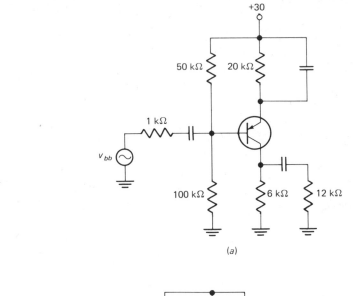

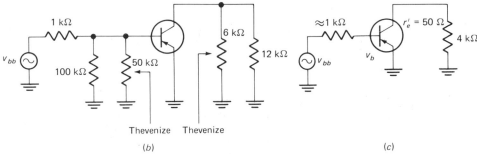

Figure 9-6.

50-kΩ base resistor, and almost all of this 10 V appears across the 20-kΩ emitter resistor (ignore V_{BE}). Therefore, the dc emitter current approximately equals 0.5 mA. And $r_e' = 50$ Ω.

Next, visualize all capacitors shorted and V_{CC} reduced to zero in Fig. 9-6a. By inverting the transistor, we get the ac equivalent circuit of Fig. 9-6b. When we Thevenize the base circuit, the 50- and 100-kΩ resistors are so much larger than the 1-kΩ resistor that the Thevenin voltage is still approximately v_{bb} and the Thevenin resistance is still close to 1 kΩ. When we Thevenize the collector circuit, we get a V_{TH} of zero and an R_{TH} of

$$R_{TH} = 6000 \parallel 12,000 = 4 \text{ kΩ}$$

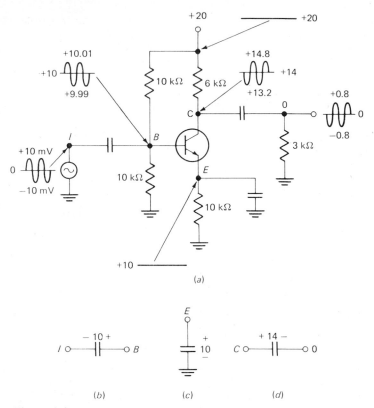

Figure 9-7.

Figure 9-6c shows the reduced amplifier. Since the *pnp* transistor has the same ideal model as an *npn* transistor, the voltage gain from base to collector equals r_C/r_e'. In this particular case,

$$A = \frac{r_C}{r_e'} = \frac{4000}{50} = 80$$

EXAMPLE 9-6
Explain the voltage waveforms shown in Fig. 9-7a.

SOLUTION
When you look at voltage waveforms with a dc-coupled oscilloscope, you will see *total* voltages, the sum of dc and ac voltage. Here is why the following waveforms appear in the amplifier of Fig. 9-7a. At point *I* (the input), an oscilloscope will show whatever the ac source voltage is; in this case, we are assuming a sine wave with a positive peak of 10 mV and a negative peak of −10 mV.

At point B (the base) the oscilloscope will show the source sine wave shifted upward by 10 V; the dc voltage from base to ground is

$$V_B = 10 \text{ V}$$

and the ac voltage from base to ground equals

$$v_b = 10 \text{ mV peak} = 0.01 \text{ V peak}$$

When we add these dc and ac voltages, we get the total base voltage shown; the maximum voltage is 10.01 V, and the minimum is 9.99 V.

At point E (the emitter) we get a straight line on an oscilloscope at +10 V ideally (9.3 V for the second approximation). There is no ac voltage at point E because the bypass capacitor ac grounds the emitter.

The 1 mA of dc emitter current produces a 6-V drop across the 6-kΩ collector resistor. This leaves a dc voltage of 14 V from collector to ground. Since $r_e' = 25$ Ω and $r_C = 2$ kΩ, the voltage gain is

$$A = \frac{r_C}{r_e'} = \frac{2000}{25} = 80$$

Therefore, the ac collector voltage equals

$$v_c = Av_b = 80(0.01) = 0.8 \text{ V peak}$$

After we add the dc and ac collector voltage, we get the inverted sine wave shown. It has a maximum value of +14.8 V and a minimum of +13.2 V.

At point O (the final output), all we see is a sine wave with a positive peak of 0.8 V and a negative peak of -0.8 V. The coupling capacitor has blocked the dc collector voltage but passed the ac collector voltage.

Finally, the V_{CC} point produces a straight line at +20 V on an oscilloscope, assuming the power-supply ripple is too small to see.

Incidentally, the capacitor voltage ratings must be greater than the dc voltages across the capacitors. Figure 9-7b through d shows each capacitor and its dc voltage. The voltage rating of each capacitor must be greater than the dc voltage across it. Furthermore, if you use electrolytic capacitors, you must connect them with the polarities shown.

9-4 *swamping the emitter diode*

The voltage gain from base to collector in a CE amplifier equals r_C/r_e'. The quantity r_e' equals 25 mV/I_E only at room temperature and for a rectangular base-emitter junction. As indicated earlier, even though the junction is not rectangular in most transistors, 25 mV/I_E is still a good approximation for r_e'. Many data sheets list h_{fe} and h_{ie}; if you see these quantities on a data sheet, use

$$r_e' \cong \frac{h_{ie}}{h_{fe}} \qquad\qquad (9\text{-}4)$$

to get an accurate value of r_e'.

Besides the differences from one transistor type to another, r_e' has a temperature dependence given by Eq. (8-6). Any change in the value of r_e' will change the voltage gain in a CE amplifier. In some applications, a change in voltage gain is acceptable. For instance, in a transistor radio you can offset changes in voltage gain by adjusting the volume control. But there are many applications where you need as stable a voltage gain as possible.

what swamping means

Many people *swamp* the emitter diode to reduce the effects of r_e'. Figure 9-8*a* shows the base-driven form and Fig. 9-8*b* shows the same circuit with the ideal

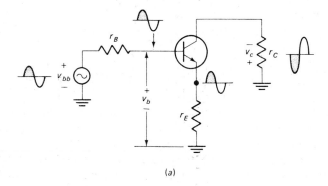

(a)

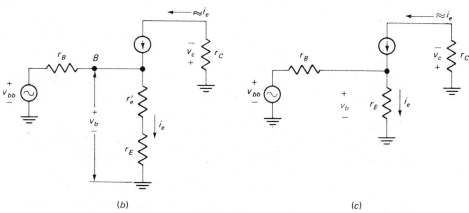

(b)   (c)

Figure 9-8. The swamped amplifier.

model for the transistor. Earlier, we proved the voltage gain from base to collector equals

$$\frac{v_c}{v_b} = \frac{r_C}{r_E + r_e'}$$

(Again, the way to remember this is by recognizing the same approximate current flows through collector and emitter circuits; therefore, the voltage ratio must equal the resistance ratio.) Swamping the emitter diode means making r_E much greater than r_e'. In this case, the voltage gain becomes

$$\frac{v_c}{v_b} = \frac{r_C}{r_E} \qquad \text{for } r_E \gg r_e' \qquad\qquad (9\text{-}5)\text{***}$$

Figure 9-8c illustrates a swamped amplifier. In a swamped amplifier, r_e' is so small compared to r_E that it produces a negligible effect on emitter current. For this reason, you can visualize the emitter branch with only r_E in it. With the same approximate current in the collector and emitter circuits, the voltage ratio v_c/v_b equals the resistance ratio r_C/r_E.

If you swamp too heavily, that is, use too large an r_E, you will get a small voltage gain. For instance, if $r_e' = 25\ \Omega, r_C = 5\ \text{k}\Omega$, and $r_E = 1\ \text{k}\Omega$, the voltage gain equals

$$\frac{v_c}{v_b} \cong \frac{r_C}{r_E + r_e'} = \frac{5000}{1000 + 25} \cong 5$$

This is heavy swamping; r_E is 40 times greater than r_e'. Since r_e' adds only 25 Ω, any changes in r_e' cause only minor changes in the denominator. The circuit is almost insensitive to changes in r_e', but the voltage gain is only 5.

The amount of swamping to use depends on the temperature range and other factors affecting the value of r_e'. The final choice of how much larger r_E should be than r_e' is a decision the designer makes in a particular application. The point is that by swamping the emitter diode, we can get a voltage gain from base to collector hardly affected by r_e'.

emitter follows base

A final point about a swamped amplifier. In Fig. 9-8a, the ac emitter-to-ground voltage equals

$$v_e = i_e r_E \qquad\qquad (9\text{-}6a)$$

As previously described, this ac voltage follows the base voltage, that is, v_e is in phase with v_b. In Fig. 9-8b, the base voltage equals

$$v_b = i_e(r_e' + r_E)$$

or
$$v_b \cong i_e r_E \qquad \text{when } r_E \gg r_e' \qquad\qquad (9\text{-}6b)$$

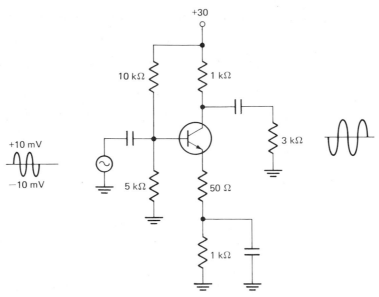

Figure 9-9.

Compare Eqs. (9-6*a*) and (9-6*b*). This comparison shows

$$v_e \cong v_b \qquad \text{when } r_E \gg r_e' \qquad (9\text{-}7)\text{***}$$

This says the ac emitter voltage approximately equals the ac base voltage in a swamped amplifier. This is important to know for troubleshooting. You can visualize this result by looking at Fig. 9-8*b*. When r_e' is much smaller than r_E, almost all of v_b appears across r_E, the ac resistance from emitter to ground.

EXAMPLE 9-7
What is the approximate ac voltage across the output 3-kΩ resistor of Fig. 9-9?

SOLUTION
First, the voltage-divider bias is about 10 V dc from base to ground. Therefore, the dc emitter current I_E approximately equals 10 mA. This means an r_e' of about 2.5 Ω.

In the ac equivalent circuit, the bottom of the 50-Ω emitter resistor is at ac ground. Therefore, $r_E = 50$ Ω, more than enough to swamp an r_e' of 2.5 Ω. So, the swamped voltage gain is

$$A = \frac{v_c}{v_b} \cong \frac{r_C}{r_E} = \frac{1000 \,\|\, 3000}{50} = 15$$

The ac voltage across the output 3-kΩ resistor is

$$v_c = Av_b = 15(10 \text{ mV}) = 150 \text{ mV}$$

The actual waveform will be a sine wave with a peak of 150 mV, and 180° out of phase with the base voltage.

Since the emitter diode is swamped, the ac voltage from emitter to ground will be a sine wave of approximately 10-mV peak, the same as the base voltage.

EXAMPLE 9-8

Suppose in wiring a circuit like Fig. 9-9 you accidentally leave out the bypass capacitor. What happens to the voltage gain?

SOLUTION

Leaving out the emitter bypass capacitor means

$$r_E = 50 + 1000 \cong 1000 \ \Omega$$

because the bottom of the 50-Ω resistor is no longer at ac ground. In a case like this, the swamping becomes enormously heavy, and the voltage gain drops to

$$\frac{v_c}{v_b} \cong \frac{1000 \parallel 3000}{1000} = 0.75$$

This tells us the ac collector voltage is less than the ac base voltage. In other words, we do not get voltage gain, but instead get *attenuation,* a loss of signal.

An open bypass capacitor in a CE amplifier like Fig. 9-7a produces the same result: a loss of signal. Remember the open bypass capacitor; it is a common trouble in discrete amplifiers.

9-5 *input impedance*

The ac source driving an amplifier must supply ac current to the amplifier. Usually, the less current the amplifier draws from the source, the better. The *input impedance* of an amplifier determines how much current the amplifier takes from the ac source. This section defines ac input impedance and works out a few basic formulas for it.

definition

In the normal frequency range of an amplifier, where coupling and bypass capacitors look like ac shorts and all other reactances are negligible, the ac input impedance is defined as

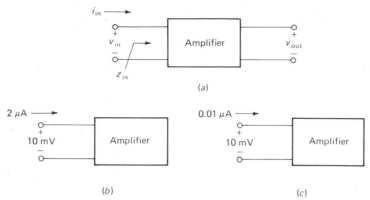

Figure 9-10. Input impedance.

$$z_{in} = \frac{v_{in}}{i_{in}} \tag{9-8}$$

where v_{in} and i_{in} are peak values, peak-to-peak values, rms values, or any consistent pair of values.

Figure 9-10a illustrates the idea of input impedance. An amplifier is inside the box. For the amplifier to work, it must have an ac voltage v_{in} across the input terminals. The ac source (not shown) delivers an ac current i_{in} to the amplifier. The ratio of v_{in} to i_{in} is the input impedance of the amplifier. For a given v_{in}, the less current the amplifier draws, the higher its input impedance.

As an example, suppose an amplifier draws 2 μA when the input voltage is 10 mV (Fig. 9-10b). Then, the ac input impedance equals

$$z_{in} = \frac{10 \text{ mV}}{2 \text{ μA}} = 5 \text{ k}\Omega$$

If another amplifier draws only 0.01 μA when 10 mV is across its input terminals (Fig. 9-10c), this second amplifier has an input impedance of

$$z_{in} = \frac{10 \text{ mV}}{0.01 \text{ μA}} = 1 \text{ M}\Omega$$

If all other characteristics are identical, the amplifier with a 1-MΩ input impedance is more desirable because it *loads* the source more lightly, that is, draws less current and absorbs less power.

two components of input impedance

In a base-driven amplifier, the ac source has to supply ac current to the base; in addition, the source has to deliver current to the biasing resistors in the

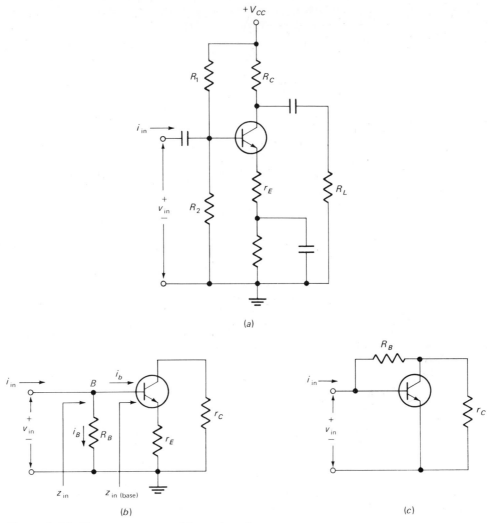

Figure 9-11. Two components of input impedance.

base circuit. For instance, Fig. 9-11a shows a base-driven amplifier, and Fig. 9-11b the ac equivalent circuit. R_B is the combined sum of R_1 and R_2, that is,

$$R_B = R_1 \parallel R_2$$

On the positive half cycle of source voltage, the input current i_{in} flows into the amplifier and splits into two components at node B. Component i_B flows down through R_B, and component i_b flows into the base (Fig. 9-11b).

Because the current splits into two parts, the input impedance of the amplifier is made up of two parts. The first is R_B and the second is $z_{in(base)}$ shown in Fig. 9-11b. Since these two impedances are in parallel, the input impedance of the amplifier is

$$z_{in} = R_B \parallel z_{in(base)} \qquad (9\text{-}9)$$

To find the input impedance of an amplifier, we need the value of R_B and $z_{in(base)}$.

How do we calculate R_B? For many amplifiers, the biasing resistors in the base are in parallel when we visualize the ac equivalent circuit. For instance, in Fig. 9-11a if R_1 and R_2 are 10 kΩ each, they appear as two parallel 10-kΩ resistors in the ac equivalent circuit. Therefore, in Fig. 9-11b,

$$R_B = 10^4 \parallel 10^4 = 5 \text{ k}\Omega$$

So whatever the amplifier, if the biasing resistors appear in parallel in the ac equivalent circuit, you combine these into a single resistance R_B as shown in Fig. 9-11b.

Collector-feedback bias is an exceptional case. When you visualize the ac equivalent circuit, the biasing resistor in the base does *not* appear across the input terminals; it still appears from collector to base as shown in Fig. 9-11c. The feedback effect described in Chap. 7 complicates the analysis slightly. For this reason, we will postpone the discussion of voltage gain and input impedance for this special circuit until Chap. 15.

For most amplifiers, R_B appears across the input terminals as shown in Fig. 9-11b; the only remaining question is how do you find $z_{in(base)}$.

impedance looking into the base

In Fig. 9-12a, the input impedance of the base is

$$z_{in(base)} = \frac{v_b}{i_b} \qquad (9\text{-}10a)$$

That is, you divide the ac base voltage by the ac base current to get the input impedance of the base. If v_b is 1 mV and i_b is 1 μA, the input impedance of the base is

$$z_{in(base)} = \frac{1 \text{ mV}}{1 \text{ } \mu\text{A}} = 1 \text{ k}\Omega$$

Equation (9-10a) is useful when you measure v_b and i_b. Often, you would prefer a formula for $z_{in(base)}$ in terms of schematic values. To get this formula, examine Fig. 9-12b, which shows the ideal model for the transistor. Since v_b appears across r_e' and r_E, we may write

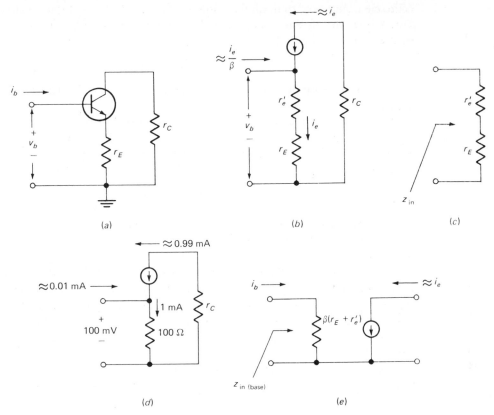

Figure 9-12. Input impedance of base.

$$z_{\text{in(base)}} = \frac{v_b}{i_b} \cong \frac{i_e(r_e' + r_E)}{i_e/\beta}$$

or

$$z_{\text{in(base)}} \cong \beta(r_E + r_e') \qquad (9\text{-}10b)\,*\!*\!*$$

This says the input impedance of the base is β times the sum of r_E and r_e'. For instance, if $\beta = 100$ and $r_E + r_e' = 100\ \Omega$, the input impedance is 10 kΩ.

This result should strike you as unusual. In fact, if you have not seen it before, it should disturb you. In Fig. 9-12b, when you look into the base, you see $r_E + r_e'$. Why isn't this the impedance? To begin with, Fig. 9-12b is *not* a one-branch circuit. Only when you have a one-branch circuit like Fig. 9-12c is the impedance equal to what you see. If the collector diode of a transistor were burned out, Fig. 9-12c would apply and the input impedance would equal $r_E + r_e'$. But when the transistor operates in the active region, the collector diode acts like a current source and the T circuit of Fig. 9-12b applies. In a T circuit like this, the input impedance is approximately β times the sum of $r_E + r_e'$.

To clear up any difficulties, look at Fig. 9-12d. The emitter branch has a 100-Ω resistor; this is not the input impedance. The ac emitter current is 1 mA; therefore, v_b equals

$$v_b = i_e(r'_e + r_E) = 0.001(100) = 100 \text{ mV}$$

The ac source does not have to supply the 1 mA of ac emitter current; all it has to supply is a much smaller ac base current. For a β of 100, the base current approximately equals 0.01 mA. The input impedance of the base is

$$z_{\text{in(base)}} = \frac{v_b}{i_b} \cong \frac{100 \text{ mV}}{0.01 \text{ mA}} = 10 \text{ k}\Omega$$

which agrees with the result you get by using

$$z_{\text{in(base)}} \cong \beta(r_E + r'_e) = 100(100) = 10 \text{ k}\Omega$$

In summary, as far as the ac source is concerned, the impedance in the emitter circuit is increased by a factor of β. For this reason, the equivalent circuit of Fig. 9-12e looks exactly the same to the source as the actual circuit of Fig. 9-12b. The input impedance of the base is $\beta(r_E + r'_e)$.

special cases

Equation (9-10b) applies to all base-driven amplifiers. For the special case of a CE amplifier, $r_E = 0$; therefore, in a CE amplifier,

$$z_{\text{in(base)}} \cong \beta r'_e \qquad (9\text{-}11a)$$

For example, if $\beta = 100$ and $r'_e = 25 \ \Omega$, $z_{\text{in(base)}}$ will equal 2.5 kΩ.

On the other hand, for the special case of a swamped amplifier, r_E is much greater than r'_e. So,

$$z_{\text{in(base)}} \cong \beta r_E \qquad (9\text{-}11b)$$

You can see right away a swamped amplifier has a larger input impedance than a CE amplifier. This is another advantage of the swamped amplifier.

EXAMPLE 9-9
What is the input impedance of the amplifier shown in Fig. 9-13a?

SOLUTION
The dc emitter current is approximately 1 mA; therefore, $r'_e = 25 \ \Omega$.

When we visualize the ac equivalent circuit and combine all resistances, we get Fig. 9-13b. In this circuit,

$$R_B = 20 \text{ k}\Omega$$

and
$$z_{\text{in(base)}} = \beta r'_e = 200(25) = 5 \text{ k}\Omega$$

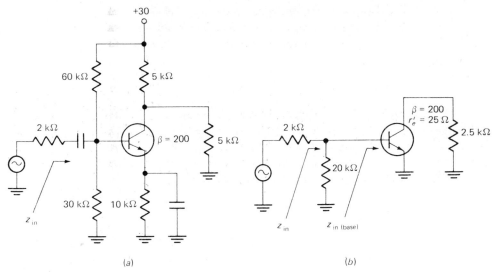

Figure 9-13.

The input impedance is

$$z_{\text{in}} = R_B \parallel z_{\text{in(base)}} = 20(10^3) \parallel 5(10^3) = 4 \text{ k}\Omega$$

EXAMPLE 9-10

What is the $z_{\text{in(base)}}$ of each stage in Fig. 9-14a? What is the voltage gain of this two-stage amplifier?

SOLUTION

The first stage has a β of 100 and an r_e' of 25 Ω. Therefore,

$$z_{\text{in(base)}} = \beta r_e' = 100 \times 25 = 2.5 \text{ k}\Omega$$

The second stage has the same β and r_e'. So, it too has a $z_{\text{in(base)}}$ of 2.5 kΩ.

Figure 9-14b shows the ac equivalent circuit of the amplifier. The voltage gain of the first stage is

$$A_1 = \frac{r_C}{r_e'} = \frac{4000 \parallel 2500}{25} = 61.5$$

The voltage gain of the second stage is

$$A_2 = \frac{r_C}{r_e'} = \frac{5000}{25} = 200$$

The overall voltage gain of the amplifier is the *product* of the individual voltage gains:

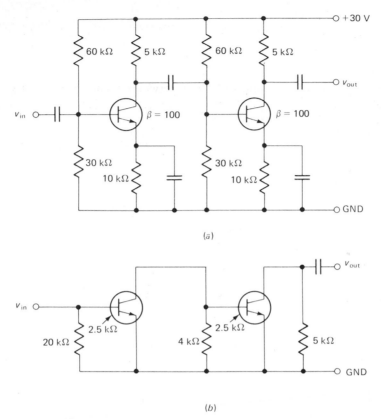

(a)

(b)

Figure 9-14. Two-stage amplifier.

$$A = A_1 A_2 = 61.5 \times 200 = 12,300$$

(*Proof:* $A = A_1 A_2 = v_2/v_{in} \times v_{out}/v_2 = v_{out}/v_{in}$.)

Note how the input impedance of the second stage is part of the load seen by the collector of the first stage. Because of the coupling capacitor in Fig. 9-14a, the first stage has a load resistance of 5 kΩ in parallel with 60 kΩ in parallel with 30 kΩ in parallel with 2.5 kΩ. These four parallel resistances are equivalent to

$$r_C = 1538 \ \Omega$$

On the other hand, the collector of the second stage sees only 5 kΩ of load resistance. This is why its gain is much higher.

EXAMPLE 9-11
Suppose we add a 220-Ω resistor between each emitter and bypass capacitor of Fig. 9-14a. Then, each stage will have $r_E = 220 \ \Omega$ in Fig. 9-14b. What is

new value of $z_{\text{in(base)}}$ for each stage? The new value of voltage gain for the amplifier?

SOLUTION

Look at Fig. 9-14b and visualize 220 Ω between each emitter and ground. The internal emitter resistance r_e' is still approximately 25 Ω because the dc emitter current decreases only slightly. Therefore,

$$z_{\text{in(base)}} = \beta(r_E + r_e') = 100(220 + 25) = 24.5 \text{ k}\Omega$$

for each stage.

The gain of the first stage is

$$A_1 = \frac{r_C}{r_E + r_e'} = \frac{4000 \,\|\, 24{,}500}{245} = 14$$

The gain of the second stage is

$$A_2 = \frac{r_C}{r_E + r_e'} = \frac{5000}{245} = 20.4$$

The overall gain is

$$A = A_1 A_2 = 14 \times 20.4 = 286$$

9-6 *source impedance*

Every ac source has some internal impedance. This impedance often is small enough to ignore, but there are many cases where we cannot neglect it.

Figure 9-15a shows an ac source driving an amplifier through source imped-ance R_S. Since the amplifier has input impedance z_{in}, we can visualize the input circuit as a voltage divider (Fig. 9-15b). Therefore, the ac voltage appearing across the input terminals of the amplifier is

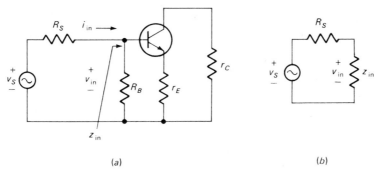

(a)   (b)

Figure 9-15. Effect of source impedance.

$$v_{in} = \frac{z_{in}}{R_S + z_{in}} v_S \qquad \text{(9-12)}***$$

As an example, if $z_{in} = 4$ kΩ and $R_S = 1$ kΩ,

$$v_{in} = \frac{4000}{1000 + 4000} v_S = 0.8 v_S$$

which tells us 80 percent of the ac source voltage appears across the input terminals of the amplifier.

With Eq. (9-12) you can calculate the input voltage to any base-driven amplifier. There is a special case worth mentioning. The input impedance z_{in} may be much larger than the source impedance R_S. In this case, Eq. (9-12) simplifies to

$$v_{in} \cong v_S \qquad \text{when } z_{in} \gg R_S \qquad \text{(9-13)}$$

This says that almost all the source voltage appears across the input terminals of the amplifier.

EXAMPLE 9-12
If $v_{in} = 1$ mV in Fig. 9-16, what does v_{out} equal?

SOLUTION
The transistor has an r_e' of 25 Ω (approximate). Since $\beta = 100$,

$$z_{in(base)} = \beta r_e' = 100 \times 25 = 2.5 \text{ kΩ}$$

The source has an internal impedance of 4 kΩ. When the signal current flows through this 4 kΩ, some of the voltage is dropped across this resistance.

How much signal reaches the base of the transistor? This is where the voltage-

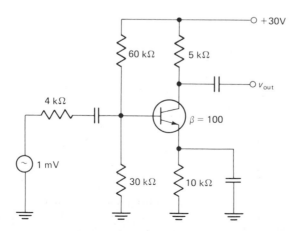

Figure 9-16.

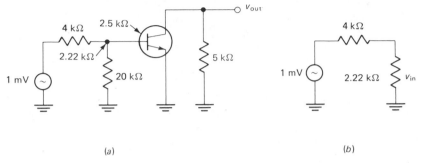

Figure 9-17.

divider theorem comes in. Figure 9-17a shows the ac equivalent circuit. The input impedance of the entire stage is the parallel of 20 kΩ and 2.5 kΩ:

$$z_{in} = 20,000 \,\|\, 2500 = 2.22 \text{ k}\Omega$$

This equivalent resistance is in series with the source resistance, as shown in Fig. 9-17b. With the voltage divider theorem,

$$v_{in} = \frac{2220}{4000 + 2220} 1 \text{ mV} = 0.357 \text{ mV}$$

Note that only 0.357 mV reaches the base of the transistor. The rest of the signal voltage (0.643 mV) is dropped across the 4 kΩ of source resistance. Whenever z_{in} is less than R_S, you lose more than half the signal across R_S.

The gain of the stage is 200. (Recall $A = r_C/r_e' = 5000/25 = 200$. See Example 9-10 if necessary.) Therefore, the output voltage of Fig. 9-16 is

$$v_{out} = Av_{in} = 200 \times 0.357 \text{ mV} = 71.4 \text{ mV}$$

9-7 *the emitter follower*

To avoid $R_S \gg z_{in}$, you sometimes have to step up the impedance level. For instance, Fig. 9-18a shows a *heavily loaded* source, one where R_L is smaller than R_S. In a case like this, most of the voltage is lost across the internal source impedance. One way to get around this is with a transformer as shown in Fig. 9-18b. Now, the input impedance seen by the source is

$$z_{in} = n^2 R_L = 10^2(100) = 10 \text{ k}\Omega$$

input impedance

Instead of a transformer, we can use an *emitter follower*. Figure 9-19 shows the *ac equivalent circuit* of an emitter follower. The ac source drives the base, and

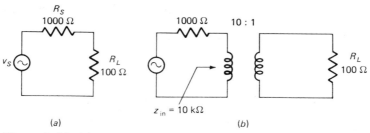

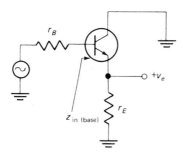

(a)

(b)

Figure 9-18. (a) *Large source impedance.* (b) *Using a transformer to step up impedance.*

Figure 9-19. Emitter follower.

the output signal is taken from the emitter. Since the collector is at ac ground, the circuit is sometimes called a *grounded-collector* or *common-collector* (CC) amplifier. As proved earlier, the input impedance of the base is

$$z_{\text{in(base)}} \cong \beta(r_E + r_e') \qquad\qquad (9\text{-}14a)\star\star\star$$

or
$$z_{\text{in(base)}} \cong \beta r_E \qquad \text{for } r_E \gg r_e' \qquad\qquad (9\text{-}14b)$$

This means the emitter follower steps up the impedance by a factor of β.

Stepping up an impedance is the main reason for using an emitter follower. Not only is it more convenient than a transformer, an emitter follower has a much better *frequency response*, that is, works over a larger frequency range.

voltage gain

With an emitter follower, the voltage gain is less than unity. We can prove this as follows. We found earlier

$$v_e = i_e r_E$$

and
$$v_b \cong i_e(r_E + r_e')$$

Therefore, the voltage gain from base to emitter is

$$A = \frac{v_e}{v_b} \cong \frac{r_E}{r_E + r_e'} \qquad\qquad (9\text{-}15a)\star\star\star$$

The denominator is greater than the numerator; so, the ratio is less than unity.
For the case of r_E much greater than r_e', we get

$$A \cong 1 \quad \text{for } r_E \gg r_e' \qquad (9\text{-}15b)$$

power gain

We do not get voltage gain with an emitter follower, but we do get *power gain*.
The output power in Fig. 9-19 is

$$p_e = i_e^2 r_E$$

and the input power to the base is

$$p_b = i_b^2 z_{\text{in(base)}} \cong i_b^2 \beta (r_E + r_e')$$

The power gain G is

$$G = \frac{p_e}{p_b} \cong \frac{i_e^2 r_E}{i_b^2 \beta (r_E + r_e')}$$

or since $i_e/i_b \cong \beta$,

$$G \cong \beta \frac{r_E}{r_E + r_e'} \qquad (9\text{-}16a)$$

The first factor is the *current gain;* the second factor is the *voltage gain*. The
product of the two gains is the power gain.

For the usual case of r_E much greater than r_e', we get

$$G \cong \beta \qquad \text{for } r_E \gg r_e' \qquad (9\text{-}16b)$$

This tells us the power gain of an emitter follower is approximately the same
as the current gain. So, if we use a transistor with a β of 100, we get a power
gain of 100.

EXAMPLE 9-13
Work out the value of v_{out} in Fig. 9-20a.

SOLUTION
The second stage is familiar from earlier examples. It has an input impedance
of approximately

$$z_{\text{in}} = 60{,}000 \,\|\, 30{,}000 \,\|\, 2500 = 2.22 \text{ k}\Omega$$

This is the R_L seen by the emitter follower. This impedance is in parallel with
the 100-kΩ emitter resistor; therefore,

$$r_E = 100{,}000 \,\|\, 2220 = 2.17 \text{ k}\Omega$$

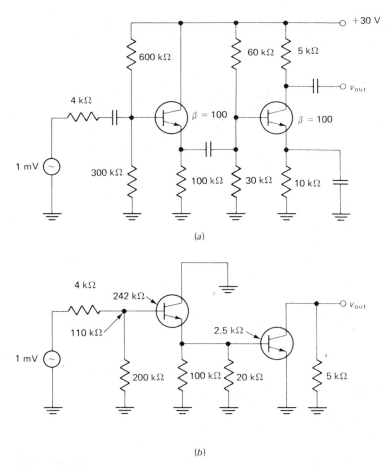

(a)

(b)

Figure 9-20.

In the first stage (the emitter follower) the dc emitter current is approximately 0.1 mA, which means $r_e' = 250\ \Omega$. Therefore, looking into the base of the emitter follower, we see an input impedance of

$$z_{\text{in(base)}} = \beta(r_E + r_e') = 100(2170 + 250) = 242\ \text{k}\Omega$$

As shown in the ac equivalent circuit (Fig. 9-20b), this 242 kΩ is in parallel with 200 kΩ, which means the source sees

$$z_{\text{in}} = 200,000 \,\|\, 242,000 = 110\ \text{k}\Omega$$

With the voltage-divider theorem, the input voltage to the first stage is

$$v_{\text{in}} = \frac{110,000}{4000 + 110,000}\ 1\ \text{mV} = 0.965\ \text{mV}$$

Almost all the source voltage reaches the base of the first stage.

The voltage gain of the emitter follower is

$$A_1 = \frac{r_E}{r_E + r_e'} = \frac{2170}{2170 + 250} = 0.897$$

The voltage gain of the second stage is still 200, found earlier. (Recall $A_2 = r_C/r_e' = 5000/25 = 200$.) The overall gain is

$$A = A_1 A_2 = 0.897 \times 200 = 179$$

The output voltage is

$$v_{\text{out}} = A v_{\text{in}} = 179 \times 0.965 \text{ mV} = 173 \text{ mV}$$

Here's the point. Without the emitter follower, the input impedance of the second stage would load down the source. The voltage divider would consist of 4 kΩ in series with 2.22 kΩ, which means most of the signal would be dropped across the 4-kΩ source resistance.

But with the emitter follower, we have stepped up the impedance level from 2.22 to 110 kΩ (this includes the biasing resistors). Therefore, almost all the source voltage reaches the base of the first stage. Since the emitter follower has a voltage gain approaching unity, almost all the signal reaches the base of the second stage. This is why the final output signal is much larger. (Compare this output with the output of Example 9-12; 173 mV versus 71.4 mV.)

9-8 *the darlington pair*

The higher the β, the higher the input impedance of the base. Many transistors have βs up to 300. With a *Darlington pair*, we can get much higher βs.

Figure 9-21a shows a Darlington pair. The collectors are connected, and the emitter of the first transistor drives the base of the second. Because of this, the overall β is

$$\beta = \beta_1 \beta_2 \qquad (9\text{-}17)$$

As an example, if $\beta_1 = 100$ and $\beta_2 = 50$, the overall β of the Darlington pair is

$$\beta = 100 \times 50 = 5000$$

By complementing and inverting the transistors, we get the *pnp* Darlington shown in Fig. 9-21b. This connection also has an overall β equal to the product of the individual βs.

Transistor manufacturers sometimes put two transistors connected as a Darlington pair inside a single transistor housing. This three-terminal device acts like a single transistor with an extremely high β. For instance, the 2N2785 is

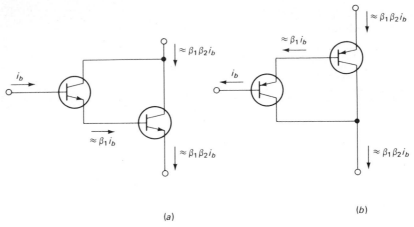

Figure 9-21. Darlington pair. (a) npn. (b) pnp.

an *npn* Darlington transistor with a minimum β of 2000 and a maximum β of 20,000.

Sometimes, the high β of a Darlington emitter follower produces a $z_{\text{in(base)}}$ over a megohm. In a case like this, you can no longer ignore r_c' in Fig. 8-10. The input impedance of the base becomes

$$z_{\text{in(base)}} = \beta(r_E + r_e') \| r_c' \qquad (9\text{-}18)$$

As discussed earlier, r_c' is usually in megohms. For this reason, r_c' represents the upper limit on the input impedance of a Darlington emitter follower. Typically, the highest input impedance you can get with bipolar transistors is a couple of megohms. (To get higher input impedances, we have to use a different device called a *field-effect* transistor.)

EXAMPLE 9-14

What is the input impedance of the first stage in Fig. 9-22 if the Darlington β is 5000 and r_c' is 2 MΩ?

SOLUTION

The second stage is familiar from earlier examples. It has a $z_{\text{in(base)}}$ of 2.5 kΩ and a stage impedance z_{in} of 2.22 kΩ. Therefore, the ac emitter resistance of the first stage is

$$r_E = 8200 \| 2220 = 1.75 \text{ k}\Omega$$

The voltage divider of the first stage produces 10 V. Because the Darlington has two transistors, two V_{BE} drops occur. Let's include these in our calculation for dc emitter current:

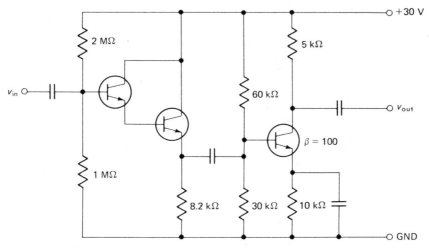

Figure 9-22.

$$I_E = \frac{V_2 - 2V_{BE}}{R_E} = \frac{10 - 1.4}{8200} = \frac{8.6}{8200} \cong 1 \text{ mA}$$

Therefore, the internal emitter resistance of the second transistor is

$$r_e' \cong 25 \ \Omega$$

This is negligible compared to $r_E = 1.75$ kΩ.

With Eq. (9-18), we can now find the impedance looking into the base of the first stage:

$$z_{in(base)} = \beta(r_E + r_e') \| r_c' = 5000(1750) \| 2{,}000{,}000$$

$$= 1.63 \text{ M}\Omega$$

When the base resistors are included, the input impedance is

$$z_{in} = R_1 \| R_2 \| z_{in(base)} = 2 \text{ M}\Omega \| 1 \text{ M}\Omega \| 1.63 \text{ M}\Omega$$

$$= 473 \text{ k}\Omega$$

Without the Darlington emitter follower, the input impedance of the amplifier would be only 2.22 kΩ, the z_{in} of the second stage. But with the Darlington emitter follower, the amplifier has an input impedance of 473 kΩ.

9-9 *types of coupling*

Figure 9-23 illustrates *resistance-capacitance (RC)* coupling, the most widely used method in discrete circuits. In this approach, the signal developed across the

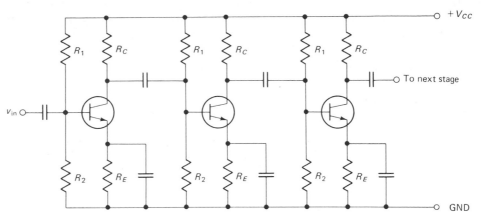

Figure 9-23. RC-*coupled amplifier.*

collector resistor of each stage is coupled into the base of the next stage. In this way, the *cascaded* (one after another) stages amplify the signal, and the overall gain equals the product of the individual gains.

The coupling capacitors transmit ac but block dc. Because of this, the stages are *isolated* as far as dc is concerned. This is necessary to prevent dc interference and shifting of the Q points. The drawback of this approach is the lower frequency limit imposed by the coupling capacitors (see Sec. 8-1 if necessary).

The bypass capacitors are needed because they bypass the emitters to ground. Without them, the voltage gain of each stage would be lost. These bypass capacitors also place a lower limit on the frequency response (Sec. 8-1).

If you are interested in amplifying ac signals with frequencies greater than 10 Hz, the *RC*-coupled amplifier is the way to go. For discrete circuits it is the most convenient and least expensive way to build a multistage amplifier.

impedance coupling

It's rarely used. Nevertheless, you sometimes see *impedance* coupling at higher frequencies. Here's the basic idea. Visualize each R_C of Fig. 9-23 replaced by an inductor. This inductor has an impedance of X_L. At high enough frequencies, X_L approaches infinity.

The advantage of impedance coupling is that no signal power is wasted in collector resistors. The disadvantage is using inductors; they cost more than resistors and their impedance drops off at lower frequencies.

transformer coupling

This was once popular at audio frequencies. But the cost and bulkiness of transformers proved to be too much of a disadvantage in most applications. Here's the idea of *transformer* coupling. Visualize each R_C of Fig. 9-23 replaced by the primary winding of a transformer; the secondary winding replaces the wire between the voltage divider and the base. In this way, the ac signal is transformer-coupled from each collector to the next base. (The coupling capacitor is removed.)

The advantage of transformer coupling is that no signal power is lost in the collector or base resistors. The disadvantages are the cost and bulkiness of transformers at audio frequencies.

The one area where transformer coupling has survived is *radio-frequency* (RF) amplifiers. Radio frequency means anything above 20 kHz. In AM radio receivers, RF ranges from 550 to 1600 kHz. In TV receivers, the RF signals have frequencies from 54 to 216 MHz (channels 2 through 13).

By shunting a capacitor across each winding, we can get resonance at a desired RF frequency. In this way, we have high gain and no signal loss in collector or base resistors. Furthermore, RF transformers are small because of the high frequencies that are involved.

direct coupling

Below 10 Hz (approximately), coupling and bypass capacitors become too large, electrically and physically. For instance, to bypass a 100-Ω emitter resistor at 10 Hz, we need about 1000 μF. The lower we go in frequency or resistance, the worse the problem gets.

To break the low-frequency barrier, we can fall back on *direct* coupling. This means designing the stages without coupling and bypass capacitors, so that dc is coupled as well as ac. In this way, there is no lower frequency limit; the amplifier enlarges signals no matter how low their frequency, including dc or zero frequency. The next section tells you more about direct coupling.

9-10 *direct coupling*

All kinds of direct-coupled designs are possible. In this section, we look at a few elementary circuits to get the idea.

one-supply circuit

Figure 9-24 is a two-stage direct-coupled amplifier; no coupling or bypass capacitors are used. Because of this, dc is amplified as well as ac. With a quiescent input voltage of +1.4 V, about 0.7 V is dropped across the first emitter diode, leaving +0.7 V across the 680-Ω resistor. This sets up approximately 1 mA of collector current. This 1 mA then produces a 27 V drop across the collector resistor. Therefore, the first collector runs at about +3 V with respect to ground.

Allowing 0.7 V for the second emitter diode, we get 2.3 V across the 2.4-

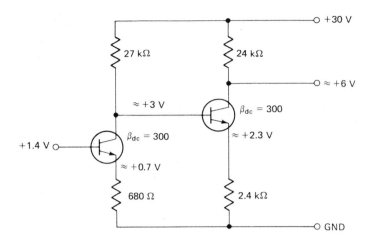

Figure 9-24. Direct-coupled amplifier.

kΩ emitter resistor. This results in approximately 1 mA of collector current, about a 24-V drop across the collector resistor, and around +6 V from the second collector to ground. Therefore, a quiescent input voltage of +1.4 V gives a quiescent output voltage of +6 V.

Because of the high β_{dc}, we can ignore the loading effect of the second base upon the first collector. Ignoring r_e', the first stage has a voltage gain of

$$A_1 = \frac{27,000}{680} = 40$$

The second stage has a voltage gain of

$$A_2 = \frac{24,000}{2400} = 10$$

And the overall gain is

$$A = A_1 A_2 = 40 \times 10 = 400$$

The two-stage circuit will amplify any change in the input voltage by a factor of 400. For instance, if the input voltage changes by +5 mV, the final output voltage changes by

$$v_{out} = A v_{in} = 400 \times 5 \text{ mV} = 2 \text{ V}$$

Because two inverting stages are involved, the final output goes from +6 to +8 V.

Here is the main disadvantage of direct coupling. Transistor characteristics like V_{BE} vary with temperature. This causes the collector currents and voltages to change. Because of the direct coupling, the voltage changes are coupled from one stage to the next, appearing at the final output as an amplified voltage change. This unwanted change is called *drift*. The trouble with drift is you can't distinguish it from a genuine change produced by the input signal.

ground-referenced input

For the two-stage amplifier of Fig. 9-24 to work properly, we need a quiescent input voltage of +1.4 V. In typical applications, it's necessary to have a *ground-referenced* input, one where the quiescent input voltage is 0 V.

Figure 9-25 shows a ground-referenced input stage. This stage is a *pnp* Darlington with the input base returned to ground through the input signal source. Because of this, the first emitter is approximately +0.7 V above ground, and the second emitter is about +1.4 V above ground. The +1.4 V biases the second stage, which operates as previously described.

The quiescent V_{CE} of the first transistor is only 0.7 V, and the quiescent V_{CE} of the second transistor is only 1.4 V. Nevertheless, both transistors are operating in the active region because the $V_{CE(sat)}$ of low-power transistors is only about 0.1 V. Furthermore, the input signal is typically in millivolts, which means these transistors continue to operate in the active region when a signal is present.

Remember the *pnp* ground-referenced input. It's used a lot in audio integrated circuits.

two-supply circuit

When a split supply is available (positive and negative voltages), we can reference both the input and output to ground. Figure 9-26 is an example. The first

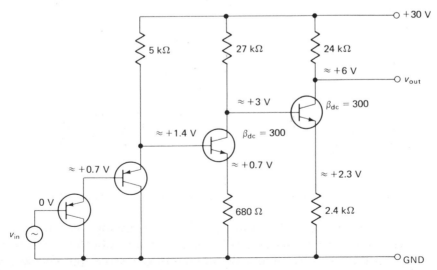

Figure 9-25. Ground-referenced input.

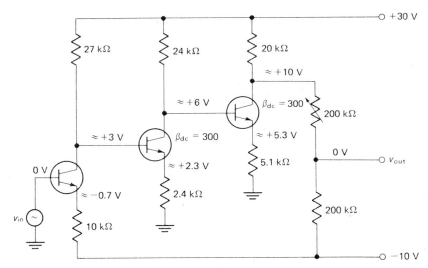

Figure 9-26. Two-supply direct-coupled amplifier.

stage is emitter-biased with an I_E around 1 mA. This produces about +3 V at the first collector. Subtracting the V_{BE} drop of the second emitter diode leaves +2.3 V at the second emitter.

The emitter current in the second stage is around 1 mA. This flows through the collector resistor, producing about +6 V from the collector to ground. The final stage has +5.3 V across the emitter resistor, which gives about 1 mA of current. Therefore, the last collector has approximately +10 V to ground.

The output voltage divider references the output to ground. When the upper resistor is adjusted to 200 kΩ, the final output voltage is approximately 0 V. The adjustment allows us to eliminate errors caused by resistor tolerances, V_{BE} differences from one transistor to the next, etc.

What is the overall voltage gain? The first stage has a gain around 3, the second stage about 10, the third stage around 4, and the voltage divider about 0.5. Therefore,

$$A = 3 \times 10 \times 4 \times 0.5 = 60$$

summary

This gives you the basic idea behind direct coupling. We leave out all coupling and bypass capacitors. This allows us to couple dc and ac from one stage to the next. Because of this, the amplifier has no lower frequency limit; it amplifies all frequencies down to zero. Herein lies the strength and weakness of direct coupling: it's good to be able to amplify very low frequencies, including dc;

it's bad, however, to amplify very slow changes in supply voltage, transistor variations, etc.

Later chapters introduce the *differential amplifier*, a two-transistor direct-coupled circuit that has become the backbone of linear integrated circuits. One reason the differential amplifier is so popular is that drift cancels out, at least partially. More is said about this later.

9-11 *the common-base amplifier*

Figure 9-27 illustrates the *grounded-base* or *common-base* (CB) amplifier. The V_{EE} supply forward-biases the emitter diode, and the V_{CC} supply reverse-biases the collector diode. The dc emitter current equals the voltage across the emitter resistor divided by the resistance. In symbols,

$$I_E = \frac{V_{EE} - V_{BE}}{R_E} \tag{9-19}$$

The collector-to-ground voltage equals the V_{CC} supply voltage minus the drop across the collector resistor:

$$V_C = V_{CC} - I_C R_C \tag{9-20}$$

Figure 9-28 shows the ac equivalent circuit. The ac output voltage is

$$v_{\text{out}} = i_c r_C$$

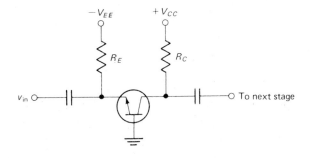

Figure 9-27. CB amplifier.

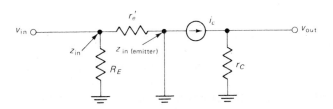

Figure 9-28.

and the ac input voltage is

$$v_{\text{in}} = i_e r_e'$$

Therefore, the voltage gain is

$$A = \frac{v_{\text{out}}}{v_{\text{in}}} = \frac{i_c r_C}{i_e r_e'}$$

Since $i_c \cong i_e$,

$$A = \frac{r_C}{r_e'} \tag{9-21}$$

This is identical to the voltage gain of a CE amplifier.

One reason why the CB amplifier is rarely used is its *low* input impedance. Looking into the emitter, an ac source sees only

$$z_{\text{in(emitter)}} = r_e' \tag{9-22}$$

The stage has an input impedance of

$$z_{\text{in}} = R_E \, \| \, r_e'$$

or
$$z_{\text{in}} \cong r_e' \tag{9-23}$$

because R_E is typically much greater than r_e'. For $I_E = 1$ mA, therefore, the approximate input impedance of a CB amplifier is only 25 Ω.

The input impedance of a CB amplifier is so low that it overloads most signal sources. In other words, the bulk of the ac signal is lost across the source impedance. Because of this, the CB amplifier is almost never used at low frequencies. it has been occasionally used in high-frequency applications above 10 MHz where low source impedances are common.

9-12 *summary*

Table 9-1 summarizes the operation of CE, CC, and CB amplifiers. A represents the voltage gain from the input transistor terminal to the output terminal. As shown, the CE and CB circuits have a voltage gain of r_C/r_e'; the CC circuit has a voltage gain of unity (ideal).

A_i is the symbol for current gain, the ratio of output current to input current. As indicated, the current gain of the CE and CC amplifiers is β; the current gain of a CB amplifier is unity.

The CE circuit has the largest power gain because G is the product of voltage gain A and current gain A_i. Also, note that the CC amplifier has the highest input impedance, whereas the CB has the lowest.

The new quantity z_{out} is the output impedance at low frequencies. This im-

Table 9-1 Ideal Approximations

	CE	CC	CB
A	r_C/r_e'	1	r_C/r_e'
A_i	β	β	1
G	$\beta r_C/r_e'$	β	r_C/r_e'
z_{in}	$\beta r_e'$	$\beta(r_E + r_e')$	r_e'
z_{out}	r_c'/β	r_e'	r_c'

pedance is identical to the Thevenin or Norton resistance of each amplifier. The derivations are too complicated for this book, but by applying Thevenin's and Norton's theorems it is possible to arrive at the ideal output impedances listed in Table 9-1. The CE amplifier has an output impedance of r_c'/β, usually more than 10,000 Ω. The CC amplifier has an output impedance of r_e', often less than 100 Ω. The CB amplifier has an output impedance of r_c', typically greater than 1 MΩ. This means the CC amplifier (emitter follower) acts approximately like a voltage source for most load resistances. The CE and CB amplifiers, on the other hand, act almost like current sources for the load resistances normally used with them.

The 2N3904 is a widely used small-signal transistor. To give you an idea of typical values, Table 9-2 lists the gains and impedances of a 2N3904 operated as a CE, CC, or CB amplifier. These quantities are based on the typical data-sheet values for a 2N3904. The CE amplifier has the best overall gain (voltage, current, and power), a medium input impedance, and a high output impedance. The CC amplifier has a low voltage gain, a high input impedance, and a low

Table 9-2 Typical Values*

	CE	CC	CB
A	100	1	100
A_i	120	120	1
G	12,000	120	100
z_{in}	3 kΩ	50 kΩ	25 Ω
z_{out}	125 kΩ	25 Ω	15 MΩ

* For a 2N3904 with $I_E = 1$ mA, $r_C = 2.5$ kΩ for CE and CB, and $r_E = 390$ Ω for CC.

output impedance. The CB amplifier has a low input impedance and a high output impedance.

self-testing review

Read each of the following and provide the missing words. Answers appear at the beginning of the next question.

1. The ac collector voltage is 180° out of phase with the ac base voltage. This _____ inversion between base and collector happens in all base-driven amplifiers. The phase of the ac emitter voltage is the same as the phase of the ac _____ voltage.

2. *(phase, base)* A voltage amplifier increases the signal voltage. The voltage gain of a base-driven amplifier equals the ac collector voltage divided by the ac _____ voltage.

3. *(base)* In a CE amplifier, ac emitter resistance r_E equals zero. Because of this, the voltage gain equals r_C divided by _____. The common-emitter amplifier is also called a grounded-emitter amplifier because the emitter is at _____ ground.

4. *(r'_e, ac)* When you look at voltage waveforms with a dc-coupled oscilloscope, you will see total voltages, that is, the sum of _____ and _____ voltages. If you switch the oscilloscope to ac input, you will see only ac voltage.

5. *(dc, ac)* Many people swamp the emitter diode to reduce the effects of _____. Swamping the emitter diode means making r_E much greater than r'_e. In this case, the voltage gain from base to collector equals r_C divided by _____.

6. *(r'_e, r_E)* In a swamped amplifier, the ac emitter voltage approximately _____ the ac base voltage. The phases of the emitter and base signals are the same because the emitter follows the base.

7. *(equals)* For the CE amplifier, $r_E = 0$; therefore, $z_{in(base)}$ equals _____ times r'_e. On the other hand, a swamped amplifier has an input impedance of β times _____.

8. *(β, r_E)* When the collector is at ac ground, the circuit is called a grounded-collector or _____ amplifier. Stepping up an impedance is the main reason for using a CC amplifier, also known as an _____ _____.

9. *(common-collector, emitter follower)* The voltage gain of an emitter follower is less than _____. If r_E is much greater than r'_e, the voltage gain approaches _____ and the input impedance is _____ times r_E. The phase of the input and output is the same.

10. *(one, one, β)* You multiply the individual βs to get the overall β of a _____ pair.

11. *(Darlington)* *RC* coupling is the most widely used in discrete circuits. Transformer coupling is still used in _____ (RF) amplifiers. Direct coupling amplifies signals no matter how low the frequency, including dc or zero frequency. The main problem with direct coupling is _____.

12. *(radio-frequency, drift)* The CB amplifier is rarely used because of its low input impedance. This input impedance overloads most signal sources, causing the bulk of the signal to be lost across the source impedance.

problems

9-1. Calculate the ac emitter current in the base-driven circuit of Fig. 9-29a.

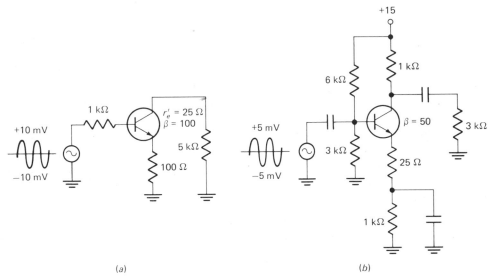

(a)   (b)

Figure 9-29.

9-2. In Fig. 9-29a, work out these ac voltages with respect to ground: v_e, v_b, and v_c. Calculate the voltage ratios v_c/v_b and v_c/v_e.

9-3. The amplifier of Fig. 9-29b has an r_e' of approximately 5 Ω. Calculate the ac emitter current. What is the value of v_e? And the voltage gain from base to collector?

9-4. Work out the voltage gain from base to collector for the CE amplifier of Fig. 9-30a.

9-5. In the *pnp* amplifier of Fig. 9-30b, how much ac collector voltage do you get if the ac source voltage is a 5-mV peak sine wave?

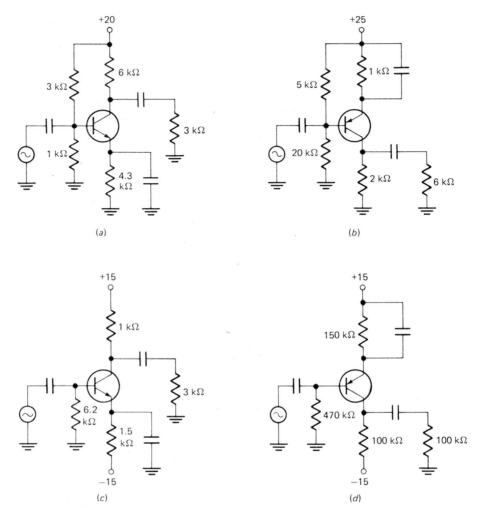

Figure 9-30.

9-6. Calculate the voltage gain for the amplifier of Fig. 9-30c.

9-7. What is the voltage gain in Fig. 9-30d?

9-8. If you connect a dc-coupled oscilloscope from the collector to ground in Fig. 9-30a, what will the total waveform be if the ac source is a 1-mV-peak sine wave?

9-9. In Fig. 9-30b, the ac source puts out a sine wave with a peak of 5 mV. Describe the total voltage waveforms you would see with a dc-coupled oscilloscope at each of these points: the base, the emitter, and the collector (all with respect to ground).

9-10. What are the minimum dc voltage ratings for the capacitors in Fig. 9-30a?

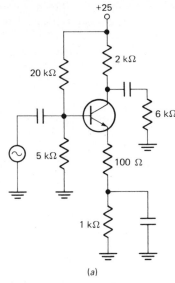

Figure 9-31.

9-11. Find the voltage gain of the swamped amplifier in Fig. 9-31a.

9-12. Calculate the voltage gain from base to collector for the swamped amplifier shown in Fig. 9-31b. If the ac source signal is a sine wave with a 0.1-V peak, what does v_c equal? And v_e?

9-13. An ac source of 2 mV peak drives the amplifier of Fig. 9-30a. What are the dc and ac voltages from the base to ground? From the emitter to ground? From the collector to ground?

9-14. In fig. 9-30b, an ac source of 5 mV peak-to-peak drives the amplifier. What are the dc and ac voltages for the following:
   (a) Base to ground
   (b) Emitter to ground
   (c) Collector to ground

9-15. If the collector coupling capacitor of Fig. 9-31b opens, what is the new value of voltage gain from base to collector?

9-16. If the collector capacitor shorts in Fig. 9-31b instead of opening, what effect will this have on the dc operation and the ac operation?

9-17. If the base coupling capacitor of Fig. 9-31b shorts, what will this do to the dc operation?

9-18. Ac current is difficult to measure. For this reason, rather than try to measure ac input current in Fig. 9-32, we can insert a *test resistor* as shown. By measuring the ac voltage on the left and right ends of this resistor, we can calculate the ac input current. Suppose the values are $v_{left} = 15$ mV, $v_{right} = 3$ mV, and $R_{test} = 1$ kΩ. Calculate the ac input current i_{in}; also, work out the input impedance z_{in}.

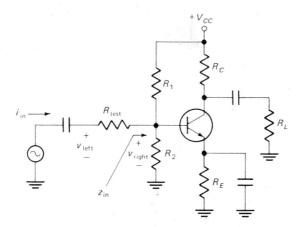

Figure 9-32.

9-19. If the transistor of Fig. 9-30a has a β of 50, what is the value of $z_{in(base)}$? Of z_{in}?

9-20. The transistor in Fig. 9-30b has a β equal to 300. Calculate $z_{in(base)}$ and z_{in}.

9-21. Suppose the β is 100 for each transistor shown in Fig. 9-31. Assume r'_e is negligible in each circuit. What is the value of $z_{in(base)}$ for each amplifier?

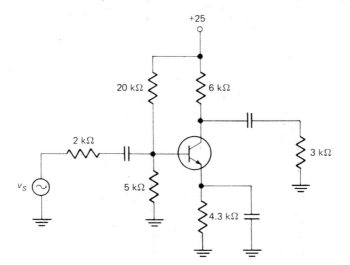

Figure 9-33.

9-22. The ac source voltage v_S of Fig. 9-33 equals 1 mV rms. If β equals 100, what does v_b equal? And v_c?

9-23. Suppose the transistor of Fig. 9-33 has a β of 200. What is the voltage gain from source to base? From base to collector? From source to collector?

9-24. What is the value of β in Fig. 9-33 that produces an impedance match between z_{in} and R_S?

9-25. Suppose the R_S of Fig. 9-33 is changed from 2 kΩ to 150 kΩ. What is the voltage gain from source to base if β equals 100? From source to collector?

9-26. If $V_{CC} = 15$ V and $r_E = 0$ in Fig. 9-34, what is the $z_{in(base)}$ of each stage? The overall voltage gain?

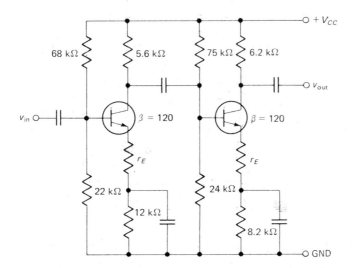

Figure 9-34.

9-27. What is the overall voltage gain in Fig. 9-34 if $V_{CC} = 20$ V and $r_E = 50$ Ω?

9-28. If $V_{CC} = 25$ V and $r_E = 100$ Ω, what is v_{out} in Fig. 9-34 if v_{in} is 2 mV?

9-29. In Fig. 9-34, $V_{CC} = 30$ V and $r_E = 220$ Ω. What does v_{out} equal if v_{in} is 5 mV?

9-30. In Fig. 9-35a, what is the value of $z_{in(base)}$? The voltage gain from base to emitter? The power gain?

9-31. In Fig. 9-35a, calculate the voltage gain from source to base and from source to emitter.

9-32. The Darlington pair of Fig. 9-35b has an overall beta of 10,000. What is the input impedance of the stage, including biasing resistors?

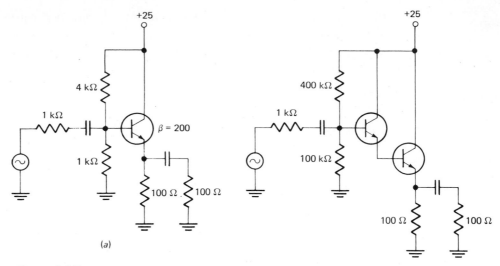

(a)

Figure 9-35.

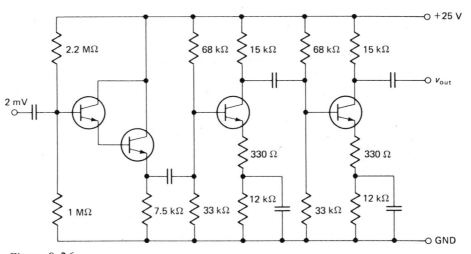

Figure 9-36.

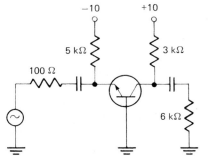

Figure 9-37.

9-33. Work out the value of v_{out} in Fig. 9-36. (Use $\beta = 100$ for all transistors.)

9-34. In Fig. 9-37, calculate the dc emitter current, the voltage gain from emitter to collector, and the input impedance looking into the emitter.

10 *class a power amplifiers*

The stages near the end in many systems are *large-signal* amplifiers, where the emphasis is on power gain. Amplifiers of this type are called *power amplifiers*. Transistors used in small-signal amplifiers are referred to as *small-signal transistors;* those used in power amplifiers are called *power transistors.* As a rule, a small-signal transistor has a power dissipation less than half a watt; a power transistor, more than half a watt.

10-1 *the* q *point*

The *quiescent* (at rest) collector current and voltage are the I_C and V_{CE} when there is no input signal. In Fig. 10-1*a*, for instance, the quiescent collector current is approximately 2 mA (ignore the V_{BE} drop). The quiescent collector-emitter voltage is the supply voltage minus the drop across the collector and emitter resistors:

$$V_{CE} \cong V_{CC} - I_C(R_C + R_E) = 30 - 0.002(10,500)$$
$$= 9 \text{ V}$$

Therefore, the coordinates of the Q point are 2 mA and 9 V, shown in Fig. 10-1*b*.

No matter what kind of bias is used, the procedure for locating the Q point is the same: you calculate I_C and V_{CE} using the *dc equivalent circuit.* Then you

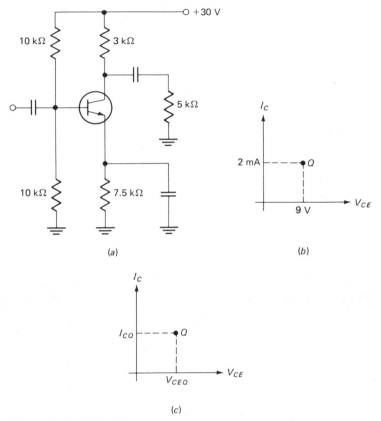

Figure 10-1. Q point.

plot the Q point with I_C on the vertical axis and V_{CE} on the horizontal axis as shown in Fig. 10-1c. To keep track of the Q point in our calculations, we will give it the subscripts indicated: I_{CQ} stands for quiescent collector current, and V_{CEQ} stands for quiescent collector-emitter voltage.

10-2 *the dc load line*

As discussed earlier, the dc load line represents all possible dc operating points. The upper end of the dc load line is called the saturation point, and the lower end is called the cutoff point. The Q point lies somewhere along the dc load line.

If the transistor of Fig. 10-1a were saturated, all supply voltage would appear across R_C and R_E, giving a current of

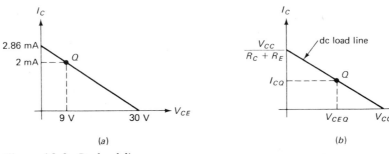

Figure 10-2. Dc load line.

$$I_{C(\text{sat})} = \frac{V_{CC}}{R_C + R_E} = \frac{30}{3000 + 7500} = 2.86 \text{ mA}$$

On the other hand, if the transistor operated in the cutoff region, ail supply voltage would appear across the collector-emitter terminals:

$$V_{CE(\text{cutoff})} = V_{CC} = 30 \text{ V}$$

Figure 10-2*a* shows the dc load line along with the *Q* point found earlier.

Figure 10-2*b* summarizes the idea of the dc load line and *Q* point for any one-supply circuit.

10-3 *the ac load line*

Every amplifier sees two loads: a dc load and an ac load. To find the saturation and cutoff points on the dc load line, you have to analyze the dc equivalent circuit. To find the saturation and cutoff points on the *ac load line,* you have to analyze the ac equivalent circuit.

where it comes from

Figure 10-3 shows the base-driven circuit analyzed in Chap. 9. In this *ac equivalent circuit,* the ac load seen by the collector is r_C and the ac load seen by the emitter is r_E. As you know, the use of coupling and bypass capacitors means r_C can be different from R_C, and r_E different from R_E.

By summing voltages around the collector-emitter loop of Fig. 10-3, we get

$$v_{ce} + i_e r_E + i_c r_C = 0$$

Since $i_c \cong i_e$, we can rewrite the equation as

$$i_c = -\frac{v_{ce}}{r_C + r_E} \tag{10-1}$$

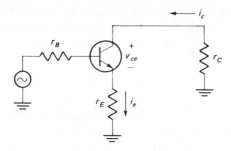

Figure 10-3. Ac equiv-
alent circuit of base-
driven amplifier.

When an ac signal drives an amplifier, it causes changes in collector current and voltage. These changes are given by

$$i_c = I_C - I_{CQ}$$

$$v_{ce} = V_{CE} - V_{CEQ}$$

Substituting these expressions into (10-1) gives

$$I_C - I_{CQ} = -\frac{V_{CE} - V_{CEQ}}{r_C + r_E} \qquad (10\text{-}2)$$

which can be rearranged as

$$I_C = -\frac{V_{CE}}{r_C + r_E} + I_{CQ} + \frac{V_{CEQ}}{r_C + r_E} \qquad (10\text{-}3)$$

This is a linear equation in I_C and V_{CE}; it is similar to

$$y = mx + b$$

As proved in basic mathematics, the graph of a linear equation is always a straight line with a slope of m and a vertical intercept of b.

Figure 10-4 shows the graph of Eq. (10-3). This line is called the *ac load line*

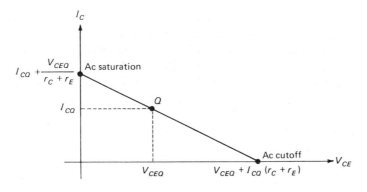

Figure 10-4. Ac load line.

because it represents all possible ac operating points. At any instant during the ac cycle, the *instantaneous* operating point lies somewhere along the ac load line, the exact point determined by the amount of change from the Q point.

large signal

The ac load line is a visual aid for understanding large-signal operation. During the positive half cycle of ac source voltage, the collector current swings from the Q point up toward saturation. On the negative half cycle of ac source voltage, the collector current swings from the Q point down toward cutoff. For a large enough ac signal, the instantaneous operating point can move all the way up

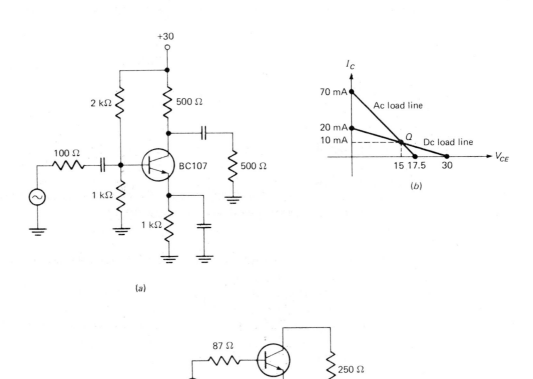

Figure 10-5. Dc and ac load line.

to saturation and all the way down to cutoff. In other words, a large-signal amplifier uses all or almost all of the active region.

EXAMPLE 10-1

The BC107 of Fig. 10-5*a* has the following maximum ratings: $I_{C(\text{max})} = 100$ mA and $BV_{CEO} = 45$ V. Show that neither of these ratings is exceeded during the ac cycle.

SOLUTION

Visualize the dc equivalent circuit and you can calculate $I_{CQ} = 10$ mA and $V_{CEQ} = 15$ V (ignore V_{BE}). Figure 10-5*b* shows the Q point.

Next, visualize the ac equivalent circuit. After Thevenizing the base and collector circuits, we get Fig. 10-5*c*. As shown, $r_E = 0$ and $r_C = 250$ Ω. Referring to the general ac load line of Fig. 10-4, we can calculate the end points on the load line as follows:

$$I_{C(\text{sat})} = I_{CQ} + \frac{V_{CEQ}}{r_C + r_E} \tag{10-4}$$

$$= 10(10^{-3}) + \frac{15}{250} = 70 \text{ mA}$$

and
$$V_{CE(\text{cutoff})} = V_{CEQ} + I_{CQ}(r_C + r_E) \tag{10-5}$$

$$= 15 + 10^{-2}(250) = 17.5 \text{ V}$$

Figure 10-5*b* shows the ac saturation current and the ac cutoff voltage. Since the BC107 has an $I_{C(\text{max})}$ rating of 100 mA, there is no danger of exceeding this rating because at most we can have 70 mA of collector current. Also, the largest collector voltage occurs at cutoff and equals 17.5 V, which is much less than the BV_{CEO} of 45 V.

Incidentally, the dc cutoff voltage equals 30 V and the dc saturation current equals 20 mA. Therefore, the dc load line is different from the ac load line (see Fig. 10-5*b*).

EXAMPLE 10-2

Show the ac load line for the emitter follower of Fig. 10-6*a*.

SOLUTION

First, get I_{CQ} and V_{CEQ}. With about 10 V dc across the 50-Ω emitter resistor, we get

$$I_{CQ} = \frac{10}{50} = 0.2 \text{ A}$$

and
$$V_{CEQ} = 20 - 10 = 10 \text{ V}$$

Figure 10-6*b* shows this Q point.

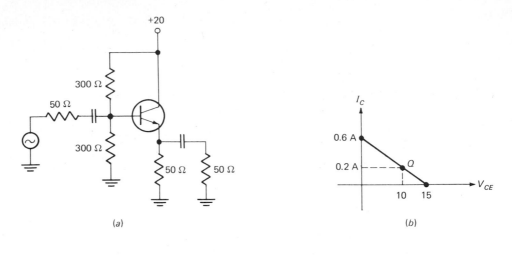

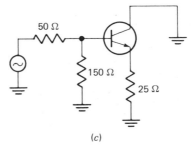

Figure 10-6.

Next, visualize the ac equivalent circuit as shown in Fig. 10-6c. Since $r_C = 0$ and $r_E = 25\ \Omega$, we calculate the ends of the load line as follows:

$$I_{C(\text{sat})} = I_{CQ} + \frac{V_{CEQ}}{r_C + r_E}$$

$$= 0.2 + \frac{10}{25} = 0.6\ \text{A}$$

and

$$V_{CE(\text{cutoff})} = V_{CEQ} + I_{CQ}(r_C + r_E)$$
$$= 10 + 0.2(25) = 15\ \text{V}$$

Figure 10-6b shows the ac load line with these end points.

EXAMPLE 10-3
Explain Fig. 10-7.

SOLUTION
The Q point has $I_{CQ} = 1$ mA and $V_{CEQ} = 7$ V. During the positive half cycle of ac source voltage, the collector current swings from 1 (point Q) to 1.5 mA

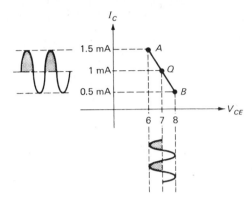

Figure 10-7. Visualizing the motion of the instantaneous operating point.

(point *A*); the collector-emitter voltage swings from 7 to 6 V. On the negative half cycle of ac source voltage, the collector current swings from 1 (point *Q*) to 0.5 mA (point *B*); the collector-emitter voltage swings from 7 to 8 V.

At any instant during the ac cycle, the coordinates of collector current and voltage are on the ac load line. Because of this, we can visualize the instantaneous operating point starting at point *Q* and moving along the ac load line toward point *A*. After peaking at *A*, the instantaneous operating point swings from *A* through *Q* to *B*. Then it returns to *Q*, where the next cycle begins.

A final point. Figure 10-7 shows only part of the ac load line. Since the slope of the line is

$$m = -0.5 \text{ mA/V}$$

the ac cutoff voltage is 9 V and the ac saturation current is 4.5 mA.

10-4 *optimum* q *point for class a*

If you *overdrive* an amplifier, the output signal will be clipped on either or both peaks. Figure 10-8*a* shows clipping on the negative half cycle of source voltage. Since the *Q* point is closer to cutoff than to saturation, the instantaneous operating point hits the cutoff point before the saturation point. Because of this, we get *cutoff clipping* as shown.

If the *Q* point is too high, that is, closer to saturation than to cutoff, we get *saturation clipping* as shown in Fig. 10-8*b*. On the positive half cycle of ac source voltage, the instantaneous operating point drives into the saturation point and results in positive clipping of the collector current.

To get the maximum unclipped signal, we can locate the *Q* point at the *center* of the ac load line (Fig. 10-8*c*). In this way, the instantaneous operating point can swing equally in both directions before clipping occurs. With the right size of input signal, we can get the maximum possible unclipped output signal.

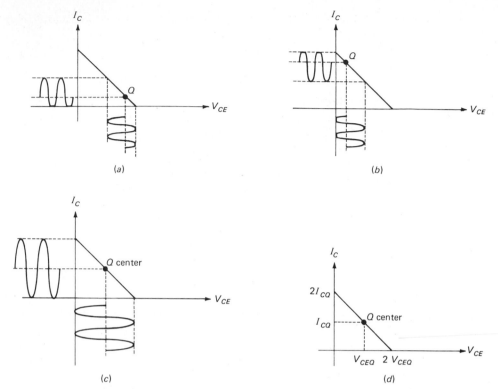

Figure 10-8. (a) *Cutoff clipping.* (b) *Saturation clipping.* (c) *Maximum unclipped signal.* (d) *Centered* Q *point.*

definition of class a

All along, we have said a transistor should stay in the active region through the ac cycle. To distinguish this operation from other kinds, we call it *class A operation.* In terms of the ac load line, class A operation means that no clipping occurs at either end of the ac load line. If we did get clipping, the operation would no longer be called class A operation.

In a class A amplifier the best place to locate the Q point is in the center of the ac load line as shown in Fig. 10-8c. The reason is clear enough. When the Q point is in the middle of the ac load line, we can get the largest possible unclipped output signal.

centered q point

Figure 10-8d shows a centered Q point. I_{CQ} and V_{CEQ} are the quiescent current and voltage. Simple geometry tells us the saturation current must be twice as large as I_{CQ} and the cutoff voltage twice as large as V_{CEQ}; otherwise, the Q point would not be centered. In symbols,

$$I_{C(\text{sat})} = 2I_{CQ} \quad (\text{centered } Q) \tag{10-6}$$

and
$$V_{CE(\text{cutoff})} = 2V_{CEQ} \quad (\text{centered } Q) \tag{10-7}$$

Since the cutoff voltage of any base-driven amplifier is given by Fig. 10-4, we can find an important relation between quiescent values and ac resistances as follows. In Fig. 10-4, the cutoff voltage is

$$V_{CE(\text{cutoff})} = V_{CEQ} + I_{CQ}(r_c + r_E) \tag{10-8}$$

Because of this, we can substitute the right member of Eq. (10-7) to get

$$2V_{CEQ} = V_{CEQ} + I_{CQ}(r_c + r_E)$$

By rearranging this, we get

$$r_c + r_E = \frac{V_{CEQ}}{I_{CQ}} \quad (\text{centered } Q) \tag{10-9}***$$

This final result is important. Here is what it tells you. *To have a centered Q point, the ac resistance of the collector and emitter circuits must equal the ratio of the quiescent collector voltage to the quiescent collector current.* You will find this equation useful in analyzing and designing power amplifiers.

EXAMPLE 10-4

Figure 10-9a is a *phase splitter* (sometimes called a phase inverter). What does it do, and is its Q point centered?

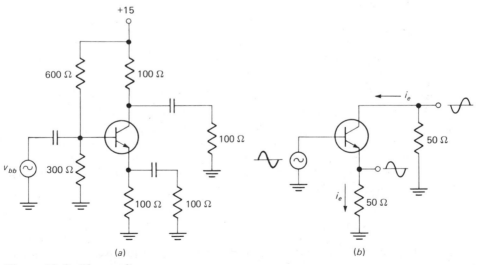

(a)   (b)

Figure 10-9. Phase splitter.

SOLUTION

First, visualize the ac equivalent circuit. After reducing, we have Fig. 10-9*b*. Since i_c closely approximates i_e, essentially the same peak voltages appear across r_C and r_E. As you recall, the collector signal is 180° out of phase with the base signal, and the emitter signal is in phase with the base signal. Therefore, the output of a phase splitter is a pair of sine waves of approximately equal amplitude but opposite phase.

Next, check if the Q point is centered. The sum of collector and emitter ac resistance is

$$r_C + r_E = 50 + 50 = 100 \ \Omega$$

When we visualize the dc equivalent circuit of Fig. 10-9*a*, we can see a base voltage of 5 V. Ideally, all this appears across the 100-Ω emitter resistor. Therefore,

$$I_{CQ} \cong I_E \cong \frac{V_E}{R_E} \cong \frac{5}{100} = 50 \ \text{mA}$$

and

$$V_{CEQ} = V_C - V_E = 10 - 5 = 5 \ \text{V}$$

The ratio is

$$\frac{V_{CEQ}}{I_{CQ}} = \frac{5}{0.05} = 100 \ \Omega$$

which does equal $r_C + r_E$. So, the phase splitter of Fig. 10-9*a* has a centered Q point.

10-5 *class a power formulas*

What is the maximum power you can get out of a class A amplifier? How much power must a transistor dissipate? How efficient is class A?

maximum output power

Figure 10-10*a* shows a base-driven circuit, and Fig. 10-10*b* is the ac load line with a centered Q point. When the ac source signal is large enough to produce the maximum unclipped output signal, we get the current and voltage waveforms of Fig. 10-10*b*.

As you can see, the ac collector current is a sine wave with a peak value of I_{CQ}. The ac collector-emitter voltage is a sine wave with a peak value of V_{CEQ}. We can work out the maximum ac output power as follows:

$$P_{O(\text{max})} = V_{RMS} I_{RMS} = \frac{V_{CEQ}}{\sqrt{2}} \frac{I_{CQ}}{\sqrt{2}}$$

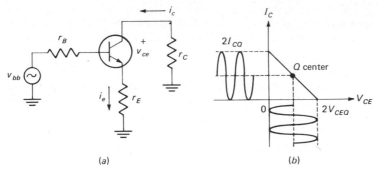

Figure 10-10. (a) *Base-driven circuit.* (b) *Centered* Q *point.*

or
$$P_{O(\text{max})} = \frac{V_{CEQ}I_{CQ}}{2} \quad \text{(centered } Q\text{)} \qquad (10\text{-}10)\text{***}$$

This says the maximum ac power delivered to r_C and r_E equals half the product of V_{CEQ} and I_{CQ}. The equation is important because it relates ac performance to dc biasing.

As an example, the phase splitter of Fig. 10-9a has $I_{CQ} = 50$ mA and $V_{CEQ} = 5$ V. When the entire ac load line is used, the ac power delivered to r_C and r_E is maximum and equals

$$P_{O(\text{max})} = \frac{V_{CEQ}I_{CQ}}{2} = \frac{5(0.05)}{2} = 125 \text{ mW}$$

power dissipation

In our discussion of the physics behind transistor action, we described how conduction-band electrons fall down the collector energy hill. As they fall, they give up energy in the form of heat. The transistor must dissipate this heat to its surroundings. During the ac cycle, the collector current and voltage change, but on the average, a fixed amount of heat is produced. In other words, the transistor has to dissipate an *average power* which we designate P_D. Data sheets always include the maximum power dissipation of the transistor type. You cannot exceed this rating without risking damage to the transistor.

The *quiescent power dissipation* of a transistor is

$$P_{DQ} = V_{CEQ}I_{CQ} \qquad (10\text{-}11)\text{***}$$

This tells us that when no ac signal is present, the transistor must dissipate power equal to the product of quiescent voltage and current. For example, in Fig. 10-9a, we found $I_{CQ} = 50$ mA and $V_{CEQ} = 5$ V. Therefore,

$$P_{DQ} = 5(0.05) = 0.25 \text{ W} = 250 \text{ mW}$$

Quiescent power represents the worst case. In other words, maximum power dissipation in a transistor occurs under no-signal conditions. In symbols,

$$P_{D(\text{max})} = P_{DQ}$$

$$(10\text{-}11a)***$$

As an example, the transistor of Fig. 10-9a has a P_{DQ} of 0.25 W. To avoid damage, the transistor needs a power rating of at least 0.25 W.

relation between output power and quiescent power

Substituting the left-hand member of Eq. (10-11) into Eq. (10-10) gives

$$P_{O(\text{max})} = \frac{P_{DQ}}{2} \quad \text{(centered } Q\text{)}$$

$$(10\text{-}11b)$$

This says the maximum ac output power is half the quiescent power. This is the best you can do with a class A amplifier, and you get this only when the Q point is centered.

As an example, if a transistor dissipates 3 W under no-signal conditions, the maximum ac output power it can deliver is 1.5 W. Conversely, if you are trying to build a class A amplifier that delivers 30 W of ac output power, you will need a transistor that can dissipate 60 W under no-signal conditions.

efficiency

The *output efficiency,* symbolized η, is defined as the ratio of ac output power to dc input power supplied to the collector-emitter circuit:

$$\eta = \frac{P_O}{P_{DC}}$$

$$(10\text{-}12)$$

Efficiency is a measure of how well an amplifier converts dc power from the supply into ac output power. If $P_{DC} = 10$ W and $P_O = 1$ W, the efficiency is 10 percent. High efficiency is important in battery-operated equipment; it means the batteries last longer.

For one-supply circuits, Eq. (10-12) becomes

$$\eta = \frac{P_O}{V_{CC} I_{CQ}}$$

$$(10\text{-}12a)$$

For two-supply circuits,

$$\eta = \frac{P_O}{(V_{CC} + V_{EE}) I_{CQ}}$$

$$(10\text{-}12b)$$

EXAMPLE 10-5
What is the efficiency of the phase splitter of Example 10-4 if the signal swings over the entire ac load line?

SOLUTION
We found $I_{CQ} = 50$ mA and $V_{CEQ} = 5$ V. The maximum output power is

$$P_{O(\text{max})} = \frac{V_{CEQ}I_{CQ}}{2} = \frac{5 \times 0.05}{2} = 125 \text{ mW}$$

The efficiency is

$$\eta = \frac{P_O}{V_{CC}I_{CQ}} = \frac{0.125}{15 \times 0.05} = 0.167$$

This says the phase splitter converts 16.7 percent of the supply power into ac output power.

Incidentally, an *RC*-coupled class A amplifier has a maximum efficiency of 25 percent. This means no more than 25 percent of supply power is converted into ac output power. (If transformer coupling is used, the efficiency can be as high as 50 percent. But transformers are rarely used except with RF signals.)

10-6 *large-signal gain and impedance*

We need to modify the approach of Chap. 9 (small-signal amplifiers) when we analyze power amplifiers. The first thing to realize is that $r'_e = 25$ mV/I_E is not useful with a power amplifier because the current and voltage swings are too large; we have to use *large-signal ac emitter resistance,* designated R'_e.

how to get R'_e

Because R'_e is a large-signal characteristic of a power transistor, we need information from the data sheet to get its value; no simple formula like 25 mV/I_E exists for the value of R'_e. Instead, we have to use the *transconductance curve* shown on the data sheet of a power transistor.

The transconductance curve is a graph of I_C versus V_{BE} similar to Fig. 10-11a. We define R'_e as the ratio of a *large change* in V_{BE} to a *large change* in I_E. In symbols,

$$R'_e = \frac{\Delta V_{BE}}{\Delta I_E} \quad \text{for large changes} \tag{10-13a}$$

Since V_{BE} and I_E are involved, R'_e is similar to r'_e, the difference being that R'_e is for large-signal operation while r'_e is for small-signal operation. As usual,

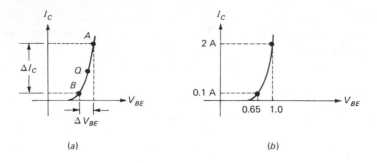

(a)   (b)

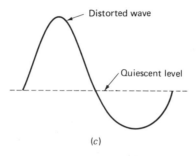

Distorted wave

Quiescent level

(c)

Figure 10-11. Bipolar transconductance curve. (a) Transconductance curve. (b) Example. (c) Nonlinear distortion.

we take advantage of the approximate equality between I_C and I_E. Because of this equality, we can rewrite Eq. (10-13a) as

$$R_e' \cong \frac{\Delta V_{BE}}{\Delta I_C} \qquad (10\text{-}13b)\text{***}$$

Equation (10-13a) gives us the physical meaning of R_e'; it is a large change in base-emitter voltage divided by a large change in emitter current. Because of this, R_e' is like an *average* ac emitter resistance that we can use for large-signal operation. Equation (10-13b), on the other hand, gives us a practical way to calculate the approximate value of R_e' using information on the data sheet of a power transistor.

As an example, Fig. 10-11b shows part of the transconductance curve of a 2N3789. Suppose we want the value of R_e' between an I_C of 0.1 and 2 A. Then, we read a change in V_{BE} of 0.35 V. And,

$$R_e' \cong \frac{\Delta V_{BE}}{\Delta I_C} = \frac{0.35}{2 - 0.1} = 0.184 \ \Omega$$

large-signal formulas

The derivations of large-signal formulas are similar to those given in Chap. 9 for small-signal amplifiers. The difference is that R_e' is used instead of r_e', and

β_{dc} instead of β. In other words, when we want a large-signal formula for gain or input impedance, all we do is replace r'_e by R'_e and β by β_{dc} in the ideal small-signal formula.

As an example, the base-driven circuit has a voltage gain from base to collector of

$$A = \frac{r_C}{r_E + r'_e}$$

for an ideal small-signal amplifier, and

$$A = \frac{r_C}{r_E + R'_e}$$

for the same amplifier used with large signals. Also, the base-driven amplifier has an input impedance of

$$z_{in(base)} = \beta(r_E + r'_e)$$

in the small-signal case, and

$$z_{in(base)} = \beta_{dc}(r_E + R'_e)$$

for the large-signal case. Finally, since power gain is the product of current gain and voltage gain,

$$G = \beta \frac{r_C}{r_E + r'_e}$$

for the small-signal case, and

$$G = \beta_{dc} \frac{r_C}{r_E + R'_e} \qquad (10\text{-}14)$$

for large signals.

This will be our approach. We will derive formulas for the *ideal small-signal* case. When we need large-signal formulas, we will replace r'_e by R'_e, and β by β_{dc}. The resulting large-signal formulas are more accurate than ideal formulas for two reasons. First, the value of R'_e is based on information from the data sheet. Second, R'_e includes the effects of base-spreading resistance r'_b because V_{BE} is the voltage from the base lead to the emitter lead.

nonlinear distortion

The transconductance curve is *nonlinear* (see Fig. 10-11a). Because of this, a large sinusoidal base voltage produces a *nonsinusoidal* collector current (see Fig. 10-11c). When this nonsinusoidal current flows through a load resistor, we get a nonsinusoidal output voltage. In other words, when a perfect sine wave goes into a large-signal amplifier, an imperfect sine wave comes out. The change in the shape of the signal is called *nonlinear distortion*. The larger the signal, the greater the distortion. (Chaps. 22 and 23 analyze nonlinear distortion.)

points A *and* B

In calculating the value of R_e' from the transconductance curve, you need to use end points A and B in Fig. 10-12b. These you get from the ac load line of the amplifier you are analyzing. Presumably, the amplifier is a power amplifier where everything is optimized for maximum power output. Because of this, the usual case is a power amplifier whose Q point is centered and whose signal swing uses most of the ac load line.

In Fig. 10-12a, we have shown the A and B points just short of saturation and cutoff. The reason is that voltage gain drops off rapidly as the instantaneous operating point approaches saturation. Also, the transconductance curve on a data sheet often does not show I_C all the way down to cutoff. For these reasons, we use A and B points that are 10 percent in from saturation and cutoff. That is, we will base our large-signal calculations on a maximum collector current of $1.9I_{CQ}$ and a minimum collector current of $0.1I_{CQ}$. If a power amplifier has an I_{CQ} of 1 A, we use a maximum of 1.9 A and a minimum of 0.1 A when we calculate R_e' in Fig. 10-12b.

The value of β_{dc} or h_{FE} to use in the large-signal formulas is the value you

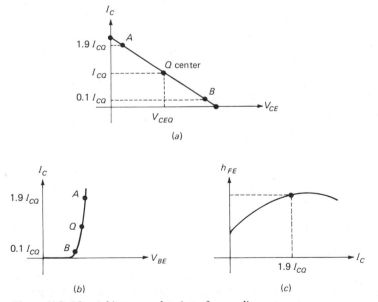

Figure 10-12. *Arbitrary end points for reading curves.*

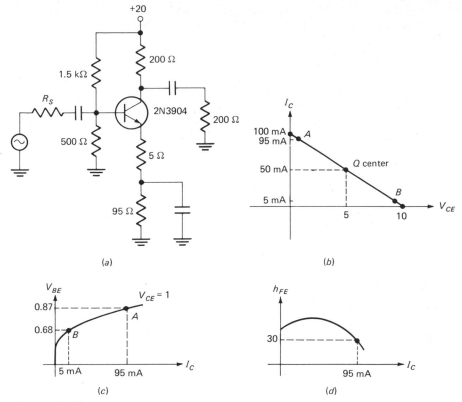

Figure 10-13.

read at $1.9I_{CQ}$. Specifically, Fig. 10-12c is similar to the graph given on a typical data sheet. After you locate $1.9I_{CQ}$, the corresponding h_{FE} is the value of β_{dc} to use in all large-signal formulas.

EXAMPLE 10-6
Calculate the large-signal voltage and power gain from base to collector in Fig. 10-13a. Also, work out the value of $z_{in(base)}$.

SOLUTION
By ideal methods of dc analysis, we can calculate $I_{CQ} \cong 50$ mA and $V_{CEQ} \cong 5$ V. Furthermore, with Eq. (10-9), we can prove the Q point is approximately centered on the load line. For this reason, we can draw the load line shown in Fig. 10-13b.

Ten percent of I_{CQ} is 5 mA. Therefore, points A and B correspond to collector currents of 95 and 5 mA. When you look at the data sheet of a 2N3904, you will find a transconductance curve like Fig. 10-13c. The V_{BE} values corresponding to I_C's of 5 and 95 mA are 0.68 and 0.87 V. So,

$$R'_e = \frac{0.87 - 0.68}{0.095 - 0.005} \cong 2.1 \ \Omega$$

The data sheet of a 2N3904 also shows h_{FE} has a value of 30 when I_C is 95 mA (Fig. 10-13d). This is the value of β_{dc} to use in all large-signal formulas.

Now, we are ready to make the calculations. The large-signal voltage gain equals

$$A \cong \frac{r_C}{r_E + R'_e} = \frac{100}{5 + 2.1} \cong 14$$

The large-signal power gain is

$$G \cong \beta_{dc} \frac{v_c}{v_b} \cong 30(14) = 420$$

And the large-signal input impedance equals

$$z_{\text{in(base)}} \cong \beta_{dc}(r_E + R'_e) \cong 30(5 + 2.1) = 213 \ \Omega$$

EXAMPLE 10-7
Calculate the power gain and input impedance of the base in Fig. 10-14a.

SOLUTION
By methods used several times before, you can calculate $I_{CQ} \cong 0.5$ A and $V_{CEQ} \cong 4$ V. Also, you will find the Q point is centered on the ac load line as shown in Fig. 10-14b.

When you examine the data sheet for an MPS-U01, you see a transconductance graph like Fig. 10-14c and a dc current-gain graph like Fig. 10-14d. With Fig. 10-14c, we get

$$R'_e = \frac{0.9 - 0.6}{0.95 - 0.05} \cong 0.33 \ \Omega$$

and with Fig. 10-14d we get $\beta_{dc} = 50$.

The large-signal power gain of the emitter follower is

$$G \cong \beta_{dc} \frac{v_c}{v_b} \cong \beta_{dc} \frac{r_E}{r_E + R'_e}$$

$$= 50 \frac{8}{8 + 0.33} = 48$$

The second factor in this product is the voltage gain, which you can see is almost unity, typical of an emitter follower.

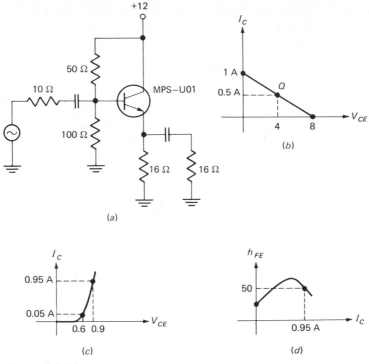

Figure 10-14.

The input impedance of the base equals

$$z_{in(base)} \cong \beta_{dc}(r_E + R'_e) = 50(8 + 0.33)$$
$$= 416 \ \Omega$$

10-7 *transistor power rating*

The temperature at the collector junction places a limit on the allowable power dissipation P_D. Depending on the transistor type, a junction temperature in the range of 150°C to 200°C will destroy the transistor. Data sheets specify this maximum junction temperature as $T_{J(max)}$. For instance, the data sheet of a 2N3904 gives a $T_{J(max)}$ of 150°C; the data sheet of a 2N3719 specifies a $T_{J(max)}$ of 200°C.

ambient temperature

The heat produced at the junction passes through the transistor *case* (metal or plastic housing) and radiates to the surrounding air. The temperature of this air, known as the *ambient* temperature, is around 25°C, but it can be much higher on hot days. Also, the ambient temperature may be much higher inside a piece of electronics equipment.

thermal resistance

The *thermal resistance* θ_{JA} is a physical property of a transistor and its case; θ_{JA} is the resistance to heat flow from the junction to the surrounding air. A low thermal resistance means it's easy for heat to flow from the junction to the surrounding air. In general, the larger the transistor case, the lower the thermal resistance. For instance, the thermal resistance of a 2N3904 is

$$\theta_{JA} = 357°\text{C/W}$$

The 2N3719 has a larger case; its thermal resistance is

$$\theta_{JA} = 175°\text{C/W}$$

This means the 2N3719 radiates heat more easily than the 2N3904.

junction temperature

A useful equation for junction temperature is

$$T_J = T_A + \theta_{JA}P_D \tag{10-15}$$

As an example, here's how to calculate the junction temperature of a 2N3904 with an ambient temperature of 25°C and a power dissipation of 0.2 W. With Eq. (10-15),

$$T_J = 25°\text{C} + (357°\text{C/W})(0.2\text{ W})$$
$$= 96.4°\text{C}$$

As another example, the junction temperature of a 2N3719 for the same ambient temperature and power dissipation is

$$T_J = 25°\text{C} + (175°\text{C/W})(0.2\text{ W})$$
$$= 60°\text{C}$$

As you can see, the lower thermal resistance of the 2N3719 results in a lower junction temperature.

derating factor

Data sheets often specify the $P_{D(\text{max})}$ of a transistor at an ambient temperature of 25°C. For instance, the 2N3904 has a $P_{D(\text{max})}$ of 350 mW for a T_A of 25°C. This means a 2N3904 used in a class A amplifier can have a quiescent power dissipation as high as 350 mW. As long as the ambient temperature is 25°C, the transistor is within its specified power rating.

What do you do if the ambient temperature is greater than 25°C? You *derate* or reduce the power rating. Data sheets specify a *derating factor D* for each degree above 25°C. The derating factor for a 2N3904 is 2.8 mW/°C. This means you reduce the power rating 2.8 mW for each degree above 25°C. As an equation,

$$P_{D(\text{max})} = P_{25} - D(T_A - 25°C) \tag{10-16}$$

where P_{25} = power rating at 25°C
D = derating factor
As an example, the 2N3904 has a P_{25} of 350 mW and a D of 2.8 mW/°C. The maximum power dissipation at an ambient temperature of 100°C is

$$P_{D(\text{max})} = 350 \text{ mW} - (2.8 \text{ mW}/°C)(100°C - 25°C)$$
$$= 140 \text{ mW}$$

Therefore, a 2N3904 can dissipate no more than 140 mW when the ambient temperature is 100°C.

heat sinks

The lower the thermal resistance, the greater the allowable power dissipation. We can reduce θ_{JA} by attaching a *heat sink* (a mass of metal) to the transistor case; the increased surface area allows heat to escape more easily and increases the allowable power dissipation. In fact, power transistors are usually mounted in thermal contact with the chassis. This way, the entire chassis becomes a heat sink. The effect is to dramatically lower the *case temperature* of the transistor.

The thermal resistance between the case and the surrounding air is designated θ_{CA}. If θ_{CA} is known, you can calculate the case temperature of a transistor with

$$T_C = T_A + \theta_{CA} P_D \tag{10-17}$$

Data sheets for power transistors usually specify the power rating for a case temperature of 25°C. Also given is a derating factor for case temperatures above 25°C. The derating equation is similar to Eq. (10-16), except we use case temperature as follows:

$$P_{D(\text{max})} = P_{25} - D(T_C - 25°C) \tag{10-18}$$

EXAMPLE 10-8
A power transistor dissipates 4 W. If $T_A = 25°C$ and $\theta_{CA} = 12°C/W$, what is the case temperature?

SOLUTION
With Eq. (10-17),

$$T_C = 25°C + (12°C/W)(4\ W) = 73°C$$

EXAMPLE 10-9
The transistor of Example 10-8 has a power rating of 10 W for a case temperature of 25°C. The derating factor is 50 mW/°C. What is the power rating of the transistor for a case temperature of 73°C?

SOLUTION
With Eq. (10-18),

$$P_{D(max)} = 10\ W - (0.05\ W/°C)(73°C - 25°C) = 7.6\ W$$

As you see, the power rating (7.6 W) is greater than the actual power dissipation (4 W in the preceding example).

self-testing review

Read each of the following and provide the missing words. Answers appear at the beginning of the next question.

1. The quiescent collector current and voltage are the I_C and V_{CE} when there is no input _____. You can determine quiescent currents and voltages from the _____ equivalent circuit. V_{CEQ} represents the collector-to-emitter voltage with _____ ac signal.

2. *(signal, dc, no)* The dc load line includes all dc collector currents and voltages that satisfy Kirchhoff's voltage law applied to the collector-emitter loop. The _____ point can be anywhere on the dc load line.

3. *(quiescent)* The horizontal intercept of the dc load line is the same as the ideal cutoff point, and the vertical intercept is identical to the _____ saturation point.

4. *(ideal)* Every amplifier stage sees two loads: a dc load and an _____ load. The ac load line includes all collector currents and voltages that satisfy Kirchhoff's voltage law applied to the ac equivalent circuit. For a large enough input signal, the instantaneous operating point moves all the way to ac _____ and all the way to ac cutoff.

5. *(ac, saturation)* Because the ac load resistance seen by the collector and emitter is usually different from the dc load resistance, the ac load line is usually different from the _____ load line.

6. *(dc)* If you overdrive an amplifier, the output signal will be _____ on either or both peaks. To get the maximum unclipped signal, we can locate the Q point at the _____ of the ac load line.

7. *(clipped, center)* In a class A amplifier with a centered Q point, the _____ power dissipation of the transistor represents the worst-case condition. That is, the transistor is safe from damage if it can dissipate P_{DQ}. The maximum efficiency for a transformerless class A amplifier is _____.

8. *(quiescent, 25 percent)* Resistance r_e' is not used when analyzing power amplifiers. Instead, we use the large-signal emitter resistance, designated _____. Also, we use _____ instead of β.

9. *(R_e', β_{dc})* The _____ resistance θ_{JA} is the resistance of heat flow from the junction to the surrounding air. A low thermal resistance means it's easy for _____ to flow from the junction to the surrounding air.

10. *(thermal, heat)* Some data sheets specify the $P_{D(max)}$ of a transistor at an ambient temperature of 25°C. If the ambient temperature is greater than 25°C, you _____ or reduce the power rating.

11. *(derate)* We can reduce θ_{JA} by attaching a heat sink to the transistor case. In fact, power transistors are usually mounted in thermal contact with the chassis. This way, the entire chassis becomes a heat sink.

problems

10-1. Plot the Q point for the amplifier of Fig. 10-15a.

10-2. Locate the I_{CQ} and V_{CEQ} for Fig. 10-15b.

10-3. What are the values of I_{CQ} and V_{CEQ} in Fig. 10-15c?

10-4. Plot the Q point for the amplifier shown in Fig. 10-15d.

10-5. What values do I_{CQ} and V_{CEQ} have in Fig. 10-15e?

10-6. Draw the dc load line for the amplifier of Fig. 10-15a.

10-7. What is the ideal dc saturation and dc cutoff point for Fig. 10-15c?

10-8. Draw the dc load line for Fig. 10-15d.

10-9. Construct the dc load line of Fig. 10-15e.

10-10. The dc cutoff voltage of an emitter-biased circuit is $V_{CC} + V_{EE}$, and the dc saturation current is $(V_{CC} + V_{EE})/(R_C + R_E)$. Draw the dc load line for the circuit of Fig. 10-15b.

10-11. The primary winding of Fig. 10-15f has a dc resistance of 100 Ω. Draw the dc load line.

10-12. Draw the ac load line for the circuit of Fig. 10-15a.

10-13. In Fig. 10-15c, what is the value of ac cutoff voltage? Of ac saturation current?

10-14. In Fig. 10-15d, what is the minimum acceptable BV_{CEO} for the transistor? The minimum acceptable $I_{C(max)}$ rating?

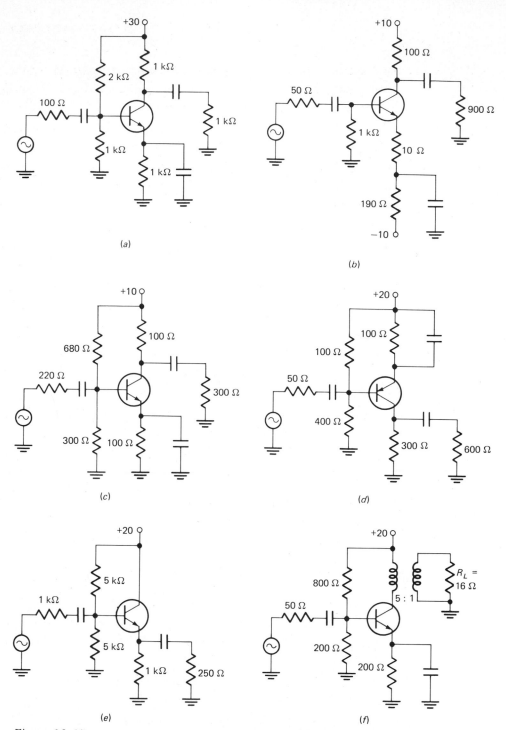

Figure 10-15.

10-15. Draw the ac load line for the emitter follower shown in Fig. 10-15e.

10-16. The transformer of Fig. 10-15f reflects an impedance of $n^2 R_L$ into the collector circuit. Draw the ac load line for the amplifier.

10-17. An amplifier has an I_{CQ} of 2 A. If $r_C + r_E = 10$ Ω, what is the voltage change when the instantaneous operating point swings from the Q point to the ac cutoff point?

10-18. Use Eq. (10-9) to check whether or not the Q point of Fig. 10-15a is centered.

10-19. Is the Q point of Fig. 10-15b centered?

10-20. Does the amplifier of Fig. 10-15c have a centered Q point?

10-21. If we change the 600-Ω resistor in Fig. 10-15d to another value, we can get a centered Q point. What value do we need?

10-22. By changing the turns ratio in Fig. 10-15f, we can get a centered Q point. What turns ratio do we need?
For all remaining problems the Q point is centered and the signal is large enough to use the entire ac load line. Neglect V_{BE}.

10-23. How much ac output power is there in Fig. 10-16a?

10-24. What is the value of ac output power in the emitter follower of Fig. 10-16b? And the value of P_{DQ}?

10-25. Calculate the ac output power for the swamped amplifier of Fig. 10-16c. What is the value of P_{DQ}?

10-26. Work out the ac output power for the phase splitter of Fig. 10-16d. What is lowest acceptable $P_{D(\max)}$ rating?

10-27. The ideal transformer of Fig. 10-16e steps up the load impedance by n^2. What is the ac power going into the primary winding? How much ac power does the 3-Ω resistor receive?

10-28. A transistor has a $P_{D(\max)}$ rating of 87 W at 25°C. If used in a class A amplifier, what is the maximum ac output power?

10-29. What $P_{D(\max)}$ rating does the transistor of Fig. 10-16a require?

10-30. What is the lowest acceptable $P_{D(\max)}$ rating for the transistor in Fig. 10-16c?

10-31. The voltage across a 16-Ω loudspeaker is 8 V. How much power does this load receive? If the amplifier driving the load is class A, what is the minimum acceptable $P_{D(\max)}$ rating for the transistor?

10-32. What is the efficiency in Fig. 10-16a?

10-33. Calculate the efficiency in Fig. 10-16b.

10-34. Work out the efficiency for Figs. 10-16c and d.

10-35. What is the efficiency for Fig. 10-16e.

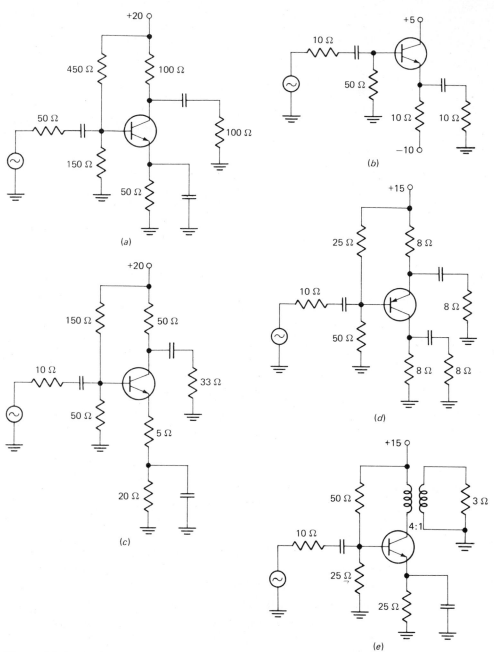

Figure 10-16.

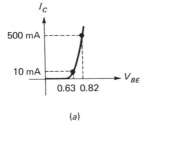

(a)

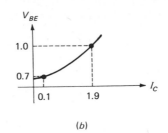

(b)

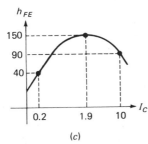

(c)

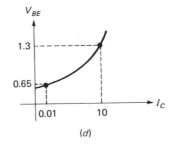

(d)

Figure 10-17.

10-36. Calculate the value of R'_e given the curve of Fig. 10-17a and the indicated points.

10-37. What is the value of R'_e between the pair of points given in Fig. 10-17b?

10-38. If $1.9I_{CQ} = 10$ A, what is β_{dc} in Fig. 10-17c?

10-39. Over the indicated range shown in Fig. 10-17d, what is the value of R'_e?
   Repeat: for all remaining problems the Q point is centered and the signal is large enough to use the entire ac load line.

10-40. The R'_e of the transistor in Fig. 10-16a is 1 Ω. Work out the large-signal voltage and power gain from base to collector using a β_{dc} of 50.

10-41. $R'_e = 2.5$ Ω and $\beta_{dc} = 50$ in Fig. 10-16a. Work out the approximate large-signal voltage and power gain from base to collector. What is the large-signal input impedance looking into the base?

10-42. $R'_e = 0.2$ Ω and $\beta_{dc} = 100$ for the transistor of Fig. 10-16d. What is the large-signal voltage gain from base to collector? From base to emitter? And what is the large-signal input impedance looking into the base?

10-43. The transistor of Fig. 10-16c has an $R'_e = 1$ Ω and a $\beta_{dc} = 75$. Calculate the large-signal voltage and power gain from base to collector, and the input impedance looking into the base. Also, what is the input impedance when the base-biasing resistors are included?

10-44. The amplifier of Fig. 10-16b has the transconductance curve given by Fig. 10-17b and the h_{FE} curve of Fig. 10-17c. Calculate the large-signal voltage and power gain from base to emitter. What is the large-signal input impedance of the base $z_{\text{in(base)}}$?

10-45. A transistor has $\theta_{JA} = 300°\text{C/W}$. If $P_D = 225$ mW and $T_A = 25°\text{C}$, what is the junction temperature?

10-46. The data sheet of a 2N4123 specifies $\theta_{JA} = 357°\text{C/W}$. What is the junction temperature if $T_A = 50°\text{C}$ and $P_D = 150$ mW?

10-47. A 2N4416 has a power rating of 300 mW for an ambient temperature of 25°C. The derating factor is 1.71 mW/°C. What is $P_{D\,\text{(max)}}$ for an ambient temperature of 90°C?

10-48. The data sheet for a 2N3055 lists a power rating of 115 W for a case temperature of 25°C. If the derating factor is 0.657 W/°C, what is $P_{D\,\text{(max)}}$ for a case temperature of 125°C?

10-49. Another useful formula for junction temperature is $T_J = T_C + \theta_{JC}P_D$. The θ_{JC} of a 2N3055 is 1.52°C/W. If $T_C = 25°\text{C}$ and $P_D = 75$ W, what does T_J equal?

10-50. A 2N497 is used in Fig. 10-16c. This transistor has a power rating of 4 W at a case temperature of 25°C. The derating factor is 22.8 mW/°C. What is the highest allowable case temperature?

10-51. Figure 10-18a shows the outline of a 2N4231, a power transistor where the collector is connected to the case to minimize thermal resistance between the collector junction and the case. The case can then be placed in thermal contact with the chassis for heat sinking. (A thin mica washer between the case and the chassis prevents a collector-to-chassis short.)

    The 2N4231 has a power rating of 35 W at a case temperature of 25°C, and the derating factor is 0.2 W/°C. What is the power rating for a case temperature of 125°C?

Collector
connected to case

Pin 1. Base
   2. Emitter
Case–collector

(a)

Metal tab

E
B
  C

(b)

Figure   10-18. (a)
Power transistor. (b)
Tab transistor.

10-52. Figure 10-18b is the outline of a D42C, a power-tab transistor. The metal tab can be fastened to the chassis for heat sinking. The D42C has a power rating of 12.5 W for a tab temperature T_T of 25°C; the derating factor is 0.1 W/°C. Given $P_{D(\text{max})} = P_{25} - D(T_T - 25°C)$, calculate the power rating for a tab temperature of 70°C.

11 *class b push-pull amplifiers*

Class A is the common way to run a transistor in linear circuits because it leads to the simplest and most stable biasing circuits. But it requires a $P_{D(\text{max})}$ rating of twice the load power; it also has a quiescent or no-signal current drain of 50 percent $I_{C(\text{sat})}$ when the Q point is centered. In the earlier stages of a system, the $P_{D(\text{max})}$ rating and no-signal current drain are usually small enough to accept. But near the end of many systems, the $P_{D(\text{max})}$ rating and the no-signal current drain become so large we can no longer use class A amplifiers.

The *class B push-pull amplifier* is a two-transistor circuit with these outstanding advantages: the $P_{D(\text{max})}$ rating drops to one-fifth of the load power and the no-signal current drain to around 1 percent of $I_{C(\text{sat})}$. The first advantage is important when large amounts of load power are needed, such as in communication transmitters. The second advantage is desirable in battery-powered systems like transistor radios.

For load power up to approximately 10 W, you can usually find a satisfactory integrated-circuit class B amplifier. Above 10 W or so, the discrete class B circuit may be used.

11-1 *the basic idea of push-pull action*

Before we discuss the idea of push-pull action, we will define other classes of operation and show the ac load line for class B circuits.

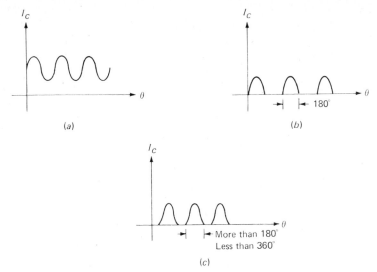

Figure 11-1. Collector-current waveforms. (a) *Class A.* (b) *Class B.* (c) *Class AB.*

class a, b, and ab

We already know the transistor of a class A amplifier remains in the active region throughout the cycle; this means the collector current of a class A amplifier flows for 360° as shown in Fig. 11-1*a*.

In a class B circuit, the transistor stays in the active region only for *half the cycle.* During the other half cycle, the transistor is cut off. This means collector current flows for 180° in each transistor of a class B circuit (Fig. 11-1*b*).

Class AB is between class A and class B. The transistor of a class AB circuit is in the active region for more than half a cycle but less than the whole cycle. In other words, collector current flows for more than 180° but less than 360° (Fig. 11-1*c*).

the ac load line for class b

Figure 11-2 shows the ac load line for one transistor in a class B push-pull circuit. Neglecting I_{CBO}, the Q point is at cutoff and has coordinates of

$$I_{CQ} = 0$$

and
$$V_{CEQ} = V_{CE(cutoff)}$$

When the ac signal comes in, the instantaneous operating point swings from Q to saturation as shown. This produces half cycles of current and voltage. By letting one transistor handle the positive half cycle and another transistor the

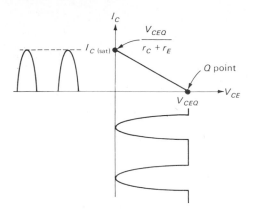

Figure 11-2. Class B load line and waveforms.

negative half cycle, we can get a final output signal that is a complete sine wave.

Especially important, we need a formula for the saturation current in a class B amplifier. Figure 10-4 showed the ac load line for any base-driven amplifier. As indicated by this figure, the saturation current is

$$I_{C(\text{sat})} = I_{CQ} + \frac{V_{CEQ}}{r_C + r_E}$$

In Fig. 11-2, I_{CQ} is zero; therefore, the saturation current in a class B circuit is

$$I_{C(\text{sat})} = \frac{V_{CEQ}}{r_C + r_E} \qquad \text{(class B)} \qquad (11\text{-}1)\text{***}$$

This is an important result because it is used in formulas for power dissipation, load power, and no-signal current drain.

an example of push-pull operation

Before discussing the complete circuit for a class B push-pull amplifier, let us get the idea of push-pull action. Figure 11-3*a* shows an *ac equivalent circuit* for an emitter follower. The dc biasing is not shown, but we will assume biasing near cutoff, that is, class B operation with an ac load line of Fig. 11-2.

During the positive half cycle of source voltage, the emitter diode is turned on, and the operating point swings from Q to saturation. During the negative half cycle of source voltage, the emitter diode is reverse-biased, and no current flows. This is why the voltage across r_E in Fig. 11-3*a* is a half-wave signal.

Next, look at the *pnp* emitter follower of Fig. 11-3*b*. Only the ac equivalent circuit is shown, and we will again assume the emitter diode is biased near cutoff. During the positive half cycle of ac source voltage, the emitter diode is

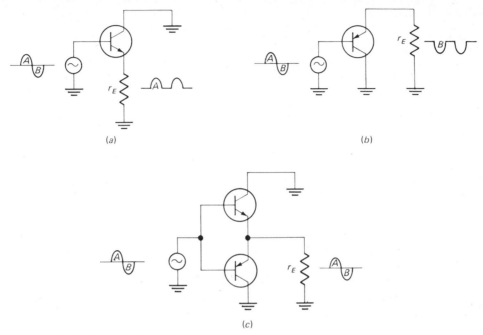

Figure 11-3. Ac equivalent circuits.

reverse-biased, and no collector current flows. But on the negative half cycle of source voltage, the emitter diode is forward-biased. Therefore, the operating point swings from Q to saturation as shown in Fig. 11-2. Because the current flows up through r_E, the voltage across r_E is negative with respect to ground. This is why the output voltage in Fig. 11-3b has only negative half cycles.

To get a push-pull circuit, we combine the two emitter followers as shown in Fig. 11-3c. The upper transistor *(npn)* takes care of the positive half cycle of source voltage; the lower transistor *(pnp)* handles the negative half cycle. In this way, the output voltage is a complete sine wave.

Other push-pull arrangements are possible. Instead of two emitter followers, we can combine two CE amplifiers; or we can use transformers in place of complementary transistors. We will discuss these alternatives later. For now, the main idea is this. In any class B push-pull amplifier, one transistor conducts during the positive half cycle of source voltage and the other during the negative half.

In Fig. 11-3c, the circuit action is complementary for each half cycle; that is, the currents and voltages of one half cycle are equal and opposite those of the other half cycle. This simplifies the ac analysis of class B push-pull circuits; it means we only have to analyze the action of one half cycle. For instance, to

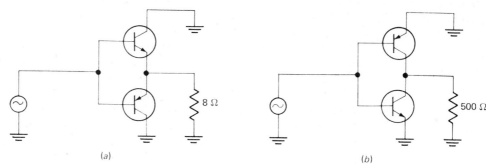

Figure 11-4.

work out the voltage gain for Fig. 11-3c, we can analyze Fig. 11-3a or b, whichever is more convenient.

EXAMPLE 11-1
Figure 11-4a shows the ac equivalent circuit for a class B push-pull emitter follower. In the dc equivalent circuit (not shown) each transistor has a $V_{CEQ} = 10$ V. Calculate the saturation current.

SOLUTION
In Fig. 11-4a, $r_C = 0$ and $r_E = 8$ Ω. With Eq. (11-1),

$$I_{C(sat)} = \frac{V_{CEQ}}{r_C + r_E} = \frac{10}{8} = 1.25 \text{ A}$$

EXAMPLE 11-2
Figure 11-4b shows the ac equivalent circuit for a class B push-pull CE amplifier. In the dc equivalent circuit (not shown) each transistor has a $V_{CEQ} = 50$ V. What is the saturation current?

SOLUTION
The lower transistor amplifies the positive half cycle of source voltage; the upper one takes care of the negative half cycle. As you can see, $r_C = 500$ Ω and $r_E = 0$. With Eq. (11-1),

$$I_{C(sat)} = \frac{V_{CEQ}}{r_C + r_E} = \frac{50}{500} = 0.1 \text{ A}$$

11-2 *distortion*

One critical thing about a class B amplifier is the Q point. In the ideal class B circuit, the Q point is at cutoff. But in a practical class B amplifier, the Q point is slightly above cutoff. This section tells you why.

crossover distortion

Figure 11-5*a* shows a class B push-pull amplifier. Suppose no bias at all is applied to the emitter diodes. Then, the incoming ac signal has to rise to about 0.7 V to overcome the barrier potential. Because of this, essentially no current flows through Q_1 when the signal is less than 0.7 V. The action on the other half cycle is complementary; the other transistor does not turn on until the negative half cycle of source voltage is more negative than approximately −0.7 V. For this reason, if no bias at all is applied to the emitter diodes, the output of a class B push-pull amplifier looks like Fig. 11-5*b*.

The signal of Fig. 11-5*b* is distorted; it no longer is a sine wave because of the clipping action between half cycles. Since this clipping occurs between the time one transistor shuts off and the other comes on, we call it *crossover distortion.*

To eliminate crossover distortion, we need to apply a slight forward bias to each emitter diode. This means locating the Q point slightly above cutoff as shown in Fig. 11-5*c*. As a guide, an I_{CQ} from 1 to 5 percent of $I_{C\text{(sat)}}$ is enough to eliminate crossover distortion, the exact value usually determined by experiment with the particular circuit. We will refer to this slight forward bias on each emitter diode as *trickle bias.*

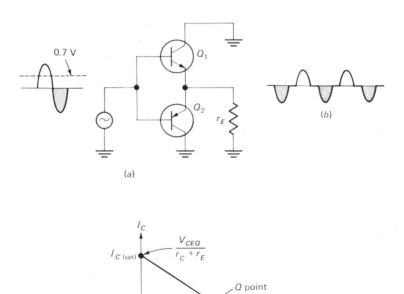

Figure 11-5. (a) *Ac equivalent circuit of emitter follower.* (b) *Crossover distortion.* (c) *Ac load line with trickle bias.*

Strictly speaking, we have class AB operation because each transistor is FR-biased for slightly more than half a cycle.

nonlinear distortion

A large-signal class A amplifier suffers from nonlinear distortion. As you recall, the nonlinear transconductance curve produces an output signal that is nonsinusoidal, elongated on one half cycle and squashed on the other. In a class B push-pull amplifier, however, both halves of the signal are identical in shape. Although some nonlinear distortion still occurs, it is less than with large-signal class A.

The reason for less nonlinear distortion with class B push-pull is that all *even harmonics* cancel out. Harmonics are multiples of the input frequency. For instance, if $f_{in} = 1$ kHz, the second harmonic is 2 kHz, the third harmonic is 3 kHz, and nth harmonic is nf_{in}. A distortionless amplifier produces no harmonics; only the original input frequency appears at the output. A large-signal class A amplifier produces all harmonics: f_{in}, $2f_{in}$, $3f_{in}$, $4f_{in}$, $5f_{in}$, and so on. A class B push-pull amplifier produces only the odd harmonics: f_{in}, $3f_{in}$, $5f_{in}$, and so on. Because of this, the overall distortion is less with class B push-pull. (Chapter 21 gets into a mathematical analysis of harmonics.)

11-3 *the current mirror*

In Fig. 11-6a, I_1 flows down through R_1 and splits into I_2 and I_B. I_2 flows through a *compensating* diode, and I_B flows into the base. Recall that the transconductance curve of a transistor is a graph of I_C versus V_{BE}. If the current-voltage curve

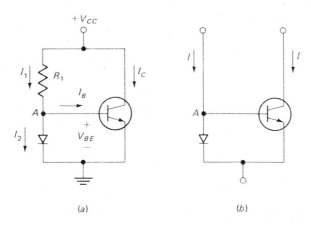

(a)

(b)

Figure 11-6. Current mirror.

of the compensating diode is identical to the transconductance curve, then the diode current equals the collector current. In symbols,

$$I_2 = I_C$$

(This is true because V_{BE} is across the compensating diode as well as the emitter diode.) By applying Kirchhoff's current law to node A, we get

$$I_1 = I_2 + I_B$$

or

$$I_1 = I_C + I_B$$

Since base current is much smaller than collector current, this becomes

$$I_1 \cong I_C$$

which is the same as

$$I_C \cong I_1 \qquad (11\text{-}2)\text{***}$$

This is important. It says collector current equals the current entering node A.

Figure 11-6b emphasizes this point. Current I into node A sets up an equal collector current I. Think of the circuit as a *mirror:* the current into node A is reflected into the collector circuit where an equal current appears. This is why the diode-transistor circuit of Fig. 11-6b is known as a *current mirror*. And the collector current is often called the *mirror current*.

Remember the current mirror. It is widely used in linear integrated circuits. In this chapter, we use it to set up the Q point of a class B push-pull amplifier.

EXAMPLE 11-3
What is the mirror current in Fig. 11-7a?

SOLUTION
Allow 0.7 V for the V_{BE} drop. Then, the current through the 10-kΩ resistor is

$$I = \frac{V_{CC} - V_{BE}}{R} = \frac{15 - 0.7}{10,000} = 1.43 \text{ mA}$$

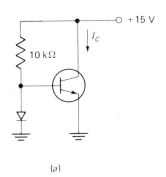
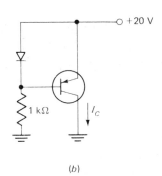

(a)   (b)

Figure 11-7. (a) npn *mirror.* (b) pnp *mirror.*

Therefore, the mirror current is

$$I_C \cong 1.43 \text{ mA}$$

EXAMPLE 11-4
Find the mirror current in Fig. 11-7*b*.

SOLUTION
The *pnp* mirror works the same as an *npn* mirror. The current through the 1-kΩ resistor is

$$I = \frac{20 - 0.7}{1000} = 19.3 \text{ mA}$$

So, the mirror current is

$$I_C \cong 19.3 \text{ mA}$$

11-4 *setting up the* q *point*

The most difficult thing about a class B amplifier is setting up a stable Q point. It is much more difficult than in a class A circuit, because V_{BE} becomes a crucial quantity in a class B circuit.

voltage-divider bias

Because the class B push-pull *emitter follower* is one of the most important class B circuits, we will concentrate on it and extend the results to other class B circuits.

Figure 11-8*a* shows *voltage-divider bias* for a class B push-pull emitter follower. The two transistors need to be *fully complementary,* meaning similar transconductance curves, maximum ratings, etc. For instance, the 2N3904 and 2N3906 are complementary, the first being an *npn* transistor and the second a *pnp;* both have almost the same maximum ratings, transconductance curves, and so on. Complementary pairs like these are commercially available for almost any application.

In Fig. 11-8*a,* the collector and emitter currents flow down through the *npn* and *pnp* transistors. Because of the series connection, the I_Cs of the transistors are equal and the V_{CE} of each transistor is half the supply voltage.

To set up trickle bias, we need to barely turn on each emitter diode. This means roughly 0.7 V across each R_2. The correct value of V_{BE} may be slightly lower or higher, depending on the transistor type, the amount of trickle current, etc. Unfortunately, the transconductance curve is so steep in the vicinity of

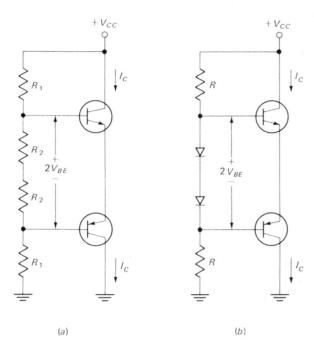

Figure 11-8. (a)
Voltage-divider bias.
(b) *Diode bias.*

(a)   (b)

turn-on that even 0.1 V can produce a huge change in collector current. Since published transconductance curves have tolerance, the correct bias voltage for a particular transistor has to be found by experiment, using an adjustable R_1 or R_2.

Voltage-divider bias like Fig. 11-8a has been used, but it's one of the worst ways to bias a class B push-pull amplifier. Any practical circuit has to contend with large changes in junction temperature. As this temperature increases, the barrier potential across the emitter diodes decreases about 2.5 mV per degree. Therefore, the fixed voltage produced by R_2 (Fig. 11-8a) will overbias the emitter diode and produce large increases in collector current when the junction temperature rises.

diode bias

Figure 11-8b shows *diode bias,* the main method used to bias class B push-pull amplifiers. The transistors are fully complementary, and the curves of compensating diodes match the transconductance curves of the transistors. Therefore, the upper half of the circuit is an *npn* mirror and the lower half is a *pnp* mirror. This combination is called a *complementary mirror.* Note that the same collector current flows through both transistors. This collector current is a reflection of the current through the biasing resistors.

Here is the reason a complementary mirror is immune to temperature changes. At 25°C, the drop across each diode is approximately 0.7 V. When the temperature increases, V_{BE} decreases 2.5 mV per degree. Since the mirror current is given by

$$I_C = \frac{V_{CC} - 2V_{BE}}{2R} \qquad (11\text{-}3)***$$

it increases only slightly as V_{BE} decreases. In fact, a high V_{CC} swamps out the changes in V_{BE}, so that I_C is virtually constant with temperature changes.

For a complementary mirror to work properly, the curves of the compensating diodes must match the transconductance curves of the transistors. In discrete circuits, this is not easy to do because of diode and transistor tolerances; so you have to settle for less than perfect mirror action. But diode bias really shines when it comes to linear integrated circuits. Since the compensating diodes and emitter diodes are on the same chip, they have almost identical characteristics; this means almost perfect mirror action over a large temperature range.

EXAMPLE 11-5
What is the trickle current in Fig. 11-8b if $V_{CC} = 20$ V and $R = 3.3$ kΩ?

SOLUTION
The current through the biasing resistors is

$$I = \frac{V_{CC} - 2V_{BE}}{2R} = \frac{20 - 1.4}{6600} = 2.82 \text{ mA}$$

Therefore, the trickle current is

$$I_C \cong 2.82 \text{ mA}$$

11-5 *a complete class b emitter follower*

Figure 11-9 shows a class B push-pull emitter follower. The diode bias sets the Q point slightly above cutoff to prevent crossover distortion. The positive half cycle of input voltage is coupled through the input capacitors, turning on the upper transistor and shutting off the lower one. The emitter voltage of the upper transistor now follows the base voltage. The output capacitor then couples the positive half cycle to the load resistance R_L.

The action is complementary on the negative half cycle. This means the upper transistor shuts off and the lower one comes on. The lower transistor now acts like an emitter follower, and the output capacitor couples the negative half cycle to the load resistance. The final signal across the load is symmetrical because the two amplified half portions are recombined.

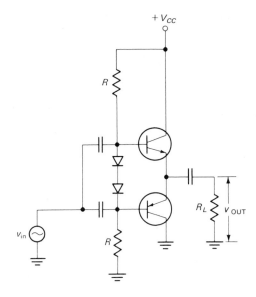

Figure 11-9. Push-pull emitter follower.

The large-signal input impedance of the active transistor is

$$z_{\text{in(base)}} = \beta_{\text{dc}}(R_L + R_e')$$ (11-4a)

Similarly, the large-signal voltage gain is

$$A = \frac{R_L}{R_L + R_e'}$$ (11-4b)

The large-signal power gain equals the product of the large-signal voltage gain and large-signal current gain:

$$G = \frac{\beta_{\text{dc}} R_L}{R_L + R_e'}$$ (11-4c)

EXAMPLE 11-6
The push-pull emitter follower of Fig. 11-10a has $\beta_{\text{dc}} = 60$ and $R_e' = 0.5 \ \Omega$. Calculate the large-signal voltage and power gain. What is the large-signal input impedance of the base?

SOLUTION
The large-signal voltage gain is

$$A = \frac{R_L}{R_L + R_e'} = \frac{8}{8 + 0.5} = 0.941$$

The large-signal power gain is

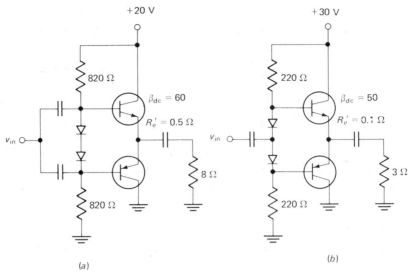

Figure 11-10.

$$G = \frac{\beta_{dc} R_L}{R_L + R_e'} = \frac{60 \times 8}{8 + 0.5} = 56.5$$

The large-signal input impedance is

$$z_{in(base)} = \beta_{dc}(R_L + R_e') = 60(8 + 0.5) = 510\ \Omega$$

EXAMPLE 11-7
As a design variation, a single-input coupling capacitor can be used, as shown in Fig. 11-10*b*. The *dynamic* resistance (same as the ac resistance) of the diodes drops very little ac signal. For this reason, almost all the input signal reaches the transistors.

Calculate $z_{in(base)}$, A, and G for $\beta_{dc} = 50$ and $R_e' = 0.1\ \Omega$.

SOLUTION

$$z_{in(base)} = \beta_{dc}(R_L + R_e') = 50(3 + 0.1) = 155\ \Omega$$

$$A = \frac{R_L}{R_L + R_e'} = \frac{3}{3 + 0.1} = 0.968$$

$$G = \frac{\beta_{dc} R_L}{R_L + R_e'} = \frac{50 \times 3}{3 + 0.1} = 48.4$$

EXAMPLE 11-8
What is the input impedance of the stage in Fig. 11-10*b*?

SOLUTION

Only one transistor is active at a time, but both biasing resistors appear in the ac equivalent circuit. Therefore,

$$z_{in} = R \parallel R \parallel z_{in(base)} = 220 \parallel 220 \parallel 155 = 64.3 \ \Omega$$

11-6 *power relations*

At the beginning of this chapter, we said a class B push-pull amplifier required a much lower $P_{D(max)}$ rating. This section discusses the $P_{D(max)}$ rating and other power relations in the class B push-pull amplifier.

output power

Figure 11-11 shows the ideal ac load line for a class B circuit. It is ideal because it neglects $V_{CE(sat)}$ and I_{CQ}. In a real amplifier, the saturation point does not quite touch the vertical axis, and the Q point is slightly above the horizontal axis.

Here is the main idea. Figure 11-11 shows the largest current and voltage waveforms we can get with one transistor of a class B push-pull circuit; the other transistor produces the dotted half cycle. For this reason, the ac output voltage has a peak value of V_{CEQ}, and the ac output current has a peak value of $I_{C(sat)}$. Therefore, the maximum ac output power is

$$P_{O(max)} = V_{RMS}I_{RMS} = \frac{V_{CEQ}}{\sqrt{2}} \ \frac{I_{C(sat)}}{\sqrt{2}}$$

or

$$P_{O(max)} = \frac{V_{CEQ}I_{C(sat)}}{2} \qquad (11\text{-}5)$$

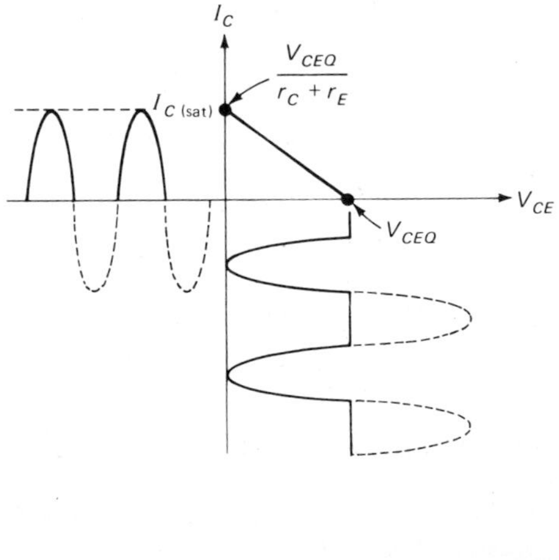

Figure 11-11. Deriving power relations for class B push-pull.

maximum efficiency

Class B push-pull amplifiers are highly efficient, one reason they are widely used as power amplifiers. In Fig. 11-9, the supply delivers the following dc power to the transistors:

$$P_{DC} = V_{CC}I_{DC} \tag{11-6a}$$

where I_{DC} is the current to the transistors averaged over one cycle. Figure 11-11 shows the current waveform. Since this is a half-wave signal,

$$I_{DC} = \frac{I_{C(\text{sat})}}{\pi}$$

Therefore, the dc power supplied to the transistors is

$$P_{DC} = \frac{V_{CC}I_{C(\text{sat})}}{\pi} \tag{11-6b}$$

In a single-supply circuit like Fig. 11-9, $V_{CEQ} = V_{CC}/2$. With Eq. (11-5), this means a single-supply class B push-pull amplifier has a maximum ac output power of

$$P_{O(\text{max})} = \frac{V_{CC}I_{C(\text{sat})}}{4}$$

When we divide this by Eq. (11-6b), we get the maximum efficiency for a class B push-pull amplifier:

$$\eta_{(\text{max})} = \frac{V_{CC}I_{C(\text{sat})}/4}{V_{CC}I_{C(\text{sat})}/\pi} = \frac{\pi}{4} = 0.785$$

which is equivalent to

$$\eta_{(\text{max})} = 78.5\% \quad \text{(class B push-pull)} \tag{11-6c}$$

(Recall that efficiency is defined as the ratio of ac output power to dc input power: $\eta = P_O/P_{DC}$.)

This is an ideal efficiency because it neglects $V_{CE(\text{sat})}$ and I_{CQ}. Nevertheless, the efficiency of a typical class B push-pull amplifier can easily be over 70 percent. The main point is this: the typical class B push-pull amplifier is far more efficient than the typical class A amplifier.

power dissipation

Under no-signal conditions, the transistors of a class B amplifier are *idling*, because only a small trickle current passes through them. In this case, the power dissipation in each transistor is negligible. But when a signal is present, large currents are in the transistors, causing significant power dissipation. The worst-case power dissipation is too complicated to derive here, but with calculus we can prove that

$$P_{D(\text{max})} = \frac{V_{CEQ}I_{C(\text{sat})}}{10} \tag{11-7}$$

With Eq. (11-5), this becomes

$$P_{D(\text{max})} = \frac{P_{O(\text{max})}}{5} \quad \text{(class B)} \tag{11-8}***$$

What does this mean? It means the power rating of each transistor in a class B push-pull amplifier must be greater than one-fifth of the maximum ac output power.

EXAMPLE 11-9

If a class B push-pull amplifier can deliver 100 W of ac output power, what is the minimum power rating of each transistor?

SOLUTION
With Eq. (11-8),

$$P_{D(\text{max})} = \frac{P_{O(\text{max})}}{5} = \frac{100\ \text{W}}{5} = 20\ \text{W}$$

The power rating of each transistor has to be at least 20 W at the highest expected junction temperature to avoid damage.

EXAMPLE 11-10

Calculate the maximum ac output power and the minimum power rating of the transistors in Fig. 11-12.

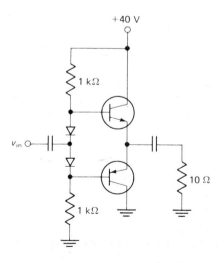

Figure 11-12.

SOLUTION

V_{CEQ} equals half the supply voltage:

$$V_{CEQ} = \frac{V_{CC}}{2} = \frac{40 \text{ V}}{2} = 20 \text{ V}$$

The ac saturation current is

$$I_{C(sat)} = \frac{V_{CEQ}}{r_C + r_E} = \frac{20 \text{ V}}{10} = 2 \text{ A}$$

The maximum ac output power is

$$P_{O(max)} = \frac{V_{CEQ} I_{C(sat)}}{2} = \frac{20 \times 2}{2} = 20 \text{ W}$$

The minimum power rating is

$$P_{D(max)} = \frac{P_{O(max)}}{5} = \frac{20 \text{ W}}{5} = 4 \text{ W}$$

EXAMPLE 11-11

The output power in Example 11-10 is ideal because it neglects $V_{CE(sat)}$ and I_{CQ} (trickle current). If the maximum ac output power is actually 18 W and the dc input power to the transistors is 25.5 W, what is the efficiency?

SOLUTION

$$\eta = \frac{P_O}{P_{DC}} = \frac{18 \text{ W}}{25.5 \text{ W}} = 0.706$$

which is the same as 70.6 percent.

EXAMPLE 11-12

A battery-powered transistor radio uses a class B emitter-follower output stage. If the ac output power is 0.5 W and the efficiency is 72 percent, how much power does the battery supply to the transistors of the output stage?

SOLUTION

Rearrange the efficiency equation to get

$$P_{DC} = \frac{P_O}{\eta} = \frac{0.5 \text{ W}}{0.72} = 0.694 \text{ W}$$

11-7 *darlington and complementary pairs*

Darlington pairs can increase the input impedance of a class B push-pull emitter follower (see Fig. 11-13a). Since the Darlington beta is much higher, the input

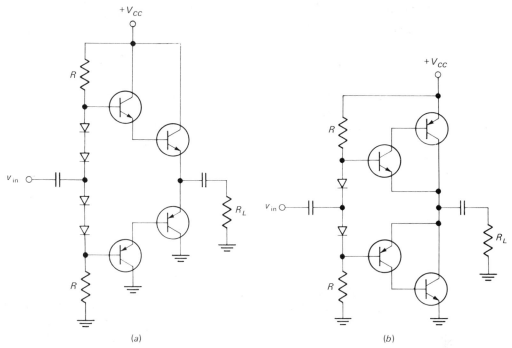

Figure 11-13. (a) *Darlington pairs.* (b) *Complementary pairs.*

impedance is much higher. But the disadvantage of using Darlingtons is four compensating diodes instead of two, which makes it more difficult to match the compensating diodes to the emitter diodes.

This is where *complementary pairs* come in. In Fig. 11-13*b*, the upper pair acts like an *npn* transistor, and the lower pair like a *pnp* transistor. The effective beta of each pair equals the product of the individual betas, the same as a Darlington. But the big advantage of complementary pairs is simpler compensation: two diodes instead of four. This means it's easier to match the compensating diodes to the emitter diodes.

11-8 *other class b push-pull amplifiers*

We have concentrated on the class B push-pull emitter follower because it is the most widely used of the class B circuits. Let's now take a brief look at other class B push-pull amplifiers.

ce push-pull amplifier

Figure 11-14*a* shows two CE amplifiers connected in a push-pull circuit. Voltage-divider bias sets up a trickle of collector current. The positive half cycle of

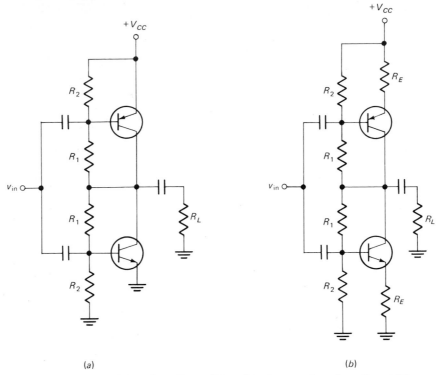

Figure 11-14. (a) *CE push-pull amplifier.* (b) *Swamped push-pull amplifier.*

input voltage turns on the lower transistor, which acts like a CE amplifier. Because of this, the voltage gain, power gain, and input impedance can be found by earlier methods of analysis. Here are the formulas of interest:

$$A \cong \frac{R_L}{R_e'}$$

$$G = \frac{\beta_{\mathrm{dc}} R_L}{R_e'}$$

and $$z_{\mathrm{in(base)}} \cong \beta_{\mathrm{dc}} R_e'$$

The advantage of a push-pull CE amplifier is voltage gain as well as power gain; the disadvantage is more nonlinear distortion. The nonlinear distortion is often bad enough that you have to swamp the emitter diodes as shown in Fig. 11-14b. By swamping the emitter diodes, you reduce the nonlinear effect the emitter diodes have on voltage gain. With heavy swamping,

$$A \cong \frac{R_L}{R_E}$$

$$G \cong \frac{\beta_{dc} R_L}{R_E}$$

and
$$z_{\text{in(base)}} \cong \beta_{dc} R_E$$

transformer-coupled push-pull amplifiers

Figure 11-15 shows a transformer-coupled push-pull CE amplifier. Diode bias sets up a trickle of collector current. Note that both transistors are *npn*. The positive half cycle of secondary voltage turns on the upper transistor, and the negative half cycle turns on the lower transistor. Since the collector currents flow in opposite directions through the output primary winding, the final signal across the load is symmetrical.

Transformers were commonly used in class B push-pull amplifiers to get the required complementary action. To work at audio frequencies, these transformers had to be bulky and expensive. The advent of complementary transistors did away with transformers in most push-pull applications. Aside from public-address systems and a few other cases, the audio transformer is now obsolete in modern solid-state amplifiers.

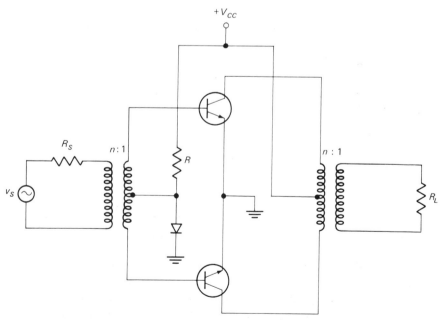

Figure 11-15. Transformer-coupled push-pull amplifier.

11-9 *a complete amplifier*

Figure 11-16 is an example of a complete amplifier. It has three stages: a small-signal class A amplifier *(Q₁)*, a large-signal class A amplifier *(Q₂)*, and a class B push-pull emitter follower. The approximate dc voltages of all nodes are listed.

The *driver stage (Q₂)* supplies the drive for the output stage. As you can see, Q_2 and its emitter resistor have replaced the biasing resistor shown earlier. This direct-coupled design is the most efficient way to connect a driver stage to the output stage.

EXAMPLE 11-13
Calculate the $P_{O(\text{max})}$ and $P_{D(\text{max})}$ for the output stage of Fig. 11-16.

SOLUTION

$$V_{CEQ} = 15 \text{ V}$$

and

$$I_{C(\text{sat})} = \frac{V_{CEQ}}{r_C + r_E} = \frac{15 \text{ V}}{100 \text{ }\Omega} = 150 \text{ mA}$$

So, the maximum output power is

$$P_{O(\text{max})} = \frac{V_{CEQ}I_{C(\text{sat})}}{2} = \frac{15 \times 0.15}{2} = 1.13 \text{ W}$$

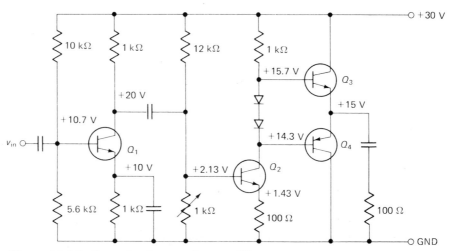

Figure 11-16. Example of a complete amplifier.

The maximum power dissipation of each transistor is

$$P_{D(max)} = \frac{P_{O(max)}}{5} = \frac{1.13 \text{ W}}{5} = 226 \text{ mW}$$

EXAMPLE 11-14

What does I_{CQ} equal in the output stage of Fig. 11-16?

SOLUTION

The emitter resistor of the driver stage has 1.43 V across it. Therefore, the emitter current in Q_2 is

$$I_E = \frac{1.43 \text{ V}}{100 \ \Omega} = 14.3 \text{ mA}$$

So, Q_2 sinks approximately 14.3 mA. This 14.3 mA flows through the diodes and the upper 1-kΩ resistor. As a result, the current mirror produces a quiescent collector current through the output transistors of

$$I_{CQ} \cong 14.3 \text{ mA}$$

EXAMPLE 11-15

Calculate the overall voltage gain of the amplifier in Fig. 11-16. Use dc betas of 100 for all transistors and an R'_e of 1 Ω for the output transistors.

SOLUTION

The first stage has a voltage gain of

$$A_1 = \frac{r_C}{r'_e} = \frac{1000 \parallel 1000 \parallel 12,000 \parallel 10,000}{2.5} = 183$$

The second stage has a voltage gain of

$$A_2 \cong \frac{r_C}{r_E} = \frac{1000 \parallel 10,000}{100} = 9.09$$

The output stage has a voltage gain of

$$A_3 = \frac{R_L}{R_L + R'_e} = \frac{100}{100 + 1} = 0.99$$

The overall voltage gain is

$$A = A_1 A_2 A_3 = 183 \times 9.09 \times 0.99 = 1650$$

EXAMPLE 11-16

What is the largest rms output signal in Fig. 11-16 without saturation or cutoff clipping? What is the input signal that just produces this maximum unclipped output?

SOLUTION

The output can swing from +15 to +30 V, and from +15 to 0 V (ideally). This is a peak-to-peak swing of 30 V, equivalent to an rms output of

$$V_{OUT} = \frac{V_{PP}}{2\sqrt{2}} = \frac{30}{2\sqrt{2}} = 10.6 \text{ V}$$

The corresponding rms input is

$$V_{IN} = \frac{V_{OUT}}{A} = \frac{10.6}{1650} = 6.42 \text{ mV}$$

self-testing review

Read each of the following and provide the missing words. Answers appear at the beginning of the next question.

1. In a class B push-pull circuit, each transistor stays in the active region only for _____ the cycle. In the emitter follower, the *npn* transistor takes care of the positive half cycle, and the _____ transistor handles the _____ half cycle.

2. *(half, pnp, negative)* To eliminate crossover distortion, we need to apply a slight _____ bias to each emitter diode. There is less nonlinear distortion with a class B push-pull amplifier because even _____ cancel out.

3. *(forward, harmonics)* A current _____ uses a compensating diode in parallel with the emitter diode. In this circuit, the current-voltage curve of the compensating diode is identical to the _____ curve.

4. *(mirror, transconductance)* The two transistors in a class B push-pull amplifier need to be fully complementary, meaning similar transconductance curves, maximum ratings, etc. Diode bias is the main method used to bias class B push-pull amplifiers. The upper half acts like an *npn* mirror and the lower half like a _____ _____.

5. *(pnp mirror)* For a complementary mirror to work properly, the curves of the _____ diodes have to match the transconductance curves of the transistors. In discrete circuits, this is not easy to do. But diode bias really shines when it comes to linear _____ circuits.

6. *(compensating, integrated)* Complementary pairs can be used instead of Darlington pairs. One advantage is simpler compensation: two _____ instead of _____.

7. *(diodes, four)* The maximum efficiency of a class B push-pull amplifier is _____ percent. Because of this, class B push-pull is much more efficient than class A.

8. *(78.5)* The driver stage supplies the drive for a class B push-pull emitter follower. This driver stage is usually direct-coupled because it's the most efficient way to drive a push-pull circuit.

problems

11-1. What is the value of $I_{C(\text{sat})}$ in Fig. 11-17?

11-2. If the 16-Ω resistor of Fig. 11-17 is changed to an 8-Ω, what is the new value of $I_{C(\text{sat})}$?

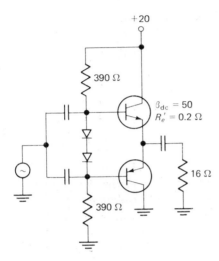

Figure 11-17.

11-3. If the V_{CC} supply of Fig. 11-17 is changed from 20 to 30 V, what will $I_{C(\text{sat})}$ equal?

11-4. A class B push-pull amplifier has an $I_{C(\text{sat})}$ of 500 mA. If I_{CQ} is 2 percent of $I_{C(\text{sat})}$, what does I_{CQ} equal? To set up this value of I_{CQ} in Fig. 11-8b, what value should R have if V_{CC} is 25 V?

11-5. In Fig. 11-17, what does I_{CQ} equal?

11-6. What is the large-signal input impedance of the base in Fig. 11-17? The voltage gain? The power gain?

11-7. In Fig. 11-18, the transistors with base-collector shorts act exactly like compensating diodes. What is the value of I_{CQ}? What does $I_{C(\text{sat})}$ equal?

11-8. If β_{dc} equals 50 in Fig. 11-18, what is the input impedance looking into the base? The voltage gain? The power gain?

11-9. What is the maximum ac output power in Fig. 11-17?

11-10. In Fig. 11-18, what is the maximum ac output power?

11-11. What is the maximum power dissipation of each transistor in Fig. 11-17?

11-12. In Fig. 11-18, what is the minimum power rating needed to avoid transistor damage?

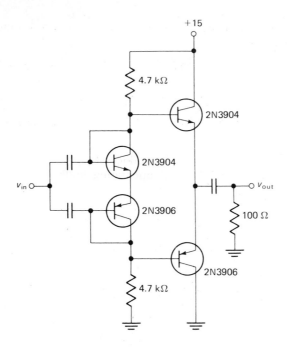

Figure 11-18. Using shorted base-collector transistors for compensating diodes.

11-13. If the efficiency is 78.5 percent, what is the maximum dc input power to the transistors of Fig. 11-17?

11-14. A class B push-pull amplifier has an ac output power of 32.4 W and a dc input power of 45.5 W. What is the efficiency?

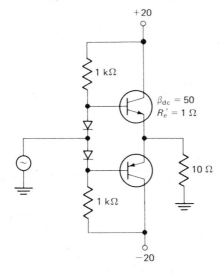

Figure 11-19. Two-supply push-pull amplifier.

11-15. To deliver a maximum ac output power of 50 W, what power rating do the transistors of a class B push-pull amplifier need?

11-16. When positive and negative power supplies are available, we can eliminate the coupling capacitors as shown in Fig. 11-19. What does V_{CEQ} equal? I_{CQ}?

11-17. What does $z_{in(base)}$ equal in Fig. 11-19? The voltage gain? The power gain?

11-18. What is the maximum ac output power in Fig. 11-19? The worst-case power dissipation in the transistors?

11-19. In Fig. 11-20, what is the value of R that produces $V_{CEQ} = 10$ V for each output transistor? (Use 0.7 V for the drop of the compensating diodes and emitter diodes.)

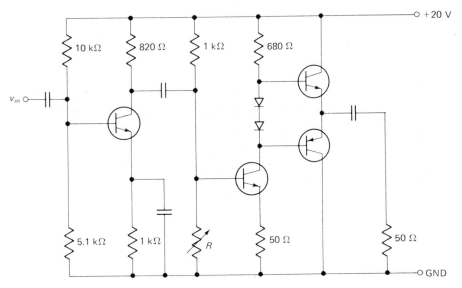

Figure 11-20.

11-20. In Fig. 11-20, the dc voltage from the output emitters to ground is +10 V. Work out the dc voltages for every node in the amplifier.

11-21. Calculate $P_{O(max)}$ and $P_{D(max)}$ for the output stage of Fig. 11-20.

11-22. What does I_{CQ} equal in the output stage of Fig. 11-20?

11-23. In Fig. 11-20, what is the largest rms output voltage without clipping? The corresponding input signal?

11-24. What is the total dc current drawn from the supply of Fig. 11-20 under no-signal conditions?

12 *class c power amplifiers*

A *class C amplifier* can deliver more load power than a class B amplifier. To amplify a sine wave, however, it has to be *tuned* to the sine-wave frequency. Because of this, the tuned class C amplifier is a *narrowband* circuit; it can amplify only the resonant frequency and those frequencies near it.

To avoid large inductors and capacitors in the resonant circuit, class C amplifiers have to operate at *radio frequencies* (above 20 kHz). Therefore, even though it is the most efficient of all classes, class C is useful only when we want to amplify a narrow band of radio frequencies (RF).

Class C power amplifiers normally use *RF power transistors;* these have characteristics optimized for RF signals. Typical $P_{D(\text{max})}$ ratings for RF power transistors are from 1 to over 75 W.

12-1 *basic class c action*

Class C means collector current flows for less than 180°. In a practical class C circuit, the current flows for much less than 180°, and looks like the narrow pulses of Fig. 12-1a. As will be discussed later, when narrow *current* pulses like these drive a high-Q resonant circuit, the *voltage* across the circuit is almost a perfect sine wave.

298

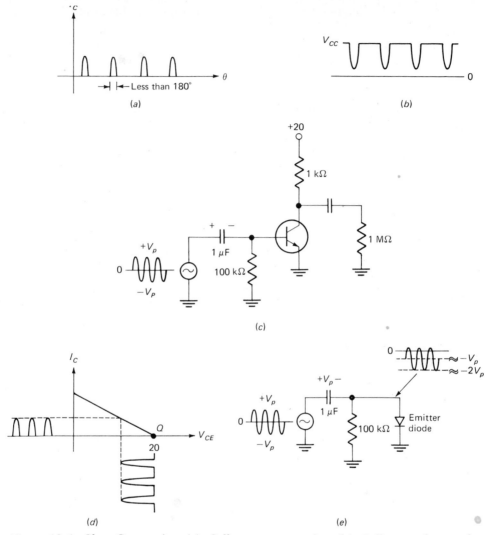

Figure 12-1. Class C operation. (a) Collector current pulses. (b) Collector voltage pulses. (c) Untuned class C amplifier. (d) Ac load line. (e) Clamping on base.

clamping

Before discussing tuned circuits, we need the basic idea of a class C *untuned* circuit (Fig. 12-1c). Without an ac input signal, no collector current flows because the emitter diode is unbiased (not even trickle bias). Therefore, the Q point is at cutoff as shown in Fig. 12-1d.

What happens when the ac signal comes in? Section 5-12 was about *clamping action*. The idea in a clamper is to charge a coupling capacitor to approximately the peak voltage of the input signal. This is precisely what we do in Fig. 12-1c. The input coupling capacitor, the 100-kΩ base resistor, and the emitter diode form a negative clamper as shown in Fig. 12-1e. The positive cycles turn on the emitter diode and charge the capacitor with the plus-minus polarity shown. On the negative half cycles, the only discharge path is through the 100-kΩ resistor. As long as the period T of the input signal is much smaller than the RC discharging time constant, the capacitor loses only a small amount of its charge. Therefore, the capacitor voltage approximately equals the peak voltage of the input signal. This produces the familiar negatively clamped waveform shown in Fig. 12-1e.

To replace the small amount of capacitor charge lost during each cycle, the base voltage swings above zero and turns on the emitter diode briefly at each positive peak. The *conduction angle* is much less than 180°. This is why the collector current waveform is a train of narrow pulses as shown in Fig. 12-1a. When these current pulses flow through the ac collector resistance, they produce the negative-going pulses of Fig. 12-1b.

In terms of the ac load line, here is what happens. At each positive peak of the base voltage, the emitter diode turns on briefly. This forces the instantaneous operating point to move from cutoff toward saturation. In the process, we get the narrow current and voltage pulses shown in Fig. 12-1d.

emitter breakdown

Now is a good time to mention BV_{EBO}, the breakdown voltage from emitter to base with an open collector (equivalent to zero collector current). Since the clamped waveform of Fig. 12-1e reaches a negative peak of approximately $-2V_P$, this is the maximum reverse voltage across the emitter diode. Unless the emitter diode can withstand this voltage, it will break down. Data sheets for RF power transistors list BV_{EBO} or its equivalent $V_{BE(max)}$ to indicate the breakdown voltage of the emitter diode. This rating has to be greater than $2V_P$ (approximately); otherwise, the emitter diode breaks down.

As an example, the data sheet of a 2N3950 gives a $V_{BE(max)}$ rating of 4 V. If we use this transistor in Fig. 12-1c, the emitter diode does not break down as long as the peak-to-peak input voltage is less than approximately 4 V. But when the peak-to-peak voltage is greater than 4 V, the emitter diode will break down.

$V_{BE(max)}$ ratings are typically less than 5 V for RF power transistors. Often, you may have an input signal whose peak-to-peak value is greater than 5 V. In a case like this, add a diode in series with the base (Fig. 12-2a) or in series with the emitter (Fig. 12-2b). On the positive half cycle, both diodes conduct and the capacitor charges as before. On the negative half cycle, the rectifier diode with its larger breakdown voltage keeps the emitter diode from breaking down. (Most rectifier diodes have breakdown voltages greater than 25 V.)

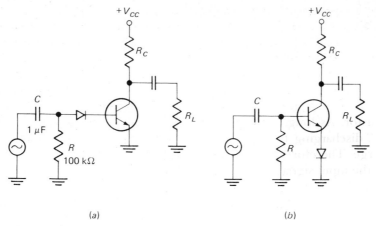

Figure 12-2. Avoiding breakdown of emitter diode.

EXAMPLE 12-1

In Fig. 12-2*a*, at what frequency does the discharge time constant equal 10 times the period?

SOLUTION

The discharge time constant equals

$$RC = 10^5(10^{-6}) = 0.1 \text{ s}$$

One-tenth of this is 0.01 s, and the frequency is

$$f = \frac{1}{0.01} = 100 \text{ Hz}$$

12-2 *other class c relations*

Figure 12-3*a* shows an untuned class C amplifier including an emitter resistor. If we Thevenize the base circuit, the Thevenin impedance is the parallel of C and R_B. Therefore, Fig. 12-3*b* is an alternative form of untuned class C amplifier; you can use the RC network of Fig. 12-3*a* or *b*, whichever is more convenient in a particular application.

saturation current and average base voltage

As already mentioned, the Q point is at cutoff as shown in Fig. 12-3*c*. Again, the general load line of Fig. 10-4 applies, so that the saturation current equals

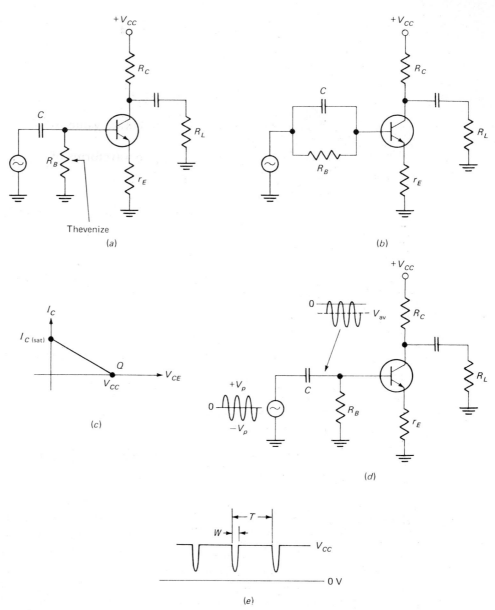

Figure 12-3. *Untuned class C amplifier.* (a) *Circuit.* (b) *Equivalent circuit.* (c) *Load line.* (d) *Waveforms.* (e) *Duty cycle.*

$$I_{C(\text{sat})} \cong I_{CQ} + \frac{V_{CEQ}}{r_C + r_E}$$

Since $I_{CQ} = 0$ and $V_{CEQ} = V_{CC}$,

$$I_{C(\text{sat})} \cong \frac{V_{CC}}{r_C + r_E} \quad \text{(untuned class C)} \tag{12-1}$$

This is a close approximation; for a more accurate value, you have to subtract $V_{CE(\text{sat})}$ from V_{CC} in the numerator.

When a signal is present, we get the voltage waveforms of Fig. 12-3d. Ideally, the capacitor voltage is

$$V_C \cong V_P$$

But to a second approximation, we must allow for the drop across the emitter diode; therefore, a more accurate value for capacitor voltage is

$$V_C \cong V_P - 0.7 \tag{12-2}$$

The base voltage is a negatively clamped waveform as shown. Ideally, the average voltage of this waveform is

$$V_{\text{av}} \cong -V_P$$

To a second approximation, the waveform actually goes above the zero level by about 0.7 V; this means the average base voltage is 0.7 V greater than $-V_P$. In symbols,

$$V_{\text{av}} \cong -V_P + 0.7 \tag{12-3}$$

base-voltage check

For class C circuits, one troubleshooting check is the average base voltage given by Eq. (12-3). As an example, if the input signal has a peak of 4 V,

$$V_{\text{av}} = -4 + 0.7 = -3.3 \text{ V}$$

With a high-impedance dc voltmeter, you can measure the dc voltage from base to ground. If you get a reading around -3.3 V, you know the clamper is working. You must use a high-impedance dc voltmeter when measuring this base voltage to avoid changing the RC time constant.

This voltmeter check is useful when you do not have an oscilloscope. But with an oscilloscope, you can look at the base-voltage waveform and see if the clamper is working properly.

duty cycle

The brief turn-on of the emitter diode at each positive peak produces narrow pulses of collector current. When this current flows through r_C, we get the negative-going voltage pulses shown in Fig. 12-3e. With pulses like these, we often refer to the *duty cycle*, defined as

$$\text{Duty cycle} = \frac{W}{T} \times 100\% \qquad (12\text{-}4)***$$

where W is the width of each pulse and T the period (see Fig. 12-3e). For instance, suppose an oscilloscope shows a pulsed waveform with each pulse being 0.2 μs wide and the period being 1.6 μs. Then, the duty cycle of the waveform is

$$\text{Duty cycle} = \frac{0.2}{1.6} \times 100\% = 12.5\%$$

The duty cycle affects transistor power dissipation. Each pulse of Fig. 12-3e appears when the transistor is conducting. Between pulses, the transistor is off. The smaller the duty cycle, the less the average power dissipation of the transistor.

EXAMPLE 12-2
Draw the ac load line for the untuned class C amplifier of Fig. 12-4a.

SOLUTION
Ac resistance r_C is the parallel of 100 Ω and 20 kΩ, which is 100 Ω to a close approximation. The ac emitter resistance r_E is adjustable from 0 to 5 Ω. Even when $r_E = 5$ Ω, $r_C + r_E$ is approximately equal to 100 Ω. With Eq. (12-1),

$$I_{C(\text{sat})} \cong \frac{V_{CC}}{r_C + r_E} \cong \frac{30}{100} = 0.3 \text{ A}$$

Incidentally, by adjusting r_E, you can change the size of the collector current pulses.

Figure 12-4b shows the ac load line. If the operating point swings over the entire load line, the current pulses will have a peak value of 0.3 A and the voltage pulses a peak of 30 V (negative-going).

EXAMPLE 12-3
The frequency of the input signal in Fig. 12-4a is 250 kHz. If the pulses of current look like Fig. 12-4c, what is the duty cycle of the current waveform?

SOLUTION
First, get the period of the input signal.

$$T = \frac{1}{f} = \frac{1}{250(10^3)} = 4 \text{ } \mu\text{s}$$

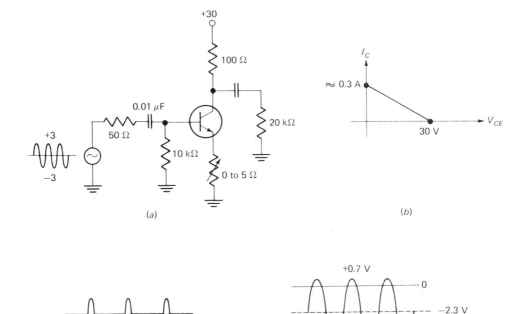

Figure 12-4.

Since the input signal produces the current waveform of Fig. 12-4c, the period of the current waveform equals 4 μs. The pulse width is given as 0.1 μs. With Eq. (12-4), the duty cycle is

$$\text{Duty cycle} = \frac{0.1}{4} \times 100\% = 2.5\%$$

EXAMPLE 12-4
If you use an oscilloscope to look at voltage waveforms, what voltage waveform will you see from base to ground in Fig. 12-4a?

SOLUTION
The input circuit is a negative clamper. Ideally, you see a negatively clamped sine wave swinging from the 0 to −6 V. To a second approximation, however, we must include the 0.7 V or thereabouts needed to turn on the emitter diode. The peak-to-peak voltage of the base waveform is still 6 V, the same as the input. But the base voltage must rise to approximately 0.7 V at each positive peak. In other words, the 6-V peak-to-peak sine wave has a positive peak of

approximately 0.7 V as shown in Fig. 12-4*d*. The average value of −2.3 V and the negative peak of −5.3 V are also shown.

12-3 *the tuned class c amplifier*

How do we get sine waves from narrow current pulses? By using a resonant circuit in the collector. Figure 12-5*a* shows one of several possible tuned circuits. As before, the base circuit acts like a negative clamper, and the emitter diode turns on briefly at each positive peak. This results in narrow collector current pulses as shown in Fig. 12-5*b*.

The *fundamental frequency* of a pulse train like Fig. 12-5*b* is defined as

$$f_1 = \frac{1}{T} \qquad (12\text{-}5)***$$

where f_1 is called the fundamental or *first harmonic*. As an example, if T equals 10 μs in Fig. 12-5*b*, the fundamental or first harmonic equals 100 kHz. This is the lowest frequency in the pulsed waveform.

When narrow current pulses like Fig. 12-5*b* drive a resonant circuit, we can

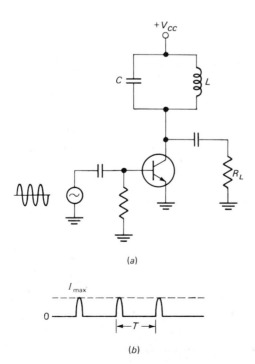

(a)

(b)

Figure 12-5. Tuned class C amplifier.

produce an almost perfect sine wave of voltage. To get a sine wave with the fundamental frequency, here is what we do. First, the resonant frequency has to equal the fundamental frequency of the pulsed waveform. Second, the tuned circuit must have a high Q (greater than 10) to get an almost perfect sine wave of voltage. As an example, if the pulse waveform of Fig. 12-5b has a period of 10 μs, the fundamental frequency is 100 kHz. By tuning the LC tank of Fig. 12-5a to 100 kHz, we can get an almost perfect sine wave of voltage across the resonant tank when the Q is greater than 10.

EXAMPLE 12-5

In the tuned class C amplifier of Fig. 12-6a, the input frequency is 1 MHz. The resonant tank is tuned to the fundamental frequency of the collector current pulses.

1. What is the value of L?

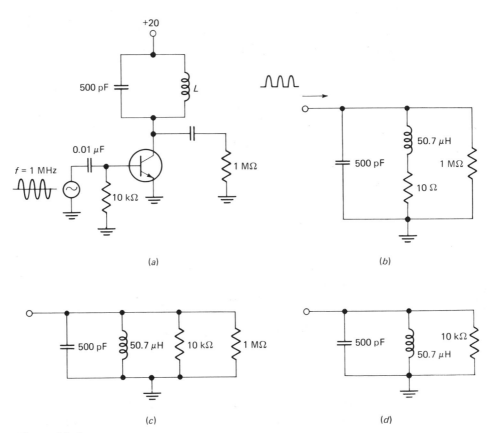

Figure 12-6.

2. If the inductor has a series resistance of 10 Ω at 1 MHz, what is the Q of the resonant circuit?

SOLUTION

As usual, the input circuit is a clamper; the clamped waveform at the base turns on the emitter diode briefly at each positive peak. The result is narrow collector current pulses. Since the input signal produces these pulses, the pulses must have the same fundamental frequency as the input signal, 1 MHz.

1. The resonant tank must be tuned to 1 MHz. Therefore, we use

$$f \cong \frac{1}{2\pi\sqrt{LC}}$$

which is the approximate formula for resonant frequency whenever Q is greater than 10. Substituting the known values gives

$$10^6 \cong \frac{1}{2\pi\sqrt{L(500)10^{-12}}}$$

Solving for L gives

$$L = 50.7 \; \mu\text{H}$$

2. Figure 12-6*b* is part of the ac equivalent circuit. We have shorted the output coupling capacitor and have reduced V_{CC} to zero. This results in a resonant tank with narrow current pulses driving it. The inductive reactance of the coil is

$$X_L = 2\pi fL = 2\pi(10^6)50.7(10^{-6}) = 318 \; \Omega$$

The Q of the coil is

$$Q_{\text{coil}} = \frac{X_L}{R_{\text{series}}} = \frac{318}{10} = 31.8$$

As discussed in basic circuit-theory books, the equivalent parallel resistance of the coil is

$$R_{\text{parallel}} \cong Q_{\text{coil}}X_L = 31.8(319) \cong 10 \; \text{k}\Omega$$

(This is a close approximation when Q is greater than 10.)

We now draw the equivalent circuit of Fig. 12-6*c*. All we have done is to represent the 10 Ω of series resistance in the coil by its parallel equivalent of 10 kΩ. In other words, because Q_{coil} is greater than 10, the circuit of Fig. 12-6*c* behaves almost the same as the circuit of Fig. 12-6*b*. Now it is plain what effect the 1-MΩ load resistor has—almost none. The 1 MΩ is so large compared to the 10 kΩ we can neglect it and use the ac equivalent circuit of Fig. 12-6*d*.

Figure 12-6*d* is the parallel resonant circuit covered in basic textbooks. Since the 1-MΩ load resistor is large enough to neglect, the *Q* of the resonant circuit equals the *Q* of the coil;

$$Q_{circuit} = Q_{coil} = 31.8$$

With a high *Q* like this, we can expect an almost perfect sine wave of voltage to appear across the tank circuit.

12-4 *tuned class c action*

Figure 12-7*a* shows a current pulse driving a resonant tank. If the pulse is narrow, the inductor looks like a high impedance and the capacitor like a low impedance. For this reason, most of the current charges the capacitor. We will assume the capacitor charges to 10 V during the pulse.

After the pulse ends, the circuit looks like Fig. 12-7*b*. The charged capacitor will discharge through the coil and the load resistor. As the current in the

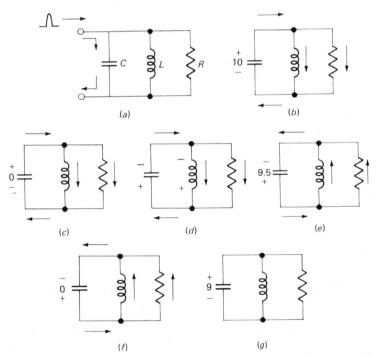

Figure 12-7. Tuned-circuit action. (a) *Initial capacitor charging.* (b) *Capacitor discharges.* (c) *Inductor current maximum.* (d) *Collapsing inductor field charges capacitor.* (e) *Capacitor discharges.* (f) *Inductor current maximum.* (g) *At end of cycle.*

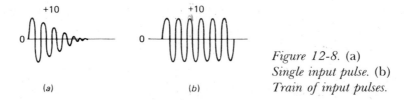

Figure 12-8. (a)
Single input pulse. (b)
Train of input pulses.

coil builds up, the capacitor voltage decreases until it reaches 0 V (Fig. 12-7c). The magnetic field around the inductor now collapses and induces a minus-plus voltage across the coil as shown in Fig. 12-6d. This forces current to flow into the capacitor and charges it with the opposite polarity shown.

After the magnetic field has completely collapsed, the inductor current drops to zero; at this instant, the capacitor has fully charged in the opposite direction. The maximum voltage on the capacitor will be less than the 10 V we started with because of the energy absorbed by the load resistor. For the sake of explanation, we will assume the capacitor voltage reaches only 9.5 V.

The capacitor cannot hold this charge; it must start to discharge through the inductor and load resistor as shown in Fig. 12-7e. The action is now complementary. That is, the magnetic field builds up with the opposite polarity. The field reaches its maximum value when the capacitor has completely discharged as shown in Fig. 12-7f. Again, the magnetic field collapses, inducing a voltage across the coil which charges the capacitor to a positive voltage. Because of energy losses in the load resistor, the maximum positive voltage on the capacitor will be less than the initial 10 V; arbitrarily, we have shown it as 9 V in Fig. 12-7g.

If only a single current pulse drives a tank circuit, the capacitor will charge and discharge repeatedly, the maximum voltage decreasing slightly each cycle, until the waveform dies out as shown in Fig. 12-8a. On the other hand, if a train of narrow current pulses drives the circuit, each pulse recharges the capacitor to the full voltage; in this way, we can get almost a perfect sine wave like Fig. 12-8b.

12-5 *power relations*

At the beginning of this chapter, we said a class C amplifier can deliver more load power than a class B amplifier. This means class C is more efficient than class B.

output power

In a tuned class C amplifier like Fig. 12-9a, the quiescent or no-signal collector voltage is

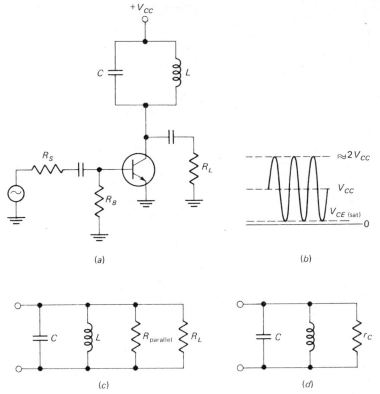

Figure 12-9. Deriving power relations.

$$V_{CEQ} \doteq V_{CC}$$

When a signal is present, it forces the total collector voltage to swing above and below this voltage. The collector voltage can drop no lower than $V_{CE(\text{sat})}$. This is why the maximum output voltage waveform of a tuned class C amplifier looks like Fig. 12-9b.

As discussed in Example 12-5, the series resistance of the inductor can be replaced by its equivalent parallel resistance as shown in Fig. 12-9c. This R_{parallel} is in shunt with R_L; the product-over-sum of these two resistances is the equivalent ac load resistance r_C shown in Fig. 12-9d.

In Fig. 12-9b, the output voltage has a peak-to-peak value of approximately $2V_{CC}$, equivalent to a peak value of V_{CC}. Therefore, the maximum ac output power is

$$P_{O(\text{max})} = \frac{V_{RMS}^2}{r_C} = \frac{(V_{CC}/\sqrt{2})^2}{r_C}$$

or
$$P_{O(\text{max})} = \frac{V_{CC}^2}{2r_C} \qquad \text{(class C)} \qquad\qquad (12\text{-}6)$$

power dissipation

The average power dissipation of the transistor depends on the duty cycle and the signal swing. It can be shown that a peak-to-peak output voltage of $2V_{CC}$ results in

$$P_D \cong 0.5\,V_{CE(\text{sat})}\,\frac{V_{CC}}{r_C} \qquad \text{(100\% load line)} \qquad (12\text{-}7a)$$

This is dissipation for maximum output signal; you get this low dissipation only when the *entire ac load line is used,* and when the *duty cycle is less than 10 percent.*

When the signal does not use the entire load line, the transistor power dissipation may be higher than the value given by Eq. (12-7a). The worst case occurs when only half the load line is used, that is, when the ac output voltage in Fig. 12-9a has a peak value of $V_{CC}/2$. In this case,

$$P_D = \frac{V_{CC}^2}{8r_C} \qquad \text{(50\% load line)} \qquad (12\text{-}7b)$$

Again, r_C is the total ac load resistance across the tank circuit; it includes the coil losses.

To take advantage of the high efficiency of a class C amplifier, you have to use most of the load line. In this way, you may approach the best-case dissipation given by Eq. (12-7a).

The ratio of the best-case dissipation to the maximum ac output power is useful. With Eqs. (12-6) and (12-7a), we get a ratio of

$$\frac{P_D}{P_{O(\text{max})}} = \frac{V_{CE(\text{sat})}}{V_{CC}}$$

or
$$P_D = \frac{V_{CE(\text{sat})}}{V_{CC}}\,P_{O(\text{max})} \qquad \text{(100\% load line)} \qquad (12\text{-}8)\text{***}$$

To do better than class B push-pull, the ratio $V_{CE(\text{sat})}/V_{CC}$ must be less than one-fifth. This is easy to accomplish. Since $V_{CE(\text{sat})}$ is typically less than 1 V, all we need is a V_{CC} of more than 5 V. Therefore, for a given transistor power rating, the tuned class C amplifier can deliver the greater ac output power.

maximum efficiency

In Fig. 12-9a, the dc input power from the supply is

$$P_{DC} = P_{O(\text{max})} + P_D$$

Therefore, the efficiency of the class C amplifier is

$$\eta = \frac{P_{O(\text{max})}}{P_{O(\text{max})} + P_D}$$

By substituting the right-hand member of Eq. (12-8) and rearranging, we get

$$\eta = \frac{V_{CC}}{V_{CC} + V_{CE(\text{sat})}} \qquad (12\text{-}9)$$

Since V_{CC} is typically much greater than $V_{CE(\text{sat})}$, the efficiency approaches 100 percent. For example, if V_{CC} is 30 V and $V_{CE(\text{sat})}$ is 1 V,

$$\eta = \frac{30}{30 + 1} = 0.968 = 96.8\%$$

In summary, transformerless class A has a maximum efficiency of 25 percent, class B of 78.5 percent, and class C near 100 percent. But remember, class C is suitable only for resonant RF applications. This is why class A and class B predominate in *audio* applications (audio means 20 to 20,000 Hz).

EXAMPLE 12-6
Calculate the maximum load power in Fig. 12-10. What $P_{D(\text{max})}$ does the transistor need for maximum output power if $V_{CE(\text{sat})}$ is 0.75 V?

SOLUTION
The peak-to-peak collector voltage in Fig. 12-10 is 100 V maximum, equivalent to a peak of 50 V. The reflected load is 200 Ω. With Eq. (12-6),

$$P_{O(\text{max})} = \frac{V_{CC}^2}{2 r_C} = \frac{50^2}{2(200)} = 6.25 \text{ W}$$

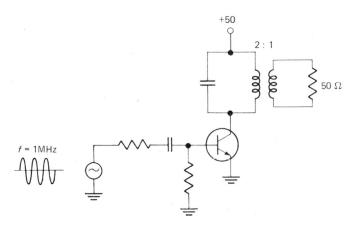

Figure 12-10.

For this ac output power, the transistor dissipation is

$$P_D = \frac{V_{CE\,(\text{sat})}}{V_{CC}}\,P_{O\,(\text{max})} = \frac{0.75}{50}\,6.25\text{ W} = 93.8\text{ mW}$$

Therefore, the transistor needs a power rating of at least 93.8 mW if the entire ac load line is used.

12-6 *frequency multipliers*

When the output signal of a tuned class C amplifier is maximum (a peak-to-peak value of approximately $2V_{CC}$), the class C amplifier is the most efficient of all classes; it delivers more load power for a given transistor dissipation than any other class. But remember, you get this efficiency only at a single frequency (or frequencies close to it).

One important application for the high efficiency of a class C amplifier is an *oscillator*. This circuit produces an output sine wave without an input signal. (A later chapter discusses oscillators.)

Another application for tuned class C circuits is with *frequency multipliers*. The idea is to tune the resonant tank to a *harmonic* or multiple of the input frequency.

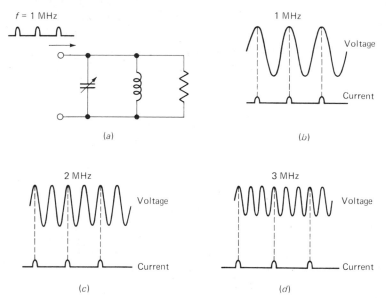

Figure 12-11. Frequency multipliers. (a) Current pulses drive resonant tank. (b) Tuned to fundamental. (c) Tuned to second harmonic. (d) Tuned to third harmonic.

Fig. 12-11a shows narrow current pulses driving a tuned circuit. The narrow current pulses have a fundamental frequency of 1 MHz. In an ordinary tuned class C amplifier, we tune the resonant circuit to the fundamental frequency. Then, each current pulse recharges the capacitor once per output cycle (Fig. 12-11b).

Suppose we change the resonant frequency of the tank to 2 MHz; this is the second multiple or second harmonic of 1 MHz. The capacitor and inductor will interact at a 2-MHz rate and produce an output voltage of 2 MHz as shown in Fig. 12-11c. Now, the 1-MHz current pulses recharge the capacitor on every other output cycle.

If we tune the tank to 3 MHz, the third harmonic of 1 MHz, we get a 3-MHz output signal (Fig. 12-11d). In this case, the 1-MHz current pulses recharge the capacitor every third output cycle. As long as the Q of the resonant circuit is high, the output voltage looks almost like a perfect sine wave.

Figure 12-12 shows one way to build a frequency multiplier. The input signal has a frequency of f; this signal is negatively clamped at the base. The resulting collector current is a train of narrow current pulses with a fundamental frequency f. By tuning the resonant tank to the nth harmonic, we get an output voltage whose frequency is nf.

In Fig. 12-12, the current pulses recharge the capacitor every nth output cycle. For this reason, the output power decreases when we tune to higher harmonics; the higher n is, the lower the output power. Because of this decreasing efficiency at higher harmonics, a tuned class C frequency multiplier like Fig. 12–12 is ordinarily used only for lower harmonics like the second or third. For n greater than 3, we usually turn to more efficient semiconductor devices like the step-recovery diode.

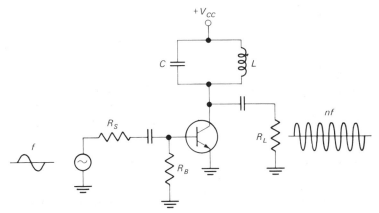

Figure 12-12. Frequency multiplier.

self-testing review

Read each of the following and provide the missing words. Answers appear at the beginning of the next question.

1. In a negative clamper, the output signal is shifted negatively. The peak-to-peak value is still $2V_P$, but the positive peak is approximately _____ and the negative peak is approximately _____.

2. *(0, $-2V_P$)* Class C means collector current flows for less than _____. In a practical class C circuit, current flows for much less than _____, and looks like narrow pulses.

3. *(180°, 180°)* The duty cycle equals W/T times 100 percent, where W is the _____ of each pulse and T is the _____. The duty cycle affects transistor _____ dissipation.

4. *(width, period, power)* The fundamental frequency of a pulse waveform is designated f_1, where f_1 is called the fundamental or _____ harmonic. When narrow current pulses drive a resonant circuit, we can get an almost perfect sine wave of voltage by tuning the resonant circuit to the fundamental _____.

5. *(first, frequency)* Transformerless class A has a maximum efficiency of _____ percent, class B push-pull of _____ percent, and class C near _____ percent. But class C is suitable only for resonant RF applications. This is why class A and class B predominate in audio applications.

6. *(25, 78.5, 100)* Another application for tuned class C amplifiers is with frequency multipliers. The idea is to tune the resonant tank to a _____ or multiple of the input frequency.

7. *(harmonic)* Because of decreasing efficiency at higher harmonics, a tuned class C frequency multiplier is ordinarily used only for lower harmonics like the second or the third.

problems

12-1. In Fig. 12-13, the period of the input signal is much smaller than the discharging time constant of the base circuit. What is the peak-to-peak voltage? Ideally, what is the maximum negative base voltage? To a second approximation, what is the largest negative base voltage? What BV_{EBO} does the transistor need?

12-2. For good clamping action in Fig. 12-13, we want the discharging time constant to be at least 10 times the period of the input signal. What is the lowest input frequency satisfying this condition?

12-3. In Fig. 12-13, what is the value of $I_{C(sat)}$? And V_{CEQ}?

12-4. If you use a high-impedance voltmeter to measure the dc voltage from base to

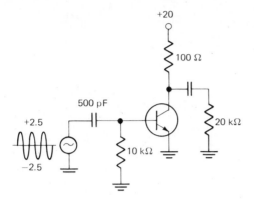

Figure 12-13.

ground in Fig. 12-13, what voltage do you read? (Give the ideal and second-approximation answers.)

12-5. If the output voltage of Fig. 12-13 has a period of 1 μs and a pulse width of 0.005 μs, what is the duty cycle of the output waveform?

12-6. Figure 12-14a shows another form of the base clamping circuit. What is the discharging time constant? For this time constant to be at least 10 times the input period, what is the lowest frequency we can use?

12-7. Sometimes the RC circuit of a clamper is moved to the emitter circuit as shown in Fig. 12-14b. After the first few input cycles, the capacitor voltage reaches approximately V_p. Between positive input peaks, the capacitor discharges through the 1-kΩ resistor. If we want the RC time constant of the emitter circuit to be at least 10 times the input period, what is lowest permitted input frequency?

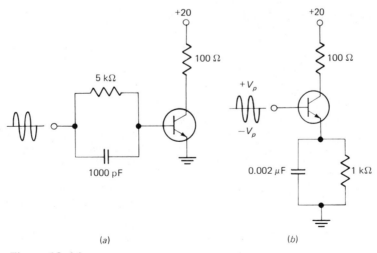

(a)   (b)

Figure 12-14.

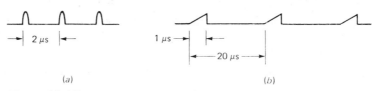

(a)   (b)

Figure 12-15.

12-8. The current waveform of Fig. 12-15*a* has a period of 2 μs. What is the fundamental frequency of this current waveform?

12-9. Figure 12-15*b* shows a current waveform made up of sawtooth pulses. The duty cycle and fundamental frequency are still defined by Eqs. (12-4) and (12-5). Calculate the duty cycle and fundamental frequency.

12-10. Figure 12-16*a* shows a tuned class C amplifier. What is the resonant frequency of the tank circuit? For the discharging tme constant of the base clamper to be 10 times the period of the input signal, what value should R have?

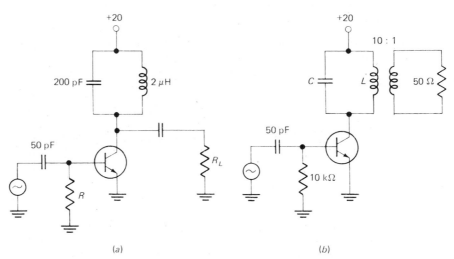

(a)   (b)

Figure 12-16.

12-11. The inductor of Fig. 12-16*a* has a series resistance of 2 Ω at the resonant frequency of the tank. What is the Q of the coil? What value does the equivalent parallel resistance of the coil have?

12-12. In Fig. 12-16*a*, the coil has a Q of 100 at the resonant frequency of the tank. What is the series resistance of this coil? The equivalent parallel resistance of the coil? If R_L equals 10 kΩ, what is the total parallel resistance across the tank circuit as resonance? The Q of the resonant circuit equals the total parallel resistance across the tank divided by X_L. What value does this Q have?

12-13. Figure 12-16*b* shows transformer coupling to the final load resistor. What is the reflected load resistance seen by the collector? If the coil has an X_L of 100 Ω and a Q_{coil} of 50, what is the Q of the entire tank circuit?

12-14. In Fig. 12-16*a*, what is the value of V_{CEQ}? What is the largest possible peak-to-peak voltage across the tank?

12-15. What is the maximum load power in Fig. 12-16*a* when r_L equals 1 kΩ?

12-16. If the r_L in Fig. 12-16*b* equals 2.5 kΩ, what is the maximum load power? How much of this total load power does the 50-Ω load resistor receive?

12-17. The transistor of Fig. 12-16*a* has a $V_{CE(sat)}$ of 0.8 V. For a low duty cycle and 100 percent load-line swing, what is the power dissipation in the transistor if the load power equals 100 mW? The efficiency?

12-18. An RF power transistor has a $P_{D(max)}$ rating of 2 W at the highest temperature it will operate at. If the transistor has a $V_{CE(sat)}$ of 1 V, what is the maximum possible load power with a V_{CC} supply of 30 V? The efficiency?

12-19. A tuned class C amplifier has a V_{CC} of 40 V, an r_L of 50 Ω, and a $V_{CE(sat)}$ of 0.6 V. How much load power can this amplifier deliver? The efficiency?

12-20. A transistor type has a $V_{CE(sat)}$ of 0.5 V and a $P_{D(max)}$ rating of 1 W. Given a V_{CC} supply of 20 V, what is the maximum load power you can get with a class A amplifier? Class B push-pull? Tuned class C?

12-21. What is the fundamental frequency of Fig. 12-15*a*? The second harmonic? The third harmonic?

12-22. What is the tenth harmonic of Fig. 12-15*b*? The fiftieth?

12-23. If we wanted to use the circuit of Fig. 12-16*a* as a *frequency doubler* ($n = 2$), what input frequency would we use?

12-24. For the circuit of Fig. 12-16*a* to act as a frequency tripler *($n = 3$)*, what input frequency do we need?

12-25. If Fig. 12-16*a* is used in an X4 frequency multiplier, what value should R have to make the discharging time constant of the base clamper ten times the period of the clamped signal?

12-26. In a tuned class C amplifier, the size of the current pulses driving the tuned circuit depends on the duty cycle. With an advanced derivation, the current pulses have an amplitude of

$$I_{max} = \frac{\pi}{4k} \frac{V_{CEQ}}{r_C + r_E}$$

where k equals W/T, the duty cycle divided by 100 percent. Calculate I_{max} for a duty cycle of 5 percent a V_{CEQ} of 10 V, and an $r_C + r_E$ of 100 Ω.

13
field-effect transistors

The bipolar transistor is the backbone of linear electronics. But in some applications the *field-effect transistor* (FET) is preferred. This chapter describes FETs and prepares us for the FET circuits of the next chapter.

13-1 *the JFET*

The *junction* FET, or JFET, is a unipolar transistor; it needs only majority carriers to work. It is much easier to understand than the bipolar transistor.

JFET regions

Figure 13-1a shows part of a JFET. The lower end is called the *source* and the upper end the *drain;* the piece of semiconductor between the source and drain is the *channel*. Since n material is used in Fig. 13-1a, the majority carriers are conduction-band electrons. Depending on voltage V_{DD} and the resistance of the channel, we get a certain amount of current.

By embedding two p regions in the sides of the channel, we get the *n-channel* JFET of Fig. 13-1b. Each of these p regions is called a *gate*. When the manufacturer connects a separate external lead to each gate, we call the device a *dual-gate* JFET. The main use of a dual-gate JFET is with a *mixer*, a special circuit to be discussed in Chapter. 23.

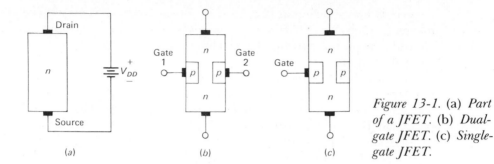

Figure 13-1. (a) *Part of a JFET.* (b) *Dual-gate JFET.* (c) *Single-gate JFET.*

Throughout this chapter we concentrate on the *single-gate* JFET, a device whose gates are *internally connected.* A single-gate JFET has only one external gate lead, as shown in Fig. 13-1*c.* When using this symbol, remember the two *p* regions have the same potential because they are internally connected.

biasing the JFET

Figure 13-2*a* shows the normal polarities for biasing an *n*-channel JFET. The idea is to apply a negative voltage between the gate and the source; this *reverse-biases* the gate. Since the gate is reverse-biased, only a negligibly small current flows in the gate lead. To a first approximation, gate current is zero.

The name *field effect* is related to the depletion layers around each *pn* junction. Figure 13-2*b* shows these depletion layers. The current from source to drain must flow through the narrow channel between depletion layers. The size of these depletion layers determines the width of the conducting channel. The more negative the gate voltage, the narrower the conducting channel becomes, because the depletion layers get closer to each other. In other words, the gate voltage *controls* the current between the source and the drain. The more negative the gate voltage, the smaller the current.

The key difference between a JFET and a bipolar transistor is this: the gate

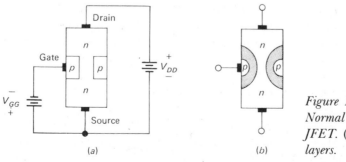

Figure 13-2. (a) *Normal biasing of a JFET.* (b) *Depletion layers.*

is reverse-biased whereas the base is forward-biased. This crucial difference means the JFET acts like a *voltage-controlled* device; ideally, input voltage alone controls the output current. This is different from a bipolar where input current controls output current.

We can summarize the first major difference between the JFET and the bipolar in terms of resistance. The input resistance of a JFET ideally approaches infinity. To a second approximation, it is well into the megohms, depending on the particular JFET type. Therefore, in applications where high input resistance is needed, the JFET is preferred.

The price we pay for larger input resistance is smaller control over output current. In other words, a JFET is less sensitive to changes in input voltage than a bipolar transistor. In almost any JFET an input change of 0.1 V produces less than a 10-mA change in output current. But in a bipolar transistor a 0.1-V change easily produces more than a 10-mA change in output current. As will be discussed later, this means a JFET has less voltage gain than a bipolar.

schematic symbol

Figure 13-3*a* shows the schematic symbol for a JFET. As a memory aid, think of the thin vertical line (Fig. 13-3*b*) as the channel; the source and drain connect to this line. Also, an arrow is on the gate; *this arrow points to the n material.* In this way, you can remember that Fig. 13-3*a* represents an *n*-channel JFET.

In many JFETs the source and drain are interchangeable; you can use either end as the source and the other end as the drain. For this reason, the JFET symbol of Fig. 13-3*a* is symmetrical; the gate points to the center of the channel.

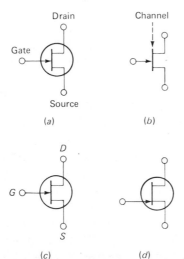

Figure 13-3. Schematic symbol for n-*channel JFET.*

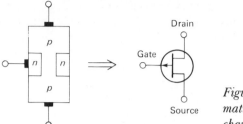

Figure 13-4. Schematic symbol for p-channel JFET.

When you use the symmetrical JFET symbol, you have to label the terminals as done in Fig. 13-3*a* or use abbreviations as shown in Fig. 13-3*c*. Or, you can use the nonsymmetrical symbol of Fig. 13-3*d;* here, the gate is drawn closer to the source.

Figure 13-4 shows the *p*-channel JFET and its schematic symbol. The *p*-channel JFET is complementary to the *n*-channel JFET. For this reason, we can concentrate on the *n*-channel JFET and extend the results to *p*-channel JFETs by the theory of complements discussed earlier.

13-2 *JFET drain curves*

Figure 13-5*a* shows a JFET with normal biasing voltages. Conventional current will flow down the channel from drain to source.

pinchoff voltage

As already discussed, the gate is reverse-biased for normal operation. Included in this is the special case of zero gate voltage. In other words, we can reduce V_{GS} to zero as shown in Fig. 13-5*b;* this special case is called the *shorted-gate* condition.

Figure 13-5*c* is the graph of drain current versus drain voltage for the shorted-gate condition. Notice the similarity to a collector curve. The drain current rises rapidly at first, but then levels off. In the region between V_P and $V_{DS(max)}$ the drain current is almost constant. When the drain voltage is too large, the JFET breaks down as shown. As with a bipolar transistor, the active region is along the almost horizontal part of the curve; in this region the JFET acts like a current source. We can symbolize this normal operating range by

$$V_P < V_{DS} < V_{DS(max)} \qquad\qquad (13\text{-}1)***$$

The *pinchoff voltage* V_P is the drain voltage above which drain current becomes essentially constant. When the drain voltage equals V_P, the channel has become

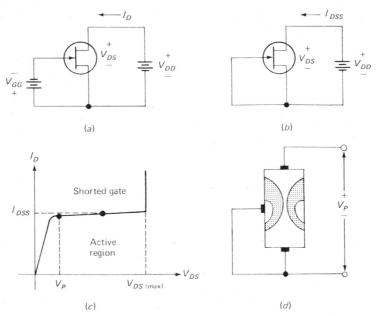

Figure 13-5. (a) *Normal bias for JFET.* (b) *Zero gate voltage.* (c) *Shorted-gate drain current.* (d) *Pinchoff.*

narrow and the depletion layers *almost touch,* as shown in Fig. 13-5*d.* The small passage between the depletion layers tends to limit the current. Further increases in drain voltage result in only a slight increase in drain current. This is why drain current is almost constant in the active region (see Fig. 13-5*c*).

shorted-gate drain current

The subscripts of I_{DSS} stand for *D*rain to *S*ource with *S*horted gate. Data sheets specify I_{DSS} at some drain voltage between pinchoff and breakdown; typically, this drain voltage is between 10 and 20 V. The important thing is this: because the curve is almost flat in the active region, I_{DSS} (no matter where measured) is a close approximation for the drain current anywhere in the active region. And, since I_{DSS} is measured under shorted-gate conditions, it is the *maximum* drain current you can get with normal operation of the JFET; all other gate voltages are negative and result in less drain current.

JFET data sheets give you the value of I_{DSS}. For instance, the data sheet of a 2N5457 lists the following typical values: $I_{DSS} = 3$ mA at $V_{DS} = 15$ V, $V_P = 2$ V, and $V_{DS(\text{max})} = 25$ V. This means the shorted-gate drain current is approximately 3 mA for any drain voltage between pinchoff (2 V) and breakdown (25

V). On a curve tracer, the highest drain curve will have approximately 3 mA throughout the almost flat part of the curve.

gate-source cutoff voltage

Drain curves look very much like collector curves. For instance, Fig. 13-6 shows drain curves for a typical JFET. The highest curve is for $V_{GS} = 0$, the shorted-gate condition. The pinchoff voltage is approximately 4 V, and the breakdown voltage is 30 V. So, the active region for this particular JFET is

$$4V < V_{DS} < 30 \text{ V}$$

As we see, I_{DSS} is 10 mA for V_{DS} of 15 V.

A negative gate voltage leads to other drain curves. A V_{GS} of -1 V drops the drain current to about 5.62 mA. A V_{GS} of -2 V reduces drain current to around 2.5 mA, and so on. The bottom curve is especially important; a V_{GS} of -4 V reduces the drain current to approximately zero. We call this voltage the *gate-source cutoff voltage* and symbolize it by $V_{GS(off)}$.

JFETs have a wide range of $V_{GS(off)}$ values. Data sheets specify $V_{GS(off)}$ at an arbitrarily small value of drain current. For instance, the data sheet of an MPF102 lists a maximum $V_{GS(off)}$ of -8 V for a drain current of 2 nA.

At $V_{GS} = V_{GS(off)}$, the depletion layers touch; this explains why the drain current is approximately zero. As we saw earlier, V_P is the value of drain voltage that pinches off the current for the shorted-gate condition. Because of this,

$$V_P = | V_{GS(off)} | \qquad (13\text{-}2)***$$

Data sheets do not list V_P, but they do give $V_{GS(off)}$, which is equivalent. For example, if a data sheet gives a $V_{GS(off)}$ of -4 V, we immediately know V_P equals 4 V.

Equation (13-1) pins down the normal range of V_{DS}. Now we can tie down

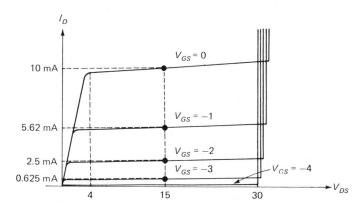

Figure 13-6. Typical set of drain curves.

the normal range of V_{GS}. Since the shorted-gate condition gives us the highest drain curve and $V_{GS(off)}$ produces the lowest drain curve, the normal range of V_{GS} is

$$V_{GS(off)} < V_{GS} < 0 \qquad (13\text{-}3a)$$

When V_{GS} is in this range, I_D must be in the interval

$$0 < I_D < I_{DSS} \qquad (13\text{-}3b)$$

As an example, in Fig. 13-6 the normal range of drain voltage is between 4 and 30 V, the normal range of gate voltage is between −4 and 0 V, and the normal range of drain current is between 0 and 10 mA.

EXAMPLE 13-1
The data sheet of a 2N5457 lists a reverse gate leakage current of 1 nA for a reverse gate voltage of 15 V. What is the resistance between the gate and the source?

SOLUTION
Use Ohm's law:

$$R_{GS} = \frac{V_{GS}}{I_G} = \frac{15 \text{ V}}{1 \text{ nA}} = 15{,}000 \text{ M}\Omega$$

Here you see the major advantage of a JFET over a bipolar transistor. The input resistance is 15,000 MΩ versus a few thousand ohms for the bipolar CE amplifier.

13-3 *JFET parameters*

As we saw earlier, the transconductance curve relates the output current to the input voltage. With a bipolar transistor the transconductance curve is the graph of I_C versus V_{BE}; with a JFET it is graph of I_D versus V_{GS}. For instance, by reading the values of I_D and V_{GS} in Fig. 13-6, we can plot the transconductance curve shown in Fig. 13-7a. In general, the transconductance curve of any JFET will look like Fig. 13-7b.

parabolic curve

The transconductance curve in Fig. 13-7b is part of a *parabola*. It can be shown that the equation of the transconductance curve is

$$I_D = I_{DSS} \left[1 - \frac{V_{GS}}{V_{GS(off)}} \right]^2 \qquad (13\text{-}4)\text{***}$$

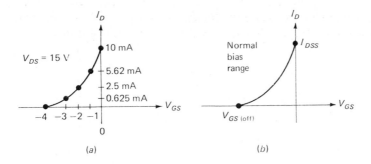

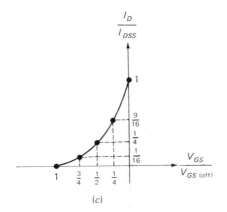

Figure 13-7. Trans-conductance curves.

As an example, suppose a JFET has an I_{DSS} of 4 mA and a $V_{GS(off)}$ of -2 V. By substitution into Eq. (13-4),

$$I_D = 0.004 \left(1 + \frac{V_{GS}}{2}\right)^2$$

With this equation we can calculate the drain current for any gate voltage in the active region.

Many data sheets do not give drain curves or transconductance curves. Instead, you get the values of I_{DSS} and $V_{GS(off)}$. By substituting these values into Eq. (13-4), you can calculate the drain current for any gate voltage.

Square law is another name for parabolic. This is why JFETs are often called square-law devices. For reasons to be discussed later, the square-law property gives the JFET another advantage over the bipolar in circuits called *mixers*.

normalized transconductance curve

We can rearrange Eq. (13-4) to get

$$\frac{I_D}{I_{DSS}} = \left[1 - \frac{V_{GS}}{V_{GS(off)}}\right]^2 \qquad (13\text{-}5)$$

By substituting 0, ¼, ½, ¾, and 1 for $V_{GS}/V_{GS(off)}$, we can calculate corresponding values of 1, $\frac{9}{16}$, ¼, $\frac{1}{16}$, and 0 for I_D/I_{DSS}. Figure 13-7c summarizes these results; it applies to all JFETs.

Here's a practical application for the curve of Fig. 13-7c. To bias a JFET near the middle of its useful current range, we need to set up an I_D of approximately half I_{DSS}. The current ratio $\frac{9}{16}$ is close to the midway point in drain current; therefore, we can set up *midpoint bias* with a V_{GS} of approximately

$$V_{GS} \cong \frac{V_{GS(off)}}{4} \quad \text{(midpoint bias)} \qquad (13\text{-}6)\text{***}$$

Given an MPF102 with a $V_{GS(off)} = -8$ V, we have to use a $V_{GS} = -2$ V to get a drain current of approximately half the maximum allowable drain current.

transconductance

The quantity g_m is called the *transconductance*, defined as

$$g_m = \frac{\Delta I_D}{\Delta V_{GS}} \quad \text{for constant } V_{DS} \qquad (13\text{-}7)$$

This says transconductance equals the change in drain current divided by the corresponding change in gate voltage. If a change in gate voltage of 0.1 V produces a change in drain current of 0.2 mA,

$$g_m = \frac{0.2 \text{ mA}}{0.1 \text{ V}} = 2(10^{-3}) \text{ S} = 2000 \text{ } \mu\text{S}$$

Note: S is the symbol for the unit "siemens," formerly referred to as "mho."

Figure 13-8 brings out the meaning of g_m in terms of the transconductance curve. To calculate g_m at any operating point, we select two nearby points like A and B on each side of the Q point. The ratio of the change in I_D to the change in V_{GS} gives us the g_m value between these two points. If we select

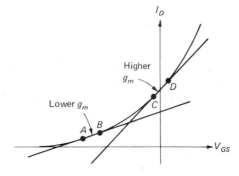

Figure 13-8. Graphical meaning of transconductance.

another pair of points further up the curve at C and D, we get more of a change in I_D for a given change in V_{GS}; therefore, g_m has a larger value further up the curve.

On a data sheet for JFETs you are usually given the value of g_m at $V_{GS} = 0$, that is, the value of g_m between points like C and D in Fig. 13-8. We will designate this value of g_m as g_{m0} to indicate it is measured at $V_{GS} = 0$. By deriving the slope of the transconductance curve at other points, we can prove any g_m equals

$$g_m = g_{m0} \left[1 - \frac{V_{GS}}{V_{GS(off)}} \right] \qquad (13\text{-}8)***$$

This equation gives g_m at any operating point in terms of the g_{m0} on a data sheet.

Incidentally, g_m is often designated as g_{fs} (forward transconductance) or y_{fs} (forward transadmittance). So if you cannot find g_m on the data sheet, look for g_{fs} or for y_{fs}. As an example, the data sheet of a 2N5951 gives $g_{fs} = 6.5$ mS at $V_{GS} = 0$; this is equivalent to a $g_{m0} = 6.5$ mS $= 6500$ μS. As another example, the 2N5457's data sheet lists $y_{fs} = 3000$ μS for $V_{GS} = 0$, equivalent to $g_{m0} = 3000$ μS.

an accurate value of $V_{GS(off)}$

With calculus, we can derive the following useful formula:

$$V_{GS(off)} = -\frac{2 I_{DSS}}{g_{m0}} \qquad (13\text{-}9)***$$

This is useful because while I_{DSS} and g_{m0} are easily measured with high accuracy $V_{GS(off)}$ is difficult to measure; Eq. (13-9) gives us a way of calculating $V_{GS(off)}$ with high accuracy.

ac drain resistance

Resistance r_{ds} is the ac resistance from the drain to the source, defined by

$$r_{ds} = \frac{\Delta V_{DS}}{\Delta I_D} \qquad \text{for constant } V_{GS} \qquad (13\text{-}10)$$

Above the pinchoff voltage, the change in I_D is small for a change in V_{DS} because the curve is almost flat; therefore, r_{ds} has large values, typically from 10 kΩ to 1 MΩ. As an example, if a change in drain voltage of 2 V produces a change in drain current of 0.02 mA,

$$r_{ds} = \frac{2V}{0.02 \text{ mA}} = 100 \text{ k}\Omega$$

Data sheets do not usually list the value of r_{ds}. Instead, they give a reciprocal specification, either g_{os} (output conductance) or y_{os} (output admittance). Drain resistance is related to these data-sheet values as follows:

$$r_{ds} = \frac{1}{g_{os}}$$

(13-10a)

and

$$r_{ds} = \frac{1}{y_{os}} \qquad \text{for low frequency}$$

(13-10b)

For instance, the data sheet of a 2N5951 gives $g_{os} = 75\ \mu$S. With Eq. (13-10a),

$$r_{ds} = \frac{1}{g_{os}} = \frac{1}{75(10^{-6})} = 13.3\ \text{k}\Omega$$

On the other hand, the 2N5457's data sheet lists $y_{os} = 50\ \mu$S. With Eq. (13-10b),

$$r_{ds} = \frac{1}{y_{os}} = \frac{1}{50(10^{-6})} = 20\ \text{k}\Omega$$

The next chapter discusses the effect of r_{ds} on the amplification of a JFET stage.

drain-source on-state resistance

In the active region, a JFET acts like a current source. But in the saturation region (drain voltage less than V_P), it acts like a resistor. Why? Because in the saturation region, a change in drain voltage produces a proportional change in drain current. This is the reason the saturation region of a JFET is often referred to as the *ohmic region.*

The resistance of a JFET operating in the ohmic region is defined as

$$r_{ds\,(\text{on})} = \frac{\Delta V_{DS}}{\Delta I_D}$$

(13-11)

For example, if a change in drain voltage of 100 mV produces a change of 0.7 mA in the ohmic region,

$$r_{ds\,(\text{on})} = \frac{100\ \text{mV}}{0.7\ \text{mA}} = 142\ \Omega$$

Chapter 14 will tell you why $r_{ds\,(\text{on})}$ is important in *choppers* and *analog switches* (two applications where the JFET is used as a switch).

EXAMPLE 13-2

A JFET has $I_{DSS} = 10$ mA and $g_{m0} = 4000\ \mu$S. Calculate $V_{GS\,(\text{off})}$. Also, work out g_m at midpoint bias.

SOLUTION

With Eq. (13-9),

$$V_{GS(\text{off})} = -\frac{2 I_{DSS}}{g_{m0}} = -\frac{2 \times 0.01}{0.004} = -5 \text{ V}$$

At midpoint bias,

$$V_{GS} = \frac{V_{GS(\text{off})}}{4} = \frac{-5 \text{ V}}{4} = -1.25 \text{ V}$$

Now use Eq. (13-8) to get

$$g_m = g_{m0}\left[1 - \frac{V_{GS}}{V_{GS(\text{off})}}\right] = 0.004\left(1 - \frac{1.25}{5}\right)$$
$$= 3000 \ \mu\text{S}$$

13-4 *depletion-enhancement MOSFETs*

The *metal-oxide semiconductor* FET or MOSFET has a source, gate, and drain; gate voltage controls the drain current. The main difference between a JFET and a MOSFET is that we can apply positive gate voltages and still have essentially zero gate current.

mosfet regions

Let's look at the parts of a MOSFET. To begin with, there's an n region with source and drain as shown in Fig. 13-9a. As before, a positive voltage applied to the drain-source terminals forces conduction-band electrons to flow from source to drain. Unlike the JFET, the MOSFET has only a single p region as shown in Fig. 13-9b. We call this region the *substrate*. This p region constricts the channel between source and drain so that only a small passage remains at the left side of Fig. 13-9b. Electrons flowing from source to drain must pass through this narrow channel.

A thin layer of metal oxide (usually silicon dioxide) is deposited over the left side of the channel as shown in Fig. 13-9c. This metal oxide is an *insulator*. Finally, a metallic gate is deposited on the insulator (Fig. 13-9d). Because the gate is insulated from the channel, a MOSFET is also known as an *insulated-gate* FET (IGFET).

depletion mode

How does the MOSFET of Fig. 13-10a work? As usual, the V_{DD} supply forces conduction-band electrons to flow from source to drain. These electrons flow through the narrow channel to the left of the p substrate.

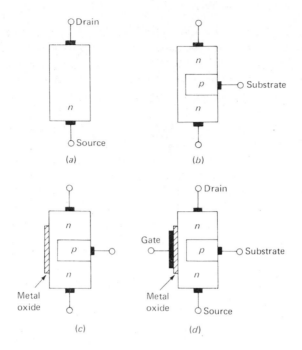

Figure 13-9. Parts of MOSFET. (a) *The* n *channel.* (b) *Adding the substrate.* (c) *Adding the metal oxide.* (d) *Adding the gate.*

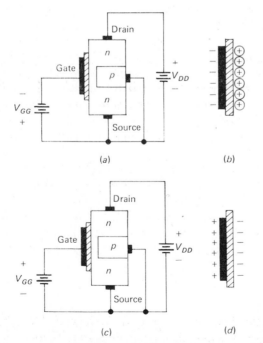

Figure 13-10. (a) *Depletion-mode operation.* (b) *Depleting the channel.* (c) *Enhancement-mode operation.* (d) *Enhancing channel conductivity.*

As before, the gate voltage can control the resistance of the n channel. But since the gate is insulated from the channel, we can apply either a positive or negative voltage to the gate. In Fig. 13-10a we have shown a negative gate voltage. The simplest way to visualize conduction in the channel is this. Think of the gate as one plate of a capacitor; the metal oxide acts like a dielectric and the n channel like the other plate. From basic theory, we know charges on a capacitor plate induce opposite charges on the other plate. Therefore, a negative gate voltage means electrons are on the gate as shown in Fig. 13-10b; these negative charges repel conduction-band electrons in the n channel, leaving a layer of positive ions in part of the channel (Fig. 13-10b). In other words, we have depleted the n channel of some of its conduction-band electrons.

The more negative the gate voltage, the greater the depletion of conduction-band electrons in the n channel. With enough negative gate voltage we can cut off the current between source and drain. Therefore, with negative gate voltage the action of a MOSFET is similar to the JFET. Because the action with a negative gate depends on depleting the channel of conduction-band electrons, we call negative-gate operation the *depletion mode*.

enhancement mode

Since the gate of a MOSFET is insulated from the channel, we can apply a positive voltage to the gate as shown in Fig. 13-10c. Again, the gate acts like a capacitor plate. But this time, the positive charges on the gate induce negative charges in the n channel (see Fig. 13-10d). These negative charges are conduction-band electrons drawn into the channel. Because these conduction-band electrons are added to those already in the channel, the total number of conduction-band electrons in the channel has increased. In other words, a positive gate voltage increases or *enhances* the conductivity of the channel. The more positive the gate voltage, the greater the conduction from source to drain.

Operation of the MOSFET with a positive gate voltage depends on enhancing channel conductivity. For this reason, positive-gate operation (Fig. 13-10c) is called the *enhancement mode*.

Because of the insulating layer, negligible gate current flows in either mode of operation. In fact, the input resistance of a MOSFET is incredibly high, typically from 10,000 to over 10,000,000 MΩ.

The device of Fig. 13-10c is an n-channel MOSFET; the complementary device is the p-channel MOSFET. We will concentrate on the n-channel MOSFET and extend the results to the p-channel MOSFET by complementing currents and voltages.

MOSFET curves

Figure 13-11a shows typical drain curves for an n-channel MOSFET. $V_{GS\text{(off)}}$ represents the negative gate voltage that cuts off drain current. For V_{GS} less

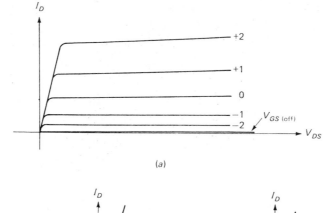

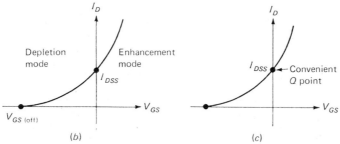

Figure 13-11. (a) *Drain curves.* (b) *Transconductance curves.* (c) *Simple bias point.*

than zero we get depletion-mode operation. On the other hand, V_{GS} greater than zero gives enhancement-mode operation.

Figure 13-11*b* is the transconductance curve of a MOSFET. I_{DSS} represents the drain-source current with a shorted gate. But now, the curve extends to the right of the origin as shown. The curve is still parabolic, and we can use the square-law equation described earlier. That is,

$$I_D = I_{DSS} \left[1 - \frac{V_{GS}}{V_{GS\,(\text{off})}} \right]^2 \tag{13-12}$$

This is identical to the square-law equation of a JFET; the value of V_{GS}, however, can be positive or negative.

MOSFETs with a transconductance curve like Fig. 13-11*c* are easier to bias than JFETs. The reason is this. If we want, we can use the Q point shown in Fig. 13-11*c*. With this Q point, $V_{GS} = 0$ and $I_D = I_{DSS}$. Setting up a V_{GS} of zero is easy; it requires no dc voltage on the gate. The resulting circuit is simple and is discussed in the next chapter.

Because the MOSFET of Fig. 13-10 can operate in either the depletion mode or the enhancement mode, we call it a *depletion-enhancement* MOSFET. Since this kind of MOSFET conducts when $V_{GS} = 0$, it also is known as a *normally on* MOSFET.

schematic symbol

Figure 13-12*a* shows the schematic symbol for a *normally on* MOSFET. The gate appears like a capacitor plate. Just to the right of the gate is the thin vertical line representing the channel; the drain lead comes out the top of the

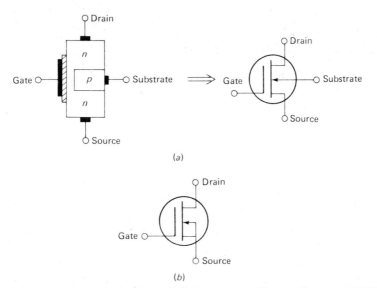

Figure 13-12. Schematic symbol for n-*channel normally on MOSFET.*

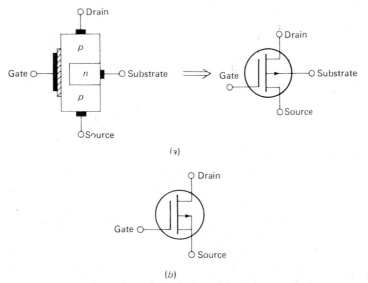

Figure 13-13. Schematic symbol for p-*channel normally on MOSFET.*

channel and the source lead connects to the bottom. The arrow is on the substrate and points to the *n* material; therefore, we have an *n*-channel MOSFET.

The manufacturer internally connects the substrate to the source; this results in a three-terminal device whose schematic symbol is shown in Fig. 13-12*b*.

Figure 13-13*a* shows a *p*-channel MOSFET and its schematic symbol. Figure 13-13*b* is the schematic symbol when the substrate is connected to the source.

13-5 *enhancement MOSFETs*

There is one other kind of MOSFET, the *enhancement-only* type. As its name implies, it operates in the enhancement mode only. This kind of MOSFET is important in digital circuits.

creating the inversion layer

Figure 13-14*a* shows the different parts of an enhancement-only MOSFET. Notice the substrate extends all the way to the metal oxide; structurally, there no longer is an *n* channel between the source and drain.

How does it work? Figure 13-14*b* shows normal biasing polarities. When $V_{GS} = 0$, the V_{DD} supply tries to force conduction-band electrons to flow from source to drain, but the *p* substrate has only a few thermally produced conduc-

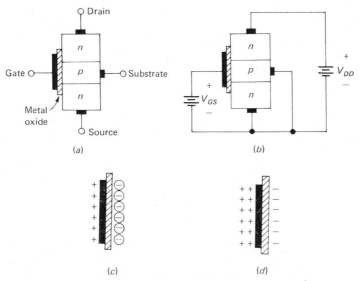

Figure 13-14. (a) *Enhancement-only MOSFET.* (b) *Normal biasing.* (c) *Creating negative ions.* (d) *Creating the* n-*type inversion layer.*

tion-band electrons. Aside from these minority carriers and some surface leakage, the current between source and drain is zero. For this reason, an enhancement-only MOSFET is also called a *normally off* MOSFET.

To get current, we have to apply enough positive voltage to the gate. The gate acts like one plate of a capacitor, the metal oxide like a dielectric, and the *p* substrate like the other plate. For lower gate voltages the positive charges in Fig. 13-14*c* induce negative charges in the *p* substrate. These charges are negative ions, produced by valence electrons filling holes in the *p* substrate. If we further increase the gate voltage, the additional positive charges on the gate can put conduction-band electrons into orbit around the negative ions (see Fig. 13-14*d*). In other words, when the gate is positive enough, it can create a thin layer of conduction-band electrons that stretch all the way from the source to the drain.

The created layer of conduction-band electrons is next to the metal oxide. This layer no longer acts like a *p*-type semiconductor. Instead, it appears like an *n*-type semiconductor. For this reason, the layer of *p* material touching the metal oxide is called an *n-type inversion layer.*

the threshold voltage

The minimum gate-source voltage that creates the *n*-type inversion layer is called the *threshold voltage* $V_{GS(th)}$. When V_{GS} is less than $V_{GS(th)}$, zero current flows from source to drain. But when V_{GS} is greater than $V_{GS(th)}$, an *n*-type inversion layer connects the source to the drain and we get current.

Threshold voltages depend on the particular type of MOSFET; $V_{GS(th)}$ can vary from less than a volt to more than 5 V. The 3N169 is an example of an enhancement-only MOSFET; it has a maximum threshold voltage of 1.5 V.

enhancement-only curves

Figure 13-15*a* shows a set of curves for an enhancement-only MOSFET. The lowest curve is the $V_{GS(th)}$ curve. For gate voltages greater than the threshold value we get the higher curves.

Figure 13-15*b* is the transconductance curve. The curve is parabolic or square-law; the vertex of the parabola is at $V_{GS(th)}$. Because of this, the equation for the parabola is different from before; it now equals

$$I_D = K[V_{GS} - V_{GS(th)}]^2 \qquad (13-13)$$

where *K* is a constant of proportionality that depends on the particular MOSFET.

Data sheets usually give the coordinates for one point on the transconductance curve as shown in Fig. 13-15*b*; after you substitute $I_{D(on)}$, $V_{GS(on)}$, and $V_{GS(th)}$ into Eq. (13-13), you can solve for the value of *K*. For instance, if an enhancement-only MOSFET has $I_{D(on)} = 8$ mA, $V_{GS(on)} = 5$ V, and $V_{GS(th)} = 3$ V, its transconduc-

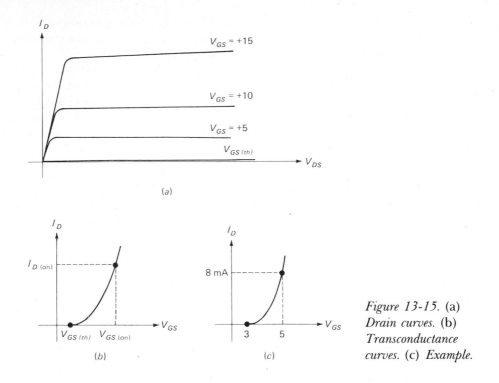

Figure 13-15. (a) *Drain curves.* (b) *Transconductance curves.* (c) *Example.*

tance curve looks like Fig. 13-15c. When we substitute these values into Eq. (13-13), we get

$$0.008 = K(5-3)^2 = 4K$$

or

$$K = 0.002$$

Therefore, the transconductance equation of Fig. 13-15c is

$$I_D = 0.002(V_{GS} - 3)^2$$

schematic symbol

When $V_{GS} = 0$, the enhancement-only MOSFET is off because no conducting channel exists between source and drain. The schematic symbol of Fig. 13-16a has a broken channel line to indicate this normally off condition. As we know, a gate voltage greater than the threshold voltage creates an n-type inversion layer that connects the source to the drain. The arrow points to this inversion layer, which acts like an n channel when the device is conducting. For this reason, we have an n-channel enhancement-only MOSFET.

Figure 13-16b shows the schematic symbol for the complementary MOSFET, the p-channel enhancement-only MOSFET.

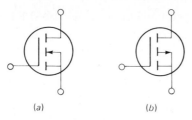

(a)   (b)

Figure 13-16. (a) n-*channel normally off MOSFET.* (b) p-*channel normally off MOSFET.*

13-6 *summary*

There are three fundamental types of FETs: the JFET, the depletion-enhancement MOSFET, and the enhancement-only MOSFET. The JFET operates only in the depletion mode (Fig. 13-17a), the depletion-enhancement MOSFET in

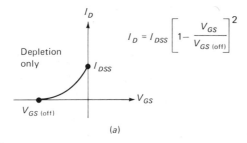

$$I_D = I_{DSS}\left[1 - \frac{V_{GS}}{V_{GS\,(off)}}\right]^2$$

(a)

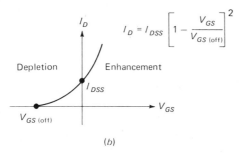

$$I_D = I_{DSS}\left[1 - \frac{V_{GS}}{V_{GS\,(off)}}\right]^2$$

(b)

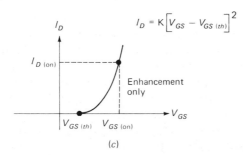

$$I_D = K\left[V_{GS} - V_{GS\,(th)}\right]^2$$

(c)

Figure 13-17. Transconductance curves. (a) *JFET.* (b) *Normally on MOSFET.* (c) *Normally off MOSFET.*

either mode (Fig. 13-17b), and the enhancement-only MOSFET only in the enhancement mode (Fig. 13-17c).

All are square-law devices, meaning their transconductance curve is parabolic. The transconductance equations of Fig. 13-17a and b are the same because each parabola has its vertex at $V_{GS(off)}$. The equation for Fig. 13-17c, however, differs because the vertex is at $V_{GS(th)}$.

Two of the main advantages of FETs are high input impedance and the square-law property. Other advantages of FETs as well as their disadvantages are discussed in the next chapter.

A final point. Data sheets usually refer to the depletion-enhancement MOSFET as a *depletion-type* MOSFET (abbreviated D MOSFET). The enhancement-only MOSFET is called an *enhancement-type* MOSFET (E MOSFET).

EXAMPLE 13-3

The data sheet for a 3N128 indicates that it's a depletion-type MOSFET with a gate leakage current of -50 pA for a gate voltage of -8 V. Calculate the gate resistance of this D MOSFET.

SOLUTION

$$R_{GS} = \frac{V_{GS}}{I_G} = \frac{8}{50(10^{-12})} = 160,000 \text{ M}\Omega$$

Gate leakage current is much smaller in a MOSFET than a JFET because of the insulated gate. Furthermore, JFET leakage current rises exponentially with temperature increase because of the reverse-biased *pn* junction; MOSFET leakage current is much less temperature-sensitive. Therefore, when you need extremely high input resistance over a large temperature range, the MOSFET is preferred to the JFET.

EXAMPLE 13-4

A 2N3796 data sheet indicates it's a D MOSFET with these maximum ratings:

1. $V_{DS(max)} = 25$ V
2. $V_{GS(max)} = \pm30$ V
3. $I_{D(max)} = 20$ mA
4. $P_{D(max)} = 300$ mW at $T_A = 25°$C; derate 1.7 mW per degree

What do these ratings mean?

SOLUTION

Ratings like these are similar to bipolar ratings, but a few comments may help.

1. $V_{DS(max)}$ is the maximum voltage you can apply between the drain and source without breakdown. You must keep V_{DS} less than $V_{DS(max)}$ for normal operation.

2. $V_{GS(max)}$ is the maximum voltage you can apply between the gate and source without destroying the thin layer of metal oxide. Breakdown of the thin insulating layer is always destructive; you can throw the MOSFET away if it happens. The polarity is immaterial; a gate voltage greater than $+30$ V or more negative than -30 V destroys the metal-oxide layer.

Aside from directly applying a V_{GS} greater than 30 V, you can destroy the insulating layer in more subtle ways. If you remove or insert a MOSFET into a circuit while the power is on, transient voltages may be large enough to ruin the MOSFET. Besides transient voltage, static voltage may destroy a MOSFET. Pick it up often and you may deposit enough static charge on the gate to exceed the $V_{GS(max)}$ rating. This is the reason MOSFETs are shipped with a wire ring around the leads. You remove the ring after the MOSFET is connected in the circuit.

3. $I_{D(max)}$ is the maximum continuous drain current you can have without risking damage; in this case, not more than 20 mA of continuous drain current should flow.

4. The $P_{D(max)}$ rating of 300 mW means the MOSFET can dissipate up to 300 mW when the ambient temperature is 25°C. For higher ambient temperatures, you subtract 1.7 mW for each degree above 25°C.

A final point. Some newer types of MOSFETs are protected by *built-in* zener diodes connected between the gate and the source. These zener diodes will break down before any damage occurs to the thin insulating layer. The disadvantage of these built-in diodes is they reduce the MOSFET's high input resistance.

self-testing review

Read each of the following and provide the missing words. Answers appear at the beginning of the next question.

1. The three parts of a JFET are the source, the _____, and the _____. The name *field effect* is related to the _____ layers around each *pn* junction. The more negative the gate voltage, the _____ the drain current.

2. *(gate, drain, depletion, smaller)* The key difference between a JFET and a bipolar transistor is this: the gate is _____-biased whereas the base is _____-biased. This crucial difference means the JFET is a _____-controlled device.

3. *(reverse, forward, voltage)* When the drain voltage equals the pinchoff voltage, the depletion layers almost _____. This keeps drain current almost constant for further increases in drain voltage.

4. *(touch)* I_{DSS} is the current from drain to source with shorted gate. Since I_{DSS} is measured with a shorted gate, it is the _____ drain current you can get with normal operation of the JFET; all other gate voltages are negative and result in _____ drain current.

5. *(maximum, less)* $V_{GS(off)}$ is the gate-source voltage that cuts off the _____ current. V_P and $V_{GS(off)}$ are _____ in value.

6. *(drain, equal)* Square law is another name for parabolic. The transconductance curve is part of a parabola. This is why JFETs are often called _____ devices.

7. *(square-law)* Transconductance equals a change in drain current divided by the corresponding change in _____ voltage. Transconductance is designated g_m. When $V_{GS} = 0$, transconductance is designated _____.

8. *(gate, g_{m0})* Because the gate is insulated from the channel, a MOSFET is also known as an insulated-gate FET. The D MOSFET can operate in either the _____ mode or the _____ mode. This type of MOSFET is also known as a normally _____ MOSFET.

9. *(depletion, enhancement, on)* The E MOSFET operates in the _____ mode only. This kind of MOSFET is important in digital circuits. It is also known as a normally _____ MOSFET.

10. *(enhancement, off)* The gate-source voltage that just turns on an E MOSFET is called the _____ voltage.

11. *(threshold)* Two of the main advantages of FETs are high input resistance and the square-law property.

problems

13-1. At room temperature the 2N4220 (an *n*-channel JFET) has a reverse gate current of 0.1 nA for a reverse gate voltage of 15 V. Calculate the resistance from gate to source.

13-2. When the ambient temperature is 150°C, a JFET has a reverse gate current of 100 nA for a reverse gate voltage of 15 V. What is the resistance from gate to source at this elevated temperature?

13-3. A JFET data sheet indicates $V_{GS(off)}$ has a minimum value of -2 V and a maximum value of -4.5 V. What is the minimum pinchoff voltage we can expect with this type of JFET? If the $V_{DS(max)}$ of this JFET is 25 V, in what range should V_{DS} be for the JFET to act almost like a current source (use the larger pinchoff voltage)?

13-4. If a JFET has the drain curves of Fig. 13-18*a*, what value does I_{DSS} have? To ensure operation on the almost flat part of the top curve, what range should V_{DS} be kept in?

13-5. A JFET has an I_{DSS} of 9 mA and a $V_{GS(off)}$ of -3 V. What is the transconductance equation for this JFET? How much drain current is there when $V_{GS} = -1.5$ V?

13-6. Write the transconductance equation for the JFET whose curve is shown by Fig. 13-18*b*. How much drain current is there when $V_{GS} = -4$ V? When $V_{GS} = -2$ V?

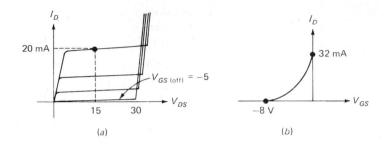

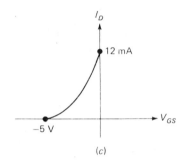

Figure 13-18.

13-7. Calculate the value of drain current for a $V_{GS} = -4.5$ V in Fig. 13-18b.

13-8. If a JFET has a transconductance curve like Fig. 13-18c, what value does I_{DSS} have? $V_{GS(off)}$? Pinchoff voltage V_P?

13-9. Write the equation for the JFET transconductance curve of Fig. 13-18c.

13-10. If a JFET has a square-law curve like Fig. 13-18c, how much drain current is there when $V_{GS} = -1$ V?

13-11. A JFET has a drain current of 5 mA. If $I_{DSS} = 10$ mA and $V_{GS(off)} = -6$ V, what is the *approximate* value of V_{GS}? Of V_P?

13-12. An n-channel depletion-type MOSEET has an I_{DSS} of 8 mA and a $V_{GS(off)}$ of -4 V. How much drain current is there when $V_{GS} = -1$ V? And when $V_{GS} = 1$ V?

13-13. A MOSFET has the transconductance curve of Fig. 13-19a. What is the value of V_P? Of I_{DSS}? Of I_D when V_{GS} is -1 V?

13-14. The normally on MOSFET of Fig. 13-19b has the transconductance curve of Fig. 13-19a. $V_{GS} = 0$. How much drain current is there? What is the value of V_{DS}? The FET is operating above the knee of the $V_{GS} = 0$ drain curve. Why?

13-15. A MOSFET has a gate leakage current of 20 pA for a gate-source voltage of 10 V. What is the resistance from gate to source?

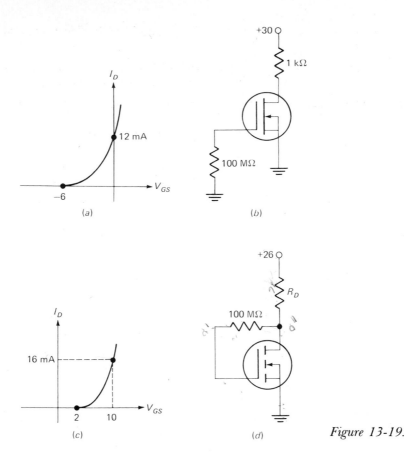

Figure 13-19.

13-16. An enhancement-type MOSFET has the transconductance curve of Fig. 13-19c. What value does the constant K in Eq. (13-13) have?

13-17. If a normally off MOSFET has the curve of Fig. 13-19c, how much drain current is there when $V_{GS} = 3$ V? $V_{GS} = 4$ V?

13-18. The MOSFET shown in Fig. 13-19d has the transconductance curve of Fig. 13-19c. Negligible gate current flows. If $V_{DS} = 10$ V, what is the value of V_{GS}? How much drain current is there? What value of R_D satisfies the given values?

13-19. The data sheet of a 2N5951 lists an I_{GS} of -200 nA for $V_{GS} = -15$ V and $T = 100°C$. Calculate the resistance between the gate and the source.

13-20. A 2N3823 has a $V_{GS(off)}$ of -8 V. What is the approximate V_{GS} for midpoint bias? To get a drain current of one-quarter I_{DSS}, what value should V_{GS} have?

13-21. When V_{GS} changes from -2.1 to -2 V, the drain current changes from 1 to 1.3 mA. What is the value of g_m in this region?

13-22. A JFET has a g_{m0} of 5000 μS and a $V_{GS(off)}$ of -4 V. Calculate g_m at -1 V and at -3 V.

13-23. A 2N3822 has a g_{m0} of 6.5 mS and a $V_{GS(off)}$ of -6 V. If it is biased at $I_D \cong 0.5 I_{DSS}$, what is the approximate value of g_m at this operating point?

13-24. A JFET has an I_{DSS} of 10 mA and a g_{m0} of 10,000 μS. Calculate its $V_{GS(off)}$.

14

FET circuit analysis

This chapter discusses dc and ac operation of FETs. After deriving bias and gain formulas, we discuss applications like _buffers, AGC amplifiers,_ and _choppers._

14-1 _self-bias_

Figure 14-1a shows _self-bias,_ the most common method for biasing a JFET. Drain current flows down through R_D and R_S, producing a drain-source voltage of

$$V_{DS} = V_{DD} - I_D(R_D + R_S) \qquad (14\text{-}1)\text{***}$$

The voltage across the source resistor is

$$V_S = I_D R_S$$

Because gate current is negligibly small, the gate terminal is at dc ground, so that

$$V_G \cong 0$$

Therefore, the difference of potential between the gate and the source is

$$V_{GS} = V_G - V_S = 0 - I_D R_S$$
or
$$V_{GS} = -I_D R_S \qquad (14\text{-}2)\text{***}$$

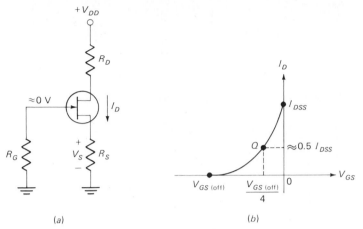

Figure 14-1. Self-bias. (a) *Circuit.* (b) *Typical* Q *point.*

This says the drop across R_S produces the biasing voltage V_{GS}. No external voltage source has to drive the gate, and this is why the circuit is known as self-bias.

Self-bias stabilizes the quiescent operating point against changes in JFET *parameters* (quantities like I_{DSS}, g_{m0}, etc.). Here's the idea. Suppose we substitute a JFET with a g_{m0} that's twice as large. Then, the drain current of Fig. 14-1*a* will try to double. But since this drain current flows through R_S, the gate-source voltage V_{GS} becomes more negative and reduces the original increase in drain current.

In Fig. 14-1*b*, a gate voltage equal to one-fourth of $V_{GS(off)}$ results in a drain current equal to half of I_{DSS} (approximately). Substituting these quantities into Eq. (14-2) and solving for R_S gives

$$R_S = \frac{-V_{GS(off)}}{2 I_{DSS}} \tag{14-3}$$

With Eq. (13-9), we can reduce the foregoing to this useful formula:

$$R_S \cong \frac{1}{g_{m0}} \qquad \text{(midpoint bias)} \tag{14-4}***$$

Given the g_{m0} of a JFET, take the reciprocal and you've got the source resistance that sets up a drain current equal to half I_{DSS}. Since g_{m0} is always accurately specified on data sheets, Eq. (14-4) gives us a fast way to set up self-bias at the midpoint of drain current.

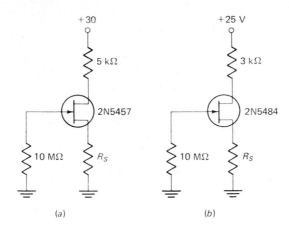

Figure 14-2.

EXAMPLE 14-1

The 2N5457 of Fig. 14-2a has g_{m0} = 5000 μS and I_{DSS} = 5 mA. What is the value of R_S that produces midpoint bias? What is the corresponding V_{GS}? The V_{DS}?

SOLUTION

$$R_S \cong \frac{1}{g_{m0}} = \frac{1}{5000(10^{-6})} = 200 \ \Omega$$

This source resistance results in a drain current of about 2.5 mA.
   The gate-source voltage is

$$V_{GS} = -I_D R_S = -2.5(10^{-3})200 = -0.5 \ \text{V}$$

The drain-source voltage is

$$V_{DS} = V_{DD} - I_D(R_D + R_S)$$
$$= 30 - 2.5(10^{-3})(5000 + 200) = 17 \ \text{V}$$

EXAMPLE 14-2

The 2N5484 of Fig. 14-2b has g_{m0} = 2.5 mS. What is the value of R_S that sets up midpoint bias?

SOLUTION

$$R_S \cong \frac{1}{g_{m0}} = \frac{1}{2.5(10^{-3})} = 400 \ \Omega$$

This resistance sets up an I_D of approximately half I_{DSS}.

EXAMPLE 14-3

The data sheet of a 2N5457 specifies a minimum g_{m0} of 1 mS and a maximum g_{m0} of 5 mS. This means that when working with thousands of 2N5457s, we will get some whose g_{m0} is as low as 1 mS and some with a g_{m0} as high as 5 mS. If a 2N5457 is used in a mass-produced self-bias circuit, what value of R_S do we need to set up midpoint bias?

SOLUTION

We have to compromise here and use an average value. As discussed in Chap. 7, whenever you get a large spread in parameter values, it's best to use a geometric average. The geometric average for transconductance is given by

$$g_{m0} = \sqrt{g_{m0(min)} \, g_{m0(max)}} \qquad (14\text{-}5)\text{***}$$

Substituting the minimum and maximum g_{m0} for a 2N5457 results in

$$g_{m0} = \sqrt{1(10^{-3})5(10^{-3})} = 2.24 \text{ mS}$$

Therefore,

$$R_S \cong \frac{1}{g_{m0}} = \frac{1}{2.24(10^{-3})} = 446 \ \Omega$$

14-2 *the self-bias graph*

With Eqs. (13-5), (13-9), and (14-2), we can derive a relation between drain current, transconductance, and the source-biasing resistor. Figure 14-3 summarizes this relation. This graph applies to all JFETs. It will help you pin down the Q point of self-biased circuits. The following examples show you how.

EXAMPLE 14-4

A self-biased circuit uses a JFET with $I_{DSS} = 10$ mA, $R_S = 100 \ \Omega$, and $g_{m0} = 3000 \ \mu$S. What does the drain current equal?

SOLUTION

$$g_{m0} R_S = 3000(10^{-6})100 = 0.3$$

In Fig. 14-3, the corresponding current ratio is

$$\frac{I_D}{I_{DSS}} = 0.78$$

Since I_{DSS} is given as 10 mA,

$$I_D = 0.78 I_{DSS} = 0.78(10 \text{ mA}) = 7.8 \text{ mA}$$

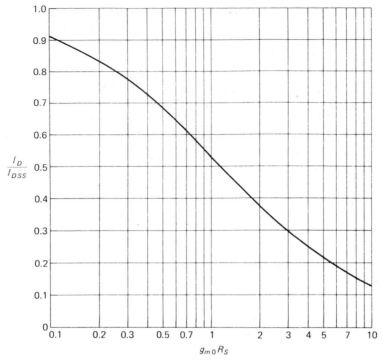

Figure 14-3. Self-bias curve.

EXAMPLE 14-5

A JFET has g_{m0} = 8000 μS. What value of R_S do we need to get an I_D of one-quarter I_{DSS}.

SOLUTION

We are given I_D / I_{DSS} = 0.25. In Fig. 14-3, read the corresponding $g_{m0} R_S$ product, which is

$$g_{m0} R_S = 4$$

The required source resistance is

$$R_S = \frac{4}{g_{m0}} = \frac{4}{8000(10^{-6})} = 500 \ \Omega$$

EXAMPLE 14-6

In Fig. 14-4, the first JFET has I_{DSS} = 8 mA and g_{m0} = 4000 μS. The second JFET has I_{DSS} = 15 mA and g_{m0} = 3300 μS. What is the drain current in each stage?

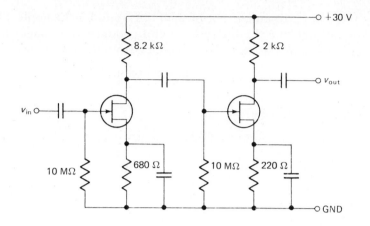

Figure 14-4.

SOLUTION

The first stage has a product of

$$g_{mo} R_S = 0.004 \times 680 = 2.72$$

With Fig. 14-3, we read a corresponding current ratio of

$$\frac{I_D}{I_{DSS}} = 0.32$$

Therefore,

$$I_D = 0.32 I_{DSS} = 0.32 \times 8 \text{ mA} = 2.56 \text{ mA}$$

The second stage has

$$g_{mo} R_S = 0.0033 \times 220 = 0.726$$

The current ratio is

$$\frac{I_D}{I_{DSS}} = 0.61$$

and the drain current is

$$I_D = 0.61 I_{DSS} = 0.61 \times 15 \text{ mA} = 9.15 \text{ mA}$$

EXAMPLE 14-7

What are the dc voltages with respect to ground for all points in Fig. 14-4?

SOLUTION

To begin with, the gate current is usually small enough to neglect. Therefore, the gate voltage to ground is approximately 0 V for both stages.

The first stage has a drain current of 2.56 mA, found in Example 14-6. This current flows through the 8.2-kΩ resistor and produces a voltage drop. The supply voltage minus this drop gives the voltage from the first drain to ground:

$$V_D = V_{DD} - I_D R_D = 30 - 0.00256(8200)$$
$$= 9 \text{ V}$$

The voltage from the source to ground is

$$V_S = I_D R_S = 0.00256 \times 680 = 1.74 \text{ V}$$

In the second stage, the drain-to-ground voltage is

$$V_D = 30 - 0.00915(2000) = 11.7 \text{ V}$$

and the source-to-ground voltage is

$$V_S = 0.00915 \times 220 = 2.01 \text{ V}$$

If you were troubleshooting an amplifier like this, a good starting point would be to check the dc voltages and make sure they agree with calculated values.

14-3 *current-source bias*

Current-source bias is the ultimate method for stabilizing drain current against variations in FET parameters.

two supplies

Figure 14-5a shows how it's done when a *split supply* is available (positive and negative supply voltages). The bipolar acts like a current source and forces the JFET to have an I_D equal to I_C.

In Fig. 14-5a, the bipolar transistor has an emitter current of

$$I_E \cong \frac{V_{EE}}{R_E}$$

The collector diode acts like a current source; therefore, it forces the drain current to approximately equal I_E. A condition that must be satisfied is

$$I_C < I_{DSS} \tag{14-6}$$

This guarantees that V_{GS} is negative.

Current-source bias like Fig. 14-5a is swamping at its best. V_{GS} and its variations are almost completely out of the picture. The only significant variable is the V_{BE} of the bipolar transistor. It varies slightly from one transistor to another, and with temperature change. But these V_{BE} changes are only a tenth volt or

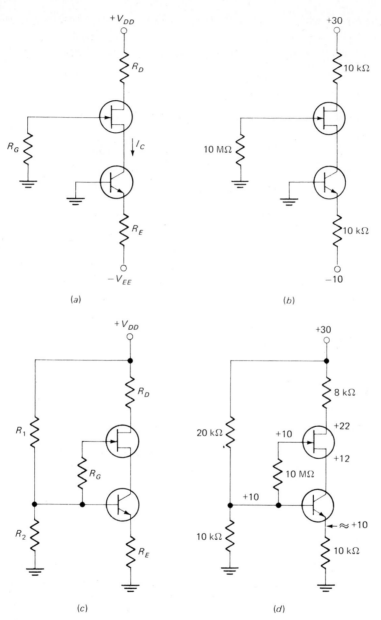

Figure 14-5. Current-source bias.

so. Therefore, with a circuit like Fig. 14-5a, we have an almost solid value of I_D.

As a concrete example, the emitter current in Fig. 14-5b is

$$I_E \cong \frac{V_{EE}}{R_E} = \frac{10}{10,000} = 1 \text{ mA}$$

This forces the drain current to approximately equal 1 mA. The drain-to-ground voltage is

$$V_D = V_{DD} - I_D R_D = 30 - 0.001(10,000) = 20 \text{ V}$$

one supply

When you do not have a negative supply voltage, you can still use current-source bias as shown in Fig. 14-5c. In this circuit the voltage divider (R_1 and R_2) sets up voltage-divider bias on the bipolar transistor. Almost all the voltage across R_2 appears across the R_E resistor. This fixes the emitter current to a value essentially independent of the JFET characteristics. Again, the collector diode acts like a current source, forcing the drain current to equal the collector current.

Especially note, in Fig. 14-5c you do not ground the bottom of R_G. You must connect it to the base of the bipolar; this is necessary to reverse-bias the collector diode.

EXAMPLE 14-8
Analyze the circuit of Fig. 14-5d.

SOLUTION
The Thevenin base voltage is 10 V. Most appears across R_E and sets up

$$I_E \cong \frac{10}{10,000} = 1 \text{ mA}$$

The collector diode now forces approximately 1 mA of current to flow through the JFET. When this flows through the drain resistor (8 kΩ), it produces an 8-V drop and makes the drain voltage 22 V with respect to ground. Since the base is 10 V to ground, the gate also must be 10 V to ground (no gate current). Assuming a V_{GS} of −2 V, the collector is 12 V with respect to ground; this is more than enough to reverse-bias the collector diode.

14-4 *biasing MOSFETs*

With D MOSFETs, V_{GS} can be positive or negative. But with E MOSFETs, V_{GS} has to be greater than $V_{GS(th)}$ to get current. Let's take a look at how to bias D and E MOSFETs.

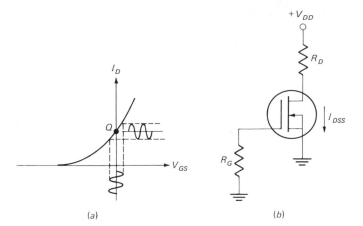

Figure 14-6. Zero bias.

(a)   (b)

zero-bias of D MOSFETs

Since a D MOSFET can operate in either the depletion or enhancement mode, we can set its Q point at $V_{GS} = 0$ as shown in Fig. 14-6a. Then, an ac input signal to the gate can produce variations above and below the Q point. Being able to use zero V_{GS} is an advantage when it comes to biasing. It permits the unique biasing circuit of Fig. 14-6b. This simple circuit has no applied gate or source voltage; therefore, $V_{GS} = 0$ and $I_D = I_{DSS}$. It follows that

$$V_{DS} = V_{DD} - I_{DSS}R_D \qquad (14\text{-}7)$$

As long as V_{DS} is greater than V_P, operation is on the almost flat part of the $V_{GS} = 0$ drain curve.

The *zero-bias* of Fig. 14-6a is unique with D MOSFETs; it will not work with a bipolar, JFET, or E MOSFET.

drain-feedback bias of E MOSFETs

Figure 14-7a shows *drain-feedback bias*, a type of bias you can use with E MOSFETs. With negligible gate current, no voltage appears across R_G; therefore, $V_{GS} = V_{DS}$. Typically, V_{DS} is kept above 10 V to ensure operation well above the pinchoff voltage. Like collector-feedback bias, the circuit of Fig. 14-7a tends to compensate for changes in FET characteristics. If I_D tries to increase for some reason, V_{DS} decreases; this reduces V_{GS}, which partially offsets the original increase in I_D.

Figure 14-7b shows the Q point on the transconductance curve. V_{GS} equals V_{DS}, and the corresponding I_D equals $I_{D(on)}$, a value of drain current well above the threshold point. To assist you, data sheets for E MOSFETs usually give a value of $I_{D(on)}$ for $V_{GS} = V_{DS}$. This helps in setting up the Q point. In design, all you do is select a value of R_D that sets up the specified V_{DS}.

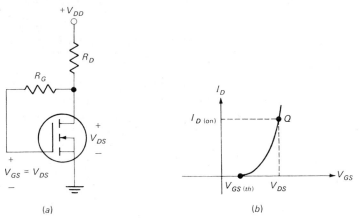

Figure 14-7. Drain-feedback bias.

For instance, suppose the data sheet of an E MOSFET specifies $I_{D(on)} = 3$ mA when $V_{GS} = V_{DS} = 10$ V. If we have a 25-V supply to work with, we can select an R_D of 5 kΩ as shown in Fig. 14-8a. When 3 mA of drain current flows, V_{DS} equals 10 V and so too does V_{GS}. Therefore, the E MOSFET is operating at its specified *on* point (Fig. 14-8b).

In a circuit like Fig. 14-8a different MOSFETs or a varying temperature may cause I_D to differ from 3 mA. But the changes are partially offset by the feedback from drain to gate. If I_D tries to increase above 3 mA, V_{DS} drops below 10 V; this lowers V_{GS}, which reduces the attempted change in I_D. The overall effect is a smaller increase in I_D than would take place without the feedback.

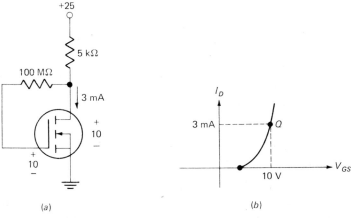

Figure 14-8.

dc amplifier

A *dc amplifier* is one that can operate all the way down to zero frequency without a loss of gain. One way to build a dc amplifier, or dc amp, is to leave out all coupling and bypass capacitors.

Figure 14-9 shows a dc amp using MOSFETs. The input stage is a D MOSFET with zero bias. The second and third stages use E MOSFETs; each gate gets its V_{GS} from the drain of the preceding stage. The design of Fig. 14-9 uses MOSFETs with drain currents of 3 mA. For this reason, each drain runs at 10 V with respect to ground. We tap the final output voltage between the 100-kΩ resistors. Since the lower resistor is returned to −10 V, the quiescent output voltage is 0 V. When an ac signal drives the amplifier, regardless of how low its frequency, we get an amplified output voltage.

There are other ways of designing dc amplifiers. The beauty of Fig. 14-9 is its simplicity.

other biasing

By examining self- and current-source bias, we can determine which MOSFETs work in. Table 14-1 summarizes the biasing circuits used with MOSFETs.

Here is what the table indicates. JFETS and D MOSFETs will work in a self-biased circuit, but the E MOSFET will not, because V_{GS} is always negative with self-bias. Also, JFETs and D MOSFETs are fine with current-source bias, but not E MOSFETs. Why? Because a positive V_{GS} would force the bipolar into saturation.

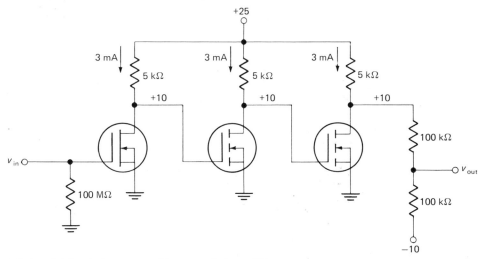

Figure 14-9. A three-stage direct-coupled amplifier.

Table 14-1 FET Biasing Circuits

	JFET	D MOSFET	E MOSFET
Self-bias	Yes	Yes	No
Current-source bias	Yes	Yes	No
Zero-bias	No	Yes	No
Drain-feedback bias	No	No	Yes

Zero-bias is not suitable with JFETs or E MOSFETs because JFETs need a reverse-biased gate and E MOSFETs need a forward-biased gate. Finally, drain-feedback bias works only with E MOSFETs.

14-5 *the common-source amplifier*

The complete ac equivalent circuit of a FET includes lead inductance, internal capacitances, and so on. We neglect the lead inductances and a few other minor effects and start with the approximate equivalent circuit shown in Fig. 14-10a. C_{gd} is the capacitance between the gate and the drain, C_{gs} the capacitance from

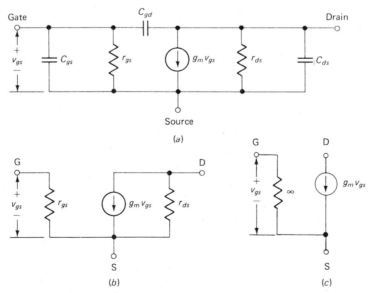

Figure 14-10. Ac equivalent circuit of FET.

gate to source, and r_{gs} the ac resistance from gate to source. On the output side, C_{ds} and r_{ds} are the ac capacitance and resistance from drain to source.

low-frequency model

At lower frequencies the X_C of each capacitor is high enough to neglect, and the equivalent circuit reduces to Fig. 14-10b. Resistance r_{gs} is hundreds of megohms, high enough to ignore in a practical analysis. The ac drain resistance r_{ds} (discussed in Chap. 13) is given by the reciprocal of g_{os} (or y_{os}), specified on data sheets. Resistance r_{ds} is almost always greater than 10 kΩ and often exceeds 100 kΩ. Because r_{ds} is large, we can ignore it in our ideal model.

Figure 14-10c shows the ideal ac equivalent circuit for JFETs and MOSFETs. The ac input voltage appears across an infinite input resistance and the drain acts like a current source with a value of

$$i_d = g_m v_{gs} \quad \text{(ideal)} \quad (14\text{-}8)$$

This equation says the ac drain current equals the transconductance times the ac gate-source voltage.

voltage gain

Ac analysis of FET amplifiers is straightforward. You start by shorting all coupling and bypass capacitors; then you reduce the dc-supply voltages to zero. You simplify this ac equivalent as much as possible by applying Thevenin's theorem and combining parallel resistances. Often, the circuit will reduce to the form shown in Fig. 14-11a.

If we replace the FET of Fig. 14-11a by its ideal model, we get Fig. 14-11b. The positive half cycle of input voltage forces the drain current to flow up through r_D; this produces the negative half cycle of output voltage. In other words, a common-source amplifier always inverts the signal.

In Fig. 14-11b,

$$v_{\text{out}} = i_d r_D = g_m v_{\text{in}} r_D$$

which rearranges into

$$\frac{v_{\text{out}}}{v_{\text{in}}} = g_m r_D$$

Therefore, the voltage gain of a common-source amplifier is

$$A = g_m r_D \quad (14\text{-}9)\text{***}$$

Sometimes the ac equivalent circuit reduces to the form shown in Fig. 14-11c. In this case, there is *local feedback,* similar to an unbypassed emitter resistor in a bipolar amplifier. The voltage gain is given by

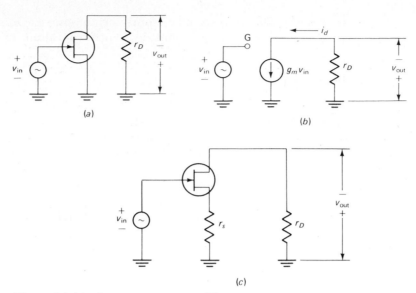

Figure 14-11. Common-source amplifier.

$$A = \frac{g_m r_D}{1 + g_m r_S} \tag{14-10}$$

effect of r_{ds}

For a common-source amplifier with $r_S = 0$, the internal drain resistance r_{ds} is in parallel with the ac load resistance r_D. Because of this, a more accurate formula for voltage gain is

$$A = g_m(r_D \| r_{ds}) \tag{14-11}$$

When r_{ds} is much greater than r_D, the parallel equivalent resistance equals r_D, and the equation simplifies to Eq. (14-9).

transconductance versus I_D/I_{DSS}

With Eqs. (13-5) and (13-8), we can derive a relation between transconductance and drain current. Figure 14-12 summarizes this relation. This graph gives the value of g_m/g_{m0} at different operating points, something we need to know when calculating voltage gain.

EXAMPLE 14-9
The JFET of Fig. 14-13a has $g_{m0} = 5000$ μS. What is the g_m at the Q point? The voltage gain?

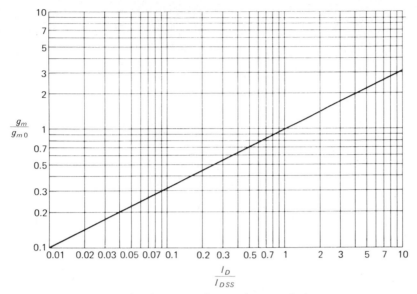

Figure 14-12. Normalized transconductance versus drain current.

SOLUTION

It's a self-biased amplifier with

$$g_{mo} R_S = 0.005 \times 400 = 2$$

In Fig. 14-3, a $g_{mo} R_S$ of 2 gives an I_D/I_{DSS} of approximately 0.38. Next, refer to Fig. 14-12 where you can see that a current ratio of 0.38 gives

$$\frac{g_m}{g_{mo}} \cong 0.62$$

or $$g_m = 0.62 g_{mo} = 0.62(5000)10^{-6} = 3100 \ \mu\text{S}$$

Now we can find the voltage gain. Visualize the ac equivalent circuit of the amplifier. Then, $r_D = 5$ kΩ. Since the source resistor is bypassed, there is no local feedback and

$$A = g_m r_D = 0.0031 \times 5000 = 15.5$$

EXAMPLE 14-10

The data sheet of a 2N5457 lists a typical y_{os} of 10 μS. If used in Fig. 14-13a, what effect does r_{ds} have on voltage gain? (Use a g_m of 3100 μS.)

SOLUTION

Resistance r_{ds} is the reciprocal of y_{os}:

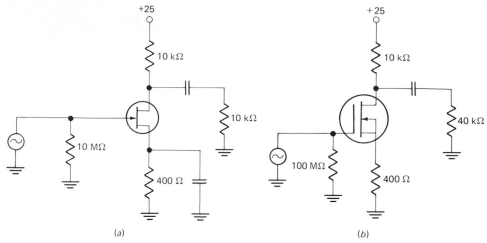

Figure 14-13.

$$r_{ds} = \frac{1}{y_{os}} = \frac{1}{10(10^{-6})} = 100 \text{ k}\Omega$$

With Eq. (14-11),

$$A = g_m(r_D \| r_{ds}) = 0.0031(5000 \| 100{,}000) = 14.8$$

EXAMPLE 14-11
The D MOSFET of Fig. 14-13*b* has a g_m of 5000 μS at its Q point. What is the ideal voltage gain of the stage?

SOLUTION
Visualize the ac equivalent circuit and you can see

$$r_D = 10{,}000 \| 40{,}000 = 8 \text{ k}\Omega$$
$$r_S = 400 \text{ }\Omega$$

With Eq. (14-10),

$$A = \frac{g_m r_D}{1 + g_m R_S} = \frac{0.005(8000)}{1 + 0.005(400)} = 13.3$$

14-6 *the common-drain amplifier*

In Fig. 14-11*c*, r_D can equal zero and the output signal can be taken from the source terminal. In this case, we have a *source follower,* analogous to the emitter follower. The voltage gain of a source follower is given by

$$A = \frac{g_m r_S}{1 + g_m r_S}$$ (14-12)***

If $g_m r_S$ is much greater than unity, this reduces to

$$A \cong 1$$

The source follower (also called the *common-drain* amplifier) acts like an emitter follower; its voltage gain is less than unity. A source follower has a very high input resistance and is often used at the front end of measuring instruments like voltmeters and oscilloscopes.

EXAMPLE 14-12

In Fig. 14-14, the 2N5457 has a g_m of 3000 μS and the 2N3906 has a β of 200. What is the voltage gain of the two stages?

SOLUTION

The overall voltage gain is the product of the individual gains:

$$A = A_1 A_2$$

The second stage has an emitter current of approximately 1 mA; therefore, its r_e' is around 25 Ω. The input resistance of the second stage is

$$z_{\text{in(stage)}} = R_1 \| R_2 \| \beta r_e'$$
$$= 20,000 \| 10,000 \| 200(25) = 2.86 \text{ k}\Omega$$

The source of the first stage sees an ac load resistance of 330 Ω in parallel with the 2.86 kΩ of the second stage. So,

$$r_S = R_S \| z_{\text{in(stage)}} = 330 \| 2860 = 296 \ \Omega$$

With Eq. (14-12), the voltage gain of the first stage is

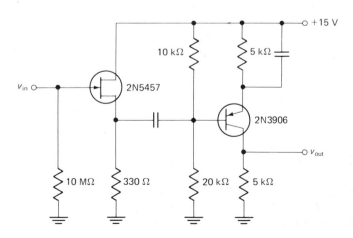

Figure 14-14. Source follower driving CE amplifier.

$$A_1 = \frac{g_m r_S}{1 + g_m r_S} = \frac{0.003(296)}{1 + 0.003(296)} = 0.47$$

The voltage gain of the second stage is

$$A_2 = \frac{r_C}{r_e'} = \frac{5000}{25} = 200$$

The total gain is

$$A = A_1 A_2 = 0.47(200) = 94$$

14-7 *the common-gate amplifier*

Figure 14-15 is the ac equivalent circuit for a *common-gate* amplifier. The output voltage is

$$v_{out} = i_d r_D = g_m v_{gs} r_D$$

The input voltage is

$$v_{in} = v_{gs}$$

Dividing v_{out} by v_{in},

$$\frac{v_{out}}{v_{in}} = \frac{g_m v_{gs} r_D}{v_{gs}}$$

or

$$A = g_m r_D \qquad (14\text{-}13)\text{***}$$

input impedance

The common-source and common-drain amplifiers have extremely high input resistance; it approaches infinity. The common-gate amplifier is different; its input resistance is low. Here's why. In Fig. 14-15,

$$i_{in} = i_d = g_m v_{gs}$$

Therefore,

$$z_{in} = \frac{v_{in}}{i_{in}} = \frac{v_{gs}}{g_m v_{gs}}$$

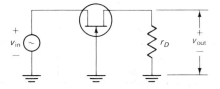

Figure 14-15. Ac equivalent of common-gate amplifier.

or
$$z_{\text{in}} = \frac{1}{g_m} \qquad\qquad (14\text{-}14)\text{***}$$

If $g_m = 5000 \ \mu\text{S}$, $z_{\text{in}} = 200 \ \Omega$.

Despite its low input resistance, the common-gate amplifier has found a few applications, the main one being a *cascode* amplifier (discussed in the next section).

summary

Table 14-2 summarizes the operation of the CS, CD, and CG amplifiers. As shown, the CS and CG amplifiers have a voltage gain of $g_m r_D$; the CD amplifier has a voltage gain less than unity.

The CS and CD amplifiers have input impedances approaching infinity, but the CG amplifier has an input impedance of only $1/g_m$.

The CS and CG circuits have output impedances of r_{ds} (typically greater than 10 kΩ), while the CD circuit has an output impedance of only $1/g_m$. This means the CD amplifier acts approximately like a voltage source for the load resistances normally used with it. The CS and CG amplifiers, on the other hand, act almost like current sources for their load resistances.

EXAMPLE 14-13

What is the voltage gain in Fig. 14-16 if the 2N5457 has a g_m of 3000 μS? The input impedance?

SOLUTION

The ac load resistance seen by the drain is

$$r_D = 5000 \,\|\, 100,000 = 4760 \ \Omega$$

The voltage gain is

$$A = g_m r_D = 0.003(4760) = 14.3$$

Table 14-2 Approximate Formulas

	Common Source	Common Drain	Common Gate
A	$g_m r_D$	$\dfrac{g_m r_S}{1 + g_m r_S}$	$g_m r_D$
z_{in}	Infinite	Infinite	$\dfrac{1}{g_m}$
z_{out}	r_{ds}	$\dfrac{1}{g_m}$	r_{ds}

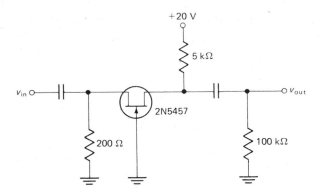

Figure 14-16.

The input impedance is

$$z_{\text{in}} = \frac{1}{g_m} = \frac{1}{0.003} = 333 \ \Omega$$

Note that in the ac equivalent circuit, the 200-Ω source resistor appears in parallel with the z_{in} of the common-gate amplifier. Therefore, the input impedance of the stage is

$$z_{\text{in(stage)}} = 333 \parallel 200 = 125 \ \Omega$$

14-8 *FET applications*

In this section, we discuss some of the applications where the FET's properties give it a clearcut advantage over the bipolar transistor.

buffer amplifier

Figure 14-17 shows a *buffer amplifier,* a stage that isolates the preceding stage from the following stage. Ideally, a buffer should have a high input impedance; this way, almost all the Thevenin voltage from stage *A* appears at the buffer input. The buffer should also have a low output impedance; this ensures that all its output reaches the input of stage *B.*

The source follower is an excellent buffer amplifier because of its high input

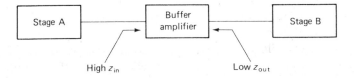

Figure 14-17. Use of buffer amplifier.

impedance (well into the megohms at low frequencies) and its low output impedance (typically a few hundred ohms). The high input impedance means light loading of the preceding stage. The low output impedance means the buffer can drive heavy loads (small load resistances).

low-noise amplifier

Noise is any unwanted disturbance superimposed upon a useful signal. Noise interferes with the information contained in the signal; the greater the noise, the less the information. For instance, the noise in television receivers produces small white or black spots on the picture; severe noise can wipe out the picture. Similarly, the noise in radio receivers produces crackling and hissing, which sometimes completely masks voice or music. Noise is independent of the signal because it exists even when the signal is off.

Chapter 23 discusses the nature of noise, how it is produced, and how to reduce it. All you need to know now is this: any electronic device produces a certain amount of noise. The FET is an outstanding low-noise device because it produces very little noise. This is especially important near the front end of receivers and other electronic equipment; subsequent stages amplify front-end noise the same as the signal. By using a FET amplifier at the front end, we get less amplified noise at the final output.

automatic gain control

When a receiver is tuned from a weak to a strong station, the loudspeaker will blare unless the volume is immediately decreased. Or the volume may change because of *fading,* a variation in signal strength caused by an electrical change in the path between the transmitting and receiving antennas. To counteract unwanted changes in volume, most receivers use *automatic gain control* (AGC).

This is where the FET comes in. As shown earlier,

$$g_m = g_{m0} \left[1 - \frac{V_{GS}}{V_{GS(off)}} \right]$$

This is a linear equation. When graphed, it results in Fig. 14-18a. For a JFET, g_m reaches a maximum value when $V_{GS} = 0$. As V_{GS} becomes more negative, the value of g_m decreases. Since a common-source amplifier has a voltage gain of

$$A = g_m r_D$$

we can *control* the voltage gain by controlling the value of g_m.

Figure 14-18b shows how it's done. A FET amplifier is near the front end of a receiver. It has a voltage gain of $g_m r_D$. Subsequent stages amplify the FET output. This amplified output goes into a peak detector that produces voltage

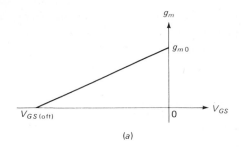

(a)

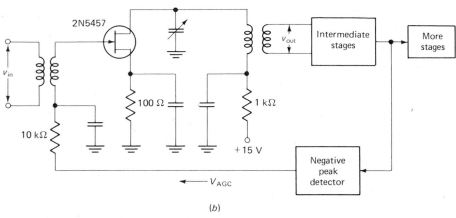

(b)

Figure 14-18. Automatic gain control.

V_{AGC}. This negative voltage returns to the FET amplifier, being applied to the gate through a 10-kΩ resistor. When the receiver is tuned from a weak to a strong station, a larger signal is peak-detected and V_{AGC} is more negative; this reduces the gain of the FET amplifier.

The overall effect of AGC is this: the final signal increases, but not nearly as much as it would without AGC. For instance, in some AGC systems an increase of 100 percent in the input signal results in an increase of less than 1 percent in the final output signal.

cascode amplifier

Figure 14-19 is an example of a *cascode amplifier,* a common-source amplifier driving a common-gate amplifier. Here's how it works. The CS amplifier has a gain of

$$A_1 = g_m r_D$$

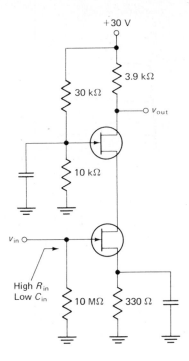

Figure 14-19. Cascode amplifier.

Resistance r_D equals $1/g_m$, the input impedance of the CG amplifier. Therefore,

$$A_1 = g_m r_D = g_m \frac{1}{g_m} = 1$$

This unity gain is important in the cascode connection because it minimizes the *Miller effect* (to be discussed in the next chapter). The CG amplifier has a gain of

$$A_2 = g_m r_D$$

So, the overall gain of the two FETs is

$$A = A_1 A_2 = g_m r_D$$

The whole purpose of the cascode connection is this: it has a low input capacitance, the direct result of minimum Miller effect. Chapter 16 will analyze the input capacitance of a cascode amplifier. For now, remember that a cascode amplifier has a high R_{in} and a low C_{in}.

analog switch

The FET is a useful switching device. Figure 14-20*a* shows a *shunt switch*. Here's the basic idea. Figure 14-20*b* is the load line of a FET used as a switch. The

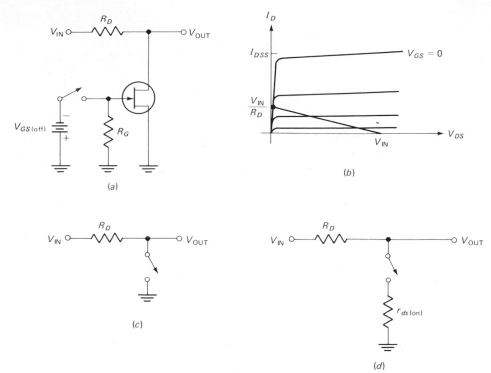

Figure 14-20. Analog switch.

cutoff voltage equals V_{IN}, and the saturation current ideally equals V_{IN}/R_D. The FET operates either at cutoff or saturation, depending on the applied gate voltage. If no voltage is applied, $V_{GS} = 0$ and the FET is saturated. In this case, it acts like a small resistance $r_{ds(on)}$, discussed in the preceding chapter. For many FETs, $r_{ds(on)}$ is less than 100 Ω; for some FETs, it's less than 10 Ω.

If a large negative voltage is applied, equal to or greater than $V_{GS(off)}$, the FET operates at the cutoff point and becomes a large resistance (well into the megohms).

To a first approximation, the shunt switch acts like the circuit of Fig. 14-20c. If the switch is open, $V_{OUT} = V_{IN}$. If the switch is closed, $V_{OUT} = 0$. To a second approximation, the shunt switch acts like Fig. 14-20d. When the switch is open, $V_{OUT} = V_{IN}$. But when it's closed, the voltage-divider theorem gives

$$V_{OUT} = \frac{r_{ds(on)}}{R_D + r_{ds(on)}} V_{IN} \qquad (14\text{-}14)$$

An *analog switch* is a device or circuit that either transmits or blocks an ac signal. The FET circuit of Fig. 14-20a is an analog switch because V_{IN} can be an ac signal.

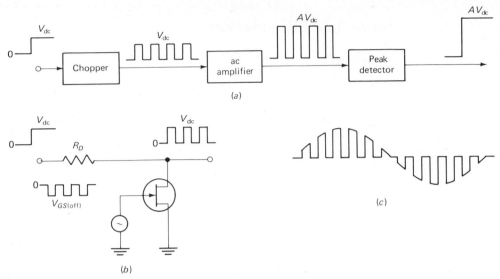

Figure 14-21. FET chopper.

FET choppers

As discussed earlier, we can build a direct-coupled amplifier by leaving out the coupling and bypass capacitors, and connecting the output of each stage directly to the input of the next stage. In this way, dc is coupled, as well as ac. But the major disadvantage of a direct coupling is drift, a slow shift in the final output voltage produced by supply and transistor variations.

Figure 14-21 shows the *chopper method* of building a dc amplifier. The input dc voltage is chopped by a switching circuit. This results in the square wave shown at the chopper output. The peak value of this square wave equals the value of V_{dc}. Because the square wave is an ac signal, we can use a conventional ac amplifier, one with coupling capacitors between the stages. The amplified output can then be peak-detected to recover the dc signal.

If we apply a square wave to the gate of a FET analog switch, it becomes a chopper (see Fig. 14-21b). The gate square wave is negative-going, swinging from 0 V to at least $V_{GS(off)}$. This alternately saturates and cuts off the FET. So, the output voltage is a square wave with a peak of V_{dc}.

If the input is a low-frequency ac signal, it gets chopped into ac waveform of Fig. 14-21c. In this way, the chopper amplifier of Fig. 14-21a can amplify either dc or low-frequency ac.

summary

Look at Table 14-3. Some of the terms are new and will be discussed in later chapters. The FET buffer has the advantage of high input impedance and low

Table 14-3 FET Applications

Application	Main Advantage	Uses
Buffer	High $z_{in'}$, low z_{out}	General purpose, measuring equipment, receivers
RF amplifier	Low noise	FM tuners, communication equipment
Mixer	Low intermodulation distortion	FM and TV receivers, communication equipment
AGC amplifier	Ease of gain control	Receivers, signal generators
Cascode amplifier	Low input capacitance	Measuring instruments, test equipment
Chopper	No drift	Dc amplifiers, guidance control systems
Voltage-variable resistor	Voltage controlled	Operational amplifiers, organ, tone controls
Low-frequency amplifier	Small coupling capacitor	Hearing aids, inductive transducers
Oscillator	Minimum frequency drift	Frequency standards, receivers
MOS digital circuit	Small size	Large-scale integration, computers, memories

output impedance. This is why the FET is the natural choice at the front end of voltmeters, oscilloscopes, etc., where you need high input resistances like 10 MΩ or more. As a guide, the input resistance looking into the gate of a JFET is from 100 to more than 10,000 MΩ. With a MOSFET the input resistance is from 10,000 to over 10,000,000 MΩ.

When used as a small-signal amplifier, the output voltage from a FET is linearly related to the input because only a small part of the square-law curve is used. But with larger signals, more of the curve is used, resulting in nonlinear distortion. This nonlinear distortion is unwanted in an amplifier. But in a *mixer,* square-law distortion has a tremendous advantage; this is why the FET is preferred to the bipolar for FM and TV mixer applications.

FETs can act as voltage-variable resistors. When a FET operates in the ohmic region, $r_{ds(on)}$ is a function of V_{GS}. This means that the FET acts like a variable resistor, controlled by its gate voltage. This can be applied to controlling the gain of *operational amplifiers.*

Oscillators are circuits that generate an output signal without an input signal. FET oscillators have the advantage of small frequency drift; they produce a sine wave whose frequency changes only slightly with change in temperature, supply voltage, and component aging.

You can expect the following usage: in discrete circuits a heavy use of bipolar transistors, a moderate use of JFETs, and a limited use of MOSFETs. On the other hand, with digital integrated circuits the bipolars and MOSFETs are heavily used; JFETs much less.

self-testing review

Read each of the following and provide the missing words. Answers appear at the beginning of the next question.

1. A common way to bias a JFET is _____. Drain current flows through the _____ resistor to produce V_S. Since the gate is approximately at dc ground, $V_{GS} = -V_S$.

2. *(self-bias, source)* Self-bias stabilizes the quiescent operating point against _____ in JFET parameters.

3. *(changes)* Given g_{m0}, take the reciprocal and you have the source resistance R_S that sets up a drain current approximately equal to _____ I_{DSS}. Whenever you get a large spread in g_{m0} for a particular JFET type, use the _____ average for transconductance.

4. *(half, geometric)* Another method of stabilizing the Q point of a JFET circuit is _____ bias. A bipolar transistor acts like a _____ source and forces the JFET to have an I_D equal to I_C.

5. *(current-source, current)* With D MOSFETs, V_{GS} can be negative or _____. But with E MOSFETs, V_{GS} has to be greater than $V_{GS(th)}$. Self-bias will work with a D MOSFET but not with an _____ MOSFET. The _____ MOSFET is the only FET that can use zero bias. The E MOSFET is the only FET that can use drain-feedback bias.

6. *(positive, E, D)* A dc amplifier is one that can operate all the way down to _____ frequency without a loss of gain. One way to build a dc amp is to leave out all coupling and bypass _____ and directly couple between stages.

7. *(zero, capacitors)* A common-source amplifier with the source terminal at ac ground has an approximate voltage gain of _____ times r_D.

8. *(g_m)* The source follower, also called a _____ amplifier, acts like an emitter follower. Its voltage gain is less than _____ and its input resistance is very _____. The source follower is often used at the front end of measuring instruments like voltmeters and oscilloscopes.

9. *(common-drain, unity, high)* The source follower is an excellent buffer amplifier because of its _____ input resistance and its _____ output resistance.

10. *(high, low)* The FET is an outstanding low-noise device because it produces very little noise. Low noise is important near the front end of electronics equipment because subsequent stages _____ front-end noise the same as the signal.

11. *(amplify)* To counteract unwanted changes in volume caused by fading or tuning, most receivers use _____ gain control. We can control the gain of a FET by changing its g_m.

12. *(automatic)* The cascode FET amplifier is a CS amplifier driving a CG amplifier. The FET is also used as an analog switch and as a chopper.

problems

14-1. A JFET has a g_{m0} of 10,000 μS. To set up midpoint bias, what value should R_S have?

14-2. The FET of Fig. 14-22a has a g_{m0} of 7000 μS. What is the approximate value of R_S for midpoint bias? If $I_{DSS} = 4$ mA, what is V_{GS} for this value of R_S? The corresponding V_{DS}?

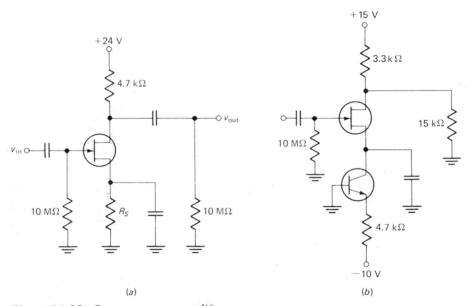

Figure 14-22. Common-source amplifiers.

14-3. If $I_G = 4$ nA in Fig. 14-22a, what is the voltage across the 10-MΩ resistor?

14-4. A 2N4222 has a minimum g_{m0} of 2500 μS and a maximum g_{m0} of 6500 μS. What is the geometric average for transconductance? What approximate R_S do you need for midpoint bias using this geometric average?

14-5. The FET of Fig. 14-22a has $I_{DSS} = 5$ mA and $g_{m0} = 2500$ μS. If $R_S = 820$ Ω, what does I_D equal? V_{GS}? V_{DS}?

14-6. A JFET has a g_{mo} of 1500 μS. What value of R_S do we need to get an I_D of three-quarters I_{DSS}?

14-7. You want to self-bias a JFET stage to $I_D/I_{DSS} = 0.8$. What is the required $g_{mo}R_S$ product? If $g_{mo} = 4500$ μS, what is the necessary R_S?

14-8. A self-biased JFET has a g_{mo} of 5000 μS; source resistance R_S equals 200 Ω. If $I_{DSS} = 2$ mA, what is the approximate value of I_D? Suppose the JFET is replaced by another JFET of the same type, except $g_{mo} = 10,000$ μS and $I_{DSS} = 3$ mA. What is the new value of I_D?

14-9. Calculate the value of I_D in Fig. 14-22b.

14-10. If a dc voltmeter is connected between the drain and ground in Fig. 14-22b, what is the approximate reading?

14-11. A self-biased p-channel JFET has minimum and maximum values as follows: g_{mo} from 2 to 6 mS, and I_{DSS} from 4 to 16 mA. Calculate the geometric averages. If R_S equals 330 Ω, what is the $g_{mo}R_S$ at the design center? The corresponding I_D?

14-12. The 2N3797 of Fig. 14-23a has these specifications: $I_{DSS(min)} = 2$ mA, $I_{DSS(typical)} = 2.9$ mA, and $I_{DSS(max)} = 6$ mA. Calculate the typical value of V_{DS}. What is the minimum possible value of V_{DS}? The maximum?

14-13. The 2N3797 of Fig. 14-23a has an I_G of 1 pA at $T = 25°C$. What is the dc voltage across the 100-MΩ resistor? When the temperature rises to $T = 150°C$, $I_G = 200$ pA. What is the new voltage across the gate resistor?

14-14. The 3N170 of Fig. 14-23b has an $I_{D(on)}$ of 10 mA when $V_{GS} = V_{DS} = 10$ V. What is the value of R_D that sets up this drain current?

14-15. Figure 14-23c shows part of an MOS digital circuit called a *flip-flop*. $V_{GS(th)} = 2$ V. When $V_{GS} = 10$ V, $I_D = 1$ mA. Suppose no drain current flows through T_1.
(a) What is the voltage from the drain of T_1 to ground?
(b) How much voltage is applied to the gate of T_2?
(c) How much voltage is on the drain of T_2?
(d) Why is T_1 off?

14-16. The g_m is 4000 μS in Fig. 14-22a. What is the voltage gain?

14-17. In Fig. 14-22b, $g_m = 3500$ μS and $g_{os} = 25$ μS. What does the voltage gain equal ideally? What does it equal when you take r_{ds} into account?

14-18. $R_S = 330$ Ω and $g_m = 3000$ μS in Fig. 14-22a. What is the voltage gain if the source bypass capacitor is not connected during production?

14-19. $I_{DSS} = 3$ mA in Fig. 14-22b. If $g_{mo} = 6500$ μS, what does A equal?

14-20. In Fig. 14-22b, $I_{DSS} = 5$ mA and $g_{mo} = 8000$ μS. What is the voltage gain?

14-21. The data sheet of a 2N3797 lists a minimum transconductance of 1500 μS and a maximum of 3000 μS. What is the minimum voltage gain in Fig. 14-23a? The maximum voltage gain?

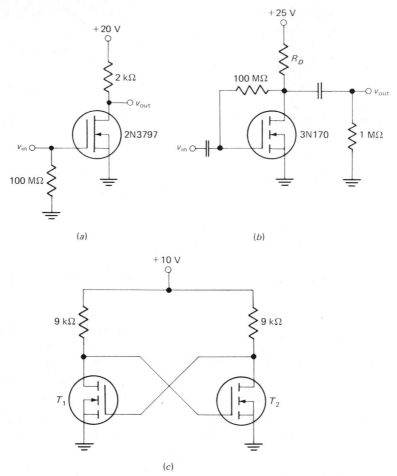

Figure 14-23. MOSFET circuits.

14-22. If $R_D = 1.5$ kΩ and $g_m = 3000$ μS in Fig. 14-23b, what is the voltage gain?

14-23. In Fig. 14-9, each MOSFET has a g_m of 2 mS. What is the voltage gain of this three-stage amplifier?

14-24. The 2N5486 of Fig. 14-24a has $g_{m0} = 6$ mS. What is the voltage gain of this source follower?

14-25. The data sheet of a 2N5486 lists the following minimum and maximum values: g_{m0} from 4 to 8 mS and I_{DSS} from 8 to 20 mA. Using geometric averages, what is the quiescent drain current in Fig. 14-24a? The dc voltage from source to ground? The voltage gain?

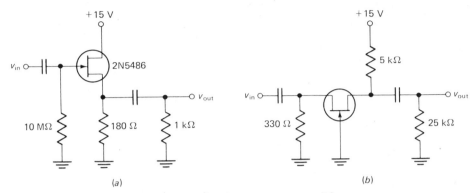

Figure 14-24. (a) *Source follower.* (b) *Common-gate amplifier.*

14-26. The FET of Fig. 14-24*b* has a g_m of 4000 μS. What is the voltage gain? The input impedance?

14-27. If g_{m0} = 5000 μS in Fig. 14-24*b*, what is the input impedance of the stage? The voltage gain?

14-28. In Fig. 14-17, stage *A* has these Thevenin values: V_{TH} = 5 V rms and R_{TH} = 100 kΩ. If the buffer amplifier has a z_{in} of 10 MΩ, how much voltage is across its input terminals? The buffer has V_{TH} = 5 V rms and R_{TH} = 100 Ω. If the z_{in} of stage *B* is 2 kΩ, what is the voltage at the input to stage *B?*

14-29. The 2N5457 has g_{m0} = 3000 μS and $V_{GS(off)}$ = −2 V in Fig. 14-18*b*. What is the value of g_m for each of these: V_{GS} = −1 V, V_{GS} = −1.5 V, and V_{GS} = −1.9 V?

14-30. Use the same g_{m0} and $V_{GS(off)}$ as the preceding problem. If the ac load resistance seen by the drain is r_D = 5 kΩ, what is the voltage gain of the AGC stage for V_{GS} = −1 V? If V_{GS} increases to −1.9 V, what is the voltage gain?

14-31. The data sheet of a 2N5457 shows g_{m0} can be as low as 1000 μS or as high as 5000 μS. Using the geometric average for g_{m0}, what is the voltage gain for the cascode amplifier if 2N5457s are used in Fig. 14-19?

14-32. In Fig. 14-20*a*, V_{IN} = 5 V and R_D = 100 kΩ. What does I_D equal with the gate switch open? With the gate switch closed? If $r_{ds(on)}$ = 50 Ω, what does V_{OUT} equal when the switch of Fig. 14-20*d* is closed?

14-33. Figure 14-25 shows a simple FET dc voltmeter. The zero adjust is set as in any voltmeter; the calibrate adjust is set periodically to give full-scale deflection when V_{in} = 2.5 V. A calibrate adjustment like this takes care of variations from one FET to another and FET aging effects.
   (*a*) The current through the 510-Ω resistor equals 4 mA. How much dc voltage is there from the source to ground?
   (*b*) If no current flows through the ammeter, what voltage does the wiper tap off the zero adjust?

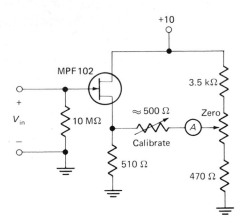

+10

MPF 102

V_{in}

+

−

10 MΩ

≈ 500 Ω

Calibrate

510 Ω

3.5 kΩ

Zero

A

470 Ω

Figure 14-25. FET voltmeter.

(c) If an input voltage of 2.5 V produces 1-mA deflection, how much deflection does 1.25 V produce (assume linearity)?

(d) The MPF102 has an I_{GSS} of 2 nA for a V_{GS} of 15 V. What is the input impedance of the voltmeter?

15 decibels, Miller's theorem, and hybrid parameters

Decibels are based on the logarithms learned in trigonometry. These logarithms simplify gain calculations. Miller's theorem applies to circuits like collector-feedback bias. When applicable, Miller's theorem can reduce a complicated circuit into two simple circuits. Hybrid parameters are easily measured transistor characteristics. When exact analysis is desired, hybrid parameters are used.

15-1 decibel power gain

Power gain G is defined as the ratio of output power to input power:

$$G = \frac{P_2}{P_1}$$

For instance, if the input power to an amplifier is 0.5 W and the output power is 15 W,

$$G = \frac{15}{0.5} = 30$$

This says the output power is 30 times greater than the input power.

Bel power gain G' is defined as the common logarithm (log to the base 10) of power gain:

$$G' = \log G \qquad (15\text{-}1)^{***}$$

If a circuit has a power gain of 100, its bel power gain is

$$G' = \log 100 = 2$$

G' is dimensionless, but to make sure it is not confused with G, we will attach the label *bel* (abbreviated B) to all answers for G'. The preceding answer therefore is written

$$G' = 2 \text{ B}$$

When an answer is in bels, we automatically know it represents the bel power gain and not the ordinary power gain.

heavily used values

Bel power gain is nothing more than the logarithm of the ordinary power gain. You can always find accurate logarithms in a table of logarithms or with a calculator. But most of the calculations in this book can be done by remembering the following approximate logarithms:

$\log 1 = 0$
$\log 2 = 0.3$
$\log 4 = 0.6$
$\log 8 = 0.9$
$\log 10 = 1$
$\log 10^2 = 2$
$\log 10^n = n$

The tens are the easiest to remember since the logarithm of 10^n equals n. The logarithms of 2, 4, and 8 are also easy when you notice the arithmetic progression 0.3, 0.6, and 0.9. (These are close approximations for the exact logarithms: 0.30103, 0.60206, and 0.90309.)

Also important to remember are these logarithmic relations:

$$\log xy = \log x + \log y$$
$$\log \frac{x}{y} = \log x - \log y$$

and

$$\log x^n = n \log x$$

All these are familiar from earlier courses. These relations along with the numerical values listed are heavily used not only in this book but throughout the electronics industry.

prefixes

0.001 volt can be written as 1 millivolt, or as 1000 microvolts, or in many other ways. So too can we express bel power gain in equivalent ways. For instance, if $G = 100$, then

$$G' = \log 100 = 2 \text{ bels (B)} = 20 \text{ decibels (dB)} = 200 \text{ centibels (cB)}$$
$$= 2000 \text{ millibels (mB)}$$

and so on. The only prefix used in practice in *deci* (one-tenth, or 10^{-1}). We abbreviate decibel as dB.

EXAMPLE 15-1
Calculate the bel power gain for each of the following: $G = 2$, 4, and 8.

SOLUTION
When $G = 2$,

$$G' = \log 2 = 0.3 \text{ B} = 3 \text{ dB}$$

When $G = 4$,

$$G' = \log 4 = 0.6 \text{ B} = 6 \text{ dB}$$

When $G = 8$,

$$G' = \log 8 = 0.9 \text{ B} = 9 \text{ dB}$$

Notice the simple progression here. When the power gain doubles, the bel power gain increases 3 dB. This is always true because if you start with G_1 and double it, you have $G = 2G_1$ and

$$G' = \log 2G_1 = \log 2 + \log G_1 = 3 \text{ dB} + \log G_1$$

The factor of 2 in front of G_1 adds 0.3 B or 3 dB to the answer.

Therefore, if we cause the power gain to double, we can make either of two equivalent statements:

1. We have doubled the power gain.
2. We have increased the bel power gain by 3 dB.

EXAMPLE 15-2
Work out the power gain for each of these: $G' = 3$ dB, 40 dB, and 43 dB.

SOLUTION
When $G' = 3$ dB or 0.3 B,

$$G = \text{antilog } 0.3 = 2$$

In other words, we work in the opposite direction, from the logarithm back to the original number.

When $G' = 40$ dB or 4 B,

$$G = \text{antilog } 4 = 10^4 = 10,000$$

When $G' = 43$ dB or 4.3 B,

$$G = \text{antilog } 4.3 = 2 \times 10^4 = 20,000$$

Alternatively, we could have converted 43 dB by realizing it is 3 dB more than 40 dB. As we saw in the preceding example, an increase of 3 dB in bel power gain means the ordinary power gain has doubled. Since a bel power gain of 40 dB is equivalent to a power gain of 10,000, a bel power gain of 43 dB is equivalent to a power gain of 20,000.

15-2 *power gain of cascaded stages*

Figure 15-1a shows blocks representing two stages in an amplifier. The input power to the first stage is p_1. The output power p_2 of the first stage goes into the second stage. The final output power is p_3.

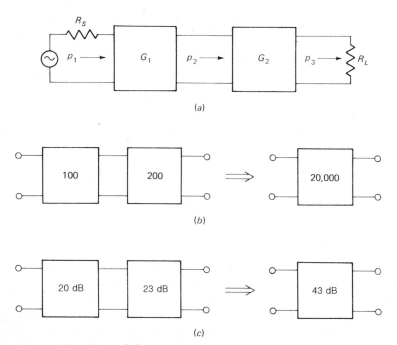

Figure 15-1. Cascaded stages.

Suppose we want to calculate the overall power gain, the ratio

$$G = \frac{p_3}{p_1}$$

There are two ways we can go about this: a direct method working with ordinary power and an indirect method using bel power gain.

direct method

The power gain of the first stage is

$$G_1 = \frac{p_2}{p_1}$$

and the power gain of the second stage is

$$G_2 = \frac{p_3}{p_2}$$

The total power gain is

$$G = \frac{p_3}{p_1}$$

which is equal to

$$G = \frac{p_3}{p_1} \frac{p_2}{p_2} = \frac{p_2}{p_1} \frac{p_3}{p_2}$$

or
$$G = G_1 G_2 \qquad \qquad (15\text{-}2)\text{***}$$

This proves the total power gain of cascaded stages equals the product of each stage gain. No matter how many stages there are, we can find the total power gain by multiplying the individual stage gains.

As an example, Fig. 15-1b shows a first-stage power gain of 100 and a second-stage power gain of 200. The total power gain is

$$G = 100 \times 200 = 20{,}000$$

indirect method

We know we can get the answer to a multiplication problem by the indirect method of adding the logarithms of factors and taking the antilog of the sum. Sometimes, this indirect method saves time and we use it.

Since bel power gain is nothing more than the logarithm of ordinary power gain, we can calculate the total power gain by adding the bel power gain of each stage and taking the antilog of the sum. Specifically, starting with G,

$$G = G_1 G_2$$

Taking the logarithm of both sides gives

$$\log G = \log G_1 G_2 = \log G_1 + \log G_2$$

With Eq. (15-1) we rewrite this as

$$G' = G_1' + G_2' \qquad\qquad (15\text{-}3)***$$

where G' is the total bel power gain, G_1' is the bel power gain of the first stage, and G_2' is the bel power gain of the second stage. This equation says the total bel power gain of two cascaded stages equals the sum of the bel power gains of each stage. It follows that no matter how many stages there are, the total bel power gain equals the sum of the individual bel power gains. Once we have found the total bel power gain, we can convert back to ordinary power gain if desired.

As an example, Fig. 15-1c shows the same two-stage amplifier as Fig. 15-1b, except the gains are expressed in decibels. The total bel power gain is

$$G' = 20 \text{ dB} + 23 \text{ dB} = 43 \text{ dB} = 4.3 \text{ B}$$

We can leave the answer like this, or convert it back to ordinary power gain as follows:

$$G = \text{antilog } G' = \text{antilog } 4.3$$
$$= 2 \times 10^4 = 20{,}000$$

An answer in decibels has the advantage of being compact and easy to write. In the foregoing example it is much easier to write 43 dB instead of 20,000 or 2×10^4.

simplified measurements

Another advantage of decibels is they simplify measurements. For example, many *microwave* instruments measure power (microwave refers to frequencies from 1000 to 100,000 MHz). The meter on such instruments often resembles Fig. 15-2a. The upper scale is marked in milliwatts. Suppose we measure the input power to a stage like Fig. 15-2b; the needle (solid line) will indicate 0.25 mW. When we measure the output power in Fig. 15-2b, the needle moves to full scale (dashed line), indicating an output power of 1 mW. Therefore, the power gain equals 4.

The lower scale is called the *dBm scale*. As shown in Fig. 15-2a, 0 dBm is equivalent to 1 mW, -3 dBm to 0.5 mW, -6 dBm to 0.25 mW, and so on. These values come about as follows. 1 mW is convenient to use as a *power reference* because power levels in microwave systems lie on both sides of this value. Suppose we let **P** stand for the *ratio of power to 1 mW*. In symbols,

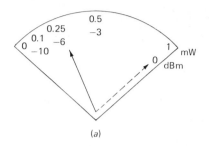

(a)

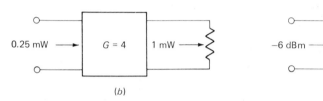

(b) (c)

Figure 15-2. Meaning of dBm.

$$\mathbf{P}=\frac{p}{p_{\text{ref}}}=\frac{p}{1\ \text{mW}}$$

Taking the logarithm of this power ratio gives

$$\mathbf{P}'=\log\frac{p}{1\ \text{mW}} \tag{15-4}$$

By working out the values of $\mathbf{P}'$ for different values of p, we can calibrate the meter face in terms of dB with respect to 1 mW.

As an example, if $p=0.5$ mW,

$$\mathbf{P}' = \log\frac{0.5\ \text{mW}}{1\ \text{mW}}=\log 0.5=\log\frac{1}{2}$$
$$=\log 1-\log 2=-0.3\ \text{B}=-3\ \text{dB}$$

To make sure we remember the milliwatt reference, we add the letter m to get

$$\mathbf{P}'=-3\ \text{dBm}$$

In this way, we can mark the lower scale in Fig. 15-2a with the correct values of dBm.

If we use the dBm scale to measure the input and output power in Fig. 15-2b, we will read −6 dBm for the input power and 0 dBm for the output power (Fig. 15-2c). Since the needle moves from −6 dBm to 0 dBm in Fig. 15-2a, the amplifier has a bel power gain of 6 dB.

simplified system design

Data sheets often specify devices in terms of their bel power gains. There is a good reason for this. When we cascade devices as shown in Fig. 15-3*a*, we can add the bel power gains to get the total bel power gain. For instance, in Fig. 15-3*b* the total bel power gain equals 46 dB, the sum of the individual bel power gains. This is usually easier and faster than multiplying ordinary power gains. To anyone designing a complicated system with many blocks, using decibels is a decided advantage because it reduces the labor of calculation.

EXAMPLE 15-3

A bipolar amplifier has a power gain of 0.5 from the signal source to base, and a power gain of 40,000 from base to collector. Calculate the individual bel power gains and the total bel power gain.

SOLUTION

The bel power gain from the signal source to base is

$$G_1' = \log 0.5 = \log \frac{1}{2} = \log 1 - \log 2$$
$$= -0.3 \text{ B} = -3 \text{ dB}$$

The bel power gain from base to collector is

$$G_2' = \log 40{,}000 = \log (4 \times 10^4) = 4.6 \text{ B} = 46 \text{ dB}$$

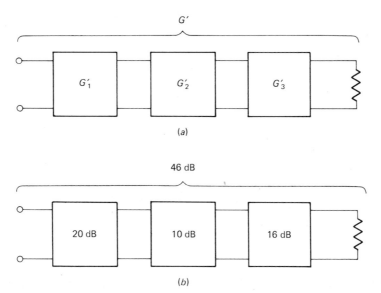

Figure 15-3. Additive property of bel power gain.

By adding the individual bel power gains, we get a total bel power gain of

$$G' = -3 \text{ dB} + 46 \text{ dB} = 43 \text{ dB}$$

15-3 *bel voltage gain*

In Fig. 15-4*a* the voltage across the amplifier input terminals equals v_1 and the voltage across the output terminals is v_2. Therefore, we have a voltage gain from input to output of

$$A = \frac{v_2}{v_1}$$

The input power to the amplifier is

$$p_1 = \frac{v_1^2}{R_1}$$

and the output power is

$$p_2 = \frac{v_2^2}{R_2}$$

So, the power gain from input to output equals

$$G = \frac{p_2}{p_1} = \frac{v_2^2/R_2}{v_1^2/R_1} = \left(\frac{v_2}{v_1}\right)^2 \frac{R_1}{R_2}$$

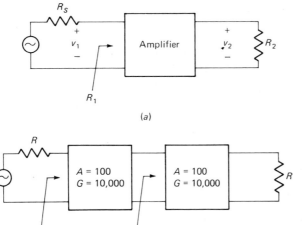

Figure 15-4. Relation of power and voltage gains.

Since the ratio v_2/v_1 is the voltage gain A, we can rewrite the power gain as

$$G = A^2 \frac{R_1}{R_2} \qquad (15\text{-}5a)$$

impedance-matched case

In many systems (microwave, telephone, etc.) we get the special case of $R_1 = R_2$. For this condition,

$$G = A^2 \qquad \text{for } R_1 = R_2 \qquad (15\text{-}5b)$$

For instance, if $A = 100$, $G = 10,000$. Figure 15-4b shows a system with these gains. Each block has an input impedance of R, and each block sees a load impedance of R. We call systems like this *impedance-matched systems;* each stage has a power gain equal to the square of its voltage gain.

Taking the logarithm of both sides of Eq. (15-5b) gives

$$\log G = \log A^2 = 2 \log A$$
or $\qquad\qquad G' = 2 \log A \qquad \text{(impedance match only)} \qquad (15\text{-}6)$

With this equation we can calculate the bel power gain without having to calculate the ordinary power gain. For example, if an impedance-matched amplifier has a voltage gain of 100, it has a bel power gain of

$$G' = 2 \log A = 2 \log 100 = 4 \text{ B} = 40 \text{ dB}$$

nonmatched case

Through the years since bel power gain was first defined, voltage gain has become more useful than power gain, at least in many branches of electronics. The reason is simple enough. It is easier to measure voltage than power. With an amplifier like Fig. 15-4a we can easily measure the input and output voltages.

Because of this, a second kind of bel gain has come into use: the *bel voltage gain.* The formula for bel voltage gain is

$$A' = 2 \log A \qquad (15\text{-}7)\text{***}$$

where A' is the bel voltage gain and A is the ordinary voltage gain. So, if $A = 100$, the bel voltage gain equals

$$A' = 2 \log A = 2 \log 100 = 4 \text{ B} = 40 \text{ dB}$$

Or, if $A = 20,000$,

$$A' = 2 \log A = 2 \log 20,000 = 2(4.3)$$
$$= 8.6 \text{ B} = 86 \text{ dB}$$

The reason for including a coefficient of 2 in Eq. (15-7) is to ensure bel power gain and bel voltage gain are equal in impedance-matched systems. In other words, with Eq. (15-7) we can rewrite Eq. (15-6)as

$$G' = A' \qquad \text{(impedance match only)} \qquad (15\text{-}8)***$$

where $G' = \log G$ and $A' = 2 \log A$. Whenever the impedances are matched, we can use G' or A' interchangeably because they are equal. For instance, an impedance-matched amplifier with a voltage gain of 100 has a power gain of 10,000; equivalently, this amplifier has a bel voltage gain of 40 dB and a bel power gain of 40 dB.

When the impedances are not matched, the bel power gain and bel voltage gain are no longer equal. Therefore, we have to calculate each by its own formula. In other words, if we want bel power gain, we have to use

$$G' = \log G$$

and for bel voltage gain, we must use

$$A' = 2 \log A$$

Both types of bel gain are widely used. Bel power gain predominates in communications, microwaves, and other systems where power is important. Bel voltage gain seems to be far in front, however, in areas of electronics where it is convenient to measure voltage rather than power.

Since bel voltage gain is based on logarithms, the additive property of cascaded stages holds. That is, the total bel voltage gain of cascaded stages equals the sum of the individual bel voltage gains. In symbols,

$$A' = A'_1 + A'_2 + \cdot \ \cdot \ \cdot \qquad (15\text{-}9)$$

Also, many voltmeters have a decibel scale. When you measure the input and output voltages of an amplifier, the difference in the dB readings is the bel voltage gain of the amplifier.

As an example, the two-stage amplifier of Fig. 15-5a has a first-stage voltage gain of 100 and a second-stage voltage gain of 200. The total voltage gain is

$$A = A_1 A_2 = 100 \times 200 = 20,000$$

The first stage has a bel voltage gain of

$$A'_1 = 2 \log 100 = 4 \text{ B} = 40 \text{ dB}$$

and the second stage,

$$A'_2 = 2 \log 200 = 2(2.3) = 4.6 \text{ B} = 46 \text{ dB}$$

The total bel voltage gain is

$$A' = 40 \text{ dB} + 46 \text{ dB} = 86 \text{ dB}$$

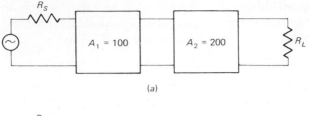

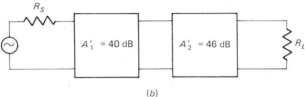

Figure 15-5. Additive property of bel voltage gain.

If we use the decibel scale of a voltmeter when measuring input and output voltages, the needle will show an increase of 40 dB for the first stage, 46 dB for the second stage, and 86 dB for the entire amplifier (see Fig. 15-5b).

EXAMPLE 15-4

The JFET in Fig. 15-6a has a g_m of 4000 μS. Calculate the bel voltage gain from gate to drain.

SOLUTION

The drain sees an ac resistance of

$$r_D = 10,000 \parallel 10,000 = 5 \text{ k}\Omega$$

The voltage gain from gate to drain is

$$A = \frac{v_d}{v_g} = g_m r_D = 0.004(5000) = 20$$

The bel voltage gain equals

$$A' = 2 \log A = 2 \log 20 = 2.6 \text{ B} = 26 \text{ dB}$$

EXAMPLE 15-5

Figure 15-6b shows a two-stage amplifier. What is the total voltage gain in decibels?

SOLUTION

We found the voltage gain of the second stage in the preceding example: 26 dB. The collector of the first stage sees approximately 10 kΩ of ac load resistance.

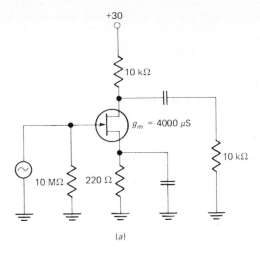

(a)

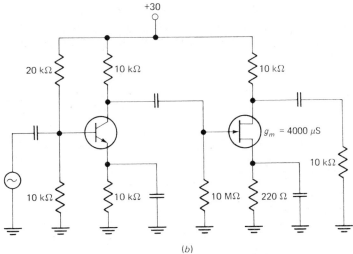

(b)

Figure 15-6.

Since the I_E of the first stage is approximately 1 mA, r_e' is about 25 Ω. The voltage gain of the first stage therefore equals

$$A_1 = \frac{r_C}{r_e'} = \frac{10,000}{25} = 400$$

And, the bel voltage gain of the first stage is

$$A_1' = 2 \log 400 = 5.2 \text{ B} = 52 \text{ dB}$$

The total bel voltage gain is the sum of individual bel voltage gains; in this case,

$$A' = A_1' + A_2' = 52 \text{ db} + 26 \text{ dB} = 78 \text{ dB}$$

EXAMPLE 15-6

What is the total bel voltage gain of the three-stage amplifier shown in Fig. 15-7?

SOLUTION

The three stages have identical biasing resistors. Approximately, 1 mA of dc emitter current flows in each transistor. Ideally, r_e' is 25 Ω in each stage. Therefore, given a β of 100 for each transistor,

$$z_{\text{in(base)}} = 100(25) = 2500 \text{ Ω}$$

in all stages.

The first collector sees a 10-kΩ resistor shunted by the input impedance of the second stage, which is

$$z_{\text{in}} = 60,000 \,\|\, 30,000 \,\|\, 2500 = 2220 \text{ Ω}$$

Therefore, the collector of the first stage sees

$$r_C = 10,000 \,\|\, 2220 = 1820 \text{ Ω}$$

And, the voltage gain from base to collector for the first stage is

$$A_1 = \frac{r_C}{r_e'} = \frac{1820}{25} \cong 73$$

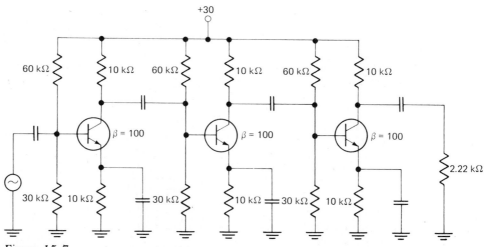

Figure 15-7.

The first-stage bel voltage gain equals

$$A_1' = 2 \log 73 = 2(1.86) \cong 3.7 \text{ B} = 37 \text{ dB}$$

The collectors of the second and third stages see exactly the same ac resistance as the collector of the first stage; therefore, each of these stages has the same bel voltage gain. The total bel voltage gain of the three-stage amplifier is the sum of the individual bel voltage gains:

$$A' = A_1' + A_2' + A_3' \cong 3(37 \text{ dB}) = 111 \text{ dB}$$

15-4 *Miller's theorem*

Figure 15-8a shows an inverting amplifier. This means the output voltage is 180° out of phase with the input voltage.

In Fig. 15-8a an impedance Z is connected from the output to the input. Many circuits have an impedance like this, and it produces a feedback effect similar to collector-feedback bias discussed earlier. *Miller's theorem* helps with circuits like these because it splits the impedance into two components as shown in Fig. 15-8b. In other words, the Miller theorem says Fig. 15-8b is equivalent to Fig. 15-8a.

To be precise, when an inverting amplifier has a voltage gain A, the Miller theorem says we can replace impedance Z by an input Miller impedance of

$$Z_{\text{in(Miller)}} = \frac{Z}{A+1} \qquad (15\text{-}10a)$$

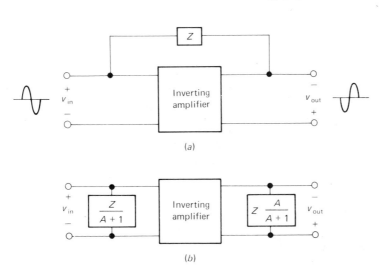

Figure 15-8. Miller theorem.

and an output Miller impedance of

$$Z_{\text{out(Miller)}} = Z\frac{A}{A+1} \tag{15-10b}$$

These Miller impedances appear across the input and output terminals as shown in Fig. 15-8b. This is useful because we can lump these Miller impedances into the input impedance and the load resistance of the amplifier. More about this later.

Z *purely resistive*

Let's make sure we understand Miller's theorem by examining two special cases. Figure 15-9a shows the first, $Z = R$. The Miller theorem says we can split this R into two parts: an input part and an output part as shown in Fig. 15-9b. In symbols,

$$R_{\text{in(Miller)}} = \frac{R}{A+1} \tag{15-11}$$

and

$$R_{\text{out(Miller)}} = R\frac{A}{A+1} \tag{15-12}$$

Here is a concrete example. Fig. 15-9c shows an inverting amplifier with a voltage gain of 99 and a resistance of 100 kΩ from input to output. With the Miller theorem we can split this into

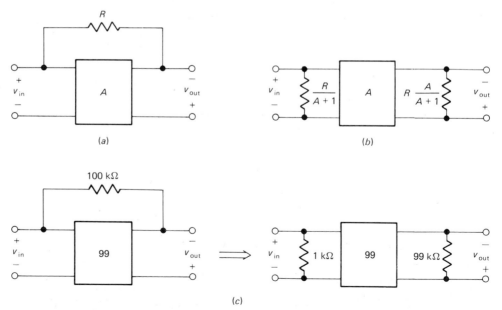

(a) (b)

(c)

Figure 15-9. Miller theorem for resistive case.

$$R_{\text{in(Miller)}} = \frac{100 \text{ k}\Omega}{99 + 1} = 1 \text{ k}\Omega$$

and
$$R_{\text{out(Miller)}} = (100 \text{ k}\Omega)\frac{99}{99 + 1} = 99 \text{ k}\Omega$$

The 1-kΩ component appears across the input, and the 99-kΩ component across the output as shown.

Z *purely capacitive*

The only other special case we are interested in is the capacitive case, $Z = X_C$ as shown in Fig. 15-10a. With the Miller theorem the reactance splits into the two components of Fig. 15-10b.

Often, we prefer to work directly with capacitance instead of reactance. Note that

$$\frac{X_C}{A + 1} = \frac{1}{2\pi f C(A + 1)} = \frac{1}{2\pi f C_{\text{in(Miller)}}}$$

where
$$C_{\text{in(Miller)}} = C(A + 1) \qquad\qquad (15\text{-}13)\ast\ast\ast$$

On the output side,

$$X_C \frac{A}{A + 1} = \frac{1}{2\pi f C}\frac{A}{A + 1} = \frac{1}{2\pi f C_{\text{out(Miller)}}}$$

where
$$C_{\text{out(Miller)}} = C\frac{A + 1}{A} \qquad\qquad (15\text{-}14)$$

Figure 15-11a and b summarize these results for the purely capacitive case. The Miller theorem says a capacitance C from input to output appears like a

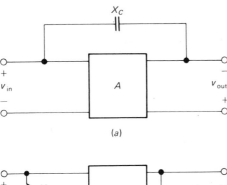

(a)

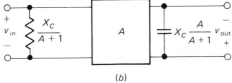

(b)

Figure 15-10. Miller theorem for capacitive case.

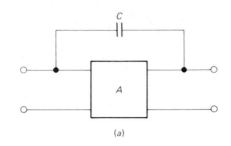

(a)

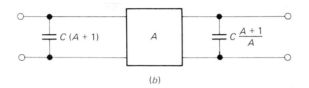

(b)

Figure 15-11. Miller theorem applied to capacitance.

much larger capacitance across the input terminals. For instance, if $C = 10$ pF and $A = 99$, then

$$C_{\text{in(Miller)}} = (10 \text{ pF})(99 + 1) = 1000 \text{ pF}$$

and

$$C_{\text{out(Miller)}} = (10 \text{ pF})\frac{99 + 1}{99} = 10.1 \text{ pF}$$

useful tool

The Miller theorem is a powerful tool. It is like the Thevenin theorem because it eliminates awkward problems and deepens our insight into circuit action. The two most important cases of the Miller theorem are the resistive case (Fig. 15-9) and the capacitive case (Fig. 15-11). Whenever an R or C connects the input and output terminals of an inverting amplifier, we can apply the Miller theorem to the ac equivalent circuit. From now on, *Millerize* a circuit means to split the R or C into its Miller components in the ac equivalent circuit.

EXAMPLE 15-7
The voltage gain from base to collector equals 200 in Fig. 15-12a. Millerize the circuit.

SOLUTION
This is the resistive case. The 1-MΩ resistor connects the input of the CE amplifier to the output. With Miller's theorem,

$$R_{\text{in(Miller)}} = \frac{R}{A + 1} = \frac{10^6}{200 + 1} \cong 5 \text{ k}\Omega$$

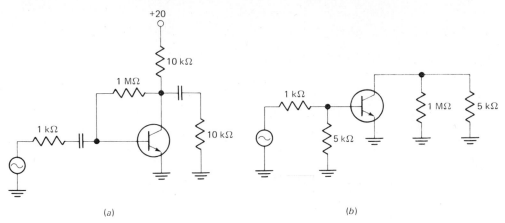

Figure 15-12.

and
$$R_{\text{out(Miller)}} = R\frac{A}{A+1} = 10^6 \frac{200}{200+1} \cong 1 \text{ M}\Omega$$

Note that the output Miller component is approximately equal to the original resistance R whenever the voltage gain is large.

Figure 15-12*b* shows the ac equivalent circuit after Millerizing. Because the 1-MΩ output Miller component is so much greater than the 5-kΩ load resistor, the ac load collector resistance is approximately 5 kΩ.

EXAMPLE 15-8

Figure 15-13*a* shows a CE amplifier with an r_e' of 25 Ω. The coupling and bypass capacitors look like ac shorts, but not the 3-pF capacitor. Millerize and simplify the ac circuit.

SOLUTION

Figure 15-13*b* shows the ac equivalent circuit. The voltage gain equals

$$A = \frac{r_C}{r_e'} = \frac{5000}{25} = 200$$

Therefore, the Miller components are

$$C_{\text{in(Miller)}} = C(A+1) = (3 \text{ pF})(200+1) \cong 600 \text{ pF}$$

and
$$C_{\text{out(Miller)}} = C\frac{A+1}{A} = 3 \text{ pF} \frac{200+1}{200} \cong 3 \text{ pF}$$

Figure 15-13*c* shows the Millerized circuit. As a final step in simplification, we can Thevenize the base circuit to the left of the input Miller capacitance. 1

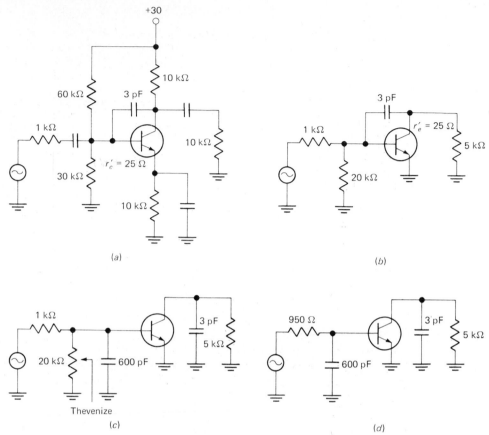

Figure 15-13. Millerizing and Thevenizing a circuit.

kΩ in parallel with 20 kΩ is approximately 950 Ω; Fig. 15-13*d* shows the Millerized and Thevenized ac equivalent for the original circuit of Fig. 15-13*a*.

15-5 *second-approximation formulas*

One of the first approximations we made was

$$\alpha \cong 1$$

With modern transistors, α is almost always greater than 0.95, and usually more than 0.98. Therefore, we expect it to have only a minor effect on voltage gain

Table 15-1 Second-approximation Formulas

	Common-emitter	Common-collector	Common-base
A	$\dfrac{\alpha r_C}{r_e' + r_b'(1-\alpha)}$	$\dfrac{r_E}{r_E + r_e' + r_b'(1-\alpha)}$	$\dfrac{\alpha r_C}{r_e' + r_b'(1-\alpha)}$
A_i	β	$\dfrac{1}{1-\alpha}$	α
z_{in}	$r_b' + \beta r_e'$	$r_b' + \beta(r_E + r_e')$	$r_e' + \dfrac{r_b'}{\beta}$
z_{out}	$\dfrac{r_c'}{\beta} + \dfrac{r_e' r_c'}{R_S + r_b' + r_e'}$	$r_e' + \dfrac{(R_S + r_b')}{\beta}$	$r_c'\left(\dfrac{R_S + r_e'}{R_S + r_b' + r_e'}\right)$

and other quantities. In this section, we will show precisely what effect α has.

Another approximation was treating r_b' as negligible. As discussed in Sec. 6-4, r_b' can get up to 1000 Ω, but usually it is from 50 to 150 Ω.

The reverse-biased collector diode has an ac resistance r_c'. This resistance is normally greater than a megohm, and we can expect it to have a small effect on voltage gain and input impedance.

By reanalyzing the CE, CC, and CB amplifiers, we can include the effects of α, r_b', and r_c' on voltage gain, current gain, input impedance, and output impedance. Table 15-1 summarizes the results.

EXAMPLE 15-9
A CE amplifier has the following: $\alpha = 0.99$, $r_C = 2$ kΩ, $r_e' = 25$ Ω, $r_b' = 100$ Ω, $\beta = 100$, $r_c' = 5$ MΩ, and $R_S = 1$ kΩ. Calculate A, A_i, z_{in}, and z_{out}.

SOLUTION
The voltage gain equals

$$A = \frac{\alpha r_C}{r_e' + r_b'(1-\alpha)} = \frac{0.99(2000)}{25 + 100(1 - 0.99)} = 76.2$$

The current gain is

$$A_i = \beta = 100$$

The input impedance equals

$$z_{\text{in}} = r_b' + \beta r_e' = 100 + 100(25) = 2.6 \text{ k}\Omega$$

The output impedance is

$$z_{out} = \frac{r'_c}{\beta} + \frac{r'_e r'_c}{R_S + r'_b + r'_e}$$

$$= \frac{5(10^6)}{100} + \frac{(25)5(10^6)}{1000 + 100 + 25} = 161 \text{ k}\Omega$$

15-6 *hybrid parameters*

Hybrid (h) parameters are easy to measure; this is the reason some data sheets specify low-frequency transistor characteristics in terms of four *h* parameters. This section tells you what *h* parameters are, and how they are related to the *r'* parameters we have been using.

what they are

The four *h* parameters of the CE connection are

h_{ie} = input impedance	$(r_C = 0)$
h_{fe} = current gain	$(r_C = 0)$
h_{re} = reverse voltage gain	$(R_S = \infty)$
h_{oe} = output admittance	$(R_S = \infty)$

The first two parameters are specified for an ac load resistance of zero, equivalent to a shorted output. The next two parameters are specified for a source resistance of infinity, equivalent to an open input.

Figure 15-14a shows how to measure h_{ie} and h_{fe}. To begin with, we use an ac short across the output. This prevents loading effects, ensuring unambiguous values for h_{ie} and h_{fe}. In other words, input impedance and current gain will vary if the ac load resistance is too large; to avoid this, we short the output terminals. In Fig. 15-14a, the ratio of input voltage to input current equals the input impedance. In symbols,

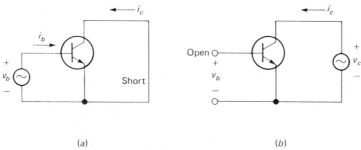

(a)   (b)

Figure 15-14. (a) *Measuring* h_{ie} *and* h_{fe}. (b) *Measuring* h_{re} *and* h_{oe}.

$$h_{ie} = \frac{v_b}{i_b} \qquad (r_C = 0)$$

The ratio of output current to input current is the current gain:

$$h_{fe} = \frac{i_c}{i_b} \qquad (r_C = 0)$$

To get the other two parameters, we use the ac equivalent of Fig. 15-14*b*. Here we drive the collector with a voltage source v_c; this forces a current i_c to flow into the collector. The base is left open; that is, $R_S = \infty$. The ratio of ac base voltage to ac collector voltage is the reverse voltage gain:

$$h_{re} = \frac{v_b}{v_c} \qquad (R_S = \infty)$$

And finally, if we take the ratio of ac collector current to ac collector voltage, we get the output admittance:

$$h_{oe} = \frac{i_c}{v_c} \qquad (R_S = \infty)$$

relation to r′ parameters

The r' parameters $(r'_e,\ r'_b,\ r'_c,\ \beta,$ and $\alpha)$ are the easiest to work with; the h parameters are the easiest to measure. Therefore, we need to know how to convert from h parameters to r' parameters. This way, when a data sheet specifies h parameters, we can get r' parameters and analyze transistor amplifiers with the methods of earlier chapters.

Table 15-2 shows the approximate relations between r' parameters and h parameters. As indicated, $\beta = h_{fe}$, $\alpha = h_{fe}/(h_{fe} + 1)$, and so on. Many data sheets do list the common-emitter h parameters, so with the formulas of Table 15-2 you can convert to r' parameters.

Table 15-2 Approximate Relations

r' Parameter	h Parameter
β	h_{fe}
α	$h_{fe}/(h_{fe} + 1)$
r'_e	h_{ie}/h_{fe}
r'_c	h_{fe}/h_{oe}
r'_b	h_{rb}/h_{ob}

The only unusual entry in Table 15-2 is

$$r'_b = \frac{h_{rb}}{h_{ob}}$$

This is the ratio of the reverse voltage gain and output admittance of a common-base circuit. In other words, the easiest and most reliable way to measure r'_b is with a common-base circuit. Reverse voltage gain h_{rb} divided by output admittance h_{ob} gives the value of r'_b.

EXAMPLE 15-10

The data sheet of a 2N3904 shows the following typical values at $I_C = 1$ mA:

$$h_{ie} = 3.5 \text{ k}\Omega$$
$$h_{fe} = 120$$
$$h_{re} = 1.3(10^{-4})$$
$$h_{oe} = 8.5 \text{ }\mu\text{S}$$

Work out the values of β, α, r'_e, and r'_c.

SOLUTION

$$\beta = h_{fe} = 120$$

$$\alpha = \frac{h_{fe}}{h_{fe} + 1} = \frac{120}{120 + 1} = 0.992$$

$$r'_e = \frac{h_{ie}}{h_{fe}} = \frac{3500}{120} = 29 \text{ }\Omega$$

$$r'_c = \frac{h_{fe}}{h_{oe}} = \frac{120}{8.5(10^{-6})} = 14.1 \text{ M}\Omega$$

EXAMPLE 15-11

The 2N1975 data sheet specifies $h_{rb} = 1.75(10^{-4})$ and $h_{ob} = 1 \text{ }\mu\text{S}$ at $I_C = 1$ mA. Calculate r'_b.

SOLUTION

$$r'_b = \frac{h_{rb}}{h_{ob}} = \frac{1.75(10^{-4})}{10^{-6}} = 175 \text{ }\Omega$$

15-7 *approximate hybrid formulas*

The h parameters are measured with shorted output and open input to avoid the ambiguous values that would result from different source and load resis-

Table 15-3 Approximate Hybrid Formulas			
	Common-Emitter	Common-Collector	Common-Base
A	$\dfrac{h_{fe}r_C}{h_{ie}}$	1	$\dfrac{h_{fe}r_C}{h_{ie}}$
A_i	h_{fe}	h_{fe}	1
z_{in}	h_{ie}	$h_{fe}r_E$	$\dfrac{h_{ie}}{h_{fe}}$
z_{out}	$\dfrac{1}{h_{oe}}$	$\dfrac{R_S + h_{ie}}{h_{fe}}$	$\dfrac{h_{fe}}{h_{oe}}$

tances. In a practical amplifier, the source resistance is not infinite, and the load resistance is not zero. Nevertheless, h parameters can be used to analyze practical amplifiers.

Table 15-3 lists the approximate formulas for the three transistor connections. If you prefer, use these formulas directly instead of converting to r' parameters. These approximate formulas are reasonably accurate for most amplifiers. If you require exact answers, you will need to use the complicated formulas given in the next section.

EXAMPLE 15-12
Using the h parameters of Example 15-10, calculate A, A_i, z_{in}, and z_{out} for a CE amplifier with $r_C = 2$ kΩ.

SOLUTION

$$A = \frac{h_{fe}r_C}{h_{ie}} = \frac{120(2000)}{3500} = 68.6$$
$$A_i = h_{fe} = 120$$
$$z_{in} = h_{ie} = 3.5 \text{ k}\Omega$$
$$z_{out} = \frac{1}{h_{oe}} = \frac{1}{8.5(10^{-6})} = 118 \text{ k}\Omega$$

EXAMPLE 15-13
The 2N3904 is used as an emitter follower with $r_E = 100$ Ω and $R_S = 1.5$ kΩ. Work out the values of z_{in} and z_{out}.

SOLUTION
With Table 15-3 and the h parameters of Example 15-10,

$$z_{in} = h_{fe}r_E = 120(100) = 12 \text{ k}\Omega$$

and
$$z_{out} = \frac{R_S + h_{ie}}{h_{fe}} = \frac{1500 + 3500}{120} = 41.7 \ \Omega$$

EXAMPLE 15-14

The 2N3904 is used in a common-base amplifier. With the h parameters of Example 15-10, calculate z_{in} and z_{out}.

SOLUTION

$$z_{in} = \frac{h_{ie}}{h_{fe}} = \frac{3500}{120} = 29 \ \Omega$$

and
$$z_{out} = \frac{h_{fe}}{h_{oe}} = \frac{120}{8.5(10^{-6})} = 14.1 \text{ M}\Omega$$

15-8 *exact hybrid formulas*

The ideal-transistor approximation is adequate for preliminary analysis and design. When more accurate answers are needed, you can use the improved approximations of Tables 15-1 through 15-3. If you need the utmost accuracy, be prepared to use complicated formulas that take everything into account.

numerical parameters

In deriving exact formulas, it's helpful to use numerical parameters h_{11}, h_{12}, h_{21}, and h_{22}. These parameters have the following meaning for any transistor connection:

h_{11} = input impedance   (shorted output)
h_{12} = reverse voltage gain   (open input)
h_{21} = current gain   (shorted output)
h_{22} = output admittance   (open input)

Table 15-4 shows how these numerical parameters are related to the h parameters of each transistor connection. For instance, for a CE amplifier

$$h_{11} = h_{ie}$$
$$h_{12} = h_{re}$$
$$h_{21} = h_{fe}$$
$$h_{22} = h_{oe}$$

Table 15-4 Hybrid Relations

Numerical	Common-Emitter	Common-Collector	Common-Base
h_{11}	h_{ie}	h_{ic}	h_{ib}
h_{12}	h_{re}	h_{rc}	h_{rb}
h_{21}	h_{fe}	h_{fc}	h_{fb}
h_{22}	h_{oe}	h_{oc}	h_{ob}

formulas

Table 15-5 lists the exact formulas for CE, CC, and CB amplifiers. All you have to do is substitute for the numerical parameters. Ac load resistance r_L depends on the transistor connection. With a CE or CB amplifier, $r_L = r_C$. With a CC connection, $r_L = r_E$.

As an example, suppose we want the voltage gain of a common-emitter amplifier. With Tables 15-4 and 15-5,

$$A = \frac{h_{fe}r_C}{h_{ie}(1 + h_{oe}r_C) - h_{re}h_{fe}r_C}$$

practical comments

The exact formulas of Table 15-5 give exact answers, provided you have the exact h parameters for the transistor being used. This is where a practical prob-

Table 15-5 Hybrid Formulas

	Exact	Approximate
A	$\dfrac{h_{21}r_L}{h_{11}(1 + h_{22}r_L) - h_{12}h_{21}r_L}$	$\dfrac{h_{21}r_L}{h_{11}}$
A_i	$\dfrac{h_{21}}{1 + h_{22}r_L}$	h_{21}
z_{in}	$h_{11} - \dfrac{h_{12}h_{21}}{h_{22} + 1/r_L}$	h_{11}
z_{out}	$\dfrac{R_S + h_{11}}{(R_S + h_{11})h_{22} - h_{12}h_{21}}$	$\dfrac{1}{h_{22}}$

lem comes in. The spread in minimum and maximum values is huge. For instance, the data sheet of a 2N3904 gives the following ranges in h parameters for $I_C = 1$ mA:

$$h_{ie} = 1 \text{ to } 10 \text{ k}\Omega$$
$$h_{re} = 0.5(10^{-4}) \text{ to } 8(10^{-4})$$
$$h_{fe} = 100 \text{ to } 400$$
$$h_{oe} = 1 \text{ to } 40 \text{ }\mu\text{S}$$

With spreads like these, exact formulas lose their appeal. When working with thousands of 2N3904s, all the formulas give us is an estimate of the voltage gain, input impedance, etc.

Here's the point. All transistors have large tolerances in their specified h parameters. For this reason, we wind up with approximate answers, so we may as well use ideal formulas because they introduce much less error than the huge tolerance of the h parameters. In other words, the ideal formulas give us typical answers, located somewhere between the minimum and maximum answers predicted by exact analysis.

EXAMPLE 15-15

A 2N3904 is used in a CE amplifier with $r_C = 2$ kΩ. With the h parameters of Example 15-10, calculate the exact voltage gain.

SOLUTION

$$A = \frac{h_{fe}r_C}{h_{ie}(1 + h_{oe}r_C) - h_{re}h_{fe}r_C}$$
$$= \frac{120(2000)}{3500 \left[1 + 8.5(10^{-6})2000\right] - 1.3(10^{-4})120(2000)}$$
$$= 68$$

self-testing review

Read each of the following and provide the missing words. Answers appear at the beginning of the next question.

1. Bel power gain is defined as the common _____ of power gain. The symbol G is used for power gain, and _____ for bel power gain. If the power gain doubles, the bel power gain increases by _____ dB.

2. *(logarithm, G', 3)* The total power gain of cascaded stages equals the product of each stage gain. The bel power gain of cascaded stages equals the _____ of each bel power gain.

3. *(sum)* The voltage gain is designated A; the bel voltage gain is designated A'. With impedance-matched systems, G' and A' are _____. When the impedances are

not matched, G' and A' are not _____. In this case, G' equals log G, and A' equals _____.

4. *(equal, equal, 2 log A)* Since bel voltage gain is based on logarithms, the additive property of cascaded stages holds. That is, the total bel voltage gain of cascaded stages equals the _____ of the individual bel voltage gains.

5. *(sum)* The Miller theorem says a capacitance C from input to output appears like a much _____ capacitance across the input terminals. This input capacitance equals C times _____.

6. *(larger, A + 1)* The four hybrid parameters of the CE connection are h_{ie}, _____, h_{re}, and h_{oe}. For $r_C = 0$, h_{ie} is the _____ impedance, and h_{fe} is the _____ gain. For $R_S = \infty$, h_{re} is the reverse voltage gain, and h_{oe} is the output _____.

7. *(hfe, input, current, admittance)* β equals _____, and _____ equals h_{ie}/h_{fe}. Other relations of lesser importance are $\alpha = h_{fe}/(h_{fe} + 1)$, and $r'_c = h_{fe}/h_{oe}$.

8. *(hfe, r'e)* For a CE amplifier, _____ equals $h_{fe}r_C/h_{ie}$, A_i equals h_{fe}, z_{in} equals _____, and z_{out} equals $1/h_{oe}$. These are approximate relations.

9. *(A, hie)* In the CC amplifier, A is approximately equal to _____, and z_{in} is approximately equal to _____.

10. *(1, hfe rE)* The exact formulas using h parameters give exact answers, provided you have the exact h parameters for the transistor being used.

problems

15-1. What is the bel power gain for each of these values of ordinary power gain: 1, 10, 100, 1000, 10,000, and 100,000?

15-2. Calculate the bel power gain for $G = 2, 20, 200$, and 2000.

15-3. Work out the bel power gain for $G = 16$, 16,000, and 1,600,000.

15-4. Convert these bel power gains to ordinary power gains: 5 B, 30 dB, 66 dB, and 89 dB.

15-5. Calculate an accurate value of bel power gain given an ordinary power gain of 3000.

15-6. A three-stage amplifier has these power gains: $G_1 = 10$, $G_2 = 100$, and $G_3 = 1000$. What is the total power gain? What is the bel power gain of each stage? The bel power gain of the three-stage amplifier?

15-7. A two-stage amplifier has these individual bel power gains: $G'_1 = 36$ dB and $G'_2 = -10$ dB. What is the total bel power gain? The total power gain?

15-8. Work out the dBm value of each of these values of power:
(a) $p = 1$ W

(b) $p = 200$ mW

(c) $p = 50$ μW

15-9. The output power of a class B push-pull amplifier has a value of 49 dBm. Work out the value in watts.

15-10. In a two-stage amplifier the first stage has a power gain of $G_1 = 40$ and the second stage a power gain of $G_2 = 10,000$. What is the total power gain? And the total bel power gain?

15-11. The power gain from a signal source to a base is 0.25; the power gain from base to collector is 5000. What is the total power gain in decibels?

15-12. An impedance-matched amplifier has a voltage gain of 100. What is the voltage gain in decibels? The power gain in decibels?

15-13. Calculate the bel voltage gain for each of these values: $A = 2$, 20, 200, and 2000.

15-14. Work out the bel voltage gain for each of the following values: $A = 8$, 800, and 800,000.

15-15. An amplifier has a voltage gain of 5000. Express this in decibels.

15-16. A two-stage amplifier has an A_1 of 100 and an A_2 of 200. What is the total voltage gain? The total bel voltage gain?

15-17. In Fig. 15-15 convert each stage voltage gain to its bel voltage gain. What is the total bel voltage gain? The total ordinary voltage gain?

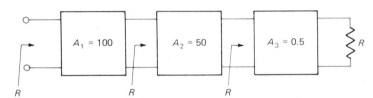

Figure 15-15.

15-18. A FET amplifier has a g_m of 2000 μS and an r_D of 10 kΩ. No swamping resistor is used. What is the voltage gain from gate to drain? And the bel voltage gain?

15-19. The FET of Fig. 15-16 has a g_m of 4000 μS. The bipolar transistor has a β of 100 and an r_e' of 25 Ω.

(a) What is the bel voltage gain of the first stage from gate to drain?

(b) What is the bel voltage gain of the second stage from base to collector?

(c) Calculate the total bel voltage gain.

15-20. Suppose a 500-Ω swamping resistor is used in the second stage of Fig. 15-16. What is the total bel voltage gain in this case?

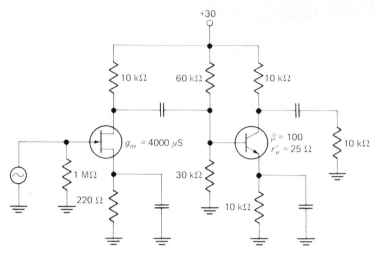

Figure 15-16.

15-21. If a swamping resistor of 500 Ω is used in each stage of Fig. 15-7, what is the total bel voltage gain?

15-22. In Fig. 15-17a what is the value of the input Miller resistance? The output Miller resistance?

15-23. What value does the input Miller resistance of Fig. 15-17b have? The output Miller resistance?

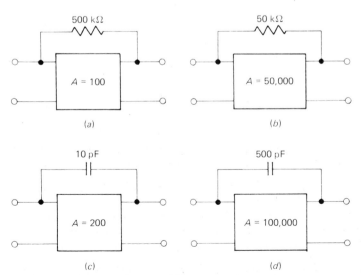

Figure 15-17. Inverting amplifiers.

15-24. Calculate the input and output Miller capacitances of Fig. 15-17c.

15-25. What are the values of input and output Miller capacitance in Fig. 15-17d?

15-26. In Fig. 15-18a the bipolar transistor has an r'_e of 5 Ω, and the voltage gain from base to collector equals 200. Calculate the input and output Miller resistances.

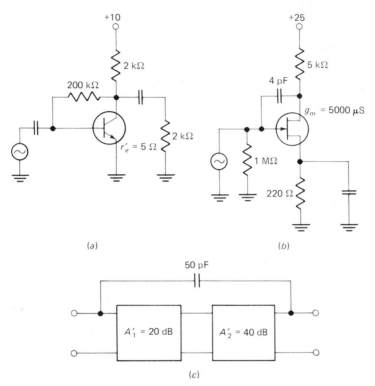

Figure 15-18.

15-27. Work out the voltage gain for the FET amplifier of Fig. 15-18b. What are the values of input and output Miller capacitance?

15-28. What is the input Miller capacitance of Fig. 15-18c.

15-29. A CE amplifier has an r_C of 10 kΩ, an α of 0.98, an r'_e of 25 Ω, an r'_b of 200 Ω, and a β of 50. What is the voltage gain from base to collector? The input impedance?

15-30. Suppose you measure these values in Fig. 15-14a: $v_b = 5$ mV, $i_b = 2$ μA, and $i_c = 300$ μA. What are the values of h_{ie} and h_{fe}?

15-31. In Fig. 15-14b, $v_b = 1$ μV, $v_c = 10$ mV, and $i_c = 0.2$ μA. What are the values of h_{re} and h_{oe}?

15-32. A data sheet gives the following h parameters at $I_C = 1$ mA: $h_{ie} = 2500$ Ω, $h_{fe} = 90$, $h_{re} = 2(10^{-4})$, and $h_{oe} = 5$ μS. Calculate α, β, r_e', and r_c'.

15-33. The data sheet of a 2N1613 gives these h parameters at $I_C = 5$ mA: $h_{rb} = 3(10^{-4})$ and $h_{ob} = 1$ μS. What is the value of r_b'?

15-34. The 2N4401 has the following typical h parameters at $I_C = 1$ mA:

$$h_{ie} = 5 \text{ k}\Omega$$
$$h_{fe} = 175$$
$$h_{re} = 0.6(10^{-4})$$
$$h_{oe} = 9 \text{ }\mu\text{S}$$

If used in a CE amplifier with $r_C = 3$ kΩ, what are the typical values of A, A_i, z_{in}, and z_{out}? (Use the approximate formulas of Table 15-3.)

15-35. Use the data of the preceding problem and work out the A, A_i, z_{in}, and z_{out} for a 2N4401 used in a CC amplifier with $r_E = 220$ Ω and $R_S = 1$ kΩ.

15-36. Repeat Prob. 15-34 for a 2N4401 used in a CB amplifier with $r_C = 2$ kΩ.

15-37. Calculate the exact values of A, A_i, z_{in}, and z_{out} for a 2N4401 used in a CE amplifier with $r_C = 2$ kΩ and $R_S = 1$ kΩ. (Use the h parameters given in Prob. 15-34.)

Table 15-6 Approximate Relations

r' Parameter	h Parameter
α	$-h_{fb}$
β	$-h_{fb}/(1 + h_{fb})$
r_e'	h_{ib}
r_c'	h_{ob}
r_b'	h_{rb}/h_{ob}

15-38. Sometimes data sheets list the common-base h parameters instead of the common-emitter h parameters. Table 15-6 gives you the relation between r' parameters and CB h parameters. The data sheet of a 2N1613 gives the following parameters at $I_C = 1$ mA: $h_{ib} = 28$ Ω, $h_{fb} = -0.98$, $h_{rb} = 0.6(10^{-4})$, and $h_{ob} = 1$ μS. Work out the values of α, β, r_e', r_c', and r_b'.

16

frequency effects

This chapter explains the action of an amplifier outside the normal frequency range where coupling and bypass capacitors no longer look like short circuits. It also takes into account the internal capacitances inside bipolars and FETs.

16-1 *the lag network*

The *lag network* of Fig. 16-1 is the key to analyzing high-frequency effects. As we will see, the output circuit of a transistor amplifier reduces to a lag network at higher frequencies.

complex voltage gain

In Fig. 16-1 a sine wave drives the lag network. By the well-known formula,

$$X_C = \frac{1}{2\pi f C}$$

Therefore, the higher the frequency, the smaller the X_C. So, we conclude the following:

1. For very low frequencies, X_C approaches infinity or an open circuit; for this case, V_{out} equals V_{in}.

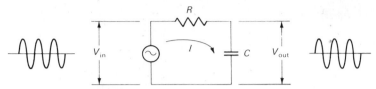

Figure 16-1. The lag network.

2. For very high frequencies, X_C goes to zero or a short circuit; in this case, V_{out} approaches zero.

We are interested in the *amplitude* and *phase angle* of the output voltage. In terms of complex numbers the output voltage equals

$$\mathbf{V}_{out} = \mathbf{I}(-jX_C) = \frac{\mathbf{V}_{in}}{R - jX_C}\,(-jX_C)$$

Dividing both sides by $\mathbf{V}_{in}$ gives the complex voltage gain from input to output:

$$\frac{\mathbf{V}_{out}}{\mathbf{V}_{in}} = \frac{-jX_C}{R - jX_C} \tag{16-1}$$

magnitude and angle

As it now stands, Eq. (16-1) is in rectangular form. To get the formulas we want, we must change it to polar form. When we do this,

$$\frac{\mathbf{V}_{out}}{\mathbf{V}_{in}} = \frac{X_C \,\underline{/-90°}}{\sqrt{R^2 + X_C^2}\,\underline{/-\arctan\ X_C/R}}$$

When an equation is in polar form like this, we can separate the magnitude and the angle to get two formulas:

$$\frac{V_{out}}{V_{in}} = \frac{X_C}{\sqrt{R^2 + X_C^2}} \tag{16-2}$$

and
$$\phi = -90° + \arctan\frac{X_C}{R} \tag{16-3}$$

These two equations are important. Equation (16-2) is the magnitude of the voltage gain. This is a pure number or magnitude; it tells nothing about the phase angle between input and output voltage. Unless otherwise indicated, when we say voltage gain, we mean *magnitude* of voltage gain.

Equation (16-3) is different. It pins down the phase angle of output voltage to input voltage. Whenever we know the values of X_C and R, we can calculate the phase angle.

critical reactance

When the frequency is zero, X_C is infinite; with Eq. (16-3),

$$\phi = -90° + \arctan \infty = 0° \qquad (f = 0)$$

At the other extreme, when the frequency approaches infinity, X_C goes to zero. With Eq. (16-3),

$$\phi = -90° + \arctan 0 = -90° \qquad (f = \infty)$$

This tells the permissible range of phase angle. In symbols,

$$\phi = 0° \text{ to } -90° \qquad \text{(lag network)}$$

The midpoint in this range is $-45°$.

The *critical reactance* is the value of X_C that makes the phase angle equal $-45°$. When we examine Eq. (16-3), we can see that ϕ will equal $-45°$ when

$$\arctan \frac{X_C}{R} = 45°$$

In turn, $\arctan X_C / R$ equals $45°$ when

$$X_C = R \qquad \text{(critical reactance)} \qquad (16\text{-}4a)***$$

Figure 16-2a shows the special case of X_C equal to R. Since X_C equals R, phase angle ϕ equals $-45°$; this is why the output waveform of Fig. 16-2a starts $45°$ after the input waveform.

At the critical reactance the voltage gain of a lag network equals

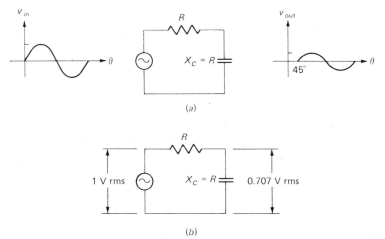

(a)

(b)

Figure 16-2. Critical reactance. (a) *Phase angle is $-45°$.* (b) *Voltage gain is 0.707.*

$$\frac{V_{\text{out}}}{V_{\text{in}}} = \frac{X_C}{\sqrt{R^2 + X_C^2}} = \frac{R}{\sqrt{R^2 + R^2}} = \frac{1}{\sqrt{2}}$$

or $\qquad\qquad A = 0.707 \qquad \text{for } X_C = R \qquad\qquad (16\text{-}4b)\text{***}$

This says the voltage gain equals 0.707 when $X_C = R$. Figure 16-2b is an example of the critical-reactance condition. Put 1 V into a lag network and you get 0.707 V out.

critical frequency

When the reactance is critical,

$$X_C = R$$

or $\qquad\qquad\qquad \dfrac{1}{2\pi fC} = R$

Solving for f gives

$$f = \frac{1}{2\pi RC}$$

We will call this value of frequency the *critical frequency* because it is the frequency that makes X_C equal to R. At this frequency, voltage gain equals 0.707.

EXAMPLE 16-1
Calculate the critical frequency for the lag network of Fig. 16-3a. Show the output waveform for an input frequency equal to the critical frequency.

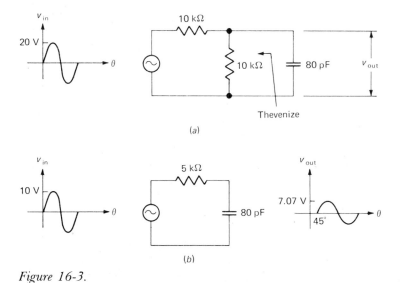

Figure 16-3.

SOLUTION

First, Thevenize the circuit. The Thevenin voltage equals half the source voltage; therefore, $V_{TH} = 10$ V peak. The Thevenin resistance is the parallel of two 10-kΩ resistances, or 5 kΩ.

Figure 16-3*b* shows the Thevenized circuit; we recognize this as a lag network. Therefore, the critical frequency equals

$$f = \frac{1}{2\pi RC} = \frac{1}{2\pi (5)10^3(80)10^{-12}} = 3.98(10^5)$$

or
$$f \cong 400 \text{ kHz}$$

At this frequency the output waveform is a sine wave with a peak of 7.07 V and a phase angle of −45° (see Fig. 16-3*b*).

16-2 *phase angle versus frequency*

To keep the critical frequency distinct from others, we will add the subscript *c*. With this notation the formula for critical frequency becomes

$$f_c = \frac{1}{2\pi RC} \tag{16-5}***$$

We can divide both sides of this equation by *f* to get

$$\frac{f_c}{f} = \frac{1}{2\pi RCf} = \frac{X_C}{R} \tag{16-6}$$

This useful ratio allows us to rewrite Eq. (16-3) as

$$\phi = -90° + \arctan\frac{f_c}{f} \tag{16-7}$$

graph of angle

We will calculate a few values of phase angle for different frequencies.

When $f = 0.1\,f_c$, $f_c/f = 10$ and Eq. (16-7) gives

$$\phi = -90° + \arctan 10 \cong -6°$$

When $f = f_c$, $f_c/f = 1$ and

$$\phi = -90° + \arctan 1 = -45°$$

When $f = 10\,f_c$, $f_c/f = 0.1$ and

$$\phi = -90° + \arctan 0.1 \cong -84°$$

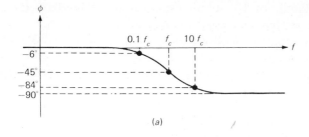

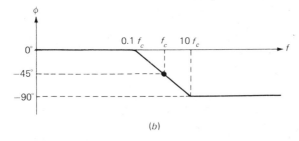

Figure 16-4. Bode plot of phase angle. (a) Exact. (b) Ideal.

To these values we can add those found earlier:

When $f = 0$, $\phi = 0$
When $f = \infty$, $\phi = -90°$

These are enough values to graph *phase angle versus frequency* (Fig. 16-4a). At very low frequencies the angle is zero. It stays near zero until we get to $0.1f_c$; at this frequency the phase angle equals $-6°$. As the frequency increases, the phase angle increases. When the input frequency equals the critical frequency, the phase angle equals $-45°$. For an input frequency of ten times the critical frequency, the phase angle is $-84°$. Further increases in frequency produce little change because the limiting value is $-90°$.

bode plot

A graph like Fig. 16-4a is called a *Bode plot* of phase angle. Knowing the phase angle is $-6°$ at $0.1f_c$ and $-84°$ at $10f_c$ is of little value except to indicate how close the phase angle is to its limiting values. Much more useful is the *ideal Bode plot* of Fig. 16-4b. This is the one to remember because it emphasizes these ideas:

1. When $f = 0.1f_c$, the phase angle is approximately zero.
2. When $f = f_c$, the phase angle is $-45°$.
3. When $f = 10f_c$, the phase angle is approximately $-90°$.

The word *decade* is used to describe changes in frequency. Decade means a *factor of 10*. So, if we say the frequency has changed by a decade, we mean it has changed by a factor of 10. As an example, if we increase the frequency from 25 to 250 Hz, we have increased it by a decade.

Another way to summarize the Bode plot of phase angle is this: at the critical frequency the phase angle equals $-45°$. A decade below the critical frequency the phase angle is approximately $0°$; a decade above the critical frequency the phase angle is approximately $-90°$.

16-3 *bel voltage gain versus frequency*

Logarithms make the analysis of lag networks easy when it comes to voltage gain. Before we take the logarithm of voltage gain, it will help if we rearrange Eq. (16-2) as follows:

$$A = \frac{V_{out}}{V_{in}} = \frac{X_C}{\sqrt{R^2 + X_C^2}} = \frac{1}{\sqrt{1 + R^2/X_C^2}}$$

With Eq. (16-6), we can write this as

$$A = \frac{1}{\sqrt{1 + f^2/f_c^2}} \qquad (16\text{-}8)$$

bel voltage gain

Now we are ready to get the bel voltage gain of a lag network. With Eq. (16-8), the bel voltage gain is

$$A' = 2 \log A = 2 \log \frac{1}{\sqrt{1 + f^2/f_c^2}}$$
$$= -2 \log \sqrt{1 + f^2/f_c^2} = -\log (1 + f^2/f_c^2)$$

Each step is based on properties of logarithms; therefore, the bel voltage gain of a lag network equals

$$A' = -\log \left(1 + \frac{f^2}{f_c^2}\right) \qquad (16\text{-}9)$$

As an example, we will calculate some bel voltage gains:

When $f = 0.1 f_c$, Eq. (16-9) gives

$$A' = -\log (1 + 0.01) = -\log 1.01$$
$$= -0.00432 \text{ B} = -0.0432 \text{ dB} \cong 0 \text{ dB}$$

When $f = f_c$,

$$A' = -\log(1+1) = -\log 2$$
$$= -0.30103 \text{ B} \cong -3 \text{ dB}$$

When $f = 10 f_c$,

$$A' = -\log(1+100) = -\log 101$$
$$= -2.00432 \text{ B} \cong -20 \text{ dB}$$

The bel voltage gains just calculated are interesting. Here is what we have found:

1. When the input frequency to a lag network is a decade below the critical frequency, the bel voltage gain is approximately zero.
2. When the input frequency equals the critical frequency, the bel voltage gain equals −3 dB (equivalent to an ordinary voltage gain of 0.707).
3. When the input frequency is a decade above the critical frequency, the bel voltage gain is approximately −20 dB.

As an example, the lag network of Fig. 16-5a has a critical frequency of 1.59 MHz. At this frequency an input sine wave with a peak of 1 V produces an output sine wave with a peak of 0.707 V; this is equivalent to the output voltage being down 3 dB from the input voltage. And at this critical frequency the output sine wave lags the input by 45° (Fig. 16-5b). If we decrease the frequency of the input signal from 1.59 MHz to 159 kHz, we have reduced frequency by a decade. In this case, the bel voltage gain is approximately 0 dB, meaning

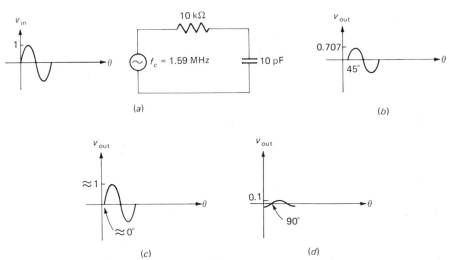

Figure 16-5. (a) *Lag network with critical frequency of 1.59 MHz.* (b) $f_{in} = 1.59$ *MHz.* (c) $f_{in} = 159$ *kHz.* (d) $f_{in} = 15.9$ *MHz.*

the output voltage almost equals the input voltage; the corresponding phase angle is approximately zero (Fig. 16-5c). On the other hand, if we increase the input frequency to 15.9 MHz (one decade above the critical frequency), the bel voltage gain is −20 dB, so that the output voltage is one-tenth of the input voltage; furthermore, the output voltage is approximately 90° behind the input voltage (Fig. 16-5d).

bode plot of bel voltage gain

In Eq. (16-9) when input frequency f is more than 10 times the critical frequency, the frequency term swamps out the 1. Because of this,

$$A' \cong -\log\frac{f^2}{f_c^2} \qquad \text{for } f \gg f_c$$

or

$$A' \cong -2\log\frac{f}{f_c} \qquad \text{for } f \gg f_c \qquad (16\text{-}10)***$$

The last equation produces an incredible simplicity when we notice the following pattern:

When $f = 10 f_c$,

$$A' = -2\log 10 = -2 \text{ B} = -20 \text{ dB}$$

When $f = 100 f_c$,

$$A' = -2\log 100 = -4 \text{ B} = -40 \text{ dB}$$

When $f = 1000 f_c$,

$$A' = -2\log 1000 = -6 \text{ B} = -60 \text{ dB}$$

and so on. In other words, *the bel voltage gain of a lag network drops 20 dB for each decade increase in frequency.*

As an example, the lag network of Fig. 16-5a has these bel voltage gains:

f	A'
15.9 MHz	−20 dB
159 MHz	−40 dB
1590 MHz	−60 dB

Note how A' drops 20 dB for each decade increase in f.

Whenever we can reduce a circuit to a lag network, we can calculate the critical frequency. Then, we immediately know the bel voltage gain is down

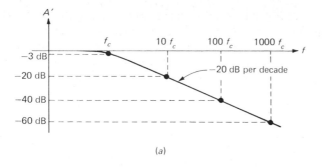

(a)

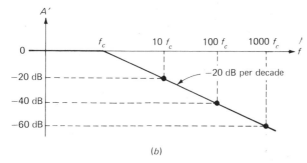

(b)

Figure 16-6. Bode plot of voltage gain. (a) Exact. (b) Ideal.

20 dB when the input frequency is one decade above the critical frequency; down 40 dB, two decades above the critical frequency; down 60 dB, three decades above the critical frequency, and so forth.

Figure 16-6*a* shows the bel voltage gain of a lag network; this is called a *Bode plot of voltage gain.* To see the graph over several decades of frequency, we have compressed the horizontal scale by showing equal distances between each decade in frequency. *Semilogarithmic* (or semilog) graph paper does the same thing; that is, the horizontal spacing is uniform between each decade. The advantage of marking off the horizontal scale in decades rather than units is this: well above the critical frequency the bel voltage gain drops off 20 dB for each decade increase in frequency; therefore, the graph of voltage gain will be a straight line with a slope of −20 dB per decade.

Figure 16-6*b* shows the *ideal Bode plot* of voltage gain. To a first approximation, we neglect the −3 dB at the critical-frequency point, and draw a straight line with a slope of −20 dB per decade. For all preliminary analysis, we use the ideal Bode plots for voltage gain and phase angle.

break frequency

Another name for the critical frequency is *break frequency,* so called because the voltage gain in Fig. 16-6*b* breaks at f_c. Beyond the break frequency the voltage gain drops off at 20 dB per decade. *Cutoff frequency* and *corner frequency* have also been used for the critical frequency.

EXAMPLE 16-2
Figure 16-7*a* shows a D MOSFET circuit. The 50 pF of capacitance is part of the next stage. Calculate the critical frequency and describe the Bode plots.

SOLUTION
The drain circuit can be reduced to a lag network as follows. Thevenize the circuit looking back into the drain. To do this, break the connection to the capacitor and calculate the Thevenin voltage and resistance. The voltage gain with no capacitor is

$$g_m r_D = 0.003(10,000) = 30$$

With a peak input of 1 mV, the Thevenin voltage at the drain is 30 mV peak.

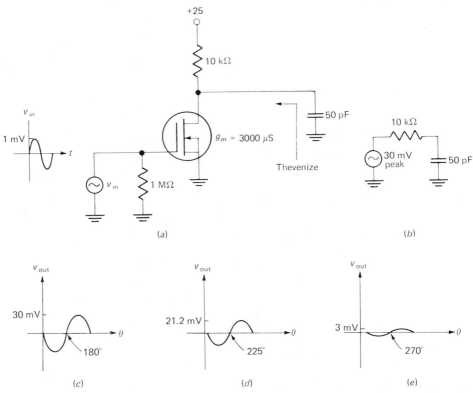

Figure 16-7. (a) *Direct-coupled FET amplifier.* (b) *Thevenized drain circuit.* (c) $f_{in} \ll f_c$. (d) $f_{in} = f_c$. (e) $f_{in} = 10f_c$.

The drain of the FET acts ideally like a current source; therefore, the Thevenin resistance equals 10 kΩ.

Figure 16-7b shows the Thevenized drain circuit connected to the 50 pF of the next stage. We recognize this as a lag network. The critical frequency of this lag network is

$$f_c = \frac{1}{2\pi RC} = \frac{1}{2\pi(10^4)50(10^{-12})} = 318 \text{ kHz}$$

Because of amplifier phase inversion, the output signal at low frequencies is 180° out of phase with the input signal. In other words, when f_{in} is at least a decade below 318 kHz, the output voltage has peak of 30 mV and lags the input by 180° (see Fig. 16-7c). When the input signal has a frequency of 318 kHz, the output voltage is down 3 dB, which is equivalent to an ordinary voltage gain of 0.707; therefore, the output voltage has a peak of

$$0.707 \times 30 \text{ mV} = 21.2 \text{ mV}$$

and lags the input by 225° (the lag network adds an additional 45° to the existing 180° as shown in Fig. 16-7d). When f_{in} is increased to 3.18 MHz (one decade above the critical frequency), the output voltage has a peak of 3 mV and lags the input by approximately 270° (Fig. 16-7e). Beyond this point, each decade increase in frequency causes the output voltage to drop 20 dB, but the output signal still lags the input by 270°.

At low frequencies the ordinary voltage gain of the amplifier is 30; this corresponds to a bel voltage gain of approximately 30 dB. Therefore, we can describe the ideal Bode plot of voltage gain as follows. The bel voltage gain is 30 dB from 0 to 318 kHz; then it decreases 20 dB for each decade increase in frequency.

The ideal Bode plot of phase angle is this. The phase angle is −180° from 0 to 31.8 kHz (one decade below the critical frequency); the phase angle equals −225° at 318 kHz; and, it equals −270° beyond 3.18 MHz (a decade above f_c).

EXAMPLE 16-3
Figure 16-8a shows the collector circuit of a transistor amplifier. The ac collector resistance is 10 kΩ. A capacitance of 50 pF is in shunt with r_C.

1. Prove that a current source driving a parallel RC circuit is equivalent to a lag network.
2. Calculate the critical frequency of the lag network.

SOLUTION

1. To prove the collector circuit is equivalent to a lag network, we Thevenize the circuit to the left of point c. This gives a Thevenin voltage of $i_c r_C$ and a Thevenin resistance of r_C.

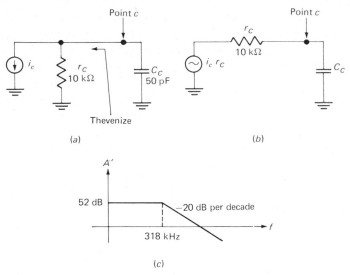

Figure 16-8.

Figure 16-8*b* shows the Thevenized collector circuit. Point *c* is still the collector node, and therefore the ac voltage from point *c* to ground is the same as the ac collector voltage in the original circuit. From now on, whenever we see a current source driving a shunt *R* and *C* (Fig. 16-8*a*), we will know it is equivalent to a lag network with the same values of *R* and *C* (Fig. 16-8*b*).

2. To get the critical frequency we use Eq. (16-5):

$$f_c = \frac{1}{2\pi RC} = \frac{1}{2\pi(10^4)50(10^{-12})} = 318 \text{ kHz}$$

This is the critical frequency of the collector lag network.

Each stage in multistage transistor amplifier has a collector lag network. The one with the *lowest* critical frequency is most important because it determines the frequency where amplifier voltage gain starts to drop off.

Incidentally, when you connect an oscilloscope from collector to ground, you will be shunting from 10 to 50 pF across r_C. If the 50 pF in Fig. 16-8*a* represents the input capacitance of an oscilloscope, this capacitance is responsible for the loss of output voltage at input frequencies above 318 kHz.

The 50 pF of capacitance may be part of the next stage. If this is the case, the input capacitance of the next stage will be causing the voltage to drop at frequencies above 318 kHz.

EXAMPLE 16-4

There are actually several lag networks in a transistor amplifier, and we will analyze them soon enough. Suppose the lag network of Fig. 16-8a is the one with the lowest critical frequency. If a CE amplifier has an r'_e of 25 Ω and its ac collector circuit looks like Fig. 16-8a, what does the ideal Bode plot of bel voltage gain look like?

SOLUTION

The voltage gain of the CE stage is

$$A = \frac{r_C}{r'_e} = \frac{10,000}{25} = 400$$

The corresponding bel voltage gain is

$$A' = 2 \log A = 2 \log 400 = 5.2 \text{ B} = 52 \text{ dB}$$

In the preceding example we calculated a critical frequency of 318 kHz. Therefore, Fig. 16-8c is the Bode plot of bel voltage gain. The bel voltage gain is 52 dB up to the break frequency; then, the bel voltage gain drops 20 dB per decade.

16-4 *dc-amplifier response*

At very low frequencies the reactances of the coupling and bypass capacitors are no longer negligible. For this reason, coupling and bypass capacitors determine the lower end of the normal frequency range. On the other hand, the lag networks in an amplifier determine the upper end of the normal frequency range. In the normal range of frequencies the amplifier has its maximum voltage gain. To keep this voltage gain distinct from voltage gain outside the normal range, we will designate it as A_{mid}. For instance, in a CE amplifier A_{mid} equals r_C/r'_e (ideally).

one dominant lag network

As mentioned earlier, we can design amplifiers with no coupling or bypass capacitors. An amplifier like this is called a *dc amplifier* (dc amp). Figure 16-9a shows the ideal Bode plot of bel voltage gain for a dc amp. As you can see, the dc amp has a bel voltage gain of A'_{mid} up to the break frequency f_2. Beyond this, the bel voltage gain drops 20 dB per decade.

The Bode plot of Fig. 16-9a assumes one dominant lag network. This is the reason the slope is −20 dB per decade. If another lag network has a critical frequency near the first, the Bode plot will break again at the critical frequency of the second lag network.

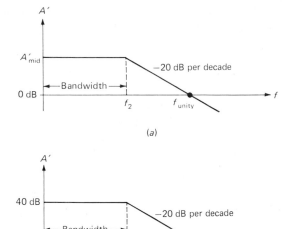

Figure 16-9. Dc-amplifier response.

In the usual dc amplifier, the bel voltage gain drops 20 dB per decade until the graph crosses the horizontal axis at f_{unity}. The subscript *unity* means the ordinary voltage gain equals unity at this frequency. For instance, Fig. 16-9b shows the Bode plot for a dc amp whose low-frequency bel voltage gain is 40 dB. The bel voltage gain breaks at 100 kHz. Ideally, the bel voltage gain drops 20 dB per decade until it crosses the horizontal axis at an f_{unity} of 10 MHz. Beyond this frequency, other lag networks will reach their critical frequencies, but we are not interested in them.

bandwidth

The *passband* of a dc amplifier is the set of frequencies from zero to break frequency f_2. Any input signal with a frequency in this set receives the full voltage gain (ideally). For instance, if a dc amplifier has a Bode plot like Fig. 16-9b, any input signal whose frequency is between 0 and 100 kHz will receive an ideal bel voltage gain of 40 dB. Signals with frequencies outside the passband of the amplifier receive less gain.

The *bandwidth B* is the numerical difference between the lowest and highest frequencies in the passband. In symbols,

$$B = f_2 - f_1 \qquad (16\text{-}11a)\texttt{***}$$

where f_1 is the lowest frequency and f_2 the highest. In a dc amplifier, f_1 is zero; therefore,

$$B = f_2 \quad (\text{dc amp}) \qquad (16\text{-}11b)\texttt{***}$$

So, for the Bode plot of Fig. 16-9b we would say the passband is the set of frequencies from 0 to 100 kHz, and the bandwidth is

$$B = 100 \text{ kHz}$$

16-5 *risetime-bandwidth relation*

In testing a dc amplifier we can vary the frequency of the input signal until the bel voltage gain is down 3 dB from its normal value; the input frequency then equals the break frequency f_2. But there is another way to find f_2.

risetime

Given an RC circuit like Fig. 16-10a, basic circuit theory tells what happens after the switch is closed. If the capacitor is initially uncharged, the voltage will rise from 0 to V. The *risetime* T_R is the amount of time it takes the voltage to go from $0.1V$ (the 10 percent point) to $0.9V$ (the 90 percent point). If it

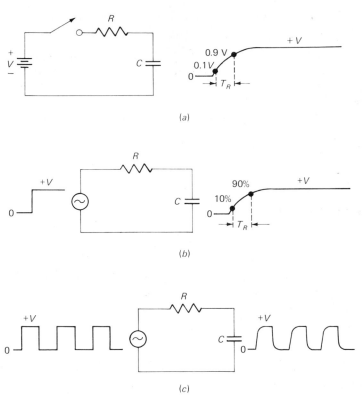

Figure 16-10. Risetime.

takes 10 μs for the exponential waveform to go from the 10 percent point to the 90 percent point, the waveform has a risetime of 10 μs.

Instead of using a switch to apply the sudden step in voltage, we can use a square-wave generator. For instance, Fig. 16-10b shows a square-wave generator driving the same RC network as before. We have shown only the leading edge of one cycle. The risetime of the output waveform will be the same as before, given the same values of R and C.

Figure 16-10c shows how several cycles would look. As we see, the input voltage changes suddenly from one voltage level to another. The output voltage takes longer to make its transitions. It cannot suddenly step because the capacitor has to charge and discharge through the resistance.

relation between T_R and RC

Basic courses prove

$$v_{\text{out}} = V(1 - \epsilon^{-t/RC})$$

where V is the total step in voltage, t is the time after the switch is closed, and RC is the time constant of the lag network. At the 10 percent point, $v_{\text{out}} = 0.1V$. At the 90 percent point, $v_{\text{out}} = 0.9V$. Substituting these voltages and solving for the time difference gives the risetime:

$$T_R = 2.2RC \qquad\qquad (16\text{-}12)\text{***}$$

As an example, if R equals 10 kΩ and C is 50 pF in Fig. 16-10a or b, then

$$RC = 10^4(50)10^{-12} = 0.5 \ \mu s$$

and the risetime of the output waveform equals

$$T_R = 2.2RC = 2.2(0.5 \ \mu s) = 1.1 \ \mu s$$

an important relation

Given a dc amplifier with one dominant lag network, we can find its break frequency with

$$f_2 = \frac{1}{2\pi RC}$$

According to Eq. (16-12), $T_R = 2.2RC$, or $RC = T_R/2.2$. When we substitute this into the formula for f_2, we get

$$f_2 = \frac{1}{2\pi T_R/2.2} = \frac{2.2}{2\pi T_R}$$

or

$$f_2 = \frac{0.35}{T_R} \qquad \text{(one lag network)} \qquad (16\text{-}13)\text{***}$$

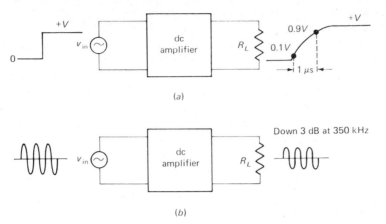

Figure 16-11. Bandwidth-risetime relation.

Why is this result important? *Because it relates sinusoidal and pulse operation of an amplifier.* This means we can test an amplifier either with an input sine wave or with an input pulse. For instance, suppose we have a dc amplifier and want its break frequency f_2. We can drive this amplifier with a step input as shown in Fig. 16-11*a*. If we measure a risetime of 1 μs, we can calculate a sinusoidal break frequency of

$$f_2 = \frac{0.35}{T_R} = \frac{0.35}{10^{-6}} = 350 \text{ kHz}$$

We do not have to measure this break frequency. But if we did set up the same dc amplifier in the circuit of Fig. 16-11*b*, we would find the output voltage is down 3 dB when the input frequency is 350 kHz.

EXAMPLE 16-5

Figure 16-12 shows a D MOSFET stage. Because of the phase inversion, the leading edge of the output falls through V volts; the trailing edge rises through V volts. In a case like this, you can measure either the falltime or the risetime since they are equal.

1. What is the risetime of the output waveform?
2. If driven by a sine wave, at what frequency is the bel voltage gain down 3 dB from its passband value?

SOLUTION

The drain ideally acts like a current source. As we saw in Example 16-3, a current source driving a parallel RC network is equivalent to a lag network.

1. The risetime of the output waveform is 2.2 times the RC time constant. So,

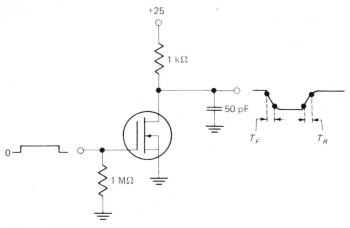

Figure 16-12.

$$T_R = 2.2\,RC = 2.2(10^3)50(10^{-12}) = 0.11\ \mu s$$

2. You can calculate f_2 in either of two ways:

$$f_2 = \frac{0.35}{T_R} = \frac{0.35}{0.11(10^{-6})} = 3.18\ \text{MHz}$$

or

$$f_2 = \frac{1}{2\pi RC} = \frac{1}{2\pi(10^3)50(10^{-12})} = 3.18\ \text{MHz}$$

16-6 *high-frequency FET analysis*

Figure 16-13a is the ac equivalent circuit of common-source FET amplifier. Resistance r_D is the ac resistance seen by the drain. Resistance R_S is the ac resistance seen by the gate when looking back to the source.

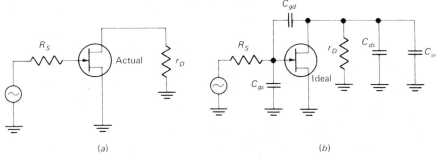

Figure 16-13. *Internal capacitances of FET.*

Ideally, the voltage gain from gate to drain is

$$A = \frac{v_d}{v_g} = g_m r_D \qquad (16\text{-}14)$$

This is the voltage gain in the midband of the amplifier. When the frequency of the input signal is high enough, however, the reactances in the amplifier are no longer negligible.

capacitances

There are capacitances in a FET amplifier that limit the high-frequency response. To begin with, the FET has internal capacitances between its three electrodes. C_{gs} is the internal capacitance between the gate and the source. C_{gd} is the capacitance between the gate and the drain and C_{ds} the capacitance between the drain and the source. Figure 16-13b shows these capacitances in the ac equivalent circuit.

If the output signal of the FET amplifier drives another amplifier stage, there is an additional capacitance C_{in} across the drain-ground terminals as shown in Fig. 16-13b. This input capacitance of the next stage includes *stray-wiring capacitance,* which is the capacitance between connecting wires and ground. As mentioned in Chap. 1, if a piece of AWG 22 wire is 1 ft long and is 1 in above a chassis, it has approximately 3.3 pF of capacitance with respect to ground. If the wire were only an inch long, it would have approximately 0.3 pF of capacitance to ground. Move this wire closer to the chassis and the capacitance increases; move it away from the chassis and the capacitance goes down. We can use 0.3 pF/in as a rough estimate for stray-wiring capacitance.

We also mentioned in Chap. 1 that a resistor has stray capacitance. As a rough estimate, we can allow 1 pF for this capacitance. When necessary, you can measure this stray capacitance with an ac bridge. Resistance r_D is the combined resistance of R_D (the drain biasing resistor) and other resistors in parallel with R_D in the ac equivalent circuit. The stray capacitance of each of these discrete resistors can be added into C_{in}.

gate lag network

To find the break frequencies of the FET amplifier of Fig. 16-13b, we have to reduce the ac circuit until we reach lag networks. The first step is to Millerize Fig. 16-13b. When we do this, we get an input Miller capacitance of

$$C_{\text{in(Miller)}} = C_{gd}(A + 1)$$

Since A is the voltage gain from the gate of the ideal FET to the collector, we can write

$$C_{\text{in(Miller)}} = C_{gd}(g_m r_D + 1)$$

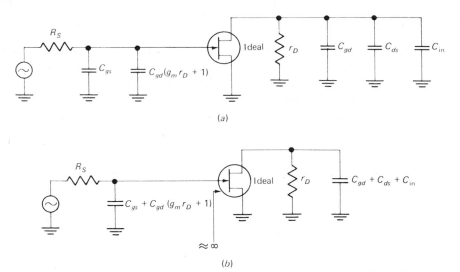

Figure 16-14. Gate and drain lag networks.

Figure 16-14a shows this input Miller capacitance.

The output Miller capacitance is

$$C_{\text{out(Miller)}} = C_{gd}\,\frac{A}{A+1} \cong C_{gd}$$

This is a reasonable approximation for A greater than 10; besides, the exact value is not important because it is only part of the total capacitance from drain to ground. Figure 16-14a shows the output Miller capacitance in shunt with C_{ds} and C_{in}.

Since the capacitances are in parallel, they can be added to get an equivalent capacitance for the gate circuit and for the drain circuit as shown in Fig. 16-14b. Looking directly into the gate, we see an almost infinite resistance; therefore, the gate circuit of Fig. 16-14b is in the form of a lag network. The critical frequency of this gate lag network is

$$f_c = \frac{1}{2\pi R_S C} \qquad \text{(gate lag network)} \qquad (16\text{-}15a)$$

where

$$C = C_{gs} + C_{gd}(g_m r_D + 1) \qquad (16\text{-}15b)$$

In the midband of the amplifier, maximum ac voltage reaches the gate (see Fig. 16-14b). But when the frequency of the input signal equals the critical frequency of the gate lag network, the ac gate voltage breaks and ideally drops 20 dB per decade.

drain lag network

Ideally, the drain acts like a current source. We already know a current source driving a parallel RC network is equivalent to a lag network (Example 16-3). Therefore the drain resistance and capacitance of Fig. 16-14b represent the R and C of a lag network. The critical frequency for this lag network is

$$f_c = \frac{1}{2\pi r_D C} \qquad \text{(drain lag network)} \qquad (16\text{-}16a)$$

where

$$C = C_{gd} + C_{ds} + C_{\text{in}} \qquad (16\text{-}16b)$$

Either the gate lag network or drain lag network dominates in Fig. 16-14b; only by a coincidence would the two lag networks have the same break frequency. Therefore, we are interested in the *lower* of the break frequencies; this break frequency determines the bandwidth. The voltage gain from signal source to final output in Fig. 16-14b equals $g_m r_D$ in the midband of the amplifier. This voltage gain breaks at the first critical frequency; it rolls off at 20 dB per decade until reaching the second break frequency; then, the voltage gain drops 40 dB per decade.

capacitances on a data sheet

Unfortunately, data sheets do not list the capacitances of Fig. 16-15a. But they do usually give information for calculating these capacitances. The manufacturer

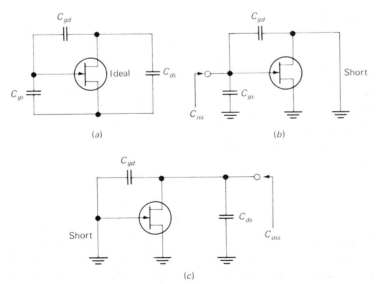

Figure 16-15. Measuring FET capacitances.

typically will measure the FET capacitances under short-circuit conditions. For instance, C_{iss} is the input capacitance with an ac short across the output as shown in Fig. 16-15b. The manufacturer measures and lists this capacitance C_{iss} on the data sheet.

In Fig. 16-15b the shorted output means C_{gd} is in parallel with C_{gs} when viewed from the input terminals. In other words,

$$C_{iss} = C_{gs} + C_{gd} \qquad (16\text{-}17a)$$

C_{oss} is the capacitance looking back into the FET with an ac short across the input terminals (Fig. 16-15c). The short across the input means C_{gd} is in parallel with C_{ds}, so that

$$C_{oss} = C_{ds} + C_{gd} \qquad (16\text{-}17b)$$

Finally, the manufacturer measures reverse capacitance C_{rss}, which equals

$$C_{rss} = C_{gd} \qquad (16\text{-}17c)$$

Solving Eqs. (16-17a) through (16-17c) simultaneously, we get these formulas:

$$C_{gd} = C_{rss} \qquad (16\text{-}18a)$$
$$C_{gs} = C_{iss} - C_{rss} \qquad (16\text{-}18b)$$
$$C_{ds} = C_{oss} - C_{rss} \qquad (16\text{-}18c)$$

With these formulas we can calculate the capacitances needed to analyze the lag networks of a FET amplifier.

A final point. Often, a data sheet will list only C_{iss} and C_{rss} values. In a case like this, you can calculate C_{gd} and C_{gs}, but not C_{ds}. However, C_{ds} remains fairly constant for most FETs. A close approximation is

$$C_{ds} \cong 1 \text{ pF}$$

Whenever C_{oss} is not given, we can use this approximation for C_{ds}. (If necessary, you can measure C_{oss} as shown in Fig. 16-15c and calculate an accurate value for C_{ds}.)

EXAMPLE 16-6
An MPF102 is an n-channel JFET. Its data sheet lists C_{iss} and C_{rss} as follows:

$$C_{iss} = 7 \text{ pF}$$
$$C_{rss} = 3 \text{ pF}$$

These are maximums or worst-case values. C_{oss} is not given. Calculate the C_{gd}, C_{gs}, and C_{ds}.

SOLUTION
With Eqs. (16-18a) and (16-18b)

$$C_{gd} = C_{rss} = 3 \text{ pF}$$
$$C_{gs} = C_{iss} - C_{rss} = 7 \text{ pF} - 3 \text{ pF} = 4 \text{ pF}$$

Since C_{oss} is not given, we approximate C_{ds} as

$$C_{ds} \cong 1 \text{ pF}$$

EXAMPLE 16-7

Calculate the upper break frequencies for the FET amplifier of Fig. 16-16a. Use a g_m of 4000 μS for the MPF102.

SOLUTION

First, visualize the ac equivalent circuit at higher frequencies. This means all coupling and bypass capacitors appear as short circuits. The MPF102 has capacitances of 4 pF, 3 pF, and 1 pF between its electrodes as shown in Fig. 16-16b; these are the capacitances we calculated in the preceding example. Also, we will allow 4 pF for stray capacitances in the drain.circuit.

The midband voltage gain is

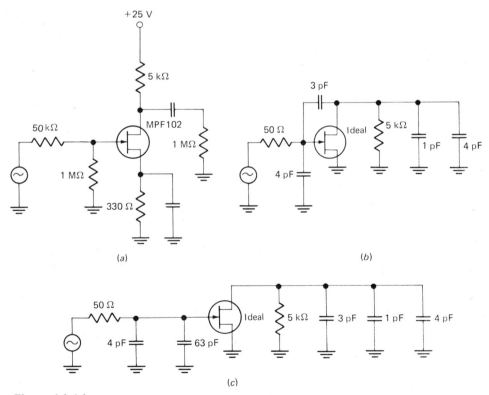

Figure 16-16.

$$A = g_m r_D = 0.004(5000) = 20$$

With this value, we can Millerize the circuit to get Fig. 16-16c. Now, we can see the two lag networks. The gate lag network has these values of R and C:

$$R = 50 \ \Omega$$
$$C = 4 \ \text{pF} + 63 \ \text{pF} = 67 \ \text{pF}$$

So, it has a critical frequency of

$$f_c = \frac{1}{2\pi RC} = \frac{1}{2\pi(50)67(10^{-12})} = 47.5 \ \text{MHz}$$

The drain lag network has

$$R = 5 \ \text{k}\Omega$$
$$C = 3 \ \text{pF} + 1 \ \text{pF} + 4 \ \text{pF} = 8 \ \text{pF}$$

and the break frequency is

$$f_c = \frac{1}{2\pi RC} = \frac{1}{2\pi(5000)8(10^{-12})} = 3.98 \ \text{MHz} \cong 4 \ \text{MHz}$$

This break frequency is lower than the other. Therefore, the drain lag network is dominant; the upper end of the passband is approximately 4 MHz.

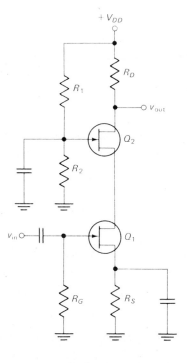

Figure 16-17.

EXAMPLE 16-8

Figure 16-17 shows a cascode amplifier. What is the input capacitance of the first stage if $C_{gd} = 3$ pF and $C_{gs} = 4$ pF?

SOLUTION

As shown earlier, the first stage has a voltage gain of unity because

$$A_1 = g_m r_D = g_m \frac{1}{g_m} = 1$$

The input Miller capacitance is

$$C_{in(Miller)} = C_{gd}(A + 1) = (3 \text{ pF})(1 + 1) = 6 \text{ pF}$$

The input capacitance of the first stage is

$$C_{in} = C_{gs} + C_{in(Miller)} = 4 \text{ pF} + 6 \text{ pF} = 10 \text{ pF}$$

The advantage of a cascode amplifier is its low input capacitance. The low gain of the first stage minimizes the input Miller capacitance. In general, the input capacitance of any cascode amplifier is

$$C_{in} = C_{gs} + 2C_{gd} \qquad (16\text{-}19)$$

16-7 *high-frequency bipolar analysis*

Figure 16-18*a* shows the ac equivalent circuit for a CE amplifier. R_S represents the Thevenin resistance looking back from base to source; it includes biasing resistors in the base circuit. Resistance r_C is the ac resistance seen by the collector. The high-frequency analysis of a bipolar transistor amplifier is so similar to the FET that we can easily derive the critical frequencies for the input and output lag networks.

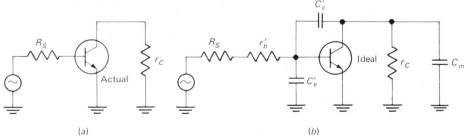

Figure 16-18. *Internal capacitances of a bipolar transistor.*

capacitances

When we first discussed the ac model of a transistor, we mentioned the internal capacitances. C'_e is the diffusion capacitance of the forward-biased emitter diode. C'_c is the transition capacitance of the reverse-biased collector diode. Because of charge storage (Sec. 4-9), C'_e is much larger than C'_c. For instance, in a 2N3904 with an I_E of 10 mA, C'_e is approximately 200 pF and C'_c is about 4 pF.

Figure 16-18b shows the ac equivalent circuit with these capacitances. The internal capacitance between the collector and the emitter (analogous to the C_{ds} in Fig. 16-13b) is small enough to neglect in a bipolar transistor. This is why we have only shown the C_{in} of the next stage. As before, we can include stray capacitance in the value of C_{in}.

Resistance r'_b is too important in the high-frequency analysis to neglect. Here is the reason. At higher frequencies where significant ac currents flow through C'_e and C'_c, additional current flows through r'_b. Because of this, there is an extra voltage drop across r'_b; this loss of signal voltage means the output voltage of the amplifier drops off.

base lag network

To find the break frequencies for the bipolar amplifier of Fig. 16-18b, we have to reduce the circuit until we have two distinct lag networks. The first step is to get the input Miller capacitance. The voltage gain from the base of the ideal transistor to the collector is

$$A = \frac{r_C}{r'_e}$$

The input Miller capacitance therefore equals

$$C_{in(Miller)} = C'_c \left(\frac{r_C}{r'_e} + 1 \right)$$

Figure 16-19a shows this input Miller capacitance.

The output Miller capacitance is approximately C'_c because the voltage gain A is ordinarily high in a CE amplifier. Figure 16-19a shows the output Miller capacitance in shunt with C_{in}.

The capacitances in Fig. 16-19a are in parallel; this allows us to add them to get an equivalent capacitance for the base circuit and for the collector circuit as shown in Fig. 16-19b. Unlike a FET, however, when we look into the base, we do not see an infinite resistance; instead, we see $\beta r'_e$. Therefore, we have to take an additional step to get to the lag networks. As proved in Sec. 9-5, we can split the ideal transistor into $\beta r'_e$ and a current source as shown in Fig. 16-19c.

In the base circuit of Fig. 16-19c the capacitance and $\beta r'_e$ are in parallel.

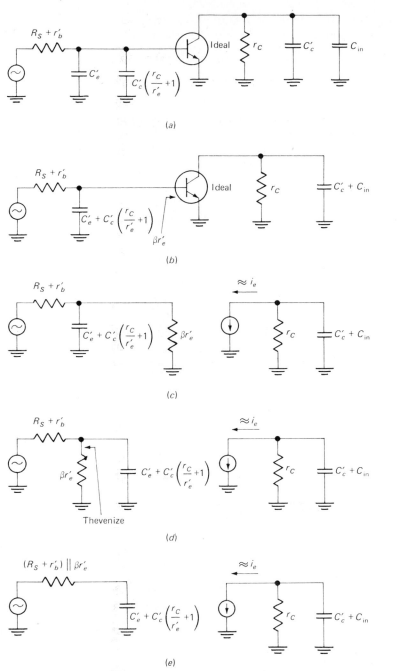

Figure 16-19. Reducing a bipolar amplifier to its base and collector lag networks.

Therefore, we can move $\beta r'_e$ to the left of the capacitance as shown in Fig. 16-19d. When we Thevenize Fig. 16-19d, we get a Thevenin resistance of

$$R_{TH} = (R_S + r_b) \| \beta r'_e$$

We don't have to bother with the Thevenin voltage because we are only after the R and C of the base lag network; this is all we need to calculate the critical frequency.

Figure 16-19e shows the final ac equivalent circuit. Aside from the complicated collection of symbols, the circuit is simple; it contains two lag networks. The base lag network has a critical frequency of

$$f_c = \frac{1}{2\pi RC} \tag{16-20a}$$

where

$$R = (R_S + r_b) \| \beta r'_e \tag{16-20b}$$

and

$$C = C'_e + C'_c \left(\frac{r_C}{r'_e} + 1 \right) \tag{16-20c}$$

If the base lag network is dominant, we calculate the break frequency of the bipolar amplifier using Eqs. (16-20a) through (16-20c).

collector lag network

Ideally, the collector acts like a current source; so, we recognize the collector circuit as a lag network. The critical frequency of this lag network is

$$f_c = \frac{1}{2\pi r_C (C'_c + C_{\text{in}})} \tag{16-20 d}$$

If the collector lag network has a lower critical frequency than the base lag network, it will determine the upper end of the amplifier passband.

capacitances on a data sheet

Again, it's a case of not getting exactly the capacitances C'_e and C'_c that we need. But enough information is usually given to calculate C'_e and C'_c. First, the collector-diode capacitance C'_c is the capacitance of the reverse-biased collector diode and depends on the dc voltage from collector to base. Data sheets usually list a value of capacitance at a specific V_{CB}, and we can use this as estimate for C'_c at other collector voltages. For instance, the 2N3904 has a C'_c of 4 pF at V_{CB} equals 5 V. If we operate at a different collector voltage, we can still use 4 pF as an estimate for C'_c.

There is no standard designation for C'_c. Data sheets may list it by any of the following equivalent symbols: C_c, C_{cb}, C_{ob}, and C_{obo}. As an example, the data sheet of a 2N2330 gives a C_{ob} of 10 pF for $V_{CB} = 2$ V. The data sheet of an FT107B lists a C_{cb} of 4 pF at $V_{CB} = 5$ V.

The C'_e we are after is not usually listed on a data sheet because it is too difficult to measure directly. Instead, the manufacturer gives a value called the *current gain-bandwidth product,* designated f_T. This particular quantity is the frequency where the current gain of a transistor drops to unity. By applying this definition of f_T to Fig. 16-19e, we can show that

$$C'_e = \frac{1}{2\pi f_T r'_e} \qquad (16\text{-}21)\text{***}$$

EXAMPLE 16-9

The data sheet of a 2N3904 gives an f_T of 300 MHz at $I_E = 10$ mA. Work out the value of C'_e.

SOLUTION

To use Eq. (16-21) we need the value of r'_e under the test conditions given on the data sheet. Since I_E is 10 mA,

$$r'_e = \frac{25 \text{ mV}}{10 \text{ mA}} = 2.5 \ \Omega$$

Therefore, C'_e equals

$$C'_e \cong \frac{1}{2\pi f_T r'_e} = \frac{1}{2\pi (300)10^6 (2.5)} = 212 \text{ pF}$$

This is the value of C'_e for an I_E of 10 mA.

EXAMPLE 16-10

Suppose we use the 2N3904 of the preceding example in a CE amplifier with the following values: $R_S = 1$ kΩ, $r_b = 100$ Ω, $\beta r'_e = 250$ Ω, $r_C = 1$ kΩ, $r'_e = 2.5$ Ω, $C'_c = 4$ pF, and $C_{in} = 5$ pF. With the C'_e found in the preceding example, work out the break frequencies of the CE amplifier.

SOLUTION

Refer to Fig. 16-19e to see how each given value fits into the lag networks. In the base lag network the equivalent R is

$$R = (R_S + r_b) \| \beta r'_e = 1100 \| 250 = 204 \ \Omega$$

The equivalent C is

$$C = C'_e + C'_c \left(\frac{r_C}{r'_e} + 1 \right) = 212 \text{ pF} + 1604 \text{ pF} \cong 1820 \text{ pF}$$

So, the critical frequency of the base lag network is

$$f_C = \frac{1}{2\pi RC} = \frac{1}{2\pi(204)1.82(10^{-9})} \cong 430 \text{ kHz}$$

In the collector circuit of Fig. 16-19e the equivalent capacitance is

$$C = C_c' + C_{in} = 4 \text{ pF} + 5 \text{ pF} = 9 \text{ pF}$$

Therefore, the critical frequency of the collector lag network is

$$f_C = \frac{1}{2\pi RC} = \frac{1}{2\pi(10^3)9(10^{-12})} = 17.7 \text{ MHz}$$

No question about which lag network is dominant. The base lag network has a much lower critical frequency; therefore, it determines the upper end of the bandwidth. The voltage gain from source to output will break first at 430 kHz; it will roll off at a rate of 20 dB per decade until 17.7 MHz; then, the voltage gain breaks again and rolls off at 40 dB per decade.

16-8 *the lead network*

Whenever you have a coupling capacitor, you have a *lead network*. Figure 16-20a shows the prototype form of a lead network. In the midband of an amplifier the coupling capacitor looks like an ac short, and the circuit acts like a voltage divider with a gain of

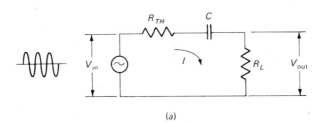

(a)

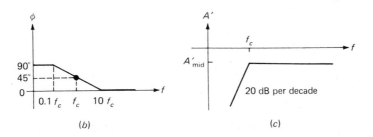

(b)   (c)

Figure 16-20. The lead network. (a) Circuit. (b) Bode plot of phase angle. (c) Bode plot of voltage gain.

$$A_{\text{mid}} = \frac{V_{\text{out}}}{V_{\text{in}}} = \frac{R_L}{R_{TH} + R_L} \qquad (16\text{-}22a)$$

At lower frequencies where X_C is no longer negligible, the current $\mathbf{I}$ equals

$$\mathbf{I} = \frac{\mathbf{V}_{\text{in}}}{R_{TH} + R_L - jX_C}$$

and the output voltage is

$$\mathbf{V}_{\text{out}} = \mathbf{I}R_L = \frac{\mathbf{V}_{\text{in}}R_L}{R_{TH} + R_L - jX_C}$$

Therefore, the complex voltage gain is

$$\mathbf{A} = \frac{\mathbf{V}_{\text{out}}}{\mathbf{V}_{\text{in}}} = \frac{R_L}{R_{TH} + R_L - jX_C} \qquad (16\text{-}22b)$$

magnitude and angle

Equation (16-22b) is the complex voltage gain. After we convert to polar form, we can split the complex voltage gain into its magnitude

$$A = \frac{R_L}{\sqrt{(R_{TH} + R_L)^2 + X_C^2}} \qquad (16\text{-}23a)$$

and its phase angle

$$\phi = \arctan \frac{X_C}{R_{TH} + R_L} \qquad (16\text{-}23b)$$

When X_C equals zero, ϕ equals zero (midband). On the other hand, when X_C approaches infinity, ϕ equals 90°. This gives us the permissible range of the phase angle:

$$0 \leqslant \phi \leqslant 90° \qquad \text{(lead network)}$$

Now we see why it is called a lead network; at lower frequencies, the output voltage leads the input voltage.

critical frequency

The critical reactance is the value of X_C that makes the phase angle equal 45°. When we examine Eq. (16-23b), it becomes apparent the condition for a 45° phase angle is

$$X_C = R_{TH} + R_L \qquad \text{(critical reactance)} \qquad (16\text{-}24a)***$$

Since $X_C = 1/2\pi fC$, we can solve for the critical frequency of a lead network to get

$$f_c = \frac{1}{2\pi(R_{TH} + R_L)C} \qquad (16\text{-}24\,b)$$

This is the frequency where the phase angle equals 45°. Also, it is the frequency where the voltage gain A is down 3 dB from its midband value.

bode plots

By working out the phase angle for $0.1f_c$, f_c, and $10f_c$, we find phase angles of 6°, 45°, and 84°. Again, the 6° and 84° indicate how close the phase angle is to ideal values of 0° and 90°. For this reason, we can show the ideal Bode plot of phase angle as given in Fig. 16-20b. At the critical frequency the output voltage leads the input by 45°. A decade below the critical frequency the output voltage leads by approximately 90°, and a decade above by approximately 0°.

In the midband, Eq. (16-22a) gives a bel voltage gain of

$$A'_{\text{mid}} = 2 \log A_{\text{mid}}$$

where A_{mid} is the ordinary voltage gain of the voltage divider formed by R_{TH} and R_L (see Fig. 16-20a). For instance, if R_{TH} equals 10 kΩ and R_L is 30 kΩ,

$$A_{\text{mid}} = \frac{30,000}{10,000 + 30,000} = 0.75$$

and

$$A'_{\text{mid}} = 2 \log 0.75 = 2 \log (7.5 \times 10^{-1})$$
$$= 2(0.875 - 1) = -0.25 \text{ B} = -2.5 \text{ dB}$$

This is the midband bel voltage gain.

When the frequency of the input signal is below the midband, the bel voltage gain drops off. At the critical frequency it is down 3 dB from the midband bel voltage gain. A decade below the critical frequency, the bel voltage gain is down 20 dB from the midband bel voltage gain. For example, if the midband bel voltage gain is −2.5 dB, at the critical frequency the bel voltage gain is −5.5 dB; a decade below the critical frequency the bel voltage gain is −22.5 dB.

Figure 16-20c shows the ideal Bode plot of bel voltage gain. In the midband the bel voltage gain equals A'_{mid}. Below the critical frequency the bel voltage gain drops 20 dB per decade. Ideally, the bel voltage gain at the critical frequency equals A'_{mid} as shown. To a second approximation it is actually down 3 dB.

EXAMPLE 16-11

In Fig. 16-21a calculate the critical frequencies of the coupling circuits. Assume the bypass capacitor looks like an ac short at these critical frequencies.

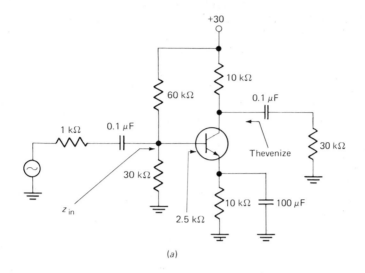

(a)

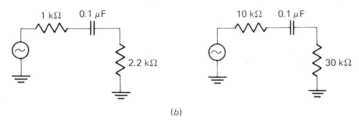

(b)

Figure 16-21.

SOLUTION

On the input side we work out the value of z_{in}.

$$z_{in} = 60,000 \parallel 30,000 \parallel 2500 \cong 2.2 \text{ k}\Omega$$

On the output side, looking back into the collector circuit, we see a Thevenin resistance of 10 kΩ (ideally). Therefore, we can show the equivalent ac circuit of Fig. 16-21b.

The base lead network has a critical frequency of

$$f_c = \frac{1}{2\pi(3.2)10^3(0.1)10^{-6}} \cong 500 \text{ Hz}$$

The collector lead network has a critical frequency of

$$f_c = \frac{1}{2\pi(40)10^3(0.1)10^{-6}} \cong 40 \text{ Hz}$$

The base lead network is the dominant one because it determines the lower end of the bandwidth.

16-9 *ac-amplifier response*

The dc amplifier has no lower critical frequency; therefore, it works all the way down to zero frequency. The *ac amplifier,* on the other hand, has lead networks that place a lower limit on the bandwidth.

bode plot of bel voltage gain

A multistage ac amplifier has many lead networks. To avoid certain problems discussed later, one of the lead networks is often made dominant so that the bel voltage gain drops 20 dB per decade until the graph crosses the horizontal axis (see Fig. 16-22*a*).

   In an ac amplifier the passband is the set of frequencies between break frequencies f_1 and f_2 shown in Fig. 16-22*a*. Any input signal with a frequency in this set ideally receives the maximum amplifier voltage gain. The bandwidth for an ac amplifier equals

$$B = f_2 - f_1 \qquad (16\text{-}25a)\ast\ast\ast$$

In an ac amplifier with no resonant circuits the ac loads on the transistors (or FETs) appear resistive over many decades of frequency. For this reason, the upper break frequency f_2 is much larger than the lower break frequency f_1. Because of this, the bandwidth of an untuned ac amplifier approximately equals

$$B \cong f_2 \qquad \text{(untuned amp)} \qquad (16\text{-}25b)$$

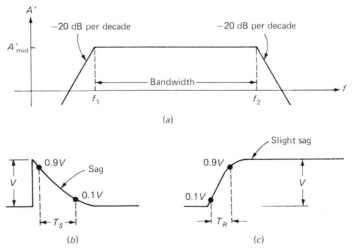

Figure 16-22. (a) *Ac-amplifier response.* (b) *Sagtime.* (c) *Risetime.*

measuring the break frequencies

If we drive an ac amplifier with a step voltage, we can measure the risetime as before and use

$$f_2 = \frac{0.35}{T_R} \qquad \text{(one lag network)} \qquad (16\text{-}26a)$$

to find the upper break frequency (see Fig. 16-22c).

By reducing the sweep speed of the oscilloscope, we can see the *sag* of Fig. 16-22b. This is caused by the dominant lead network, specifically by the charging of the dominant coupling capacitor. The *sagtime* T_S is the time between the 90 and 10 percent points shown in Fig. 16-22b. By a derivation almost identical to that given earlier for Eq. (16-26a),

$$f_1 = \frac{0.35}{T_S} \qquad \text{(one lead network)} \qquad (16\text{-}26b)\textbf{***}$$

EXAMPLE 16-12

Suppose you measure a sagtime of 7 ms when you drive an ac amplifier with a step voltage. What is the lower break frequency of the amplifier? (Assume one dominant lead network.)

SOLUTION
With Eq. (16-26b),

$$f_1 = \frac{0.35}{0.007} = 50 \text{ Hz}$$

16-10 *the bypass capacitor*

At lower frequencies the coupling or bypass capacitor may dominate. In this section we assume the bypass capacitor causes the response to break before the coupling capacitor does.

bipolar case

Figure 16-23a shows a base-driven prototype. When we Thevenize the circuit to the left of the bypass capacitor, we get the equivalent circuit of Fig. 16-23b. In the midband the capacitor looks like an ac short. But at lower frequencies the reactance is no longer negligible. Since R_E is usually negligibly large, the break frequency is

$$f_c \cong \frac{1}{2\pi (r_E + r_e' + R_S/\beta)C_E} \qquad (16\text{-}27)$$

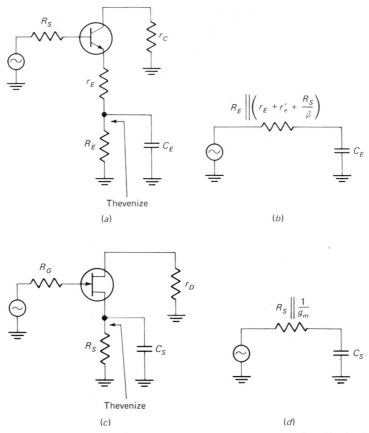

Figure 16-23. *Bypass capacitor.* **(a)** *Bipolar circuit.* **(b)** *Equivalent bipolar circuit.* **(c)** *FET circuit.* **(d)** *Equivalent FET circuit.*

In this equation, R_S includes the shunting effects of biasing resistors seen when looking from the base to the source.

When analyzing an amplifier, you calculate the break frequencies produced by each bypass capacitor and coupling capacitor. The highest break frequency determines the lower end of the passband.

An advantage of collector-feedback bias is that the emitter is at dc ground; therefore, it needs no bypass capacitor.

FET case

Figure 16-23c shows the gate-driven FET prototype. The Thevenin resistance to the left of the bypass capacitor is R_S in parallel with $1/g_m$ (Fig. 16-23d).

The break frequency produced by the bypass network is

$$f_c = \frac{1}{2\pi(R_S \parallel 1/g_m)C_S} \qquad (16\text{-}28)$$

EXAMPLE 16-13
Calculate the break frequency produced by the bypass capacitor of Fig. 16-21a. Use an r'_e of 25 Ω and a β of 100.

SOLUTION
Resistance r_E is zero and R_S is approximately 1000 Ω. With Eq. (16-27),

$$f_c = \frac{1}{2\pi(25 + 1000/100)10^{-4}} \cong 45 \text{ Hz}$$

EXAMPLE 16-14
A FET amplifier has an R_S of 1 kΩ, a g_m of 5000 μS, and a C_S of 10 μF. Calculate the break frequency.

SOLUTION
The reciprocal of g_m is 1/0.005, or 200 Ω. With Eq. (16-28),

$$f_c = \frac{1}{2\pi(1000 \parallel 200)10(10^{-6})} \cong 95 \text{ Hz}$$

self-testing review

Read each of the following and provide the missing words. Answers appear at the beginning of the next question.

1. In a lag network, the output voltage appears across a capacitor. This output has an amplitude and a _____ angle. The critical reactance in a lag network is the value of X_C that makes the phase angle equal to _____. This occurs when _____ equals R.

2. *(phase, − 45°, X_C)* The critical frequency equals the reciprocal of 2π times _____. At this frequency, the voltage gain equals _____ and the phase angle of the output is _____ with respect to the input. As the frequency approaches infinity, the phase angle approaches _____.

3. *(RC, 0.707, −45°, −90°)* Here is a summary of the ideal Bode plot of phase angle: At the critical frequency the phase angle of a lag network is _____; one decade below f_c the phase angle is approximately _____; one decade above f_c the phase angle is approximately _____.

4. *(−45°, 0°, −90°)* The bel voltage gain of a lag network is down 20 dB when the input frequency is one _____ above the critical frequency; down _____, two decades above f_c; down 60 dB, _____ decades above f_c, and so on. In a

Bode plot, this means bel voltage gain drops off 20 dB per _____ above the break frequency.

5. *(decade, 40 dB, three, decade)* Each stage in a multistage transistor amplifier has a collector lag network. The one with the _____ critical frequency is most important because it determines the frequency where the amplifier voltage gain starts to drop off. Critical frequency is also known as break frequency, _____ frequency, and cutoff frequency.

6. *(lowest, corner)* In a dc amplifier with one lag network, the bel voltage gain drops _____ per decade above the critical frequency until the Bode plot crosses the horizontal axis at _____. At this frequency the bel voltage gain is 0 dB; the ordinary voltage gain is _____.

7. *(20 dB, f_{unity}, unity)* The bandwidth of amplifier is the numerical difference between the _____ and the _____ frequencies in the passband. In a dc amplifier, f_1 is zero; therefore, $B = f_2$.

8. *(lowest, highest)* When a voltage step drives a lag network, the output increases exponentially from zero to the final value. The risetime T_R is the time it takes for the output voltage to rise from the _____ percent point to the _____ percent point. The bandwidth of a dc amp equals 0.35 divided by the _____; this is for one lag network.

9. *(10, 90, risetime)* In a single-stage FET amplifier, there is a gate lag network and a _____ lag network. Each has a break frequency. The voltage gain first breaks at the _____ break frequency. It rolls off 20 dB per decade until reaching the second break frequency. Then, the voltage gain drops _____ per decade.

10. *(drain, lower, 40 dB)* The advantage of the cascode amplifier is its low input capacitance. The low gain of the first stage minimizes the input _____ capacitance.

11. *(Miller)* The current gain-bandwidth product of a bipolar transistor is designated _____. This is the frequency where the current gain of the transistor drops to _____.

12. *(f_T, unity)* A coupling capacitor and resistor form a lead network. At the critical frequency, the output voltage leads the input by _____. A decade below the critical frequency, the output voltage leads by approximately _____, and a decade above by approximately _____.

13. *(45°, 90°, 0°)* The ac amplifier has lead networks that place a lower frequency limit on the passband. The lead network with the _____ break frequency is the most important because it determines where the voltage gain first breaks.

14. *(highest)* An advantage of collector-feedback bias is that the emitter is at dc ground; therefore, it needs no bypass capacitor between the emitter and ground.

problems

16-1. The capacitance C in a lag network equals 100 pF. Calculate the critical frequency for each of these values of R: 100 Ω, 10 kΩ, and 1 MΩ.

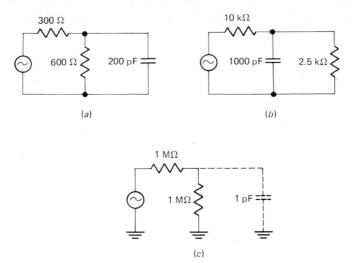

Figure 16-24.

16-2. Calculate the critical frequency for the lag network of Fig. 16-24a.

16-3. Work out the critical frequency for the circuit shown in Fig. 16-24b.

16-4. Figure 16-24c shows a voltage divider with a 1-MΩ resistors. The 1-pF capacitance represents the stray capacitance of the megohm resistor on the right. What critical frequency does the lag network have?

16-5. The FET of Fig. 16-25a has a g_m of 8000 μS. What is the voltage gain for low frequencies where X_C is negligible? At what frequency is the voltage gain down 3 dB? At what frequency is the phase shift of the lag network equal to −45°? What is the total phase shift from gate to drain at the critical frequency?

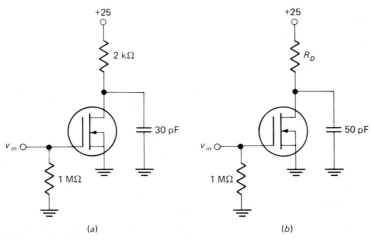

Figure 16-25.

16-6. The g_m of the FET in Fig. 16-25b has a value of 5000 μS. If R_D equals 5 kΩ, what is the voltage gain for an input sine wave whose frequency is 10 kHz? What is the critical frequency?

16-7. The FET in Fig. 16-25b has a g_m of 5000 μS. Calculate the voltage gain and critical frequency for each of these:

(a) $R_D = 1$ kΩ
(b) $R_D = 2$ kΩ
(c) $R_D = 4$ kΩ

Is the voltage gain directly or inversely proportional to the critical frequency?

16-8. Sketch the ideal Bode plot of bel voltage gain for the amplifier of Fig. 16-25a using a g_m of 10,000 μS. Also show the ideal Bode plot of phase angle.

16-9. Figure 16-26a represents the collector circuit of a transistor. Calculate the break frequency of this lag network. What happens to the break frequency if we double the resistance?

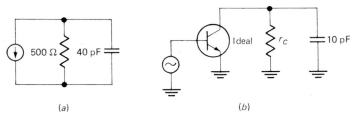

(a)   (b)

Figure 16-26.

16-10. The 40 pF in Fig. 16-26a is part of the circuit. If we connect an oscilloscope with an input capacitance of 50 pF across the 500-Ω resistor, what will the new critical frequency be? What is the critical frequency of the circuit when the oscilloscope is disconnected?

16-11. In Fig. 16-26b the only frequency effect we will consider is the collector lag network made up of r_C and the 10-pF capacitance. If r_e' is 25 Ω, what is the voltage gain v_c/v_b and the critical frequency for each of these:

(a) $r_C = 1$ kΩ
(b) $r_C = 2$ kΩ
(c) $r_C = 10$ kΩ

Which of these r_C values produces the maximum voltage gain? Which is the one with the highest critical frequency?

16-12. Sketch the ideal Bode plot of phase angle and bel voltage gain for the collector lag network of Fig. 16-26b with these values: $r_e' = 25$ Ω and $r_C = 5$ kΩ.

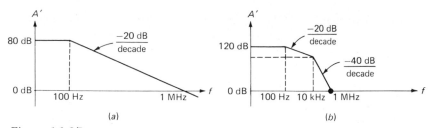

Figure 16-27.

16-13. For the Bode plot of Fig. 16-27a what is the midband bel voltage gain? What is the bandwidth? What is the ideal bel voltage gain at a frequency of 10 kHz? What is the phase angle at 1 MHz?

16-14. Figure 16-27b shows an ideal Bode plot of bel voltage gain. At 10 kHz the gain breaks and then drops off 40 dB per decade. What is the ideal bel voltage gain at 10 kHz? At 100 Hz?

16-15. Given a Bode plot like Fig. 16-27b, what is the phase angle at 100 Hz? Phase angles are additive; what is the phase angle at 10 kHz? At 1 MHz?

16-16. An *octave* is a *factor of* 2 in frequency, similar to a decade being a factor of 10. Saying the frequency increases an octave is equivalent to saying it doubles. Use Eq. (16-10) to prove the bel voltage gain drops 6 dB per octave when $f \gg f_c$. (Instead of 20 dB per decade, some people use the equivalent rate of 6 dB per octave.)

16-17. The amplifier of Fig. 16-28a has an A_{mid} of 100. If V_{in} equals 20 mV, what is the output voltage at the 100 percent point? The 90 percent point? What is the upper break frequency of the amplifier?

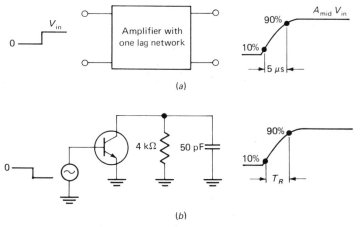

Figure 16-28.

16-18. The negative input step voltage produces a positive-going output in the ac equivalent circuit of Fig. 16-28b. The only frequency effect to consider is the lag network in the collector. What risetime does the output waveform have?

16-19. A dc amplifier has a bel voltage gain of 60 dB and a break frequency of 10 kHz. If the bel voltage gain drops off 20 dB per decade until it equals 0 dB, what is the risetime of the output when a step input is used? If the dc amplifier is modified to get a bel voltage gain of 40 dB and a break frequency of 100 kHz, what will the new risetime be?

16-20. You have two data sheets for different dc amplifiers. The first shows a break frequency of 1 MHz. The second lists a risetime of 1 μs. Which amplifier has the greater bandwidth?

16-21. The data sheet for a 2N3797 lists these maximum values of capacitance: $C_{iss} = 8$ pF and $C_{rss} = 0.8$ pF. Calculate C_{gd} and C_{gs}.

16-22. The FET of Fig. 16-29 has these capacitances: C_{gs} is 6 pF, C_{gd} is 4 pF, and C_{ds} is 1 pF. Neglect all other capacitances. Calculate the critical frequencies of the gate and drain lag networks.

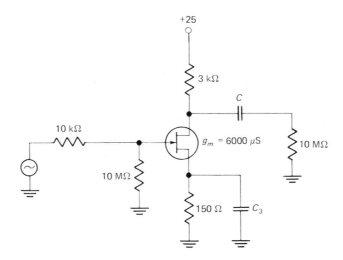

Figure 16-29.

16-23. Same data as in the preceding problem, but add a stray capacitance of 4 pF across the output. Work out the critical frequencies.

16-24. Suppose we want the amplifier of Fig. 16-29 to have an f_2 break frequency of 20 kHz. One way to get this is to shunt a capacitor across the output. Neglect all other capacitances and calculate the required C.

16-25. A 2N3300 has an f_T of 250 MHz for an I_E of 50 mA. Calculate the value of C_e' for this value of emitter current.

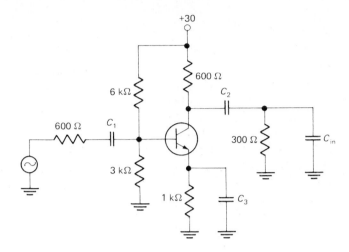

Figure 16-30.

16-26. The bipolar transistor of Fig. 16-30 has these values: $r_b' = 50$ Ω, $\beta = 80$, $r_e' = 2.5$ Ω, $f_T = 200$ MHz, $C_{ob} = 3$ pF, and $C_{in} = 50$ pF. What is the break frequency of the base lag network? Of the collector lag network?

16-27. The bipolar transistor of Fig. 16-30 has these values: $r_b' = 75$ Ω, $\beta = 20$, $r_e' = 1$ Ω, $f_T = 500$ MHz, $C_{ob} = 2$ pF, and $C_{in} = 100$ pF. Calculate the break frequencies for the base and collector lag networks.

16-28. $C_1 = 1$ μF and $C_2 = 4$ μF in Fig. 16-30. What are the break frequencies of the input and output lead networks for a $\beta r_e'$ of 300 Ω?

16-29. An ac amplifier has a sagtime of 2 ms. If it has one dominant lead network, what is the break frequency of this network?

16-30. C_3 equals 200 μF in Fig. 16-30. What break frequency does the bypass capacitor produce for a β of 50 and an r_e' of 2.5 Ω?

17 *integrated circuits*

As mentioned in Chap. 3, a manufacturer can produce an integrated circuit (IC) on a chip. ICs eliminate much of the drudgery in electronics work; you can find an IC for almost any routine application.

17-1 *making an IC*

Though we are mainly interested in using ICs, we should know something about how they are made.

the p *substrate*

First, the manufacturer produces a *p* crystal several inches long and 1 to 2 in in diameter (Fig. 17-1*a*). This is sliced into many thin *wafers* like Fig. 17-1*b*. One side of the wafer is lapped and polished to get rid of surface imperfections. This wafer is called the *p substrate;* it will be used as a chassis for the integrated components.

the epitaxial n *layer*

Next, the wafers are put in a furnace. A gas mixture of silicon atoms and pentavalent atoms passes over the wafers. This forms a thin layer of *n*-type semiconductor

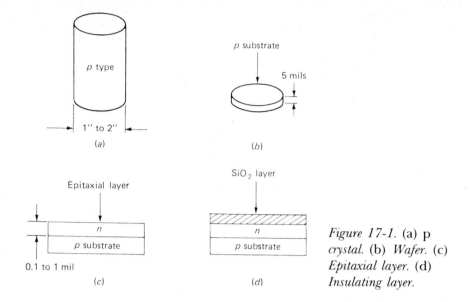

Figure 17-1. (a) p *crystal.* (b) *Wafer.* (c) *Epitaxial layer.* (d) *Insulating layer.*

on the heated surface of the substrate (see Fig. 17-1*c*). We call this thin layer an *epitaxial layer.* As shown in Fig. 17-1*c,* the epitaxial layer is about 0.1 to 1 mil thick.

the insulating layer

To prevent contamination of the epitaxial layer, pure oxygen is blown over the surface. The oxygen atoms combine with the silicon atoms to form the layer of silicon dioxide (SiO_2) shown in Fig. 17-1*d*. This glasslike layer of SiO_2 seals off the surface and prevents further chemical reactions; sealing off the surface like this is known as *passivation.*

chips

Visualize the wafer subdivided into the areas shown in Fig. 17-2. Each of these areas will be a separate chip after the wafer is cut. But before the wafer is cut,

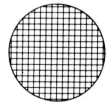

Figure 17-2. Cutting the wafer into chips.

the manufacturer produces hundreds of circuits on the wafer, one on each area of Fig. 17-2. This simultaneous mass production is the reason for the low cost of integrated circuits.

forming a transistor

Here is how an integrated transistor is formed. Part of the SiO_2 layer is etched off, exposing the epitaxial layer (see Fig. 17-3a). The wafer is then put into a furnace and trivalent atoms are diffused into the epitaxial layer. The concentration of trivalent atoms is enough to change the exposed epitaxial layer from n-type semiconductor to p-type. Therefore, we get an *island* of n material under the SiO_2 layer (Fig. 17-3b).

Oxygen is again blown over the wafer to form the complete SiO_2 layer shown in Fig. 17-3c. A hole is now etched in the center of the SiO_2 layer; this exposes

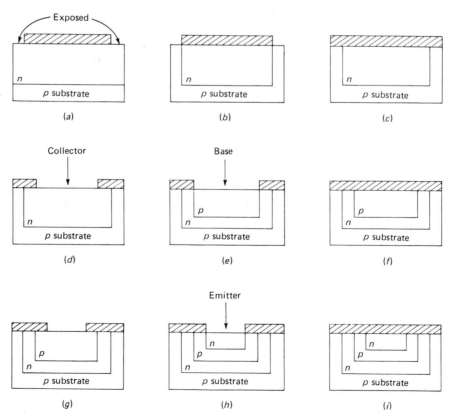

Figure 17-3. Steps in making a transistor.

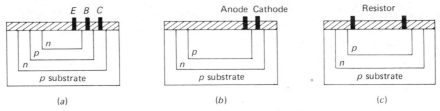

Figure 17-4. Integrated components. (a) *Transistor.* (b) *Diode.* (c) *Resistor.*

the *n*-epitaxial layer (Fig. 17-3*d*). The hole in the SiO₂ layer is called a *window*. We are now looking down at what will be the collector of the transistor.

To get the base, we pass trivalent atoms through the window; these impurities diffuse into the epitaxial layer and form an island of *p*-type material (Fig. 17-3*e*). Then, the SiO₂ layer is reformed by passing oxygen over the wafer (Fig. 17-3*f*).

To form the emitter, we etch a window in the SiO₂ layer and expose the *p* island (Fig. 17-3*g*). By diffusing pentavalent atoms into the *p* island, we can form the small *n* island shown in Fig. 17-3*h*. We then passivate the structure by blowing oxygen over the wafer (Fig. 17-3*i*).

IC components

By etching windows in the SiO₂ layer, we can deposit metal to make electrical contact with the emitter, base, and collector. In this way, we have the integrated transistor of Fig. 17-4*a*.

To get a diode, we follow the same steps up to the point where the *p* island has been formed and sealed off (Fig. 17-3*f*). Then, we etch windows to expose the *p* and *n* islands. By depositing metal through these windows, we make electrical contact with the cathode and anode of the integrated diode (Fig. 17-4*b*).

By etching two windows above the *p* island of Fig. 17-3*f*, we can make metallic contact with the *p* island; this gives us an integrated resistor (Fig. 17-4*c*).

Transistors, diodes, and resistors are easy to integrate on a chip. For this reason, almost all integrated circuits use these components. Inductors and large capacitors are not practical to integrate on the surface of a chip.

a simple example

To give you an idea of how a circuit is produced, look at the simple three-component circuit of Fig. 17-5*a*. To integrate this, we would simultaneously produce hundreds of circuits like this on a wafer. Each chip area would resemble Fig. 17-5*b*. The diode and resistor would be formed at the point mentioned earlier. At a later step, the emitter of the transistor would be formed. Then,

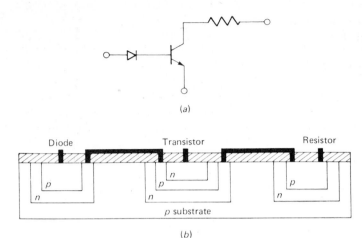

Figure 17-5. Simple integrated circuit.

we would etch windows and deposit metal to connect the diode, transistor, and resistor as shown in Fig. 17-5*b*.

Regardless of how complicated a circuit may be, it is mainly a process of etching windows, forming *p* and *n* islands, and connecting the integrated components.

The *p* substrate *isolates* the integrated components from each other. In Fig. 17-5*b*, depletion layers exist between the *p* substrate and the three *n* islands touching it. Because the depletion layers have essentially no current carriers, the integrated components are insulated from each other. This kind of insulation is known as *depletion-layer* or *diode isolation*.

monolithic ICs

The integrated circuits we have described are called *monolithic* ICs. The word "monolithic" is from Greek and means "one stone." The word is apt because the components are part of one chip.

Monolithic ICs are by far the most common, but there are other kinds. *Thin-film* and *thick-film* ICs are larger than monolithic ICs but smaller than discrete circuits. With a thin- or thick-film IC, the passive components like resistors and capacitors are integrated simultaneously on a substrate. Then, *discrete* active components like transistors and diodes are connected to form a complete circuit. Therefore, commercially available thin- and thick-film circuits are combinations of integrated and discrete components.

Hybrid ICs either combine two or more monolithic ICs in one package or monolithic ICs with thin- or thick-film circuits.

SSI, MSI, and LSI

Figure 17-5*b* is an example of *small-scale integration* (SSI); only a few components have been integrated to form the complete circuit. As a guide, SSI refers to ICs with less than 12 integrated components.

Medium-scale integration (MSI) refers to ICs that have from 12 to 100 integrated components per chip. *Large-scale integration* (LSI) refers to more than a hundred components. As mentioned earlier, it takes fewer steps to make an integrated MOSFET. Furthermore, a manufacturer can produce more MOSFETs on a chip than bipolar transistors. For this reason, MOS/LSI has become the largest segment of the LSI market.

17-2 *the differential amplifier*

Transistors, diodes, and resistors are the only practical components in a monolithic IC. Capacitances have been integrated but usually are less than 50 pF. Therefore, IC designers cannot use coupling and bypass capacitors like a discrete-circuit designer. Instead, the stages of a monolithic IC have to be direct-coupled.

The *differential amplifier* (diff amp) is heavily used in linear ICs. A diff amp uses no coupling or bypass capacitors; all it requires are resistors and transistors, both easily integrated on a chip.

tail current

Figure 17-6*a* shows the basic form of a diff amp. There are two input signals and one output signal. Note that the output signal is the voltage between the collectors. Ideally, the circuit is symmetrical; each half is identical to the other half. Monolithic ICs can approach this symmetry because the integrated components are on the same chip and have virtually identical characteristics.

A diff amp is sometimes called a "long-tail pair" because it consists of a pair of identical transistors connected to a common emitter resistor (the tail). The current through this common resistor is known as the *tail current* I_T.

The diff amp of Fig. 17-6*a* uses emitter bias, discussed earlier. As you recall, the key to analyzing emitter-biased circuits is to remember that the top of the emitter resistor is an *approximate ground point*. The dc equivalent circuit of Fig. 17-6*b* emphasizes this idea. Because of the ground, almost all the V_{EE} supply voltage appears across R_E. Therefore, the dc tail current is

$$I_T \cong \frac{V_{EE}}{R_E} \qquad (17\text{-}1)***$$

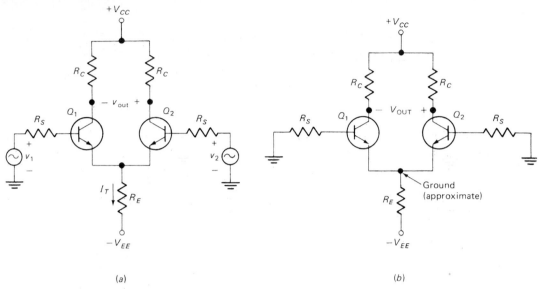

Figure 17-6. The differential amplifier.

When both transistors are identical, the tail current divides equally between them. That is, the dc emitter current in each transistor is half the tail current:

$$I_E = \frac{I_T}{2} \qquad (17\text{-}1a)***$$

As usual, I_C equals I_E to a close approximation:

$$I_C \cong I_E \qquad (17\text{-}1b)$$

other quantities

In Fig. 17-6b, the dc voltage from the Q_1 collector to ground equals the V_{CC} supply voltage minus the drop across the collector resistor:

$$V_{C1} = V_{CC} - I_C R_C \qquad (17\text{-}2a)***$$

Similarly, the dc voltage from the Q_2 collector to ground is

$$V_{C2} = V_{CC} - I_C R_C \qquad (17\text{-}2b)***$$

When the transistors and collector resistors are identical, V_{C1} equals V_{C2}, so there is no voltage between the collectors. In other words, the dc output voltage is zero:

$$V_{OUT} = 0 \qquad (17\text{-}2c)$$

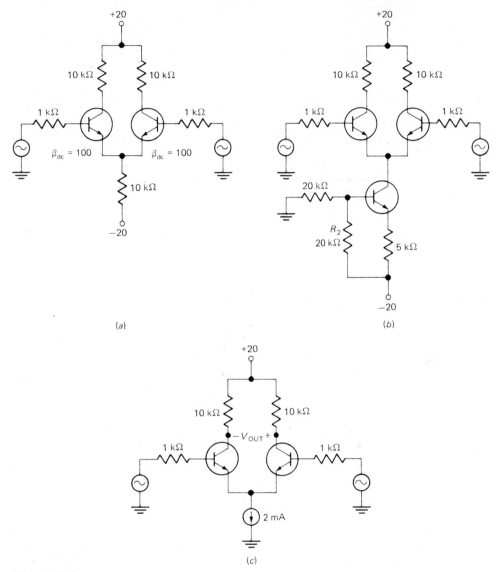

Figure 17-7.

EXAMPLE 17-1

Calculate the dc emitter current in each transistor of Fig. 17-7*a*.

SOLUTION

Visualize the emitters at dc ground. Then, all the V_{EE} supply voltage is across the emitter resistor. This gives a tail current of

$$I_T \cong \frac{V_{EE}}{R_E} = \frac{20\text{ V}}{10\text{ k}\Omega} = 2\text{ mA}$$

If the transistors are identical, each gets half of this tail current:

$$I_E = \frac{I_T}{2} \cong \frac{2\text{ mA}}{2} = 1\text{ mA}$$

This is the approximate value of dc emitter current in each transistor.

When high accuracy is needed, you can calculate the tail current with this formula:

$$I_T = \frac{V_{EE} - V_{BE}}{R_E + R_S/\beta_{dc}}$$

(See Example 7-15 if interested in the proof.) With the component values given in Fig. 17-7a,

$$I_T = \frac{20 - 0.7}{10,000 + 10} = 1.93\text{ mA}$$

So, each emitter gets half of this:

$$I_E = \frac{I_T}{2} = \frac{1.93\text{ mA}}{2} = 0.965\text{ mA}$$

EXAMPLE 17-2

We can use current-source bias to set up the dc tail current as shown in Fig. 17-7b. Calculate the dc currents.

SOLUTION

The bottom transistor uses voltage-divider bias. The voltage across R_2 is approximately 10 V. Most of this 10 V appears across the 5-kΩ resistor and produces a dc tail current of about

$$I_T \cong \frac{10}{5000} = 2\text{ mA}$$

Because the two halves of the diff amp are identical, the tail current splits so that each gets 1 mA.

Since the lower transistor acts like a current source, we may draw the equivalent circuit of Fig. 17-7c; this is simpler and emphasizes that a constant current is being forced through the diff-amp transistors. Current-source bias is used a great deal in integrated circuits.

EXAMPLE 17-3

Calculate the dc collector-to-ground voltages in Fig. 17-7c. Also, work out the dc voltage between collectors.

SOLUTION

In Fig. 17-7*c* each transistor has a dc collector-to-ground voltage of

$$V_C = V_{CC} - I_C R_C \cong 20 - 0.001(10,000) = 10 \text{ V}$$

This is what you measure with a dc voltmeter between each collector and ground.

The dc output voltage in Fig. 17-7*c* equals the difference of two equal collector-to-ground voltages; therefore,

$$V_{\text{OUT}} = 0 \text{ V dc}$$

17-3 *ac analysis of a diff amp*

Now that we have the main idea of dc action, let's find out what an ac signal does. In Fig. 17-8*a*, the output is taken between collectors. If the two halves of the diff amp are identical, the output voltage is zero when v_1 and v_2 are zero. If v_1 and v_2 change by exactly the same amount, the output voltage remains zero because of the symmetry. Only when there is a *difference* between v_1 and v_2 do we get an output voltage. When v_1 is more positive than v_2, more collector current flows through the transistor on the left; because of this, the output voltage has the polarity shown in Fig. 17-8*a*.

superposition theorem

In a good diff-amp design, V_{EE} and R_E set up an almost constant tail current and we can visualize the diff amp by the equivalent circuit of Fig. 17-8*b*. If

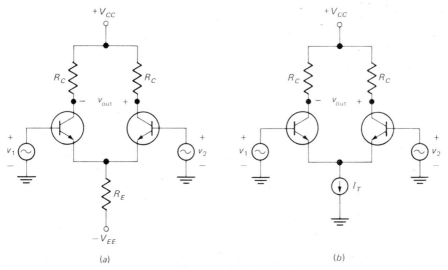

Figure 17-8. Diff-amp circuits.

current-source bias is used, Fig. 17-8b becomes a highly accurate equivalent circuit.

To get the ac equivalent circuit, visualize all dc sources of Fig. 17-8b reduced to zero. This is equivalent to ac grounding the V_{CC} supply point and opening the I_T current source. Since two ac sources drive the diff amp, we can use the superposition theorem to find the ac output voltage. This means taking one ac source at a time, finding its contribution to the ac output, and then algebraically adding the separate contributions.

Start with v_1 active and v_2 off, as shown in Fig. 17-9a. Q_1 acts like a CE amplifier; therefore, an inverted sinusoid appears at its collector. This ac voltage is between the Q_1 collector and ground. The ac emitter current from Q_1 has to flow into the emitter of Q_2 because the I_T current source is open to ac. Q_2 therefore acts like a CB amplifier, and an in-phase signal appears at its collector. The ac output voltage is taken between the collectors. Because it is the algebraic difference of two equal and opposite sine waves, the output has twice the amplitude (see Fig. 17-9b). So, the contribution of the first source acting alone is

$$v_{\text{out}(1)} = Av_1 \qquad\qquad (17\text{-}3a)$$

where $A \cong R_C/r'_e$.

Next, v_2 is made active and v_1 off (Fig. 17-9c). This time, Q_2 acts like a CE amplifier and Q_1 like a CB amplifier. An in-phase signal appears at the Q_1 collector and an inverted signal at the Q_2 collector. The algebraic difference of these two collector voltages is the ac output (see Fig. 17-9d). This ac output voltage is the second contribution:

$$v_{\text{out}(2)} = -Av_2 \qquad\qquad (17\text{-}3b)$$

where $A \cong R_C/r'_e$. This is the ac output produced by v_2 alone. The minus sign means the contribution has the opposite polarity from that of Fig. 17-9b.

The superposition theorem tells us that the ac output voltage when both sources act simultaneously is the algebraic sum of the two separate contributions:

$$v_{\text{out}} = v_{\text{out}(1)} + v_{\text{out}(2)}$$

or
$$v_{\text{out}} = Av_1 - Av_2 \qquad \text{(two inputs)} \qquad (17\text{-}4)***$$

where A approximately equals R_C/r'_e. Therefore, when we want the ac voltage out of a diff amp, we multiply each input voltage by A and take the difference.

input impedance

Each source acting alone sees an input impedance. In Fig. 17-9a, the v_1 source sees an input impedance of

$$z_{\text{in}} = \beta(r_E + r'_e)$$

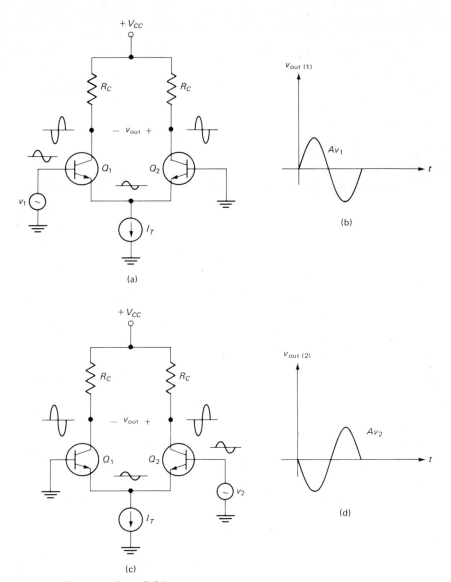

Figure 17-9. Single-ended inputs.

Since Q_2 acts like a CB stage, $r_E = r_e'$ and

$$z_{in} = 2\beta r_e' \qquad (17\text{-}5)\;\text{***}$$

Similarly, the v_2 source sees an input impedance of $2\beta r_e'$.

inverting and noninverting inputs

In the diff amp of Fig. 17-10*a* the output voltage can drive another stage, possibly a diff amp. As we saw, a positive v_1 acting alone produces a positive output voltage. For this reason, the v_1 input is the *noninverting input*. On the other hand, a positive v_2 acting alone produces a negative output voltage. This is why the v_2 input is called the *inverting* input.

differential input

Besides not having coupling and bypass capacitors, the diff amp has other advantages. If we like, we can ground one input and drive the other input. This special case is called *single-ended input* (similar to Fig. 17-9*a*).

Or, we can drive the diff amp with a signal between the bases as shown in Fig. 17-10*b*. This is known as *double-ended* or *differential input*. When we have a differential input, v_{in} is the difference of v_1 and v_2. In symbols,

$$v_{in} = v_1 - v_2$$

With this viewpoint, Eq. (17-4) becomes

$$v_{out} = Av_1 - Av_2 = A(v_1 - v_2)$$

or $\qquad v_{out} = Av_{in} \qquad$ (differential input) $\qquad\qquad$ (17-6)***

where A equals R_C/r_e'.

If we have two separate signal sources, we think of the output voltage as $Av_1 - Av_2$. On the other hand, if we have only one signal source driving the diff amp in the differential mode, we think of the output as Av_{in}.

common-mode input

Figure 17-10*c* illustrates the *common-mode input;* the same signal is applied to both inputs. If each half of the diff amp is identical, the ac output voltage will equal zero. About the only time we deliberately use a common-mode input signal is when we are testing the diff amp to see how well balanced the two halves are.

The *common-mode rejection ratio* CMRR is defined as

$$\text{CMRR} = \frac{Av_{in(CM)}}{v_{out(CM)}} \qquad\qquad (17\text{-}7a)$$

where the numerator is calculated and the denominator is measured. As an example, suppose $v_{in(CM)}$ is 1 V in Fig. 17-10*c*. Ideally, we should get nothing out, but there may be a small output signal because of nonsymmetry. Suppose A is 100 and $v_{out(CM)}$ is 0.01 V. Then, with Eq. (17-7*a*),

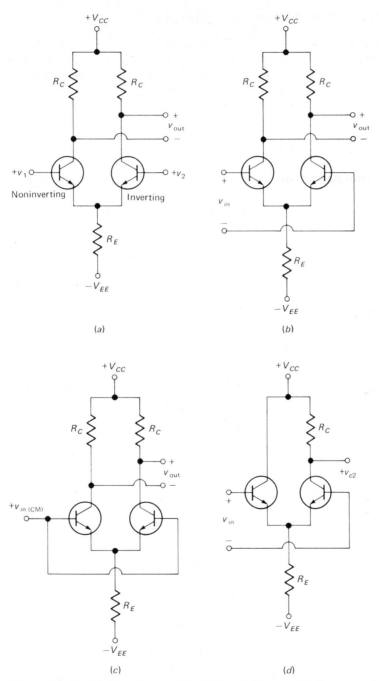

Figure 17-10. (a) *Two inputs.* (b) *Differential input.* (c) *Common-mode input.* (d) *Single-ended output.*

$$\text{CMRR} = \frac{100(1\ \text{V})}{0.01\ \text{V}} = 10,000$$

On a data sheet, CMRR is given in decibels. So, in this particular example,

$$\text{CMRR}' = 2 \log 10,000 = 8\ \text{B} = 80\ \text{dB}$$

The larger the value of CMRR, the better the diff amp. Ideally, a common-mode input should produce zero output voltage; therefore, the ideal diff amp has a CMRR of infinity.

We can rearrange Eq. (17-7a) to get

$$v_{\text{out(CM)}} = \frac{A v_{\text{in(CM)}}}{\text{CMRR}} \qquad (17\text{-}7b)$$

With this equation we can calculate how much common-mode output voltage occurs for common-mode input signals. Most forms of interference, static, induced voltages, etc., drive a diff amp in the common mode. A good diff amp has a large CMRR value, which means it's virtually free of interference signals.

single-ended output

A diff amp with a *single-ended output* (Fig. 17-10d) is sometimes used in the later stages of an amplifier. Since only half the available output voltage is used, the voltage gain drops in half:

$$v_{c2} = \frac{A}{2} v_{\text{in}} \qquad (17\text{-}8)$$

where A equals R_C/r_e'.

EXAMPLE 17-4
Calculate the approximate output voltage for each diff amp in Fig. 17-11.

SOLUTION
In all diff amps,

$$\frac{V_{EE}}{R_E} \cong \frac{10}{5000} = 2\ \text{mA}$$

Therefore, each transistor has a dc emitter current of about 1 mA; this means an r_e' of approximately 25 Ω. The ideal voltage gain is

$$A \cong \frac{R_C}{r_e'} = \frac{10,000}{25} = 400$$

In Fig. 17-11a we are using a single-ended noninverting input. Since v_2 is zero, Eq. (17-4) gives

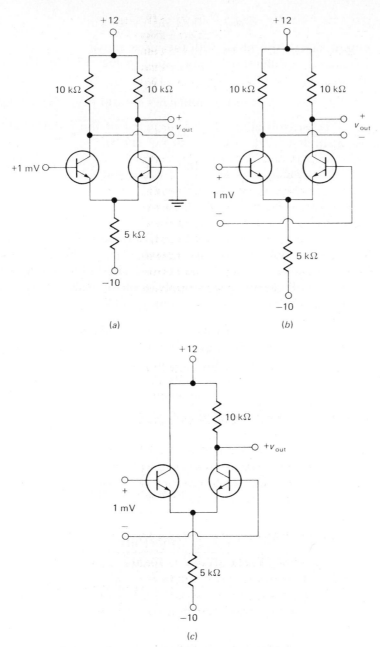

(a)

(b)

(c)

Figure 17-11.

$$v_{\text{out}} = Av_1 \cong 400(1 \text{ mV}) = 400 \text{ mV}$$

This ac output signal is in phase with the input signal.

We are applying a differential input signal in Fig. 17-11b. Therefore, using Eq. (17-6), we get

$$v_{\text{out}} = Av_{\text{in}} \cong 400(1 \text{ mV}) = 400 \text{ mV}$$

In Fig. 17-11c we still apply the input signal in the differential mode, but now we use a single-ended output. With Eq. (17-8),

$$v_{\text{out}} = \frac{A}{2} v_{\text{in}} \cong \frac{400}{2} 1 \text{ mV} = 200 \text{ mV}$$

EXAMPLE 17-5

Suppose you connect the output of Fig. 17-11a to the input of Fig. 17-11b. Then, the double-ended output of the first diff amp is the differential input to the second diff amp. Not only will you get the desired input signal to the second diff amp, you also will get the power-supply ripple on each collector. This ripple goes into the second diff amp in the common mode.

1. If the differential input signal to the second diff amp is 1 mV, how much output voltage is there? Use a gain of 400 for the second stage.
2. The ripple into each base of the second diff amp equals 1 mV. How much output ripple is there if CMRR equals 10,000?
3. Calculate the signal-to-ripple ratio at the input to the second diff amp and at the output.

SOLUTION

1. The output voltage of the second diff amp is

$$v_{\text{out}} = Av_{\text{in}} = 400(1 \text{ mV}) = 400 \text{ mV}$$

2. With Eq. (17-7b),

$$v_{\text{out(CM)}} = \frac{Av_{\text{in(CM)}}}{\text{CMRR}} = \frac{400(1 \text{ mV})}{10,000} = 0.04 \text{ mV}$$

3. At the input to the second diff amp, the signal-to-ripple ratio is

$$\frac{v_s}{v_r} = \frac{1 \text{ mV}}{1 \text{ mV}} = 1 \qquad \text{(input)}$$

At the output of the second diff amp, the ratio is

$$\frac{v_s}{v_r} = \frac{400 \text{ mV}}{0.04 \text{ mV}} = 10,000 \qquad \text{(output)}$$

This final result shows how effectively the diff amp discriminates against ripple in the power supply. The input signal and ripple are equal going into the second diff amp, but the output signal is 10,000 times greater than the output ripple.

As mentioned, most forms of interference are applied with equal intensity to both sides of the diff amp, that is, in the common mode. For instance, externally produced noise from electric motors, neon signs, ignition systems, lighting, etc., induces common-mode voltages. The signal-to-noise ratio (common-mode noise) will therefore be improved by a factor of CMRR.

17-4 *cascaded diff amps*

Figure 17-12 will give you an idea of how diff amps can be cascaded. The first diff amp has noninverting and inverting inputs. We can drive this diff amp with two separate single-ended inputs or with one differential input. The second diff amp uses Darlington pairs to prevent excessive loading of the first diff-amp stage. The single-ended output of the second diff amp drives a Darlington emitter follower which produces the final output voltage.

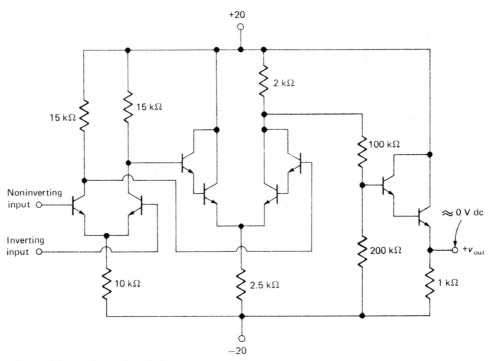

Figure 17-12. Cascaded diff amps.

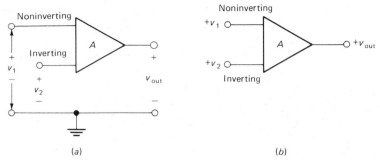

Figure 17-13. Schematic symbols.

Because no coupling or bypass capacitors are used, Fig. 17-12 is a dc amplifier. As a result, Fig. 17-12 can amplify signals with frequencies all the way down to zero. The midband voltage gain of the circuit is high, better than 10,000, because each diff amp has a voltage gain over 100.

Drift is much smaller in Fig. 17-12 than with the direct-coupled amplifiers of Sec. 9-10. Why? Because transistors on the same chip have almost identical changes with temperature. In a diff amp, these changes cancel since they are equivalent to a common-mode signal. Therefore, the drift at the final output of Fig. 17-12 is much smaller because of the diff amps.

schematic symbol

Given an amplifier like Fig. 17-12, we can save time by using a schematic symbol for the entire amplifier. Figure 17-13a shows a simple way to represent an amplifier with two inputs and one output. A is the unloaded voltage gain; this is the gain we get when no load resistor is used, or equivalently, the voltage gain when R_L is much greater than the Thevenin output impedance of the amplifier. The input and output voltages are with respect to the ground line.

Most of the time we don't bother drawing the ground line; we simply draw the schematic symbol of an amplifier as shown in Fig. 17-13b. Whenever you see this symbol, remember the voltages are with respect to ground.

Figure 17-14 shows the most widely used symbol. A positive input signal to the noninverting input produces a positive output voltage. For this reason, we mark the *noninverting input* with a *plus sign*. On the other hand, a positive input

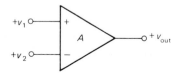

Figure 17-14. The most common symbol.

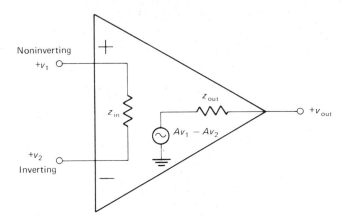

Figure 17-15. Input impedance and output circuit.

signal on the inverting input produces a negative output voltage; this is why the *inverting input* is marked with a *minus sign.*

input impedance and Thevenin circuit

In Fig. 17-12 the input impedance equals $2\beta r_e'$ for a single-ended or differential input signal. By using Darlington pairs or FETs, we can step up this input impedance to high values. In any case, an amplifier has a z_{in} between its terminals as shown in Fig. 17-15.

Theoretically, we can Thevenize the output of any linear circuit. Therefore, we can visualize the output side of the amplifier as a Thevenin voltage source of

$$v_{TH} = Av_1 - Av_2$$

and a Thevenin impedance of z_{out} shown in Fig. 17-15.

Figure 17-15 is important because it summarizes important values on data sheets, namely, z_{in}, z_{out}, and A. We need these values to analyze amplifier action under loaded conditions. For convenience, we don't normally show z_{in}, z_{out}, and A in schematic diagrams. Nevertheless, whenever you see a symbol like Fig. 17-14, remember there is a z_{in} between the two input terminals and a Thevenin circuit looking back into the output terminal; the Thevenin output voltage is $Av_1 - Av_2$, and the Thevenin output impedance is z_{out}.

EXAMPLE 17-6
Verify the peak output voltage in Fig. 17-16.

SOLUTION
The input impedance is 500 kΩ. Since the source impedance is only 1 kΩ, essentially all of the 1-mV source signal reaches the noninverting input terminal, that is, $v_1 = 1$ mV.

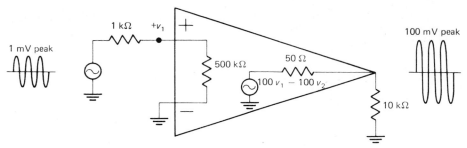

Figure 17-16.

The Thevenin or unloaded voltage gain is 100. Since there is no inverting input, the Thevenin output voltage is

$$v_{TH} = Av_1 = 100(1 \text{ mV}) = 100 \text{ mV}$$

The Thevenin output impedance is 50 Ω. With a load impedance of 10 kΩ, less than 1 percent of the signal is dropped across z_{out}. Therefore, the peak load voltage is almost equal to 100 mV. This output signal is in phase with the input signal.

17-5 *the operational amplifier*

About a third of all linear ICs are *operational amplifiers* (op amps). An op amp is a *high-gain dc amplifier* usable from 0 to over 1 MHz. By connecting external resistors to an op amp, you can adjust the voltage gain and bandwidth to your requirements. There are over 2000 types of commercially available op amps. Almost all are monolithic ICs with room-temperature dissipations under a watt. Whenever you need voltage gain, check available op amps. In many cases, an op amp will satisfy your requirements.

If you look at the schematic of a typical op amp, you will find many of the stages are diff amps; this allows operation down to zero frequency and provides common-mode rejection. Most op amps have two inputs and one output like Fig. 17-15.

first-generation op amps

The first quality IC op amp came out in 1965; it was the famous Fairchild μA709. Similar op amps followed: Motorola's MC1709, National Semiconductor's LM709, Texas Instruments' SN72709, and others. The last three digits in each of these types are 709. All these op amps behave the same; that is, they all have the same specifications. For this reason, we will refer to them as the 709.

The 709 family has different models like the 709, 709A, 709B, and 709C. For the same manufacturer, all these have the same schematic diagram, but the tolerances are larger for later members of the family; at the same time, the price of an op amp decreases. For instance, the 709 has the best tolerance and costs the most. At the other extreme, the 709C has the worst tolerance but costs the least.

When you look at the schematic diagram of a 709, be sure to look at the labels on each input terminal; these labels give you a great deal of information. Even better, if the data sheet includes a *pin diagram* like Fig. 17-17a or b, you can learn a lot about the op amp. For instance, pins 2 and 3 are the inverting and noninverting inputs. Pins 4 and 7 are for supply voltages. Pin 6 is the output, while pins 1 and 5 refer to lag networks to be externally connected.

The 709C has these typical values: $z_{in} = 250$ kΩ, $z_{out} = 150$ Ω, and $A = 45,000$ (approximately 93 dB). These values immediately tell us the 709C has a high input impedance, a low output impedance, and a high voltage gain.

The 709 typifies the first generation of op amps. The advantages are high input impedance, low output impedance, and high voltage gain. The disadvan-

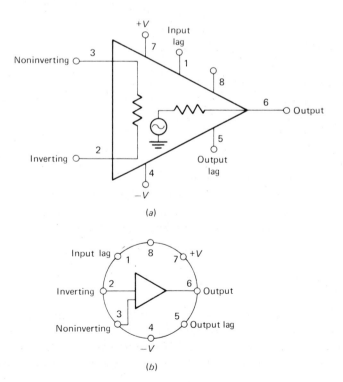

Figure 17-17. Pin diagrams.

tages include possible *latch-up,* no *short-circuit protection,* and the need for *external lag networks.* Latch-up means the output voltage can be latched or stuck at some value, regardless of the value of input voltage; latch-up may occur for large values of common-mode input voltage. No short-circuit protection means that accidentally shorting the output terminals may destroy the op amp; this can easily happen with a 709 and other first-generation op amps. Finally, having to add external lag networks means extra work; ideally, these should be integrated in the op amp.

second generation

The 741 is typical of second-generation op amps. It has no latch-up problems, includes short-circuit protection, and has its own integrated lag network. Unlike the 709, which has no internal capacitors, the 741 has an integrated 30-pF MOS capacitor. This capacitor is part of a lag network that rolls off the bel voltage gain at a rate of 20 dB per decade. The importance of this is discussed later.

Other improvements in the 741 are its higher input impedance (over a megohm), larger voltage gain (200,000), and lower output impedance (75 Ω). Because it is inexpensive and easy to use, the 741 has become one of the most widely used op amps.

later generations

The 709 and 741 are industry standards, used for general-purpose applications. For special applications, however, we need newer devices in which one or more specifications have been sharply improved. For instance, if we want high input impedance, *BIFET* op amps are the way to go. (BIFET stands for bipolar transistors and JFETs on the same chip.) These special op amps use JFETs for the input stage, followed by bipolar stages. This combination gives us the high input impedance associated with JFETs and the high voltage gain of bipolars. Other improved op amps are available for critical applications requiring low drift, low power consumption, high load current, etc.

Here are the main points to remember:

1. An op amp uses direct coupling, which gives voltage gain down to zero frequency.
2. It has high input impedance, low output impedance, and high voltage gain.
3. Op amps are not meant to be used *open loop;* you have to connect external components to get normal operation. (The only exception is the *comparator,* discussed later.)

17-6 *a simplified op-amp circuit*

Figure 17-18 is a simplified schematic diagram for a typical op amp. This circuit is equivalent to the 741 and many later-generation op amps. To pin down some important ideas, we will analyze how this circuit works.

input stage

Q_{13} and Q_{14} are a current mirror. Therefore, Q_{14} sources tail current to the input diff amp (Q_1 and Q_2). The diff amp drives a current mirror, consisting of Q_3 and Q_4. An input signal V_{IN} produces an amplified current out of this mirror which goes into the base of Q_5.

second and third stages

The second stage is an emitter follower (Q_5). It steps up the input impedance of the third stage (Q_6) by a factor of β. Note that Q_6 is the driver for the output stage. Incidentally, the + sign on the collector of Q_5 means it's connected to the positive V_{CC} supply; similarly, the minus signs at the bottoms of R_2 and R_3 mean these are connected to the negative V_{EE} supply.

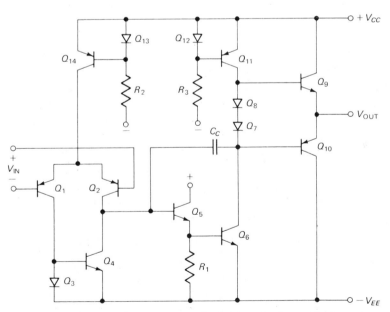

Figure 17-18. Simplified schematic for 741 and other typical op amps.

output stage

The last stage is a class B push-pull emitter follower (Q_9 and Q_{10}). Because of the split supply (equal positive and negative voltages), the quiescent output is ideally 0 V. Q_{11} is part of a current mirror that sources current through the compensating diodes (Q_7 and Q_8). Q_{12} is the input half of the mirror, and biasing resistor R_3 sets up the desired mirror current.

compensating capacitor

C_C is called a *compensating* capacitor. Because of the Miller effect, this small capacitor (typically 30 pF) has a pronounced effect on the frequency response. C_C is part of a base lag network that rolls off the bel voltage gain at a rate of 20 db per decade (equivalent to 6 dB per octave). This is necessary to prevent *oscillations* (an unwanted signal produced by the amplifier). Later sections tell you more about the compensating capacitor and oscillations.

active loading

All CE stages discussed up to now have used a *passive* load (a resistor). In Fig. 17-18, we have an example of a CE stage (Q_6) driving an *active* load (Q_{11}). Because Q_{11} is part of a current mirror, it sources a *fixed* current; therefore, the amplified signal out of Q_6 sees only the high output impedance associated with the Q_{11} current source. Because of this, any signal current out of Q_6 is forced into the final output transistors (Q_9 or Q_{10}), whichever is conducting.

Active loading (using transistors for loads instead of resistors) is very popular in integrated circuits because it is easier and less expensive to fabricate transistors on a chip than resistors. MOS digital integrated circuits use active loading almost exclusively; in these ICs, one MOSFET is the active load for another.

input impedance

Recall that the input impedance of a diff amp is

$$z_{\text{in}} = 2\beta r_e'$$

With a small tail current in the input diff amp, an op amp can have a fairly high input impedance. For instance, the input diff amp of a 741 has a tail current of approximately 15 μA. Since each emitter gets half of this,

$$r_e' = \frac{25 \text{ mV}}{I_E} = \frac{25 \text{ mV}}{7.5 \text{ μA}} = 3.33 \text{ k}\Omega$$

The input transistors have typical βs of 300; therefore,

$$z_{in} = 2\beta r_e' = 2 \times 300 \times 3{,}330 = 2 \text{ M}\Omega$$

This agrees with the data-sheet value for a 741.

summary

The circuit of Fig. 17-18 is relatively simple compared to the actual schematic diagram of an op amp. Nevertheless, this equivalent circuit is tremendous first step toward understanding and using op amps.

A final point. The diodes of Fig. 17-18 are actually transistors connected as diodes (see Prob. 11-7 if necessary). By shorting the collector and base together, the transistor acts like a diode. This is the easiest way to fabricate a matching diode.

17-7 *op-amp specifications*

To enhance our understanding of op amps, we need to discuss the more important specifications that appear on a data sheet.

input bias current

The op amp of Fig. 17-18 is equivalent to what is *inside* the IC package or housing. For the circuit to work, you need to connect V_{CC} and V_{EE} supplies. But that's not all. You also have to connect external dc returns for the *floating* input bases. In other words, the Q_1 and Q_2 base currents have to flow to ground to complete the circuit, because the other ends of the power supplies are grounded.

The signal sources driving the op amp provide the dc returns unless they are capacitively coupled; in this case, you have to add dc returns (described in Sec. 5-14). Either way, the two base currents of the input diff amp must flow through *external* resistances. These base currents are close in value, but not necessarily equal. When slightly unequal base currents flow through external dc returns, they produce a small differential input voltage or unbalance; this is a false input signal. The smaller the base currents are, the better, because the unbalance is minimized.

The *input bias current* shown on data sheets is the *average* of the two input currents. It tells you approximately what each input current is. As a guide, the smaller the input bias current, the smaller the possible unbalance. The 741 has an input bias current of 80 nA, which is acceptable in many applications. But in critical applications, a later-generation op amp may be preferred. For example, a BIFET op amp like the 357 has an input bias current of only 30 pA.

input offset current

The *input offset current* is the *difference* between the two input currents. The 741 has an input offset current of 20 nA. When working with 741s, we may find 20 nA more current in one base than the other. These unequal base currents produce a false differential input signal when they flow through dc returns. Again, the general guide is this: the smaller the input offset current, the better.

input offset voltage

The *input offset voltage* is the differential input voltage needed to null or zero the quiescent output voltage. For example, a 741 has a worst-case input offset voltage of 5 mV. When using 741s, we need to apply up to 5 mV differential input to zero the output voltage.

cmrr and output voltage swing

The common-mode rejection ratio was defined earlier. For a 741, CMRR′ = 90 dB. If the signal/common-mode ratio is unity at the input to a 741, the signal/common-mode ratio will be 90 dB at the output. In other words, the common-mode signal will be approximately 30,000 times smaller than the signal at the output.

Output voltage swing V_O is the maximum possible swing in output voltage. Typically, it is within 1 or 2 V of the supply voltages. For instance, when V_{CC} = +15 V and V_{EE} = −15 V, a 741 has V_O = ±14 V. This means the output voltage can swing up to +14 V or down to −14 V. V_O is slightly less than the supply voltages because of V_{BE} and other drops in later stages of an op amp.

17-8 *slew rate and power bandwidth*

Among all specifications affecting the ac operation of an op amp, *slew rate* is the most important. Why? Because it places a severe limit on large-signal operation.

basic idea

Circuit theory tells us the charging current in a capacitor is given by

$$i = C\frac{dv}{dt}$$

where i = current into capacitor
C = capacitance
dv/dt = rate of capacitor voltage change

We can rearrange this basic equation to get

$$\frac{dv}{dt} = \frac{i}{C}$$

This says the rate of voltage change equals the charging current divided by the capacitance.

The greater the charging current, the faster the capacitor charges. If for any reason the charging current is limited to a maximum value, so too is the rate of voltage change *limited to a maximum value.*

Figure 17-19*a* brings out the idea of current limiting and its effect on output voltage. A current of I_{MAX} charges the capacitor. Because this current is constant, the capacitor voltage increases linearly as shown in Fig. 17-19*b*. The rate of voltage change with respect to time is

$$\frac{dv_{out}}{dt} = \frac{I_{MAX}}{C_C} \qquad\qquad (17\text{-}9)$$

As an example, if $I_{MAX} = 60 \ \mu A$ and $C_C = 30$ pF (see Fig. 17-19*c*), the maximum rate of voltage change is

$$\frac{dv_{out}}{dt} = \frac{60 \ \mu A}{30 \ pF} = 2 \ V/\mu s$$

This says the output voltage across the capacitor changes at a maximum rate of 2 V per microsecond (Fig. 17-19*d*). The voltage *cannot change faster than this* unless we can increase I_{MAX} or decrease C_C.

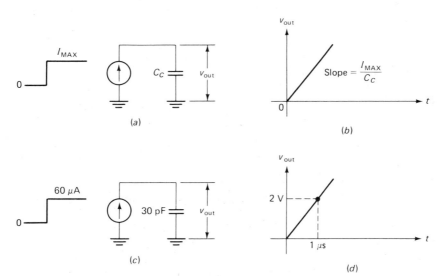

Figure 17-19. Slew rate equals maximum charging current divided by capacitance.

Slew rate is defined as the maximum rate of output voltage change. Because of this, we can rewrite Eq. (17-9) as

$$S_R = \frac{I_{\text{MAX}}}{C_C} \qquad (17\text{-}10)$$

In Fig. 17-19a, slew rate pins down the rate at which the output voltage can change. If $I_{\text{MAX}} = 60$ μA and $C_C = 30$ pF, the circuit can slew no faster than 2 V per microsecond.

slew-rate distortion

When a large positive step of input voltage drives the op amp of Fig. 17-18, Q_1 saturates and Q_2 cuts off. Therefore, all the tail current I_T passes through Q_1 and Q_3. Because of the current mirror, the current through Q_4 equals I_T. Since Q_2 is cut off, all the tail current passes on to the next stage. Initially, all of it goes to C_C (Q_5 base current is negligible).

As C_C charges, the output voltage rises. Assuming unity voltage gain for the output stage, the rate of output voltage change is equal to the rate of voltage change across C_C. With Eq. (17-10), the maximum rate of output voltage change is

$$S_R = \frac{I_T}{C_C}$$

This says the output voltage can change no faster than the ratio of I_T to C_C.

Here's an example. The input stage of a 741 has $I_T = 15$ μA; $C_C = 30$ pF. Therefore, the slew rate of a 741 is

$$S_R = \frac{15 \ \mu\text{A}}{30 \ \text{pF}} = 0.5 \ \text{V}/\mu\text{s}$$

This is the ultimate speed of a 741; its output voltage can change no faster than 0.5 V per microsecond. Figures 17-20a and b show what's going on. If we overdrive a 741 with a large step input (Fig. 17-20a), the output slews as shown in Fig. 17-20b. It takes 20 μs for the output voltage to change from 0

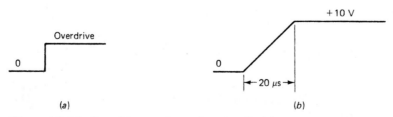

Figure 17-20. Overdrive produces slew-rate limiting.

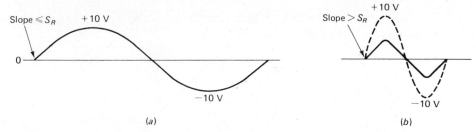

Figure 17-21. Slew-rate distortion of sine wave.

to 10 V (nominal output voltage swing). It is impossible for the output of a 741 to change faster than this.

We can also get slew-rate limiting with a sinusoidal signal. Figure 17-21*a* shows the maximum sinusoidal output from a 741 with $V_O = \pm 10$ V. As long as the initial slope of the sine wave is less than or equal to S_R, there is no slew-rate limiting. But when the initial slope of the sine wave is greater than S_R, we get the slew-rate distortion shown in Fig. 17-21*b*. The output begins to look triangular; the higher the frequency, the smaller the swing and the more triangular the waveform.

power bandwidth

Slew-rate distortion of a sine wave starts at the point where the initial slope of the sine wave equals the slew rate of the op amp. With calculus, we can derive this useful equation:

$$f_{\text{max}} = \frac{S_R}{2\pi V_P} \tag{17-11}$$

where f_{max} = highest undistorted frequency
$\quad\quad S_R$ = slew rate of op amp
$\quad\quad V_P$ = peak of output sine wave
As an example, if a 741 has $V_O = 10$ V and $S_R = 0.5$ V/µs, the maximum frequency for large-signal operation is

$$f_{\text{max}} = \frac{0.5 \text{ V/µs}}{2\pi \times 10 \text{ V}} = 7.96 \text{ kHz}$$

Frequency f_{max} is often called the *power bandwidth* of an op amp. We have just found that the 10-V power bandwidth of a 741 is approximately 8 kHz. This means the undistorted bandwidth for large-signal operation is 8 kHz. Try to amplify higher frequencies of the same peak value and you will get slew-rate distortion.

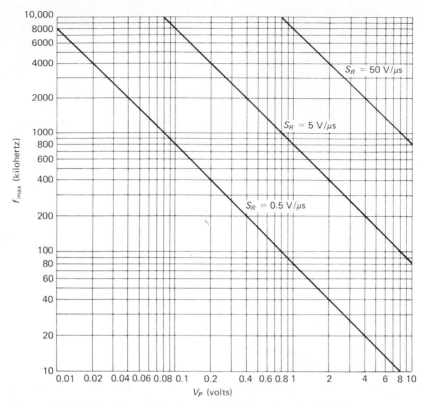

Figure 17-22. Trading off peak amplitude for power bandwidth.

tradeoff

One way to increase the power bandwidth is to accept less than maximum output swing. Figure 17-22 is a graph of Eq. (17-11) for three different slew rates. By trading off amplitude for frequency, we can improve the power bandwidth. For instance, if peak amplitudes of 1 V are acceptable in an application, the power bandwidth of a 741 increases to 80 kHz (the bottom curve). If an op amp has a slew rate of 50 V/μs (top curve), its 10-V power bandwidth is 800 kHz, and its 1-V power bandwidth is 8 MHz.

17-9 *other linear ICs*

This section briefly examines other linear ICs. Our survey is not complete because there are many special-purpose ICs; we cover the main types.

audio amplifiers

Preamplifiers (preamps) are audio amplifiers with less than 50 mW of output power. Preamps are optimized for low noise because they are used at the front end of audio systems where they amplify weak signals from phonograph cartridges, magnetic tape heads, microphones, etc.

An example of an IC preamp is the LM381, a low-noise *dual* preamplifier. Each amplifier is completely independent of the other. The LM381 has a voltage gain of 112 dB, a 10-V power bandwidth of 75 kHz; it operates from a positive supply of 9 to 40 V. Its input impedance is 100 kΩ, and its output impedance is 150 Ω. The LM381's input stage is a diff amp, which allows differential or single-ended input.

Medium-level audio amplifiers have output powers from 50 to 500 mW. These are useful near the output end of small audio systems like transistor radios or signal generators. An example is the MFC4000P with an output power of 250 mW.

Audio power amplifiers deliver more than 500 mW of output power. They are used in phonograph amplifiers, intercoms, AM-FM radios, etc. The LM380 is an example. It delivers 5 W of output power. The voltage gain is 34 dB.

Figure 17-23 shows a simplified schematic diagram of the 380. The input diff amp uses *pnp* ground-referenced inputs (see Sec. 9-10 if necessary). Because

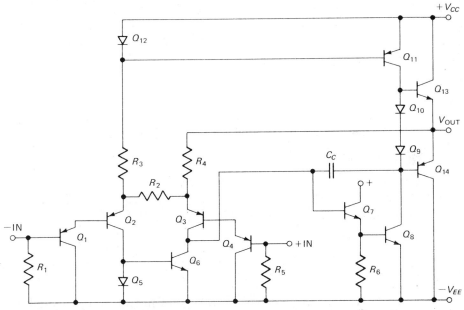

Figure 17-23. Simplified schematic for the LM380 and similar audio ICs.

of this, the signal can be directly coupled, which is an advantage with transducers. The diff amp drives a current mirror $(Q_5$ and $Q_6)$. The output of the mirror goes to an emitter follower (Q_7) and a CE driver (Q_8). The output stage is a class B push-pull emitter follower $(Q_{13}$ and $Q_{14})$.

There is an internal compensating capacitor of 10 pF that rolls off the bel voltage gain at a rate of 20 dB per decade. This capacitor produces a slew rate of approximately 5 V/μs.

wideband amplifiers

A *wideband amplifier* has a flat response (constant bel gain) over a very broad range of frequencies. It also is known as a video amp, broadband amp, or linear-pulse amp. Wideband amps are not necessarily dc amps, but they often do have a response that extends down to zero frequency. They are used in applications where the range of input frequencies is very large. For instance, many oscilloscopes handle frequencies from 0 to over 10 MHz; instruments like these use wideband amps to increase the signal strength before applying it to the cathode-ray tube. As another example, the video section of a television receiver uses a wideband amp that can handle frequencies from near zero to about 4 MHz.

IC wideband amps have voltage gains and bandwidths you can adjust by connecting different external resistors. For instance, the μA702 has a bel voltage gain of 40 dB and a break frequency of 5 MHz; by changing external components, you can get useful gain out to 30 MHz. The MC1553 has a bel voltage gain of 52 dB and a bandwidth of 20 MHz; these are adjustable by changing external components. The μA733 has a very wide bandwidth; it can be set up to give 20 dB gain and a bandwidth of 120 MHz.

rf and if amplifiers

A radio-frequency (RF) amplifier is usually the first stage in a radio or television receiver; intermediate-frequency (IF) amplifiers typically are the middle stages. The basic idea is this: RF and IF amplifiers are tuned or resonant so that they amplify only a *narrow band* of frequencies; this allows us to separate the signals from different radio or television stations.

As mentioned, inductors and large capacitors are impractical to integrate on a chip. For this reason, you have to add L's and C's externally to get a resonant circuit.

voltage regulators

Chapter 5 discussed rectifiers and power supplies. After filtering, we have a dc voltage with ripple. This dc voltage is proportional to the line voltage, that is, it will change 10 percent if the line voltage changes 10 percent. In many applications, a 10 percent change in dc voltage is too much. For this reason, we can use a *voltage regulator*, a device that delivers an almost constant dc output voltage even though the input voltage changes 10 percent or thereabouts. The typical IC regulator can hold the output dc voltage to within 0.01 percent for normal changes in line voltage and load resistance. Other features include positive or negative output, adjustable output voltage, and short-circuit protection.

self-testing review

Read each of the following and provide the missing words. Answers appear at the beginning of the next question.

1. Monolithic ICs are the most common. The components are part of one _____. Transistors, diodes, and resistors are easy to fabricate in a monolithic IC, but _____ and large _____ are not practical.

2. *(chip, inductors, capacitors)* The current through the common emitter resistor of a diff amp is called the _____ current. The top of the emitter resistor is an approximate _____ point. Sometimes a _____ source produces the tail current.

3. *(tail, ground, current)* The voltage gain of a diff amp equals R_C/r'_e, and the input impedance is $2\beta r'_e$. A signal driving both inputs at the same time is called a _____ signal. CMRR indicates how well this signal is rejected.

4. *(common-mode)* An op amp is a high-gain dc amplifier usable from 0 to over 1 MHz. By connecting external resistors, you can adjust the voltage _____ and bandwidth to your requirements. The 709 and _____ are industry standards, used for general-purpose applications. BIFET op amps use _____ for the input stage and bipolar transistors for later stages.

5. *(gain, 741, JFETs)* An op amp uses _____ coupling, which gives it voltage gain down to zero frequency. It has high input impedance, low output impedance, and high _____ gain. It is not meant to be used open loop; you have to connect external components to get normal operation.

6. *(direct, voltage)* C_C is called a _____ capacitor. Because of the Miller effect, it has a pronounced effect on frequency response. It is part of a base lag network that rolls off the bel voltage gain at a rate of 20 dB per decade. A resistor is a passive load; a _____ is an active load.

7. *(compensating, transistor)* Output voltage swing V_O is the _____ possible swing in output voltage. Typically, it is within 1 or 2 V of the supply voltages. Other specifications include input bias current, input offset current, and input _____ voltage.

8. *(maximum, offset)* The maximum rate of output voltage change is called the _____ rate. In a 741 and similar op amps, it equals the tail current divided by C_C. A sine wave with slew-rate distortion looks triangular.

9. *(slew)* The _____ bandwidth is the highest frequency an op amp can process without slew-rate distortion. One way to increase the power bandwidth is to accept less than maximum output swing.

10. *(power)* Other linear ICs include audio amplifiers, wideband amplifiers, RF and IF amplifiers, and voltage regulators.

problems

17-1. What is the dc emitter current in each transistor of Fig. 17-24a?

17-2. Calculate the dc voltage from each collector to ground in Fig. 17-24a.

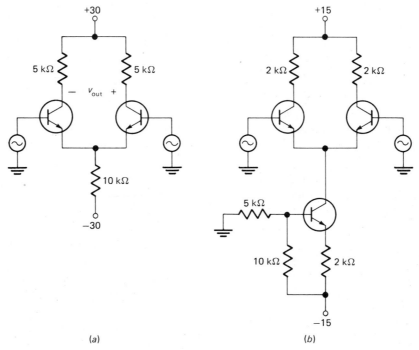

(a)   (b)

Figure 17-24.

17-3. If each source in Fig. 17-24a has an R_S of 1 kΩ and if β equals 100, what is the value of dc emitter current (second approximation)?

17-4. What is the ideal value of dc emitter current in the diff amp of Fig. 17-24b?

17-5. Calculate the collector-to-ground voltage in Fig. 17-24b for the two upper transistors.

17-6. In Fig. 17-24*b* suppose the collector resistors have a 1 percent tolerance. What is maximum dc output voltage if the two transistors are identical?

17-7. A sine wave with a peak of 1 mV drives the single-ended diff amp of Fig. 17-25*a*. How much output voltage is there?

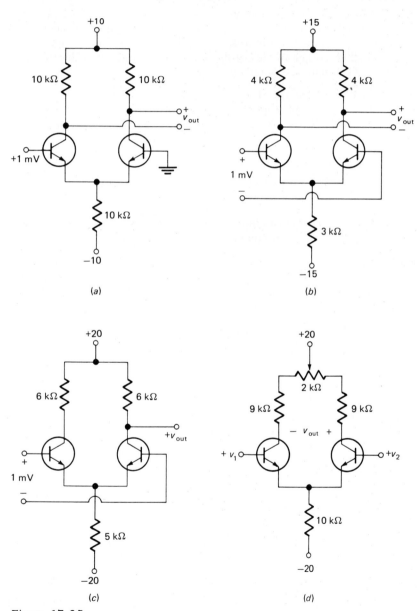

Figure 17-25.

17-8. What is the voltage gain A for the diff amp of Fig. 17-25b? With a differential input of 1 mV, what is the output voltage?

17-9. A sine wave with a peak of 1 mV drives the diff amp shown in Fig. 17-25c. How much ac output voltage is there? How much dc output voltage?

17-10. When you build a discrete diff amp, the two transistors almost never have close enough characteristics to get a reasonable balance between the two halves of the diff amp. For this reason, a potentiometer can be added as shown in Fig. 17-25d. To get an idea of the adjustment range, treat the transistors as identical and calculate the range of v_{out}.

17-11. In Fig. 17-25d the dc output voltage equals zero when the wiper is near the center of the potentiometer. What is the approximate value of voltage gain A? If v_1 equals 101 mV and v_2 equals 100 mV, how much output voltage is there?

17-12. What is the input impedance of the diff amp of Fig. 17-25a for a β of 75?

17-13. The common-mode input signal of 1 V in Fig. 17-26a produces an ac output voltage of 2 mV. Calculate the common-mode rejection ratio. Express the answer in decibels.

17-14. If a diff amp has a common-mode rejection ratio of 60 dB and a voltage gain of 200, how much common-mode output voltage do you get with a common-mode input voltage of 10 mV?

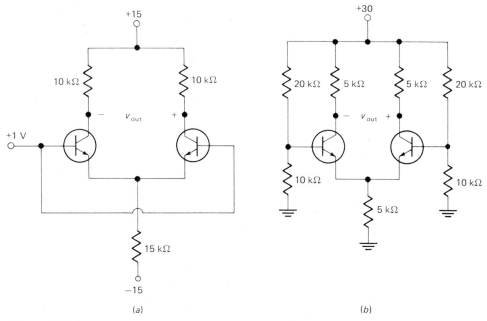

Figure 17-26.

17-15. Figure 17-26*b* shows a diff amp using a single supply. Treat the voltage across the base-emitter diodes as zero and calculate the dc emitter current in each transistor.

17-16. Suppose the diff amp of Fig. 17-26*b* has a CMRR' of 100 dB. The power supply delivers 30 V dc plus a ripple of 30 mV rms. How much common-mode output ripple is there?

17-17. The amplifier of Fig. 17-27*a* has a z_{in} of 1 MΩ, a z_{out} of 100 Ω, and an A of 10,000. Calculate the approximate output voltage.

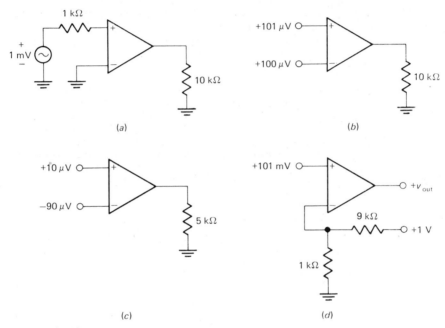

Figure 17-27.

17-18. As shown in Fig. 17-27*b*, the noninverting input equals 101 μV and the inverting input equals 100 μV. If A' equals 92 dB and z_{out} is 10 Ω, how much output voltage is there?

17-19. A positive voltage is applied to the noninverting input and a negative voltage to the inverting input in Fig. 17-27*c*. If A equals 100,000 and z_{out} is 10 Ω, how much output voltage is there?

17-20. In Fig. 17-27*d*, z_{in} is negligibly large, z_{out} negligibly small, and A equals 10,000. A sine wave with a peak of 101 mV is applied to the noninverting input. Another sine wave with exactly the same phase but a peak of 1 V is applied to the voltage divider driving the inverting input. What is the value of the output voltage?

17-21. The input base currents of an op amp are $I_{B1} = 50$ nA and $I_{B2} = 40$ nA. What does the input bias current equal? The input offset current?

17-22. A capacitor has a constant charging current of 1 mA. If the capacitance is 50 pF, what is the rate of voltage change with respect to time?

17-23. A 100-pF capacitor has a maximum charging current of 150 μA. What is the slew rate?

17-24. An op amp has a slew rate of 35 V/μs. How long will it take the output to change from 0 to 15 V?

17-25. The input stage of an op amp like Fig. 17-18 has $I_T = 100$ μA. If $C_C = 30$ pF, what is the slew rate?

17-26. An op amp has a slew rate of 2 V/μs. If the peak output is 12 V, what is the power bandwidth?

17-27. An op amp data sheet gives a 15-V power bandwidth of 25 kHz. What is the slew rate?

17-28. If $V_{CC} = V_{EE} = 15$ V and $R_2 = 1$ MΩ in Fig. 17-18, what does I_T equal in the diff-amp stage?

17-29. A 318 has a slew rate of 50 V/μs. Refer to Fig. 17-22 to determine the power bandwidth for
(a) $V_P = 10$ V
(b) $V_P = 4$ V
(c) $V_P = 2$ V

18 *negative feedback*

In a feedback control system, the output is sampled and a fraction of it is sent back to the input. This returning signal combines with the original input, producing remarkable changes in system performance. *Negative feedback* means the returning signal has a phase that opposes the input signal. The advantages of negative feedback are stabilizing the gain, improving the input and output impedances, reducing nonlinear distortion, and increasing bandwidth.

18-1 *the four basic feedback connections*

A feedback amplifier has two parts: an amplifier and a feedback circuit. Depending on the output connection, output voltage or current drives the feedback circuit. The feedback circuit returns a signal to the input which modifies the overall action of the system. The main purpose of the feedback is to allow input to precisely control the value of output.

There are four basic feedback connections. Figure 18-1a shows an amplifier and feedback circuit connected series-parallel (SP): the inputs of the amplifier and feedback circuit are in *series*, but the outputs are in *parallel*. With this connection, the controlling input variable is v_{in}, and the controlled output variable is v_{out}. The whole purpose of SP feedback is to produce a highly stable and precise ratio of v_{out}/v_{in}.

Figure 18-1b shows another way to connect the amplifier and feedback circuit.

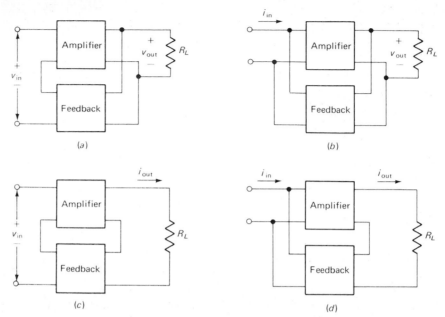

Figure 18-1. Four feedback connections. (a) *SP.* (b) *PP.* (c) *SS.* (d) *PS.*

This is called parallel-parallel (PP) feedback because the inputs are in *parallel* and the outputs are in *parallel*. As will be proved later, the controlling input variable is i_{in}, and the controlled output variable is v_{out}. In other words, PP feedback produces a highly stable and precise ratio of v_{out}/i_{in}.

Figure 18-1c illustrates the third basic type of feedback: series-series (SS). Here the inputs of the amplifier and feedback circuit are in *series,* and the outputs are in *series.* The controlling input variable is v_{in}, and the controlled output variable is i_{out}. The purpose of SS feedback is to get an extremely accurate ratio of i_{out}/v_{in}.

Figure 18-1d shows the last basic feedback connection: parallel-series (PS). The amplifier and feedback inputs are in *parallel,* while the outputs are in *series.* With this feedback, we get a highly stable ratio of i_{out}/i_{in}.

Table 18-1 summarizes these four types. The most widely used are SP and PP feedback; therefore, the remainder of the chapter will emphasize these two. But always remember that four distinct kinds of feedback exist, each with special properties ideally suited to certain applications. For instance, the next section will show you why SP feedback leads to the ideal voltage amplifier, one that has extremely high input impedance, extremely low output impedance, and a precise voltage gain. A later section will tell you why PP feedback results in an ideal current-to-voltage converter, a circuit with extremely low input impedance, extremely low output impedance, and a precise ratio of v_{out}/i_{in}. The

Table 18-1 The Four Types of Feedback

	SP	PP	SS	PS
Input connection	Series	Parallel	Series	Parallel
Input signal	v_{in}	i_{in}	v_{in}	i_{in}
Output connection	Parallel	Parallel	Series	Series
Output signal	v_{out}	v_{out}	i_{out}	i_{out}
Stabilized ratio	v_{out}/v_{in}	v_{out}/i_{in}	i_{out}/v_{in}	i_{out}/i_{in}

final section of this chapter will summarize the properties of all four feedback connections.

EXAMPLE 18-1

Figure 18-2a shows an SP negative-feedback amplifier. A stands for the voltage gain of the internal amplifier, B for the voltage gain of the feedback circuit, and A_{SP} for the voltage gain of the overall feedback amplifier. Calculate the values of A, B, and A_{SP}.

SOLUTION
Since 1 mV drives the internal amplifier and 10 V come out,

$$A = \frac{10 \text{ V}}{1 \text{ mV}} = 10,000$$

10 V drive the feedback circuit and 200 mV come out. Therefore, the voltage gain of the feedback circuit is

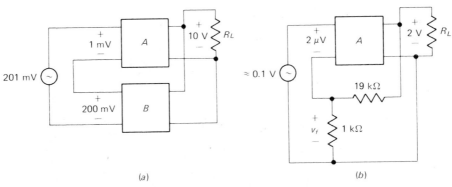

(a)   (b)

Figure 18-2.

$$B = \frac{200 \text{ mV}}{10 \text{ V}} = 0.02$$

In any practical negative-feedback circuit, B is always between 0 and 1.

The input signal driving the feedback amplifier is 201 mV. So, the overall voltage gain is

$$A_{SP} = \frac{v_{\text{out}}}{v_{\text{in}}} = \frac{10 \text{ V}}{201 \text{ mV}} = 49.8$$

EXAMPLE 18-2

The feedback circuit of a typical SP feedback amplifier is usually a *voltage divider*. Because of this, the feedback signal is a voltage equal to a fraction of the output voltage.

In Fig. 18-2*b*, calculate B, v_f, A, and A_{SP}.

SOLUTION

The gain of the voltage divider is

$$B = \frac{v_f}{v_{\text{out}}} = \frac{1000}{19,000 + 1000} = 0.05$$

The feedback voltage is

$$v_f = B v_{\text{out}} = 0.05 \times 2 \text{ V} = 0.1 \text{ V}$$

The internal gain is

$$A = \frac{2 \text{ V}}{2 \ \mu\text{V}} = 1,000,000$$

The overall system gain is

$$A_{SP} \cong \frac{2 \text{ V}}{0.1 \text{ V}} = 20$$

18-2 *sp closed-loop voltage gain*

Figure 18-3 is an SP feedback amplifier. In an SP feedback control system, the input to the internal amplifier is called the *error voltage*. This voltage is the difference between the input signal and the feedback signal. In a typical control system, the error voltage approaches zero; the reason is that internal gain A is very high, from 10,000 to more than 1,000,000. Since the output voltage swing of most amplifiers is in the vicinity of 10 V, the error voltage is somewhere in the microvolt region at low frequencies.

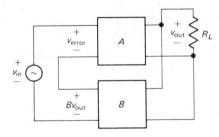

Figure 18-3. SP negative feedback.

Compared to all other signal voltages in the circuit, therefore, v_{error} is *insignificantly small.* To emphasize this important idea, we can write

$$v_{error} \cong 0 \qquad (18\text{-}1)***$$

why feedback stabilizes gain

Here's why the overall voltage gain of an SP feedback amplifier is stable or accurately fixed in value. Suppose internal gain A increases because of temperature change or some other reason. The output voltage will rise. This means more voltage is fed back to the input. This feedback signal Bv_{out} returns with a phase that opposes the input signal v_{in}, causing v_{error} to decrease. The reduced error voltage to the internal amplifier almost completely offsets the original increase in voltage gain A. The result is that v_{out} hardly increases at all. For instance, if A increases 10 percent, v_{error} decreases almost 10 percent, so that v_{out} shows only the slightest increase.

A similar argument applies to a decrease in gain A. If A decreases for any reason, the output voltage decreases. In turn, feedback voltage Bv_{out} decreases, causing v_{error} to increase. This increase in v_{error} almost completely nullifies the decrease in A. As a result, the output voltage shows only the slightest decrease.

This gives you the basic idea of what negative feedback does. Attempted changes in output voltage are fed back to the input, causing the error voltage to change in the opposite direction. The overall effect is that output voltage is virtually independent of changes in the internal gain A.

mathematical analysis

Now for the mathematical proof. Applying Kirchhoff's voltage law to the input side of Fig. 18-3 gives

$$v_{error} = v_{in} - Bv_{out} \qquad (18\text{-}2)$$

Furthermore, the output voltage equals

$$v_{out} = Av_{error}$$

With Eq. (18-2), this becomes

$$v_{out} = A(v_{in} - Bv_{out})$$

By expanding and rearranging, we get

$$\frac{v_{out}}{v_{in}} = \frac{A}{1 + AB}$$

or

$$A_{SP} = \frac{A}{1 + AB} \qquad (18\text{-}3)$$

For SP feedback to be effective, the designer deliberately makes the product of AB much greater than unity. Because of this design condition, Eq. (18-3) reduces as follows:

$$A_{SP} = \frac{A}{1 + AB} \cong \frac{A}{AB}$$

Then, the A cancels, giving

$$A_{SP} \cong \frac{1}{B} \qquad (18\text{-}4)\text{***}$$

Why is this result so important? Because it says the voltage gain with feedback equals the reciprocal of B, the voltage gain of the feedback circuit. As mentioned earlier, the feedback circuit is usually a voltage divider, a circuit we can make with *precision* resistors. This means B can be an accurate and stable value. Because of this, the voltage gain of a feedback amplifier becomes a rock-solid value equal to the reciprocal of B.

A final point. The internal gain A is called the *open-loop gain* because it is the gain we would get if the feedback path were opened. On the other hand, the overall system gain A_{SP} is called the *closed-loop gain* because it is the gain we get when there's a closed loop or signal path all the way around the circuit.

EXAMPLE 18-3
Figure 18-4 shows the most widely used form of an SP feedback amplifier. The input signal drives the noninverting input of an op amp. The op amp provides the internal gain A; resistors R_1 and R_2 form the feedback voltage divider. Since the returning signal drives the inverting input, it opposes the input signal. In other words, the feedback is negative.

If the 741 has an open-loop gain of 200,000, what is the closed-loop gain? If $v_{in} = 1$ mV, what do the output and error voltages equal?

SOLUTION
The gain of the voltage divider is

$$B = \frac{R_1}{R_1 + R_2} = \frac{2000}{100,000} = 0.02$$

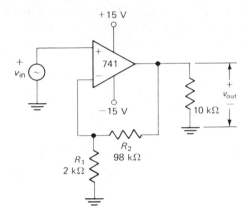

Figure 18-4.

The closed loop gain is

$$A_{SP} \cong \frac{1}{B} = \frac{1}{0.02} = 50$$

This is extremely accurate. As proof, we can calculate the exact gain with Eq. (18-3):

$$A_{SP} = \frac{A}{1 + AB} = \frac{200,000}{1 + 200,000(0.02)}$$

$$= 49.9875 \cong 50$$

If $v_{in} = 1$ mV, the output voltage is

$$v_{out} = A_{SP}v_{in} = 50 \times 1 \text{ mV} = 50 \text{ mV}$$

The error voltage is

$$v_{error} = \frac{v_{out}}{A} = \frac{50 \text{ mV}}{200,000} = 0.25 \ \mu V$$

Notice how small the error voltage is. This confirms Eq. (18-1), which says the error voltage is small compared to the other signal voltages in the circuit.

EXAMPLE 18-4
The 741 of Fig. 18-4 is replaced by another 741 whose open-loop gain is 100,000. Recalculate the exact value of A_{SP}. Also, what are the new values of v_{out} and v_{error} for $v_{in} = 1$ mV?

SOLUTION

$$A_{SP} = \frac{A}{1 + AB} = \frac{100,000}{1 + 100,000(0.02)}$$

$$= 49.975 \cong 50$$

The closed-loop gain is still extremely close to 50, despite the open-loop gain dropping in half. This is what SP feedback is all about. You trade off extremely high open-loop gain and its instability for a much lower closed-loop gain that is very stable in value.

Since $A_{SP} \cong 50$, v_{out} is still approximately 50 mV. But the error voltage changes to

$$v_{error} = \frac{v_{out}}{A} = \frac{50 \text{ mV}}{100,000} = 0.5 \ \mu V$$

Compared to the preceding example, the error voltage has increased from 0.25 to 0.5 μV.

Do you understand what has happened? When the open-loop gain drops in half, the error voltage doubles; therefore, the output voltage remains at approximately 50 mV. This echoes our earlier explanation of negative feedback. Attempted changes in output voltage are fed back to the input, causing the error voltage to change in the opposite direction. As a result, the output voltage remains essentially constant and independent of changes in open-loop gain.

A noninverting amplifier like Fig. 18-4 is used a lot. Since $B = R_1/(R_1 + R_2)$,

$$A_{SP} = \frac{R_2}{R_1} + 1 \qquad\qquad (18\text{-}5)***$$

Remember this formula. It's convenient to use.

18-3 *sp input and output impedances*

Figure 18-5 shows an SP negative-feedback amplifier. The internal amplifier has an input impedance of z_{in} and an output impedance of z_{out}. The overall amplifier, however, has an input impedance of $z_{in(SP)}$ and an output impedance of $z_{out(SP)}$.

why input impedance increases

Suppose v_{in} increases in Fig. 18-5; then, i_{in} increases. This produces a greater error voltage and a greater output voltage. As a result, the feedback voltage Bv_{out} increases, opposing the original increase in i_{in}. This means SP negative feedback *increases* the input impedance seen by the source.

why output impedance decreases

Suppose R_L decreases in Fig. 18-5. Then, i_{out} increases, producing a larger drop across the Thevenin impedance z_{out} of the internal amplifier. This results

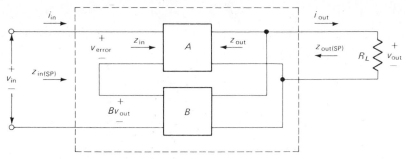

Figure 18-5. SP input and output impedances.

in a lower output voltage v_{out}. But then the feedback voltage Bv_{out} also decreases. This causes v_{error} to increase and almost completely offset the original decrease in output voltage. In other words, even though R_L decreases, the output voltage shows almost no decrease at all, compared to what it would do without feedback. This is equivalent to saying that SP negative feedback *decreases* the output impedance.

mathematical analysis

In Fig. 18-5,

$$v_{in} = v_{error} + Bv_{out}$$
or
$$v_{in} = v_{error} + ABv_{error} = (1 + AB)v_{error}$$

Because $v_{error} = i_{in}z_{in}$, the equation becomes

$$v_{in} = (1 + AB)i_{in}z_{in}$$
or
$$\frac{v_{in}}{i_{in}} = (1 + AB)z_{in}$$

But $z_{in(SP)}$ is the overall input impedance, so this equation becomes

$$z_{in(SP)} = (1 + AB)z_{in} \qquad (18\text{-}6)\text{***}$$

Since AB is much greater than unity, the input impedance with feedback is much greater than the input impedance of the internal amplifier.

By a similar derivation, we can prove the following:

$$z_{out(SP)} = \frac{z_{out}}{1 + AB} \qquad (18\text{-}7)\text{***}$$

This says the output impedance with feedback equals the Thevenin impedance of the internal amplifier divided by the factor $1 + AB$. In other words, $z_{out(SP)}$ is much smaller than z_{out}.

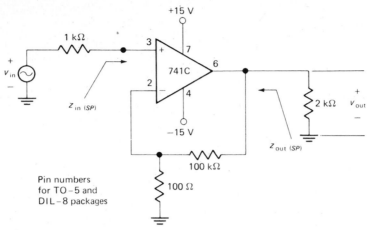

Figure 18-6.

EXAMPLE 18-5

The 741C of Fig. 18-6 has $A = 100{,}000$, $z_{in} = 1$ MΩ, and $z_{out} = 300$ Ω. Calculate $z_{in(SP)}$ and $z_{out(SP)}$.

SOLUTION

In Fig. 18-6,

$$B = \frac{R_1}{R_1 + R_2} = \frac{100}{100{,}100} \cong 0.001$$

and

$$1 + AB = 1 + 100{,}000(0.001) \cong 100$$

Then,

$$z_{in(SP)} = (1 + AB)z_{in} \cong 100 \times 1 \text{ MΩ}$$
$$= 100 \text{ MΩ}$$

and

$$z_{out(SP)} = \frac{z_{out}}{1 + AB} \cong \frac{300 \text{ Ω}}{100} = 3 \text{ Ω}$$

EXAMPLE 18-6

Figure 18-7 is a *voltage follower*. Calculate its A_{SP}. Also, what do $z_{in(SP)}$ and $z_{out(SP)}$ equal?

SOLUTION

R_1 is infinite and R_2 is zero; therefore, $B = 1$. This is massive negative feedback, the most you can have. In this case,

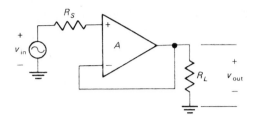

Figure 18-7. The voltage follower.

$$A_{SP} \cong \frac{1}{B} = \frac{1}{1} = 1$$

The closed-loop voltage gain equals unity to a very close approximation.

The input impedance of a voltage follower is

$$z_{in(SP)} = (1 + AB)z_{in} = (1 + A)z_{in}$$
$$\cong Az_{in}$$

because A is much greater than unity. Likewise,

$$z_{out(SP)} = \frac{z_{out}}{1 + AB} + \frac{z_{out}}{1 + A}$$
$$\cong \frac{z_{out}}{A}$$

For typical op amps, $z_{in(SP)}$ approaches infinity and $z_{out(SP)}$ approaches zero.

A voltage follower is almost an ideal buffer amplifier because of its high input impedance, low output impedance, and unity gain. Figure 18-7 is an ac equivalent circuit; it does not show the supply voltages. Whenever V_{CC} and V_{EE} are omitted from a circuit diagram, remember that the actual circuit does include them.

18-4 *other benefits of sp negative feedback*

By sacrificing open-loop gain, we can greatly improve the qualities of a closed-loop amplifier. As shown earlier, the gain stability improves, the input impedance increases, and the output impedance decreases.

why negative feedback reduces nonlinear distortion

A large-signal stage has nonlinear distortion because its voltage gain changes at different points in the cycle. For instance, the positive half cycle may be elongated, and the negative half cycle may be compressed. Change in the internal gain is the cause of nonlinear distortion.

SP negative feedback *reduces nonlinear distortion* because it eliminates the effects of changing internal gain. As long as the open-loop gain is much greater than the closed-loop gain, the output voltage is almost unaffected by changes in the internal gain. In other words, the nonlinearity of the internal amplifier produces negligible distortion of the signal.

By a mathematical analysis given later,

$$D_{SP} = \frac{D}{1 + AB} \qquad (18\text{-}8)\text{***}$$

where D_{SP} = distortion of overall amplifier
$\quad\quad D$ = distortion of internal amplifier
This says SP negative feedback reduces nonlinear distortion by a factor of $1 + AB$.

bandwidth increases

SP negative feedback increases the *bandwidth*. Here's why. The open-loop gain is much greater than the closed-loop gain. When the internal gain is down 3 dB, the closed-loop gain still closely equals $1/B$. This means the bandwidth of the overall amplifier is greater than the bandwidth of the internal amplifier.

It can be shown that

$$f_{1(SP)} = \frac{f_1}{1 + AB} \qquad (18\text{-}9)\text{***}$$

where $f_{1(SP)}$ = lower break frequency of overall amplifier
$\quad\quad f_1$ = lower break frequency of internal amplifier
This says SP negative feedback decreases the lower break frequency by a factor of $1 + AB$.

On the other hand, the upper break frequency increases by the same factor:

$$f_{2(SP)} = (1 + AB)\, f_2 \qquad (18\text{-}10)\text{***}$$

where $f_{2(SP)}$ = upper break frequency of overall amplifier
$\quad\quad f_2$ = upper break frequency of internal amplifier
The next section continues the discussion of *bandwidth stretching*—decreasing the lower break frequency and increasing the upper break frequency.

sacrifice factor

The closed-loop voltage gain of an SP negative-feedback amplifier is

$$A_{SP} = \frac{A}{1 + AB}$$

which rearranges to

$$1 + AB = \frac{A}{A_{SP}}$$

The factor $1 + AB$ is called the *sacrifice factor* because it is the ratio of open-loop gain to closed-loop gain. The sacrifice factor indicates how much internal gain is sacrificed to improve the qualities of the overall amplifier.

As an example, if $A = 100{,}000$ and $A_{SP} = 100$, the sacrifice factor is

$$1 + AB = \frac{A}{A_{SP}} = \frac{100{,}000}{100} = 1000$$

This means the closed-loop gain is smaller than the open-loop gain by a factor of 1000, equivalent to 60 dB. In exchange for this sacrifice, however, we improve gain stability, input impedance, and other characteristics by a factor of 1000.

From now on, S will represent the sacrifice factor. In symbols,

$$S = 1 + AB \qquad\qquad (18\text{-}11)***$$

Remember: the easiest way to calculate S is with

$$S = \frac{A}{A_{SP}} \qquad\qquad (18\text{-}12)***$$

Table 18-2 summarizes the effects of SP negative feedback. For your convenience, the formulas are given as they first appeared, and are also rewritten in terms of S.

EXAMPLE 18-7

An SP negative feedback amplifier has a midband internal gain of 120 dB and a closed-loop gain of 40 dB. What is the sacrifice factor in the midband of the amplifier? Express the answer in decibels and as an ordinary number.

Table 18-2 SP Negative Feedback*

Quantity	Effect	Original formula	Alternative
Voltage gain	Decreases	$1/B$	$1/B$
Input impedance	Increases	$(1 + AB)z_{in}$	Sz_{in}
Output impedance	Decreases	$z_{out}/(1 + AB)$	z_{out}/S
Distortion	Decreases	$D/(1 + AB)$	D/S
Lower break frequency	Decreases	$f_1/(1 + AB)$	f_1/S
Upper break frequency	Increases	$(1 + AB)f_2$	Sf_2

* In these formulas, A and S are calculated inside the midband of the internal amplifier. In other words, $A = A_{mid}$ and $S = S_{mid}$, discussed in Sec. 18–5.

SOLUTION
Since

$$S = \frac{A}{A_{SP}}$$

the log of both sides gives

$$\log S = \log A - \log A_{SP}$$

Now it's clear the decibel equivalent is

$$S' = A' - A'_{SP} = 120 \text{ dB} - 40 \text{ dB}$$
$$= 80 \text{ dB}$$

The antilog gives

$$S = 10,000$$

EXAMPLE 18-8
In the preceding example, what is the sacrifice factor one decade above the upper break frequency of the internal amplifier, assuming a rolloff of 20 dB/ decade?

SOLUTION
One decade above the f_2 of the internal amplifier, the internal gain has dropped 20 dB from the midband value; therefore,

$$A' = 100 \text{ dB}$$

The new value of sacrifice factor is

$$S' = 100 \text{ dB} - 40 \text{ dB} = 60 \text{ dB}$$
or $$S = 1000$$

The point of the example is this: the sacrifice factor decreases outside the midband of the internal amplifier because the internal gain decreases. When using the formulas of Table 18-2, remember to use midband values for A, A_{SP}, and S. In fact, from now on, we will use A_{mid}, $A_{SP(mid)}$, and S_{mid} for the midband values. The next section tells you more about this.

18-5 *the bandwidth of an sp feedback amplifier*

The internal amplifier has an upper break frequency f_2 produced by its dominant lag network. The upper break frequency of the entire feedback amplifier, however, is much higher.

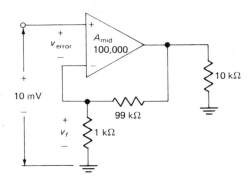

Figure 18-8.

a simple explanation

In Fig. 18-8, the internal gain in the midband is 100,000. Because of the external feedback resistors, the closed-loop gain equals 100. With a 10-mV input, the output equals 1 V.

As the frequency of the input signal increases, we eventually reach the break frequency f_2 of the *internal amplifier*. At this frequency the internal gain A is down 3 dB; in this case, A equals 70,700. The sacrifice factor is still high and equals 707. Because of this, the closed-loop voltage gain closely equals 100 and the output voltage still equals 1 V. The internal gain is so high to begin with that even though it has gone down 3 dB, it is high enough to ensure a large value of S.

As described earlier, as long as A is large, the only significant voltage change is in v_{error}. If A goes down by a factor of 2, v_{error} increases by a factor of almost 2; this compensates for the change in A.

With increasing frequency the internal gain A keeps decreasing until S is no longer large. When this happens, we begin to notice a decrease in v_{out}. When v_{out} is down 3 dB, we have reached the break frequency of the feedback amplifier. To keep this frequency distinct from others, we designate it as $f_{2(SP)}$.

relation between break frequencies

By an advanced derivation, the break frequencies are related as follows:

$$f_{2(SP)} = S_{mid}f_2 \qquad (18\text{-}13)\text{***}$$

where

$$S_{mid} = \frac{A_{mid}}{A_{SP(mid)}} \qquad (18\text{-}14)$$

A_{mid} and $A_{SP(mid)}$ are the voltage gains well inside the bandwidth of the internal amplifier. As an example, in Fig. 18-8 A_{mid} is 100,000 and $A_{SP(mid)}$ equals 100; therefore,

$$S_{\text{mid}} = \frac{100,000}{100} = 1000$$

If the internal amplifier has an f_2 of 200 Hz, Eq. (18-13) gives

$$f_{2(SP)} = 1000(200\text{ Hz}) = 200\text{ kHz}$$

What it means is this. When you sacrifice gain by a factor S_{mid}, you *stretch* the bandwidth of the feedback amplifier by the same factor. In the foregoing example we started with an open-loop gain of 100,000 and a bandwidth of 200 Hz; we wind up with a closed-loop gain of 100 and a bandwidth of 200 kHz.

special condition

A crucial condition must be satisfied when you apply Eq. (18-13). The internal amplifier must have *one dominant lag network* so that the voltage gain A decreases 20 dB per decade above f_2. This 20-dB rolloff must continue to the break frequency $f_{2(SP)}$. In other words, the next important lag network in the internal amplifier must have a break frequency greater than $f_{2(SP)}$. Unless this condition is satisfied, the neat relation given by Eq. (18-13) does not apply.

EXAMPLE 18-9
The data sheet of a 741C shows a bel graph like Fig. 18-9*a*. If the 741C is used in the SP negative-feedback amplifier of Fig. 18-9*c*, calculate the break frequency of the feedback amplifier.

SOLUTION
In Fig. 18-9*a* the midband voltage gain of the 741C is 100 dB, equivalent to an ordinary gain of

$$A_{\text{mid}} = 100,000$$

The break frequency of the 741C is

$$f_2 = 10\text{ Hz}$$

The bel graph indicates a rolloff of 20 dB per decade, meaning the 741C has one dominant lag network (integrated on the chip). This rolloff is constant out to the unity-gain frequency of

$$f_{\text{unity}} = 1\text{ MHz}$$

In the feedback amplifier of Fig. 18-9*c* the midband feedback gain equals

$$A_{SP(\text{mid})} = \frac{R_2}{R_1} + 1 = 9 + 1 = 10$$

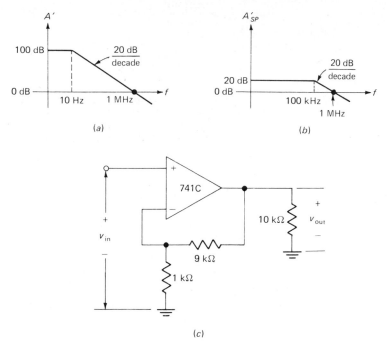

Figure 18-9.

which is equivalent to 20 dB. The sacrifice factor equals

$$S_{mid} = \frac{A_{mid}}{A_{SP(mid)}} = \frac{100,000}{10} = 10,000$$

With Eq. (18-13),

$$f_{2(SP)} = S_{mid}f_2 = 10,000(10 \text{ Hz}) = 100 \text{ kHz}$$

So, an S factor of 10,000 stretches the bandwidth by a factor of 10,000.

Figure 18-9b shows the bel voltage gain of the feedback amplifier. In the midband of the feedback amplifier, the bel gain is 20 dB. The break frequency of the feedback amplifier is 100 kHz. The rolloff rate is 20 dB per decade; therefore, the unity-gain frequency is still 1 MHz.

EXAMPLE 18-10
By changing the ratio of R_2 to R_1 in Fig. 18-9c, we get different values of $A_{SP(mid)}$. Calculate the $f_{2(SP)}$ for each of the following values: $A_{SP(mid)} = 1000, 100, 10,$ and 1.

SOLUTION

Since A_{mid} is 100,000, Eq. (18-13) gives

$$f_{2(SP)} = \frac{100,000}{A_{SP(\text{mid})}} f_2$$

We can use this to calculate $f_{2(SP)}$ for different $A_{SP(\text{mid})}$. When $A_{SP(\text{mid})}$ is 1000,

$$f_{2(SP)} = \frac{100,000}{1000} \, 10 \text{ Hz} = 1 \text{ kHz}$$

When $A_{SP(\text{mid})}$ equals 100,

$$f_{2(SP)} = \frac{100,000}{100} \, 10 \text{ Hz} = 10 \text{ kHz}$$

When $A_{SP(\text{mid})}$ is 10,

$$f_{2(SP)} = \frac{100,000}{10} \, 10 \text{ Hz} = 100 \text{ kHz}$$

When $A_{SP(\text{mid})}$ equals 1,

$$f_{2(SP)} = \frac{100,000}{1} \, 10 \text{ Hz} = 1 \text{ MHz}$$

The point is this. We can trade off closed-loop gain for bandwidth. By changing the external resistors in Fig. 18-9c, we can tailor a voltage amplifier for a particular application. The values in this example give these choices of gain-bandwidth:

$$A_{SP(\text{mid})} = 1000, \ f_{2(SP)} = 1 \text{ kHz}$$
$$A_{SP(\text{mid})} = 100, \ f_{2(SP)} = 10 \text{ kHz}$$
$$A_{SP(\text{mid})} = 10, \ f_{2(SP)} = 100 \text{ kHz}$$
$$A_{SP(\text{mid})} = 1, \ f_{2(SP)} = 1 \text{ MHz}$$

Notice that the product of gain and break frequency is a constant value of 1 MHz.

18-6 *the gain-bandwidth product*

The preceding example suggests the idea of the *gain-bandwidth product*. Since

$$f_{2(SP)} = S_{\text{mid}} f_2 = \frac{A_{\text{mid}}}{A_{SP(\text{mid})}} f_2 \qquad (18\text{-}15)$$

we can multiply both sides by $A_{SP(\text{mid})}$ to get

$$A_{SP(\text{mid})} f_{2(SP)} = A_{\text{mid}} f_2 \qquad (18\text{-}16)\text{***}$$

The right-hand side of this equation is the product of internal gain and the internal break frequency. For instance, the typical 741C has an A_{mid} of 100,000 and an f_2 of 10 Hz; therefore, it has a gain-bandwidth product of

$$A_{mid}f_2 = 100,000(10 \text{ Hz}) = 1 \text{ MHz}$$

Since A_{mid} is dimensionless and f_2 is in hertz, the gain-bandwidth product always has the dimensions of hertz.

gain-bandwidth product is constant

The left-hand side of Eq. (18-16) is the product of closed-loop gain and break frequency. Therefore, no matter what the values of R_1 and R_2, the product of $A_{SP(mid)}$ and $f_{2(SP)}$ must *equal the gain-bandwidth product* of the internal amplifier. If the internal amplifier has a gain-bandwidth product of 1 MHz, the product of $A_{SP(mid)}$ and $f_{2(SP)}$ must equal 1 MHz. Example 18-10 illustrated this; each pair of $A_{SP(mid)}$ and $f_{2(SP)}$ had a product of 1 MHz.

Equation (18-16) is often summarized by saying the *gain-bandwidth product is a constant.* In other words, in Eq. (18-16) if the right-hand product is a constant, so too is the left-hand product. For this reason, even though $A_{SP(mid)}$ and $f_{2(SP)}$ change when we change external resistors, the product of these two quantities remains constant for a given internal amplifier.

gain-bandwidth product equals unity-gain frequency

Figure 18-10 summarizes the idea of constant gain-bandwidth product. $A'_{SP(mid)}$ can have any value between 0 dB and A'_{mid}; for the special case of 0 dB, $A_{SP(mid)}$ equals unity, and Eq. (18-16) reduces to

$$f_{unity} = A_{mid}f_2 \qquad\qquad (18\text{-}17)\textbf{***}$$

This says the unity-gain frequency equals the gain-bandwidth product.

Figure 18-10 illustrates all the key ideas behind the gain-bandwidth product. The open-loop gain breaks at f_2; it drops 20 dB per decade. The closed-loop gain breaks at $f_{2(SP)}$; it too drops 20dB per decade. Both graphs superimpose beyond the $f_{2(SP)}$ break point until they cross the axis at a frequency of f_{unity}.

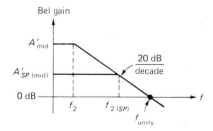

Figure 18-10. Open-loop and closed-loop response.

The gain-bandwidth product gives us a fast way of comparing amplifiers; the greater the gain-bandwidth product, the higher we can go in frequency and still have usable gain. The 318, for example, with its gain-bandwidth product of 15 MHz provides gain out to higher frequencies than a 741C with a gain-bandwidth product of only 1 MHz.

slew rate and power bandwidth

SP negative feedback has no effect on slew rate or power bandwidth. Until the output voltage has had a chance to change, there is no feedback signal and no benefits of negative feedback. In other words, slew rate and power bandwidth are the same with or without feedback.

The discussion of gain-bandwidth product has assumed small-signal operation. This means output peak values less than the limit given by Eq. (17-11). To put it another way, the power bandwidth still determines the frequency where slew-rate limiting begins. As long as $f_{2(SP)}$ is less than f_{max}, you have small-signal operation and no slew-rate distortion.

EXAMPLE 18-11
Figure 18-11a shows the bel voltage gain of an op amp. What does the gain-bandwidth product equal?

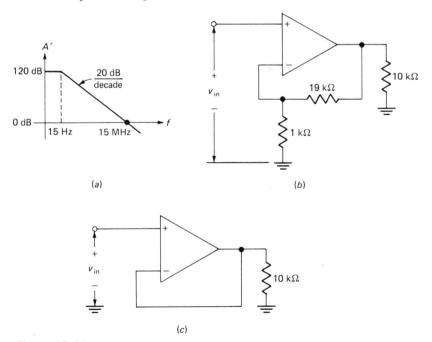

(a)

(b)

(c)

Figure 18-11.

SOLUTION
We can get this in either of two ways. First, the midband gain is 120 dB, equivalent to

$$A_{mid} = 1,000,000$$

The open-loop break frequency is

$$f_2 = 15 \text{ Hz}$$

Therefore, the open-loop gain-bandwidth product equals

$$A_{mid}f_2 = 1,000,000(15 \text{ Hz}) = 15 \text{ MHz}$$

Second, we can get the same answer by reading the unity-gain frequency; this is how you normally would do it given the data sheet of an op amp. In Fig. 18-11a,

$$f_{unity} = 15 \text{ MHz}$$

which equals the gain-bandwidth product.

EXAMPLE 18-12
If an op amp with the bel graph of Fig. 18-11a is used in Fig. 18-11b, what is the break frequency of the feedback amplifier?

SOLUTION
The closed-loop voltage gain is

$$A_{SP(mid)} = \frac{R_2}{R_1} + 1 = 19 + 1 = 20$$

Since f_{unity} equals 15 MHz, the gain-bandwidth product equals 15 MHz. With Eq. (18-16),

$$f_{2(SP)} = \frac{A_{mid}f_2}{A_{SP(mid)}} = \frac{f_{unity}}{A_{SP(mid)}}$$

$$= \frac{15 \text{ MHz}}{20} = 750 \text{ kHz}$$

EXAMPLE 18-13
Suppose an op amp with the bel graph of Fig. 18-11a is used in the voltage follower of Fig. 18-11c. What is the break frequency of the voltage follower?

SOLUTION
A voltage follower has unity gain. Because of this, the break frequency equals f_{unity}; in this case $f_{2(SP)}$ equals 15 MHz.

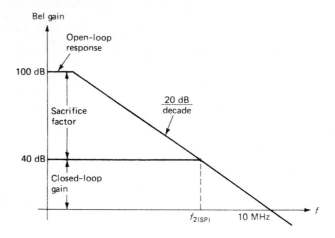

Figure 18-12.

EXAMPLE 18-14

Figure 18-12 is the bel graph of an SP negative-feedback amplifier. What does the midband sacrifice factor equal? The closed-loop break frequency? The open-loop break frequency?

SOLUTION

The sacrifice factor in the midband is

$$S'_{mid} = 100 \text{ dB} - 40 \text{ dB} = 60 \text{ db}$$

equivalent to

$$S_{mid} = 1000$$

Since the rolloff rate is 20 db per decade, the closed-loop break frequency is two decades below f_{unity}:

$$f_{2(SP)} = 100 \text{ kHz}$$

Figure 18-12 also makes it clear that the open-loop break frequency is five decades below f_{unity}, or three decades below $f_{2(SP)}$; therefore,

$$f_2 = 100 \text{ Hz}$$

18-7 *pp negative feedback*

Figure 18-13a shows a *parallel-parallel* (PP) negative-feedback amplifier. The input voltage drives the *inverting* terminal of the amplifier; the noninverting terminal is grounded. The amplified-and-inverted output signal is then applied to the right end of a feedback resistor *R*.

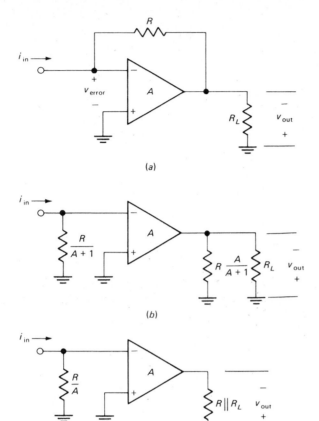

(a)

(b)

(c)

*Figure 18-13. PP neg-
ative feedback.*

output voltage

The easiest way to find what the circuit does is to Millerize it, that is, split R
into its two Miller components as shown in Fig. 18-13b. If the amplifier is a
discrete circuit, we can expect a gain of 100 or more; if the amplifier is an op
amp, the gain will be better than 10,000. For this reason, $A + 1$ is closely
equal to A, and we can use the simplified circuit of Fig. 18-13c.

In Fig. 18-13c, R/A is in parallel with the z_{in} of the internal amplifier. Therefore,
the input impedance of the PP feedback amplifier is

$$z_{in(PP)} = \frac{R}{A} \,\Big\|\, z_{in}$$

In a typical PP negative feedback amplifier, R/A will be much smaller than z_{in}
so that

$$z_{\text{in}(PP)} \cong \frac{R}{A} \qquad\qquad (18\text{-}18)\text{***}$$

In other words, negligible current flows into the internal amplifier; almost all of i_{in} flows through the input Miller resistance.

The error voltage equals

$$v_{\text{error}} = i_{\text{in}} z_{\text{in}(PP)} \cong i_{\text{in}} \frac{R}{A}$$

and the output voltage is

$$v_{\text{out}} = A v_{\text{error}} \cong A i_{\text{in}} \frac{R}{A}$$

or
$$v_{\text{out}} \cong R i_{\text{in}} \qquad\qquad (18\text{-}19)\text{***}$$

Here's how to remember the foregoing equation. In Fig. 18-13a, almost all input current i_{in} flows through the feedback resistor R. Because the error voltage approaches zero, Kirchhoff's voltage law tells us the output voltage approximately equals the voltage across the feedback resistor. The voltage across the feedback resistor is $R i_{\text{in}}$; therefore,

$$v_{\text{out}} \cong R i_{\text{in}}$$

This result is valid for all inverting op-amp circuits where the noninverting input is grounded. Remember it because we will use it a great deal in a later chapter.

virtual ground

When an op amp is used, the input impedance of a PP negative-feedback amplifier is

$$z_{\text{in}(PP)} \cong \frac{R}{A}$$

Since A is very large, $z_{\text{in}(PP)}$ approaches zero:

$$z_{\text{in}(PP)} \cong 0$$

This is one crucial difference between SP and PP negative feedback. SP input impedance approaches infinity, whereas PP input impedance approaches zero.

Because of the very low input impedance, the inverting input of Fig. 18-13a is called a *virtual ground*. This means the inverting input acts like a ground point in the sense that the error voltage is approximately zero. On the other hand, the inverting input is not truly a ground point because it does not sink the input current; rather, almost all the input current flows through the feedback resistor R because of the Miller effect.

The virtual-ground concept simplifies the analysis of inverting op-amp circuits discussed in a later chapter. It's important to remember these two ideas about virtual ground:

1. The inverting input is at ground potential (ideally).
2. All the input current flows through the feedback resistor (ideally).

output impedance

Like SP feedback, PP feedback makes the Thevenin output impedance approach zero. In other words, the output circuit tends to act like a perfect voltage source. The value of $z_{out(PP)}$ depends on the source impedance R_S connected to the inverting input. By a derivation similar to that given for SP negative feedback,

$$z_{out(PP)} = \frac{z_{out}}{1 + AB} \tag{18-20}$$

where

$$B = \frac{R_S}{R + R_S}$$

The denominator of Eq. (18-20) is so large in the typical PP feedback amplifier that $z_{out(PP)}$ approaches zero, much the same as in the SP negative-feedback amplifier.

the ideal current-to-voltage converter

The PP negative-feedback amplifier approaches the *perfect current-to-voltage converter*, a device with *zero input impedance, zero output impedance*, and a *fixed relation between i_{in} and v_{out}*. Whenever we see a PP negative-feedback amplifier, we can visualize it as the perfect current-to-voltage converter shown in Fig. 18-14a. The output voltage equals R times the input current.

How can we use PP negative feedback? A moving-coil ammeter has an internal resistance that may disturb the circuit whose current is being measured. For instance, an ammeter with a full-scale deflection of 50 μA typically has a resistance of 2 kΩ, far from ideal; when an ammeter like this is placed in series with a branch, it adds 2 kΩ of resistance to the branch. With PP negative feedback, we can build an *electronic ammeter whose resistance approaches zero*.

Figure 18-14b shows a complete circuit for such an ammeter. The typical gain of a 741C is 100,000. Therefore,

$$z_{in(PP)} = \frac{R}{A} = \frac{100,000}{100,000} = 1 \ \Omega$$

and

$$v_{out} = Ri_{in} = 10^5 i_{in}$$

The first equation tells us that we will add only 1 Ω of resistance to the branch whose current is being measured. The second equation tells us how sensitive

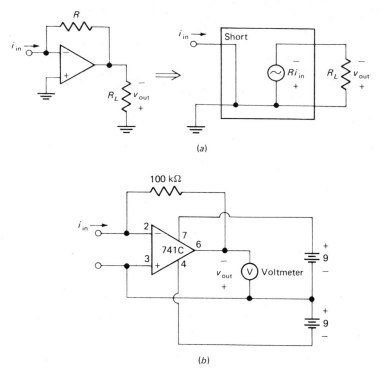

Figure 18-14. (a) *Current-to-voltage converter.* (b) *Electronic ammeter.*

the ammeter is; if i_{in} equals 50 μA, v_{out} equals 5 V, enough voltage to measure easily with an inexpensive voltmeter. Therefore, even though it uses inexpensive parts, the electronic ammeter of Fig. 18-14*b* is superior to a moving-coil ammeter as far as input impedance is concerned.

EXAMPLE 18-15

Suppose we have an oscilloscope with an input sensitivity of 10 mV/cm. By connecting the current-to-voltage converter of Fig. 18-15*a* to the vertical input, we can measure current. If we want 1 μA of input current to produce 1 cm of vertical deflection, what value should R have?

SOLUTION

To get 1 cm of deflection, 1 μA must produce 10 mV. With Eq. (18-19),

$$R = \frac{v_{out}}{i_{in}} = \frac{10(10^{-3})}{10^{-6}} = 10 \text{ k}\Omega$$

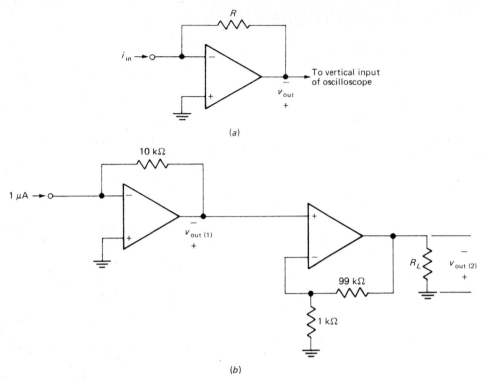

Figure 18-15. (a) *Using an oscilloscope to measure current.* (b) *Current-to-voltage converter drives voltage amplifier.*

Therefore, a 10-kΩ resistor in Fig. 18-15a allows us to measure microamp currents with an oscilloscope.

EXAMPLE 18-16
If 1 μA of input current i_{in} drives the system of Fig. 18-15b, what is the value of the output voltage?

SOLUTION
The first stage is a current-to-voltage converter with an output voltage of

$$v_{out(1)} = Ri_{in} = 10^4(10^{-6}) = 10 \text{ mV}$$

The second stage is a voltage amplifier with a gain of 100; therefore, the output voltage of the system is

$$v_{out(2)} = Av_{in} = 100(10 \text{ mV}) = 1 \text{ V}$$

18-8 *ss negative feedback*

Figure 18-16*a* shows a *series-series* (SS) negative-feedback amplifier. The input voltage drives the *noninverting* input terminal. The feedback voltage drives the inverting input terminal.

output current

What is the formula for i_{out} in Fig. 18-16*a*? First, note that i_{out} flows through the load resistor R_L and through the feedback resistor R. Therefore,

$$v_f = i_{\text{out}} R$$

As we saw with SP feedback, a series input connection means v_f approximately equals v_{in}. Therefore,

$$v_{\text{in}} \cong i_{\text{out}} R$$

or
$$i_{\text{out}} \cong \frac{v_{\text{in}}}{R} \qquad (18\text{-}21)\text{***}$$

Here's how to remember this result. In Fig. 18-16*a*, v_{error} approaches zero. To satisfy Kirchhoff's voltage law, therefore, approximately all the input voltage v_{in} appears across R. Since i_{out} flows through R, $i_{\text{out}} \cong v_{\text{in}}/R$.

input impedance

Because of the series input connection, the input impedance approaches infinity. By a straightforward derivation like the one for SP feedback, we can prove

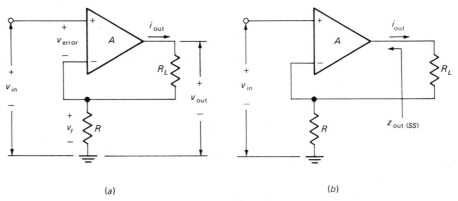

(a)　　　　　　　　　　(b)

Figure 18-16. SS negative feedback.

$$z_{in(SS)} \cong (1 + AB)z_{in} \qquad (18\text{-}22)$$

where B equals $R/(R_L + R)$.

output impedance

Look at Fig. (18-16b). Ideally, the output current of an SS negative-feedback amplifier is independent of the value of R_L. In other words, in Fig. 18-16b we can change R_L, but the value of i_{out} remains constant because it equals v_{in}/R (ideally). This implies the output of an SS feedback amplifier acts like a current source.

The output impedance in Fig. 18-16b is

$$z_{out(SS)} = AR \qquad (18\text{-}23)$$

For instance, if A is 10,000 and R is 100 Ω, $z_{out(SS)}$ equals 1 MΩ.

the ideal voltage-to-current converter

Here is what counts. An SS negative-feedback amplifier ideally behaves like a *perfect voltage-to-current* converter, a device with *infinite input impedance, infinite output impedance,* and a *fixed relation between v_{in} and i_{out}.* Figure 18-17a illustrates the idea. Apply an input voltage v_{in} to the converter and out comes a current of v_{in}/R. Since R can be a precision resistor, the output current is precisely fixed by the ratio v_{in}/R.

Where can you use a voltage-to-current converter? Figure 18-17b shows one application, a sensitive dc voltmeter. The load is an ammeter with a full-scale deflection of 100 μA. When v_{in} is 1 mV,

$$i_{out} = \frac{v_{in}}{R} = \frac{10^{-3}}{10} = 100 \ \mu A$$

This tells us we get full-scale deflection of the ammeter with only 1-mV input. In this way, we can accurately measure very small dc voltages.

If we change the 10-Ω to a 100-Ω resistor, it will take a 10-mV input to get full-scale deflection. If we use a 1000-Ω resistor, it takes 100-mV input to get full-scale deflection, etc. By using a switch and precison resistors between 10 Ω and 100 kΩ, we will have a sensitive dc voltmeter with full-scale voltage ranges from 1 mV to 10 V.

The 741C data sheet indicates we can null the output with a 10-kΩ potentiometer between pins 1 and 5, as shown in Fig. 18-17b. This is how we can zero the voltmeter before taking a reading.

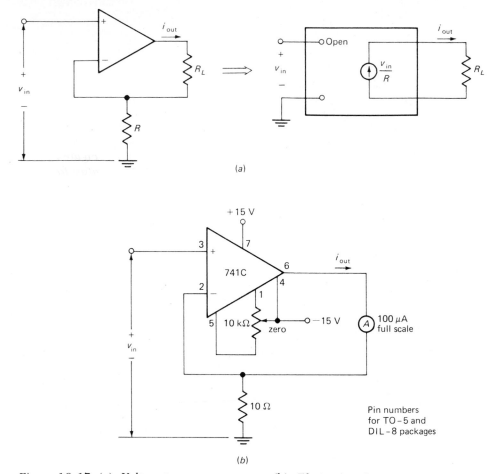

Figure 18-17. (a) Voltage-to-current converter. (b) Electronic voltmeter.

18-9 *ps feedback*

The fourth basic form of feedback is the *parallel-series* (PS) connection shown in Fig. 18-18. The input current drives the *inverting* terminal; the noninverting terminal is grounded.

Because of the Miller effect, the inverting input of Fig. 18-18 is a virtual ground. In other words, $z_{in\,(PS)}$ approaches zero. Also, the series output connection tends to act like a perfect current source; that is, $z_{out\,(PS)}$ approaches infinity.

With PS negative feedback, the input variable is i_{in} and the output variable

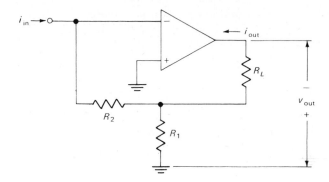

Figure 18-18. PS negative feedback.

is i_{out}. Because of the virtual ground, we can apply the current-divider theorem to Fig. 18-18 to get the PS current gain:

$$A_i = \frac{R_2}{R_1} + 1 \qquad (18\text{-}24)$$

This is identical to the formula for SP voltage gain, except the feedback-stabilized ratio is current gain i_{out}/i_{in}.

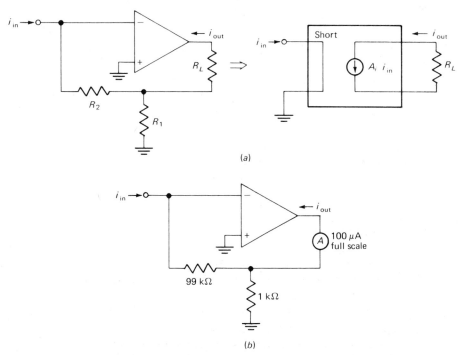

Figure 18-19. (a) *Current amplifier.* (b) *Sensitive electronic ammeter.*

The PS negative-feedback amplifier acts approximately like a perfect current amplifier, a circuit with zero input impedance, infinite output impedance, and a precise current gain as shown in Fig. 18-19a. Inject a current i_{in} and out comes a current of

$$i_{out} = A_i i_{in}$$

EXAMPLE 18-17

Figure 18-19b shows part of a sensitive dc ammeter. How much input current do we need to get full-scale deflection?

SOLUTION

The current gain of the PS negative-feedback amplifier is

$$A_i = \frac{R_2}{R_1} + 1 = 99 + 1 = 100$$

Since the ammeter has a full-scale current of 100 μA, it takes only 1 μA of input current to produce full-scale deflection.

18-10 *summary*

A parallel-output connection is often referred to as *voltage feedback* because the output voltage of the amplifier drives the feedback circuit. Voltage feedback stabilizes the output voltage against changes in open-loop gain, internal impedance, etc. In other words, voltage feedback tends to make the output circuit act like a perfect voltage source.

Table 18-3 Feedback Summary*

	SP	PP	SS	PS
Stable ratio	v_{out}/v_{in}	v_{out}/i_{in}	i_{out}/v_{in}	i_{out}/i_{in}
Input impedance	Increases	Decreases	Increases	Decreases
Output impedance	Decreases	Decreases	Increases	Increases
Distortion	Decreases	Decreases	Decreases	Decreases
Bandwidth	Increases	Increases	Increases	Increases
Equivalent circuit	Voltage amplifier	Current-to-voltage converter	Voltage-to-current converter	Current amplifier

* Common equivalent names are as follows: SP is voltage-series, PP is voltage-shunt, SS is current-series, and PS is current-shunt.

On the other hand, a series-output connection is often referred to as *current feedback* because the output current drives the feedback circuit. Current feedback stabilizes the output current against changes in open-loop gain, internal impedance, etc. To put it another way, current feedback makes the output circuit act approximately like a perfect current source.

The returning feedback signal can be applied in series or in parallel with the input. As shown earlier, the series-input connection produces an input impedance approaching infinity. With this connection, the controlling input variable is voltage v_{in}. On the other hand, a parallel-input connection results in an input impedance approaching zero; in this case, the controlling input variable is current i_{in}.

Table 18-3 summarizes the highlights of the four basic kinds of negative feedback. As indicated, SP negative feedback stabilizes the ratio v_{out}/v_{in}, PP negative feedback stabilizes the ratio v_{out}/i_{in}, SS negative feedback stabilizes i_{out}/v_{in}, and PS negative feedback stabilizes i_{out}/i_{in}.

self-testing review

Read each of the following and provide the missing words. Answers appear at the beginning of the next question.

1. There are _____ basic feedback connections. The most widely used are SP and PP negative feedback. The feedback circuit in an SP negative-feedback amplifier is usually a _____ divider.

2. *(four, voltage)* The error voltage is insignificantly _____. If the internal gain of an SP negative-feedback amplifier decreases, the error voltage will _____. The overall effect is that output voltage is virtually independent of changes in the open-loop gain.

3. *(small, increase)* The closed-loop gain of an SP negative-feedback amplifier approximately equals $1/B$. Because of this, changes in open-loop gain have almost no effect on the _____ gain. SP input impedance is very _____, but SP output impedance is very _____.

4. *(closed-loop, large, small)* Other benefits of negative feedback are reduction of nonlinear _____ and a stretching of the _____. The sacrifice factor is the ratio of open-loop gain to closed-loop gain.

5. *(distortion, bandwidth)* The gain-bandwidth product is a _____, provided the open-loop gains rolls off at a rate of 20 dB per decade out to f_{unity}. The gain-bandwidth product _____ f_{unity}.

6. *(constant, equals)* The input signal drives the _____ input of a PP negative-feedback amplifier. PP input impedance approaches _____, and PP output impedance approaches _____. The inverting input acts like a virtual ground. Ideally, this means the inverting input is at ground potential, but all the input current flows through the feedback resistor.

7. *(inverting, zero, zero)* The PP negative-feedback amplifier acts like a current-to-voltage converter, a circuit with very _____ input impedance, very _____ output impedance, and a fixed ratio of v_{out}/i_{in}.

8. *(low, low)* Input voltage drives the noninverting terminal of an SS negative-feedback amplifier. This feedback amplifier is equivalent to a voltage-to-current converter, a circuit with very _____ input impedance, very _____ output impedance, and a fixed ratio of i_{out}/v_{in}.

9. *(high, high)* The PS negative-feedback amplifier acts like a _____ amplifier. It has very _____ input impedance and very high output impedance.

10. *(current, low)* A parallel-output connection is often referred to as _____ feedback, because the output voltage drives the feedback circuit. SP and PP are examples of this kind of feedback.

11. *(voltage)* A series-output connection is called current feedback because output current drives the feedback circuit.

problems

18-1. In Fig. 18-20a calculate the gain A of the internal amplifier, the gain B of the feedback circuit, and the gain A_{SP} of the feedback amplifier.

18-2. Calculate the sacrifice factor for the feedback amplifier of Fig. 18-20a.

18-3. How much output voltage is there in Fig. 18-20b? If the internal amplifier is a 741C with a typical gain of 100,000, what is the value of S?

18-4. If the internal amplifier of Fig. 18-20b has a gain of 1,000,000, what are the approximate values of v_{error} and v_f? If the gain decreases to 100,000, what are the approximate values of v_{error} and v_f?

18-5. What is the closed-loop voltage gain in Fig. 18-20c?

18-6. Calculate the output voltage in Fig. 18-20c. If the z_{in} of the internal amplifier equals 10 kΩ and A is 50,000, what is the value of $z_{in\,(SP)}$?

18-7. The internal gain A of Fig. 18-20b equals 1,000,000 and the z_{in} is 100 kΩ. Calculate $z_{in\,(SP)}$.

18-8. Suppose the data sheet of an op amp lists an open-loop voltage gain of 100,000 and a z_{in} of 500 kΩ. If this op amp is used in Fig. 18-20b what is the value of $z_{in\,(SP)}$?

18-9. An op amp has an internal gain A of 200,000 and an output impedance z_{out} of 1 kΩ.
   (a) What is the value of $z_{out(SP)}$ when this op amp is used in Fig. 18-20b?
   (b) If used in Fig. 18-20c, what is the value of $z_{out\,(SP)}$?

18-10. Figure 18-21 is a two-stage bipolar amplifier with SP negative feedback. The feedback circuit consists of 9.8 kΩ and 200 Ω. What is the closed-loop gain? If

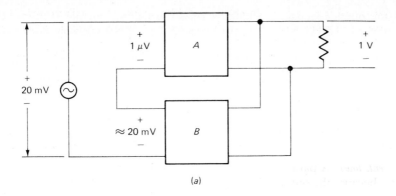

(a)

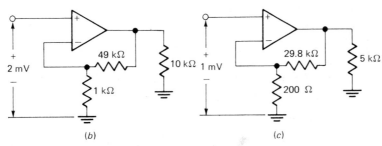

(b)   (c)

Figure 18-20.

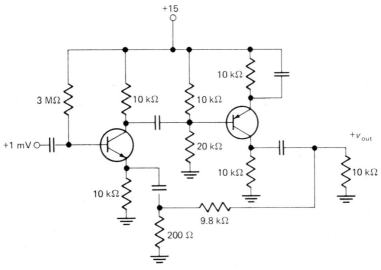

Figure 18-21.

the open-loop gain is 10,000, what is the sacrifice factor? If the z_{in} of the first stage is 2.5 kΩ, what does $z_{in\,(SP)}$ equal? If z_{out} is 10 kΩ, what does $z_{out\,(SP)}$ equal?

18-11. An input current of 1 mA drives the current-to-voltage converter of Fig. 18-22a. What is the output voltage for each position of the switch?

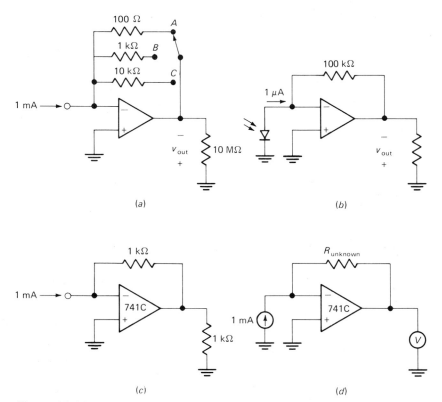

Figure 18-22.

18-12. Figure 18-22b shows a *photodiode* (the two arrows represent light) driving a current-to-voltage converter. With 1 μA coming out of the photodiode, how much output voltage is there?

18-13. The 741C of Fig. 18-22c has a typical internal gain A of 100,000 and a z_{out} of 300 Ω. What is the value of $z_{in\,(PP)}$? If the source supplying the 1 mA has an impedance R_S of 1 MΩ, what is the value of $z_{out\,(PP)}$?

18-14. A 1-mA current source drives the current-to-voltage converter of Fig. 18-22d. A voltmeter across the output has ranges of 1 mV, 10 mV, and 100 mV (full scale). We can use the circuit as an electronic ohmmeter. What is the value of $R_{unknown}$ that produces full-scale deflection for each given voltage range? If we change the current source to 1 μA, what is the value of $R_{unknown}$ that produces full-scale deflection for each range?

18-15. Many transducers are resistive types; a nonelectrical input quantity changes their resistance. A carbon microphone is an example of a resistive transducer; the sound-wave input produces changes in resistance. Other examples are strain gages, thermistors, and photoresistors. Suppose we use a resistive transducer in the circuit of Fig. 18-23; we can convert the changes in resistance to changes in output voltage.

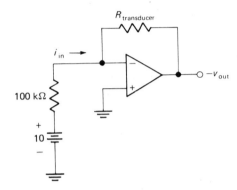

Figure 18-23.

(a) What is the value of i_{in}?
(b) If the quiescent value of the transducer resistance is 1 kΩ, what is the quiescent output voltage?
(c) If the nonelectrical quantity driving the transducer produces a change of $\pm$100 Ω in the transducer resistance, how much output ac voltage is there?

18-16. Figure 18-24a shows a sensitive dc voltmeter. Calculate the input dc voltages that produce full-scale deflection of the ammeter in each switch position.

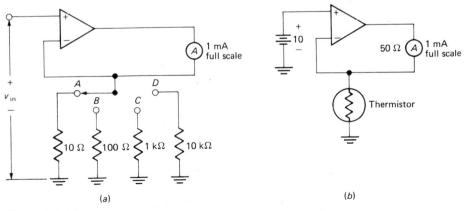

(a)

(b)

Figure 18-24.

18-17. The ammeter of Fig. 18-24a has a resistance of 40 Ω. If the internal amplifier has a gain A of 1,000,000 and a z_{in} of 100 kΩ, what is the value of $z_{in\ (SS)}$ in position A of the switch?

18-18. If the internal amplifier of Fig. 18-24a has a gain A of 100,000, what is the minimum value of $z_{out\ (SS)}$?

18-19. Figure 18-24b shows an electronic thermometer. At 0°C, the thermistor has a resistance of 20 kΩ. The resistance decreases 200 Ω for each degree rise, so that $R_{thermistor}$ equals 19.8 kΩ, 19.6 kΩ, 19.4 kΩ, and so on for T equals 1°C, 2°C, 3°C, etc. What does the ammeter read at 0°C? At 25°C? At 50°C?

18-20. The meter resistance in Fig. 18-25 is 50 Ω. The internal amplifier has a gain of 1,000,000. Calculate the value of i_{out}.

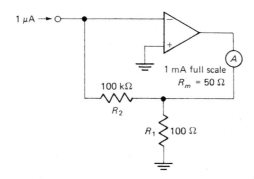

Figure 18-25.

18-21. To get a current gain of 200 in Fig. 18-25 what value should we use for R_1 if R_2 is kept at 100 Ω?

18-22. An SP feedback amplifier has a sacrifice factor of 1000. If the internal amplifier has a break frequency f_2 of 10 Hz, what is the feedback break frequency $f_{2\ (SP)}$?

18-23. The op amp of Fig. 18-26 has an internal gain A of 1,000,000 and a break frequency of 10 Hz. Work out the value of $f_{2\ (SP)}$ for each position of the switch.

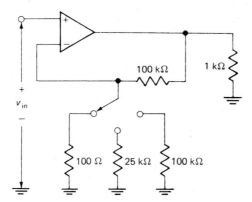

Figure 18-26.

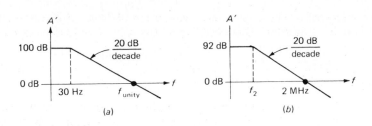

Figure 18-27.

18-24. Calculate the gain-bandwidth product for each of these:
  (a) $A_{mid} = 50,000$ and $f_2 = 100$ Hz
  (b) $A'_{mid} = 100$ dB and $f_2 = 20$ Hz
  (c) $A'_{mid} = 120$ dB and $f_2 = 15$ Hz

18-25. What is the value of f_{unity} for each amplifier given in the preceding problem if the internal gain rolls off 20 dB per decade until it crosses the horizontal axis?

18-26. If an amplifier has the bel voltage gain shown in Fig. 18-27a, what does its gain-bandwidth product equal? What is the value of f_{unity}?

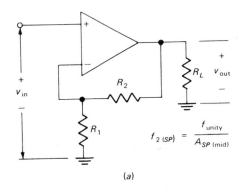

(a)

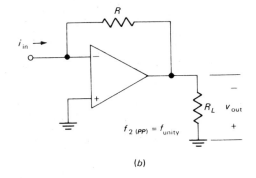

(b)

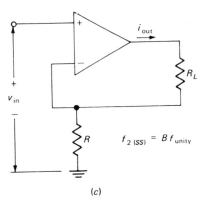

(c)

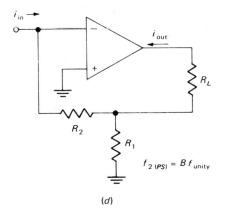

(d)

Figure 18-28.

18-27. For an amplifier whose bel voltage gain looks like Fig. 18-27b, what is the value of its internal break frequency f_2? The value of its gain-bandwidth product?

18-28. If a voltage follower uses an internal amplifier with the bel voltage gain shown in Fig. 18-27b, what will the $f_{2(SP)}$ equal?

18-29. The SP feedback amplifier of Fig. 18-20b uses an internal amplifier whose bel voltage gain looks like Fig. 18-27b. What is the $f_{2(SP)}$ of the feedback amplifier?

18-30. Figure 18-28 shows the four feedback connections and the closed-loop break frequency of each. If the internal amplifier of Fig. 18-28b has the bel graph of Fig. 18-27b, what does $f_{2(PP)}$ equal?

18-31. For the SS negative feedback shown in Fig. 18-28c, the value of B is given by

$$B = \frac{R}{R + R_L}$$

If the ammeter of Fig. 18-24a has a resistance of 50 Ω and the open-loop gain looks like Fig. 18-27b, what does $f_{2(SS)}$ equal in position A?

18-32. With the PS negative feedback of Fig. 18-28d,

$$B = \frac{R_1 \parallel R_2}{R_L + R_1 \parallel R_2}$$

What does $f_{2(PS)}$ equal in Fig. 18-25 if the op amp has the bel graph of Fig. 18-27b?

19 *positive feedback*

The main use of positive feedback is in *oscillators,* circuits that generate an output signal without an input signal. In an oscillator, part of the output is fed back to the input; this feedback signal is the only input to the internal amplifier.

19-1 *the basic idea*

Before mathematical analysis, let us get the main idea of how an oscillator produces an output signal without an external input signal. To begin with, Fig. 19-1a shows a voltage source v driving the error terminals of the amplifier. The amplified signal Av drives the feedback circuit to produce feedback voltage ABv. This voltage returns to point x. If the phase shift through the amplifier and feedback circuit is correct, the signal at point x will be exactly in phase with the signal driving the error terminals of the amplifier. Stated another way, if the phase shift around the entire loop is $0°$, we have *positive feedback.*

loop gain AB

The value of AB (the *loop gain*) is important. We will explain the action of an oscillator in a moment. For now, assume we connect points x and y, and remove voltage source v; the feedback signal drives the error terminals of the amplifier (Fig. 19-1b). If AB is less than unity, ABv is less than v, and the output signal

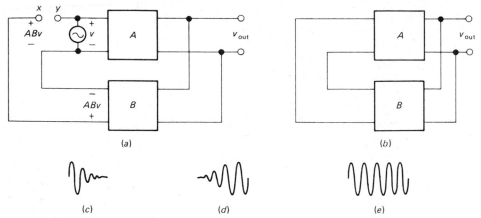

Figure 19-1. Positive feedback.

will die out as shown in Fig. 19-1c because we are not returning enough voltage. On the other hand, if AB is greater than unity, ABv is greater than v, and the output voltage builds up (Fig. 19-1d). Finally, if AB equals unity, there will be no change in output voltage; we will get a steady output like Fig. 19-1e.

In a practical oscillator the value of loop gain AB is greater than unity when the power is first turned on. A small starting voltage is applied to the error terminals, and the output voltage builds up as shown in Fig. 19-1d. After the output voltage reaches a desired level, the value of AB automatically decreases to unity and the output amplitude remains constant (Fig. 19-1e).

the starting voltage

The starting voltage of an oscillator is built into every resistor inside the oscillator. As will be discussed later, every resistor generates noise voltages; these voltages are produced by the random motion of electrons in the resistor. The motion is so random it contains sinusoidal frequencies to over 1000 GHz (10^{12} Hz). In other words, each resistor acts like a small voltage source producing essentially all frequencies.

In Fig. 19-1b here is what happens. When you first turn on the power, the only signals in the system are noise voltages. These signals are very small in amplitude. All are amplified and appear at the output terminals. The amplified noise now drives the feedback circuit, usually a resonant circuit. Because of this, feedback voltage ABv will be maximum at the resonant frequency of the feedback circuit; furthermore, the phase is correct for positive feedback only at the resonant frequency.

In other words, the amplified noise is filtered so that only one sinusoidal component returns with exactly the right phase for positive feedback. When loop gain AB is greater than unity, the oscillations build up at this frequency (Fig. 19-1d). After a suitable level is reached, AB decreases to unity and we get a constant-amplitude output signal (Fig. 19-1e).

AB *decreases to unity*

There are two ways for AB to decrease to unity. First, the increasing signal will eventually force the output stage to clip; when this happens, the value of A decreases. The harder the clipping, the lower the voltage gain. In this way, the gain decreases to whatever value is needed to make AB equal to unity.

Second, we can put something in the feedback circuit to reduce the gain of the feedback circuit. Often, this something is a nonlinear resistor that reduces B when the output signal has reached the desired value. In this way, the oscillator automatically makes AB equal unity after the oscillations have built up.

summary

Here are the key ideas behind any feedback oscillator:

1. Initially, loop gain AB must be greater than unity at the frequency where the loop phase shift is 0°.
2. After the desired level is reached, AB must decrease to unity by reducing either A or B.

19-2 *the phase-shift oscillator*

Figure 19-2 is a *phase-shift oscillator*. The amplifier has 180° of phase shift because the signal drives the inverting input. The amplifier output is fed back to three cascaded lead networks. As you recall, a lead network produces a phase shift between 0° and 90°, depending on the frequency. Therefore, at a particular frequency the total phase shift of the three lead networks equals 180°. As a result, the phase shift around the entire loop will be 360°, equivalent to 0°. If AB is greater than unity at this particular frequency, oscillations can start.

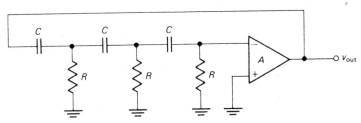

Figure 19-2. Phase-shift oscillator.

A straightforward but complicated analysis of the three lead networks leads to these formulas. The frequency of oscillation is given by

$$f_0 = \frac{1}{2\pi RC\sqrt{6}} \qquad\qquad (19\text{-}1)\text{***}$$

and the voltage gain of the lead networks is

$$B = \frac{1}{29}$$

So, if A is greater than 29, oscillations can start.

Phase-shift oscillators can also work with three lag networks. Oscillations occur when the total phase shift of the three lag networks equals 180°.

EXAMPLE 19-1
Figure 19-3 shows a FET phase-shift oscillator. At what frequency does the circuit oscillate?

SOLUTION
Use Eq. (19-1):

$$f_0 = \frac{1}{2\pi RC\sqrt{6}} = \frac{1}{2\pi(10^6)68(10^{-12})\sqrt{6}}$$
$$= 956 \text{ Hz}$$

With the adjustable source resistor reduced to zero, the FET has a maximum voltage gain of

$$A = g_{m0}R_L = 0.005 \times 10,000 = 50$$

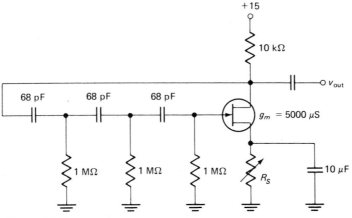

Figure 19-3.

As the source resistance is increased, g_m decreases, and so does voltage gain. Therefore, we can adjust the resistance to get a gain of slightly more than 29. This allows oscillations to start and prevents heavy distortion from excessive clipping.

19-3 *the Wien-bridge oscillator*

The *Wien-bridge oscillator* is the standard oscillator circuit for all low frequencies in the range of 5 Hz to about 1 MHz. It is almost always used in commercial audio generators and is usually preferred for other low-frequency applications.

lead-lag network

To understand how a Wien-bridge oscillator works, we have to discuss the *lead-lag network* of Fig. 19-4. With this feedback circuit, the phase angle leads for low frequencies and lags for high frequencies. Especially important, there is one frequency where the phase shift exactly equals 0°; this important property allows the lead-lag network to determine the frequency of oscillation.

At very low frequencies the series capacitor of Fig. 19-4a looks open and we get no output. At very high frequencies the shunt capacitor looks shorted and we get no output. In between these extremes the output voltage reaches a maximum value (see Fig. 19-4b). As can be proved, the frequency f_0 equals

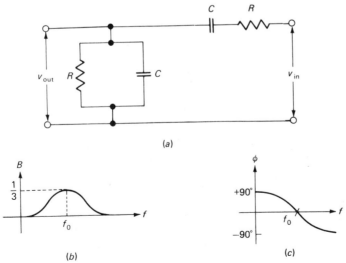

Figure 19-4. Lead-lag network.

$$f_0 = \frac{1}{2\pi RC} \qquad \text{(resonant frequency)} \qquad \text{(19-2)}\ast\ast\ast$$

Figure 19-4c shows the phase angle of output voltage with respect to the input voltage. As we see, at very low frequencies the phase angle is positive and the circuit acts like a lead network. On the other hand, at very high frequencies the phase angle is negative and the circuit acts like a lag network. The frequency f_0 produces zero degrees phase shift.

Therefore, the lead-lag network of Fig. 19-4a acts like a resonant circuit; the voltage gain reaches a maximum at f_0, and the phase angle goes to zero at f_0. For this reason, f_0 is the *resonant frequency* of the lead-lag network. The lead-lag network is the key to how a *Wien-bridge oscillator* works.

the basic idea

Figure 19-5a shows a Wien-bridge oscillator; it uses positive and negative feedback. The positive feedback is through the lead-lag network to the noninverting input. The negative feedback is through the voltage divider to the inverting input.

When we first apply power, the *tungsten lamp* has low resistance and we do not get much negative feedback. For this reason, AB is greater than unity and oscillations can build up at the resonant frequency f_0. As the oscillations build up, the tungsten lamp heats, and this increases its resistance. At a desired output level, the tungsten lamp has a resistance of R'. As we will prove, AB equals unity for this value.

initial conditions

At first, the tungsten lamp has a resistance *less* than R'. The gain from the noninverting input to the output equals A_{SP} therefore,

$$A_{SP} = \frac{R_2}{R_1} + 1 > 3$$

This gain is greater than 3 because R_2 equals $2R'$ and R_1 is less than R'. As shown earlier, the voltage gain B of the lead-lag network is 1/3 at the resonant frequency. Therefore, the voltage gain around the positive-feedback loop is initially greater than unity.

As the output voltage builds up, the resistance of the lamp increases as shown in Fig. 19-5b. At a certain voltage V' the tungsten lamp will have a resistance of R'. At this point, the gain from the noninverting input to the output is

$$A_{SP} = \frac{2R'}{R'} + 1 = 3$$

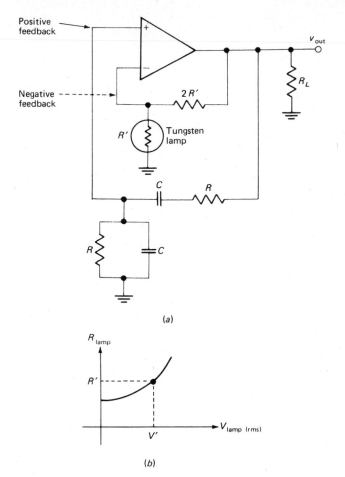

Figure 19-5. Wien-bridge oscillator.

and the gain of the positive feedback loop becomes unity. When this happens, the output amplitude levels off and becomes constant. (In a practical oscillator the tungsten lamp does not become luminescent; this would waste signal power.)

amplifier phase shift

In a Wien-bridge oscillator, the phase shift of the lead-lag network equals zero when the oscillations have a frequency of

$$f_0 = \frac{1}{2\pi RC}$$

Because of this, we can adjust the frequency by varying the value of R or C. This assumes the phase shift of the amplifier is negligibly small. Stated another

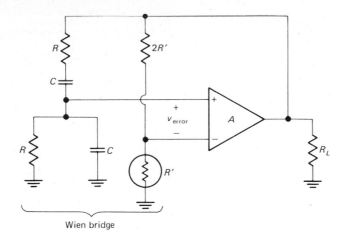

Wien bridge

Figure 19-6.

way, the amplifier must have an SP break frequency well above the resonant frequency f_0. In this way, the amplifier introduces no additional phase shift. If the amplifier did introduce phase shift, the neat calibration formula $f_0 = 1/2\pi RC$ would no longer hold.

why called a Wien-bridge oscillator

Figure 19-6 shows another way to draw the oscillator. The lead-lag network is the left side of a bridge and the voltage-divider is the right side. This particular bridge is known as a *Wien bridge* and is used in applications besides oscillators. The error voltage is the output of the bridge. When the bridge approaches balance, the error voltage approaches zero.

The Wien bridge is an example of a *notch filter*, a circuit with zero output at one particular frequency. For the Wien bridge the notch frequency equals

$$f_0 = \frac{1}{2\pi RC}$$

Ideally, we get oscillations at f_0. But actually, we have to have some error voltage no matter how high the internal gain A. In the Wien-bridge oscillator the error voltage is so small the bridge is essentially balanced.

EXAMPLE 19-2
Calculate the minimum and maximum frequency in the Wien-bridge oscillator of Fig. 19-7*a*.

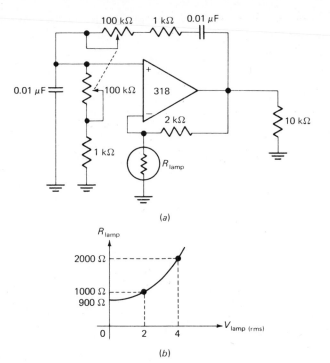

(a)

(b)

Figure 19-7.

SOLUTION

The ganged rheostats can vary from 0 to 100 kΩ; therefore, the value of R goes from 1 to 101 kΩ. The minimum frequency of oscillation is

$$f_0 = \frac{1}{2\pi RC} = \frac{1}{2\pi(101)10^3(0.01)10^{-6}} = 158 \text{ Hz}$$

and the maximum frequency is

$$f_0 = \frac{1}{2\pi RC} = \frac{1}{2\pi(10^3)10^{-8}} \cong 16 \text{ kHz}$$

EXAMPLE 19-3

Figure 19-7b shows the lamp resistance of Fig. 19-7a. Calculate the output voltage.

SOLUTION

In Fig. 19-7a the output amplitude becomes constant when the lamp resistance equals 1kΩ. In Fig. 19-7b this means the lamp voltage is 2 V rms. The current flowing through the lamp also flows through the 2-kΩ resistor, which means a

4-V rms signal across the 2-kΩ resistor; therefore, the output voltage equals the sum of 4 V and 2V, or

$$V_{out} = 6 \text{ V rms}$$

19-4 *lc oscillators*

Although superb at low frequencies, the Wien-bridge oscillator is not suited to high frequencies (well above 1 MHz). Its R's and C's become too small and the amplifier phase shift is a problem. *LC oscillators* are used at frequencies from less than 1 to over 500 MHz. With an amplifier and LC resonant circuit we can feed back a signal with the right amplitude and phase to sustain oscillations.

ce oscillators

Figure 19-8a shows the ac equivalent circuit of a *Hartley oscillator*. When the LC tank is resonant, it appears purely resistive to the collector, so we get 180° phase shift because of inversion from base to collector. There is an additional 180° phase shift because of the tapped inductor as the feedback signal returns to the base. For this reason, the phase shift around the loop is 0° at the resonant frequency. The Hartley oscillator is used a great deal in transistor radios and other entertainment receivers.

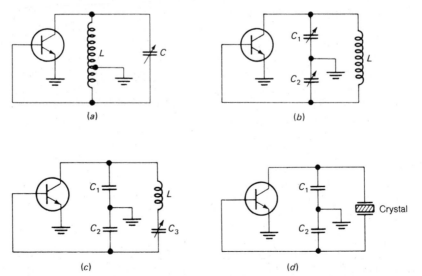

Figure 19-8. Ac forms of common oscillators. (a) *Hartley.* (b) *Colpitts.* (c) *Clapp.* (d) *Crystal.*

The *Colpitts oscillator* of Fig. 19-8*b* is a superb circuit, widely used in commercial signal generators above 1 MHz. This kind of oscillator relies on a capacitive tap rather than an inductive tap; the voltage developed across C_2 is used to drive the base. By gang-tuning the two capacitors we can vary the frequency of oscillation.

Figure 19-8*c* is the ac equivalent circuit for a *Clapp oscillator.* It resembles a Colpitts because of the capacitive tap. The inductive branch, however, has a capacitor C_3. We will show later that a Clapp oscillator produces a more stable frequency than a Colpitts oscillator.

When accuracy and stability of the frequency are important, a *quartz-crystal oscillator* is used. In the ac equivalent circuit of Fig. 19-8*d* the feedback signal comes from a capacitive tap. As discussed later, the crystal acts like a large inductor in series with a small capacitor (similar to the Clapp).

resonant frequency

Oscillators operate class A, B, or C. As discussed in Chap. 12, when the collector current is not sinusoidal, you need a high-Q tank circuit to produce a sinusoidal output voltage. For this reason, most oscillators use tank circuits with a Q greater than 10. Basic circuit books prove this formula for the resonant frequency of a parallel tank circuit:

$$f_0 = \frac{1}{2\pi\sqrt{LC}} \sqrt{\frac{Q^2}{1 + Q^2}} \qquad (19\text{-}3a)$$

With this equation we can calculate the exact resonant frequency. Most of the time, however, we can use the approximation

$$f_0 \cong \frac{1}{2\pi\sqrt{LC}} \qquad \text{for } Q \gg 1 \qquad (19\text{-}3b)\text{***}$$

This is accurate to better than 1 percent when Q is greater than 10.

Given one of the ac oscillator prototypes, what values of L and C do you use in Eq. (19-3)? Use the inductance and capacitance the *circulating* tank current flows through. For instance, in the Colpitts tank of Fig. 19-9*a* the circulating

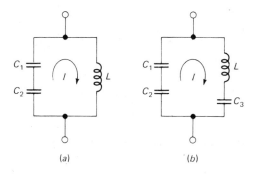

(a)   (b)

Figure 19-9. Circulating current.

or loop current I flows through L and C_1 *in series with* C_2. The equivalent capacitance is

$$C = \frac{C_1 C_2}{C_1 + C_2} \qquad \text{(Colpitts)} \tag{19-4a}$$

So, if C_1 and C_2 are 100 pF each, you would use 50 pF in Eq. (19-3).

As another example, the Clapp-oscillator tank of Fig. 19-9b has a circulating current I that flows through the three capacitors in series. Therefore, the equivalent capacitance to use in the formula for resonant frequency is

$$C = \frac{1}{1/C_1 + 1/C_2 + 1/C_3} \qquad \text{(Clapp)} \tag{19-4b}$$

As discussed later, in a Clapp oscillator C_3 is deliberately made *much smaller* than C_1 and C_2. Since the capacitors are in series with respect to circulating current, C_3 *dominates*, that is,

$$C \cong C_3 \qquad \text{(Clapp)} \tag{19-4c}$$

For instance, if $C_1 = 1000$ pF, $C_2 = 5000$ pF, and $C_3 = 50$ pF,

$$C = \frac{10^{-12}}{1/1000 + 1/5000 + 1/50} \cong 50 \text{ pF}$$

starting conditions

The required starting condition for an oscillator is

$$AB > 1$$

at the resonant frequency of the tank circuit. This is equivalent to

$$A > \frac{1}{B}$$

After you have reduced an oscillator circuit to its ac form, you have to calculate the small-signal gain A and the feedback gain B. In the circuits of Fig. 19-8 the voltage gain is

$$A \cong \frac{r_C}{r_e'} \qquad \text{(small-signal gain)} \tag{19-5}$$

The value of B depends on the particular ac circuit. In any case, the ratio of r_C/r_e' must be greater than $1/B$, or the oscillator will not start.

Figure 19-10a shows a Colpitts tank circuit. The output voltage is across C_1 and the feedback voltage is across C_2. Since the circulating current is the same for both capacitors,

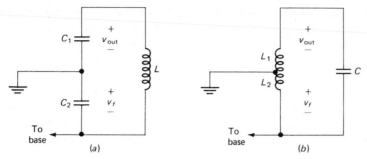

Figure 19-10. Feedback voltages.

$$\frac{v_f}{v_{out}} = \frac{X_{C2}}{X_{C1}} = \frac{1/2\pi f C_2}{1/2\pi f C_1}$$

or

$$B = \frac{C_1}{C_2}$$

and the starting condition is

$$\frac{r_C}{r_e'} > \frac{C_2}{C_1} \tag{19-6}$$

This condition applies to the Clapp and crystal oscillators as well, because they too use a capacitive voltage divider to develop the feedback signal.

Similarly, in the Hartley tank circuit of Fig. 19-10b,

$$\frac{v_f}{v_{out}} = \frac{X_{L2}}{X_{L1}} = \frac{2\pi f L_2}{2\pi f L_1}$$

or

$$B = \frac{L_2}{L_1}$$

and the starting condition is

$$\frac{r_C}{r_e'} > \frac{L_1}{L_2} \quad \text{(Hartley)} \tag{19-7}$$

output voltage

The exact formula for output voltage depends on the class of operation, the current gain-bandwidth product f_T, and other factors.

Light feedback (small B) results in class A operation. With light feedback the value of A is only slightly larger than $1/B$. When you first turn on the power, the oscillations build up and the signal swings over more and more of the ac load line. With this increased signal swing, the operation changes from small signal to large signal. As described in Chap. 10, *large-signal voltage gain is less than small-signal voltage gain;* therefore, with light feedback the value of *AB* can decrease to unity without clipping. As an example, we can get light-feedback oscillations with a Colpitts oscillator by using a large C_2/C_1 ratio, provided *AB* is initially greater than unity. With increasing signal swing, *AB* drops to unity before clipping occurs.

With *heavy feedback* (large *B*), clipping occurs at either or both peaks, depending on the oscillator circuit, the amount of feedback, and other factors. This reduces the gain and decreases *AB* to unity. If the feedback is too heavy, you lose some output signal.

In practice, you can adjust the amount of feedback to maximize the output voltage. For instance, with a Colpitts oscillator you increase C_2/C_1 until you get maximum output voltage.

EXAMPLE 19-4
Figure 19-11 is a complete Colpitts oscillator. The **RF** *choke* ideally looks like a dc short and an ac open. It prevents loss of signal power that would occur if you used a resistor. The final 1-kΩ load may be a discrete load or the input resistance of another stage.

What is the frequency of oscillation?

SOLUTION
The series capacitance is

$$C = \frac{C_1 C_2}{C_1 + C_2} = 909 \text{ pF}$$

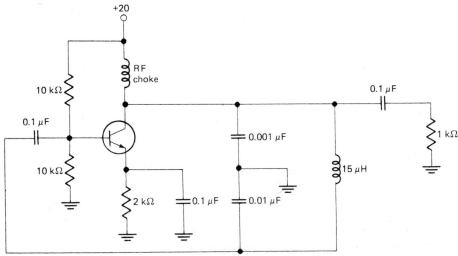

Figure 19-11. Colpitts oscillator.

The inductance is 15 μH; therefore, the frequency of oscillation is

$$f_0 = \frac{1}{2\pi\sqrt{LC}} = \frac{1}{2\pi\sqrt{15(10^{-6})909(10^{-12})}}$$
$$= 1.36 \text{ MHz}$$

EXAMPLE 19-5

If 50 pF is added in series with the 15-μH inductor of Fig. 19-11, the circuit becomes a Clapp oscillator. What is the frequency of oscillation?

SOLUTION

The added capacitor C_3 is only 50 pF; therefore,

$$C = \frac{1}{1/C_1 + 1/C_2 + 1/C_3} \cong C_3 = 50 \text{ pF}$$

The approximate oscillation frequency is

$$f_0 = \frac{1}{2\pi\sqrt{LC_3}} = \frac{1}{2\pi\sqrt{15(10^{-6})50(10^{-12})}}$$
$$= 5.81 \text{ MHz}$$

Why is this frequency more stable than in a Colpitts oscillator? Because C_1 and C_2 are shunted by stray capacitances, transistor capacitances, and so on. Even though these additional capacitances are small, they affect the values of C_1 and C_2 slightly. In a Colpitts oscillator, therefore, the exact frequency depends on these extra capacitances. But in a Clapp oscillator C_3 is more important than C_1 and C_2. For this reason, the oscillation frequency of a Clapp oscillator is more stable and accurate than in a Colpitts oscillator. This is why the Clapp oscillator is occasionally used instead of the Colpitts.

19-5 *quartz crystals*

Some crystals found in nature exhibit the *piezoelectric effect;* when you apply an ac voltage across them, they vibrate at the frequency of the applied voltage. Conversely, if you mechanically force them to vibrate, they generate an ac voltage. The main substances that produce this piezoelectric effect are *quartz, Rochelle salts,* and *tourmaline.*

Rochelle salts have the greatest piezoelectric activity: for a given ac voltage, they vibrate more than quartz or tourmaline. Mechanically, they are the weakest; they break easily. Rochelle salts have been used to make microphones, phonograph pickups, headsets, and loudspeakers.

Tourmaline shows the least piezoelectric activity but is the strongest of the

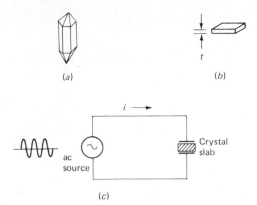

Figure 19-12. Quartz crystal.

three. It is also the most expensive. It is occasionally used at very high frequencies.

Quartz is a compromise between the piezoelectric activity of Rochelle salts and the strength of tourmaline. Because it is inexpensive and readily available in nature, quartz is widely used for RF oscillators and filters.

The natural shape of quartz is a hexagonal prism with pyramids at the ends (see Fig. 19-12a). To get a usable crystal out of this, we have to slice a rectangular slab out of the natural crystal. Figure 19-12b shows this slab with a thickness t. The number of slabs we can get from a natural crystal depends on the size of the slabs and the angle of the cut.

There are a number of different ways to cut the natural crystal; these cuts have names like the X cut, Y cut, XY cut, and AT cut. For our purposes, all we need to know is the cuts have different piezoelectric properties. (Manufacturers' catalogs are usually the best source of information on different cuts and their properties.)

For use in electronic circuits the slab must be mounted between two metal plates as shown in Fig. 19-12c. In this circuit the amount of crystal vibration depends on the frequency of the applied voltage. By changing the frequency we can find resonant frequencies where the crystal vibrations reach a maximum. Since the energy for the vibrations must be supplied by the ac source, the ac current maximizes at each resonant frequency.

fundamental and overtones

Most of the time, the crystal is cut and mounted to vibrate best at one of its resonant frequencies, usually the *fundamental* or lowest frequency. Higher resonant frequencies, called *overtones,* are almost exact multiples of the fundamental frequency. As an example, a crystal with a fundamental frequency of 1 MHz has a first overtone of approximately 2 MHz, a second overtone of approximately 3 MHz, and so on.

The formula for the fundamental frequency of a crystal is

$$f = \frac{K}{t} \tag{19-8}$$

where K is a constant that depends on the cut and other factors, and t is the thickness of the crystal. As we see, the fundamental frequency is inversely proportional to thickness. For this reason, there is a practical limit on how high we can go in frequency. The thinner the crystal, the more fragile it becomes and the more likely it is to break because of vibrations.

Quartz crystals work well up to 10 MHz on the fundamental frequency. To reach higher frequencies we can use a crystal mounted to vibrate on overtones; in this way, we can reach frequencies up to 100 MHz. Occasionally, the more expensive but stronger tourmaline is used at higher frequencies.

ac equivalent circuit

What does the crystal look like as far as the ac source is concerned? When the mounted crystal of Fig. 19-13a is not vibrating, it is equivalent to a capacitance C_m because it has two metal plates separated by a dielectric. C_m is known as the *mounting capacitance*.

However, when the crystal is vibrating, it looks like a tuned circuit. Figure 19-13b shows the ac equivalent circuit of a crystal vibrating at or near its fundamental frequency. Typical values of L are in henrys, C_s in fractions of a picofarad, R in hundreds of ohms, and C_m in picofarads. As an example, here are the values for one available crystal: $L = 3$ H, $C_s = 0.05$ pF, $R = 2000$ Ω, and $C_m = 10$ pF. Among other things, the cut, thickness, and mounting of the slab affect these values.

The outstanding feature of crystals compared with discrete LC tank circuits is their incredibly high Q. For the values just given, we can calculate a Q over

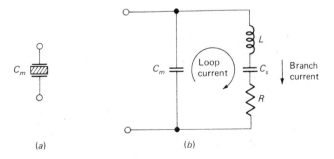

Figure 19-13. Crystal and equivalent circuit.

3000. Q's can easily be over 10,000. On the other hand, a discrete LC tank circuit seldom has a Q over 100. The extremely high Q of a crystal leads to oscillators with very stable values of frequency.

series and parallel resonance

Besides the Q, L, C_s, R, and C_m of the crystal, there are two other characteristics we should know about. The *series resonant frequency* f_s of a crystal is the resonant frequency of the *LCR branch* in Fig. 19-13b. At this frequency the branch current reaches a maximum value because L resonates with C_s. The formula for this series resonant frequency is

$$f_s = \frac{1}{2\pi\sqrt{LC_s}} \tag{19-9}$$

The *parallel resonant frequency* f_p of the crystal is the frequency where the circulating or *loop* current of Fig. 19-13b reaches a maximum value. Since this loop current must flow through the series combination of C_s and C_m, the equivalent C is

$$C_{\text{loop}} = \frac{C_m C_s}{C_m + C_s} \tag{19-10}$$

and the parallel resonant frequency is

$$f_p = \frac{1}{2\pi\sqrt{LC_{\text{loop}}}} \tag{19-11}$$

Two capacitances in series always produce a capacitance smaller than either; therefore, C_{loop} is less than C_s, and f_p is greater than f_s.

In any crystal, C_s is much smaller than C_m. For instance, with the values given earlier, C_s was 0.05 pF and C_m was 10 pF. Because of this, Eq. (19-10) gives a value of C_{loop} just slightly less than C_s. In turn, this means f_p is only slightly more than f_s. When you use a crystal in an oscillator circuit like Fig. 19-14, the additional circuit capacitances appear in shunt with C_m. Because of this, *the oscillation frequency will lie between f_s and f_p.* This is the advantage of knowing

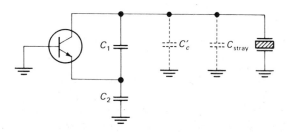

Figure 19-14. Effect of transistor and stray capacitance.

the values of f_s and f_p; they set the lower and upper limits on the frequency of a crystal oscillator.

crystal stability

The frequency of an oscillator tends to change slightly with time; this *drift* is produced by temperature, aging, and other causes. In a crystal oscillator the frequency drift with time is very small, typically less than 1 part in 10^6 (0.0001 percent) per day. Stability like this is important in *electronic wristwatches*; they use quartz-crystal oscillators as the basic timing device.

By using crystal oscillators in precision temperature-controlled ovens, crystal oscillators have been built with frequency drifts less than 1 part in 10^{10} per day. Stability like this is needed in *frequency and time standards*. To give you an idea of how precise 1 part in 10^{10} is, a clock with this drift will take 300 years to gain or lose 1 s.

EXAMPLE 19-6

A crystal has these values: $L = 3$ H, $C_s = 0.05$ pF, $R = 2000$ Ω, and $C_m = 10$ pF. Calculate the f_s and f_p of the crystal to three significant digits.

SOLUTION
With Eq. (19-9),

$$f_s = \frac{1}{2\pi\sqrt{LC_s}} = \frac{1}{2\pi\sqrt{3(0.05)10^{-12}}} = 411 \text{ kHz}$$

With Eq. (19-10),

$$C_{\text{loop}} = \frac{(10 \text{ pF})(0.05 \text{ pF})}{10 \text{ pF} + 0.05 \text{ pF}} = 0.0498 \text{ pF}$$

With Eq. (19-11),

$$f_p = \frac{1}{2\pi\sqrt{LC_{\text{loop}}}} = \frac{1}{2\pi\sqrt{3(0.0498)10^{-12}}} = 412 \text{ kHz}$$

If this crystal is used in an oscillator, the frequency of oscillation must lie between 411 and 412 kHz.

19-6 *unwanted oscillations in amplifiers*

When you build an amplifier with several stages of gain, you can easily get unwanted oscillations. This section tells why they occur and how to avoid them.

low-frequency oscillations

Look at Fig. 19-15a. There is no feedback path from output to input; therefore, the circuit cannot possibly oscillate. Right? Wrong! There are subtle feedback paths that can make any high-gain amplifier produce unwanted oscillations.

Motorboating is a putt-putt sound from a loudspeaker connected to an amplifier

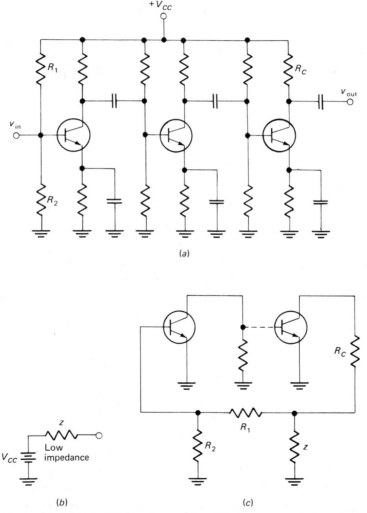

Figure 19-15. (a) *High-gain amplifier.* (b) *Power-supply impedance.* (c) *Current feedback through supply.*

like Fig. 19-15a. This sound represents very low-frequency oscillations, like a few hertz. The feedback path is *by way of the power supply*. Ideally, the power supply looks like a perfect ac short to ground. But to a better approximation the supply has an impedance z as shown in Fig. 19-15b. For example, if the supply has an output filter capacitor, the capacitive reactance increases as frequency decreases. Because of the nonzero supply impedance at low frequencies, we can get current feedback through the supply impedance and the voltage-divider bias of the first stage.

Figure 19-15c illustrates this current feedback from the last stage to the first stage. The ac current out of the last stage flows through R_C and through the supply impedance z. The voltage across z can then drive the R_1-R_2 voltage divider of the first stage. The frequency of oscillation is determined by the lead networks in the amplifier and the reactance of the power supply. At some frequency below the midband of the amplifier, the loop phase shift is 0°. If AB is greater than unity at this frequency, motorboating occurs.

What is the cure for motorboating? Use a *regulated* power supply (Chap. 20). This kind of supply has an internal impedance under 0.1 Ω (some as low as 0.0005 Ω). The current feedback is then too small for oscillations.

high-frequency oscillations

You can get unwanted oscillations above the midband of the amplifier. The electric or magnetic fields around the last stage may induce feedback voltages in an earlier stage with the right phase for oscillations. If AB is greater than unity at the frequency where this happens, you will get oscillations.

Figure 19-16a illustrates the idea. The output circuit acts like one plate of a capacitor, the input circuit like the other plate. This capacitance between output and input is small (usually under 1 pF); but at a high frequency, the capacitance may feed back enough signal to produce oscillations.

Magnetic coupling is also possible. The output wire labeled *primary* in Fig. 19-16a can act like the primary winding of a transformer; the input wire labeled *secondary* can act like a secondary winding. As a result, ac current in the primary can induce a voltage in the secondary. If the feedback signal is strong enough and the phase correct, we get oscillations.

What is the cure for high-frequency oscillations? One of the approaches is to increase the distance between stages; this cuts down the coupling between them. If this is not practical, you can enclose each stage in a *shield* or metallic container (see Fig. 19-16b). Shielding like this is common in many high-frequency applications; it blocks high-frequency electric and magnetic fields. If only the capacitive coupling is a problem, a *baffle shield* (a metallic plate) between stages may eliminate high-frequency oscillations (Fig. 19-16c).

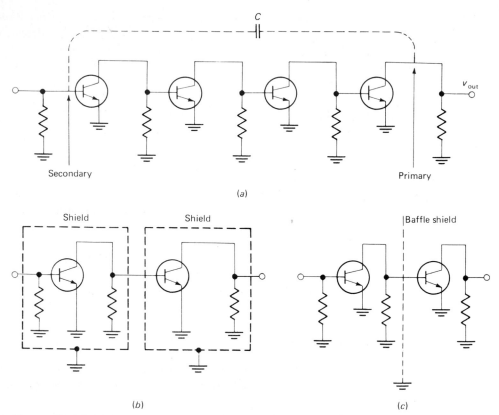

Figure 19-16. *Unwanted coupling from output to input.* (a) *Circuit.* (b) *Shielding each stage.* (c) *Baffle shield.*

ground loops

Another subtle cause of high-frequency oscillations is a *ground loop,* a difference of potential between two ground points. In Fig. 9-16a, all ac grounds are *ideally* at the same potential. But in reality, the chassis or whatever serves as ground has some nonzero impedance that increases with frequency. Therefore, if ac ground currents from the last stage happen to flow through part of the chassis being used by an earlier stage, we can get enough unwanted positive feedback to cause oscillations.

The solution to a ground-loop problem is proper layout of the stages to prevent ac ground currents of later stages from flowing through ground paths of earlier stages. One way to accomplish this is the *single-point* ground system; this means returning all stage grounds to a single point. In this way, there can be no difference of potential between two ground points.

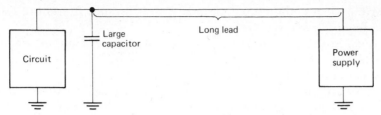

Figure 19-17. Capacitor bypasses lead inductance of supply line.

supply bypassing

Watch out for lead inductance between the power supply and the circuit (see Fig. 19-17). A long lead may have enough inductance to result in current feedback at high frequencies. The solution is to add a large bypass capacitor across the circuit, as shown in Fig. 19-17.

Bypassing the supplies is almost always essential with ICs. Depending on the particular IC, bypass capacitors from about 0.1 to over 1 μF may be needed to prevent oscillations. These bypass capacitors should be located as close as possible to the IC.

negative-feedback amplifiers

Figure 19-18*a* shows a three-stage negative-feedback amplifier. In the midband of the internal amplifier the phase shift is 180° because there is an odd number

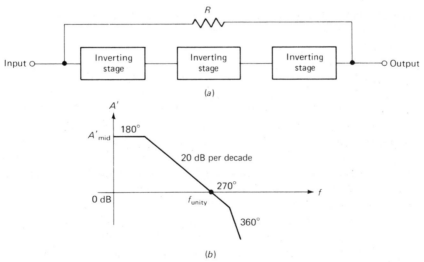

Figure 19-18. (a) Feedback around three inverting stages. (b) One dominant lag network to get unity-gain stability.

of inverting stages; therefore, the phase shift around the entire loop is 180° and the feedback is negative.

But outside the midband the internal lag networks of the stages produce additional phase shifts. Because of this, at some high frequency the phase shift around the entire loop is 0° and the feedback becomes positive. In other words, Fig. 19-18a acts like a phase-shift oscillator using lag networks. At some high frequency the three lag networks can produce a phase shift of 180° (60° each if the networks are identical).

The only way to prevent oscillations outside the midband is to make sure loop gain AB is less than unity when the phase shift reaches 0°. The safest and most widely used method is this: make one of the lag networks dominant enough to produce a 20-dB rolloff until the loop gain crosses the 0-dB axis. A 20-dB rolloff at the horizontal crossing means only one lag network is operating beyond its break frequency; all others are still operating below. This implies the loop phase shift is around 270° at f_{unity}, which makes it impossible to have oscillations.

Figure 19-18b illustrates the idea of one dominant lag network. In the midband the gain is high and the phase shift is 180°. One of the stages has a dominant lag network, so the gain breaks at a low frequency and rolls off at 20 dB per decade. A decade above this break frequency the phase shift is 270°. It stays at approximately 270° until the gain crosses the horizontal axis at f_{unity}. Beyond this point, oscillations are impossible because the loop gain must be less than unity, no matter what the value of B.

With many later-generation op amps the dominant lag network is integrated on the chip and automatically provides the 20-dB rolloff until f_{unity} is reached. For instance, the 741 uses a 30-pF compensating capacitor that is part of a Miller lag network; the gain breaks at 10 Hz and rolls off at 20 dB per decade until an f_{unity} of 1 MHz is reached.

With uncompensated op amps like the 709, you have to add external resistors and capacitors to get this 20-dB rolloff; the manufacturer's data sheet tells you what sizes of R and C to use.

19-7 *thyristors*

Thyristors are four-layer semiconductor devices that use internal positive feedback to produce *latching* action. This section is about two of the most widely used thyristors: the four-layer diode and the silicon controlled rectifier. As an aid to understanding these devices, we begin our discussion with the *ideal latch*.

ideal latch

Figure 19-19a is an ideal latch, a special way to connect complementary transistors. The Q_1 collector drives the Q_2 base, and the Q_2 collector drives the Q_1

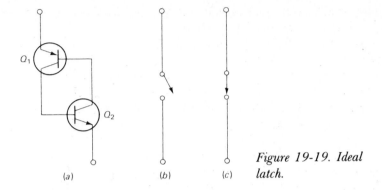

(a) (b) (c)

Figure 19-19. Ideal latch.

base. Therefore, we have direct-coupled feedback. Furthermore, the feedback is positive because a change in current at any point in the loop is amplified and returned to the starting point with the same phase.

An ideal latch can be in either of two states: *open* (Fig. 19-19*b*) or *closed* (Fig. 19-19*c*). If it's in the open position, it stays open until an input current forces it to close. If it's in the closed position, it stays there until an input current forces it to open.

One way to close a latch is by *triggering,* shown in Fig. 19-20*a*. At point A in time, a positive pulse hits the Q_2 base. This trigger momentarily forward-biases the Q_2 base. Because of the large positive feedback, the returning amplified current is approximately β^2 times the original input current. Since the Q_1 collector now supplies the Q_2 base current, the trigger voltage is no longer needed. We call this *regenerative* feedback because once started, the action sustains itself.

The regenerative feedback quickly drives both transistors into saturation, at

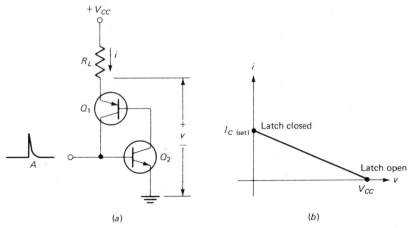

(a) (b)

Figure 19-20. Latch circuit.

which point loop gain drops to unity. The transistors will now remain saturated indefinitely (the upper end of Fig. 19-20 b).

One way to open a latch is by applying a negative trigger to the Q_2 base, which pulls Q_2 out of saturation. Once this happens, regeneration takes over and quickly drives the transistors to the cutoff point shown in Fig. 19-20 b.

Another way to open a latch is by *low-current dropout.* This means reducing the V_{CC} supply voltage almost to zero, at which point the transistors come out of saturation and regeneration drives them into cutoff.

four-layer diode

Figure 19-21 a shows a *four-layer diode* (also called a Shockley diode). This device is classified as a diode because it has two external leads. Because of its four doped regions, it's often called a *pnpn* diode. The easiest way to understand how it works is as follows. Visualize the device separated into two halves as shown in Fig. 19-21 b. This means it's equivalent to a latch (Fig. 19-21 c). Schematic diagrams use the symbol of Fig. 19-21 d for the four-layer diode.

The only way to close a four-layer diode is with *breakover.* This means raising the supply voltage enough to break down either collector diode. Once this happens, the increased current starts the regenerative action and both transistors saturate. Ideally, the four-layer diode then appears shorted or closed.

The only way to open a four-layer diode is by low-current dropout. As described earlier, this means reducing the supply voltage almost to zero, at which point the transistors come out of saturation and regeneration opens the latch. The internal transistors come out of saturation when the current is reduced to a low value called the *holding current.*

For instance, a 1N5158 is a four-layer diode with a breakover voltage of 10 V and a holding current of 4 mA. To close the diode, the supply voltage has to be increased to at least 10 V. Once the diode has closed, you then have to reduce the supply voltage until the current drops below 4 mA; then, the diode will open.

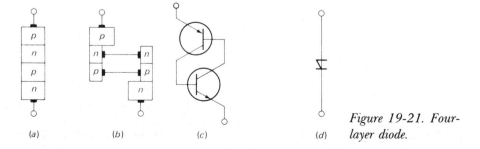

(a)   *(b)*   *(c)*   *(d)*

Figure 19-21. Four-layer diode.

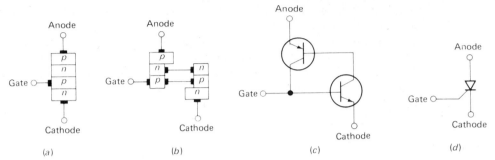

Figure 19-22. Silicon controlled rectifier.

silicon controlled rectifier

Figure 19-22*a* shows the *silicon controlled rectifier*. We can visualize the four doped regions separated into two transistors as shown in Fig. 19-22*b*. This means the silicon controlled rectifier (SCR) is equivalent to a latch with an external trigger input to the lower base (Fig. 19-22*c*). Schematic diagrams use the symbol of Fig. 19-22*d*.

Commercially available SCRs have breakover voltages from about 50 to 500 V. This means they can block a forward voltage (remain open) as long as the supply voltage is less than the breakover voltage. The SCR remains open until a positive trigger hits the gate input. Then, the SCR latches and remains closed even though the trigger is removed.

Once latched, the SCR remains latched indefinitely. The only way to open it is to reduce its current below the holding current. The action is then similar to a four-layer diode; the low-current dropout drives both internal transistors to cutoff and the SCR opens.

As an example, the 2N4444 has a forward blocking voltage of 600 V, a trigger current of 30 mA, and a holding current of 10 mA. This means it can withstand a forward voltage of 600 V without breaking over. It takes a trigger current of 30 mA to close the SCR. And you have to reduce the SCR current to 10 mA to open the SCR.

EXAMPLE 19-7
Figure 19-23*a* shows a *sawtooth generator*. Describe the circuit action.

SOLUTION
Without the four-layer diode, the capacitor voltage would charge exponentially to +50 V (the dashed line of Fig. 19-23*b*). But with the diode in the circuit, breakover occurs at +10 V. At this instant, the diode closes and discharges the capacitor, producing the *flyback* portion of the sawtooth wave. At some point

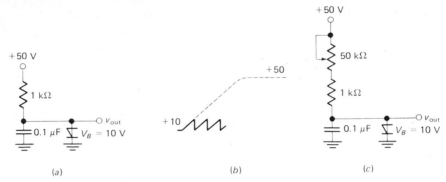

Figure 19-23. Sawtooth generator.

along the flyback, the current drops under the holding value and the diode opens. The next cycle then begins.

This is an example of a *relaxation oscillator,* a circuit that generates an output signal whose frequency depends on the charging or discharging of a capacitor or inductor. If we increase the RC time constant in Fig. 19-23a, it takes longer to reach the breakover point; therefore, the frequency is lower. For instance, a circuit like Fig. 19-23c gives us a 50:1 range in frequency.

EXAMPLE 19-8
Figure 19-24a shows a simple way to build an SCR *crowbar.* Describe the circuit action.

SOLUTION
The zener diode is open as long as the voltage out of the supply is 20 V. Initially, the SCR is open because it has received no trigger; therefore, 20 V appears across the load resistor.

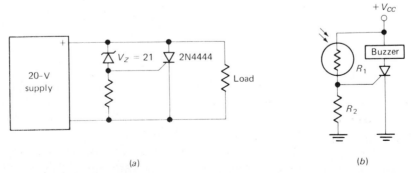

Figure 19-24. SCR circuits.

If something goes wrong with the supply and its voltage tries to rise above 21 V, the zener diode breaks down and delivers a trigger to the SCR. Immediately, the SCR latches and shuts down the power supply. The action is the same as throwing a crowbar across the load terminals. Because the SCR turn-on is very fast (1 μs for a 2N4444), the load is immediately protected against the damaging effects of a large overvoltage.

Crowbarring, though a drastic form of protection, is necessary with many ICs; they can't take much overvoltage. Rather than destroy expensive ICs, therefore, we can use an SCR crowbar to short the load terminals at the first sign of overvoltage. Power supplies with an SCR crowbar need another form of protection called *current-limiting,* which prevents excessive current from damaging the power supply. (Chapter 20 shows you how current limiting works.)

EXAMPLE 19-9

Figure 19-24*b* is an *overlight detector.* How does it work?

SOLUTION

R_1 is a *photoresistor,* a device whose resistance decreases with increasing light intensity. When no light strikes R_1, its resistance is high and the voltage across R_2 is insufficient to trigger the SCR. But when R_1 is in strong light, its resistance is low and we get enough voltage across R_2 to trigger the SCR. When this happens, the buzzer sounds the alarm. Even if the strong light disappears, the latched SCR keeps the buzzer on.

Instead of a buzzer, we could use a LED. If excessive light hits R_1, even temporarily, the LED will remain on as a record of this event.

19-8 *the unijunction transistor*

The *unijunction transistor* (UJT) has two doped regions with three external leads: the emitter, base 1, and base 2, as shown in Fig. 19-25*a.* Figure 19-25*b* is the equivalent circuit, and Fig. 19-25*c* is the schematic symbol.

The intrinsic standoff ratio is defined as

$$\eta = \frac{R_1}{R_1 + R_2} \qquad (19\text{-}12)$$

The typical range of η is 0.5 to 0.8. For instance, a 2N2646 has an η of 0.65. If this UJT has 10 V between B_1 and B_2, then 6.5 V appears across R_1. The voltage across R_1 is known as the *standoff voltage.*

Here's how a UJT works. The latch of Fig. 19-25*b* appears open as long as the input voltage to the emitter *(E)* is less than the standoff voltage. When the input emitter voltage is greater than the standoff voltage, however, the latch

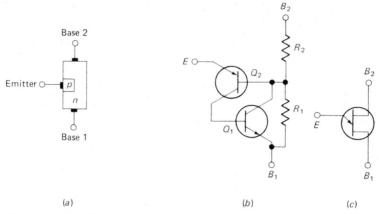

            (a)   (b)   (c)

Figure 19-25. Unijunction transistor.

closes. It then remains closed as long as the latch current (emitter current) is greater than the holding current (also called the valley current).

    A *programmable* UJT (usually designated PUT) allows us to control the standoff ratio. PUTs are useful in building voltage-controlled oscillators (described in Chap. 21).

EXAMPLE 19-10
Describe the circuit action of Fig. 19-26.

SOLUTION
This is a UJT relaxation oscillator. The capacitor charges toward V_{CC}. As soon as the capacitor voltage exceeds the standoff voltage, however, the UJT latches. This discharges the capacitor because of the low resistance between the emitter and ground. As the capacitor voltage flys back to zero, low-current dropout takes place and the UJT latch opens. The next cycle then begins. By varying R, we can control the frequency of the output sawtooth.

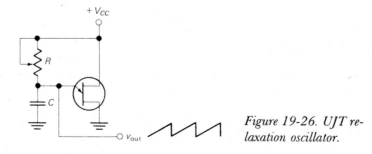

Figure 19-26. UJT relaxation oscillator.

self-testing review

Read each of the following and provide the missing words. Answers appear at the beginning of the next question.

1. The main use of positive feedback is in _____, circuits that generate an output signal without an input signal. Initially, loop gain *AB* must be greater than _____ at the frequency where the loop phase shift is 0°. After the desired level is reached, *AB* must decrease to unity.

2. *(oscillators, unity)* In a phase-shift oscillator the amplifier has a phase shift of 180°. At a particular frequency the total phase shift of the three lead networks is _____, which results in a phase shift around the entire loop of 360°. Phase-shift oscillators can also work with three _____ networks.

3. *(180°, lag)* The _____ oscillator is the standard oscillator circuit for all low frequencies from about 5 Hz to 1 MHz. With the lead-lag network there is one frequency where the phase shift is exactly 0°; this is called the _____ frequency.

4. *(Wien-bridge, resonant)* The Hartley oscillator uses a tapped inductor. The Colpitts oscillator relies on a _____ tap rather than an inductive tap. The Clapp oscillator produces a more stable _____ than the Colpitts. When the utmost accuracy and frequency stability are needed, a quartz-crystal oscillator is used.

5. *(capacitive, frequency)* When a quartz crystal is vibrating, it looks like a tuned circuit. The outstanding advantage of crystals compared with discrete *LC* tank circuits is the incredibly high _____, which can be over 10,000.

6. *(Q)* Motorboating is a putt-putt sound representing very low-frequency oscillations. The _____ is by way of the power supply. The cure for motorboating is to use a regulated power supply. Its extremely low impedance prevents oscillations.

7. *(feedback)* Electric or magnetic coupling can produce high-frequency _____. To prevent this, we can increase the distance between stages, or we can enclose each stage in a _____.

8. *(oscillations, shield)* Another cause of high-frequency oscillations is a _____ loop, a difference of potential between two ground points. One way to eliminate this problem is to use a single-point ground system.

9. *(ground)* Bypassing the power supplies is essential with IC circuits. Capacitors from 0.1 to 1 μF are usually adequate to prevent power-supply _____. These bypass capacitors should be mounted as close as possible to the IC.

10. *(feedback)* Outside the midband of a negative-feedback amplifier, the internal lag networks produce additional phase shift. At some high frequency the phase shift around the entire loop is 0° and the feedback becomes _____.

11. *(positive)* The only way to prevent oscillations outside the midband of a negative-feedback amplifier is to make the loop gain less than _____ when the phase shift reaches _____.

12. *(unity, 0°)* Thyristors are four-layer devices that use positive feedback to produce latching action. The only way to close a four-layer diode is with _____. The only way to open it is with low-current dropout.

13. *(breakover)* An SCR remains open until a trigger hits the gate. It then latches and remains closed even though the trigger is removed. The only way to open an SCR is to reduce its current below the _____ current.

14. *(holding)* The UJT closes when the emitter voltage exceeds the intrinsic standoff voltage. It then remains latched until the emitter current is reduced to less than the valley current.

problems

19-1. A phase-shift oscillator has three lead networks with $R = 500$ kΩ and $C = 50$ pF. What is the frequency of oscillation?

19-2. The Wien-bridge oscillator of Fig. 19-27a uses a lamp with the characteristics of Fig. 19-27b. How much output voltage is there?

19-3. Position D in Fig. 19-27a is the highest frequency range of the oscillator. We can vary the frequency by the ganged rheostats. What are the minimum and maximum frequencies of oscillation on this range?

19-4. Calculate the minimum and maximum frequency of oscillation for each position of the ganged switch of Fig. 19-27a.

19-5. To change the output voltage of Fig. 19-27a to a value of 6 V rms, what change can you make?

19-6. In Fig. 19-27a the break frequency of the amplifier with negative feedback is at least one decade above the highest frequency of oscillation. What is the break frequency?

19-7. Figure 19-28 shows one way to build a *phase-shift oscillator*. At the frequency of oscillation, the phase shift of the 741C is $-270°$; this is the sum of 180° plus an additional $-90°$ produced by the internal lag network of a 741C. Each external lag network (10 kΩ and 0.001 μF) produces $-45°$; in this way, the loop phase shift is $-360°$, or equivalently, 0°.

In other words, we get oscillations at the break frequency of the external lag networks. Calculate the approximate frequency using $1/2\pi RC$. (The 5-kΩ rheostat has nothing to with frequency; it allows you to change the output amplitude.)

19-8. What is the approximate value of dc emitter current in Fig. 19-29? The dc voltage from collector to emitter?

19-9. What is the approximate frequency of oscillation in Fig. 19-29? The value of B? For the oscillator to start, what is the minimum value of A?

19-10. A crystal has a fundamental frequency of 5 MHz. What is the approximate value of the first overtone frequency? The second overtone? The third?

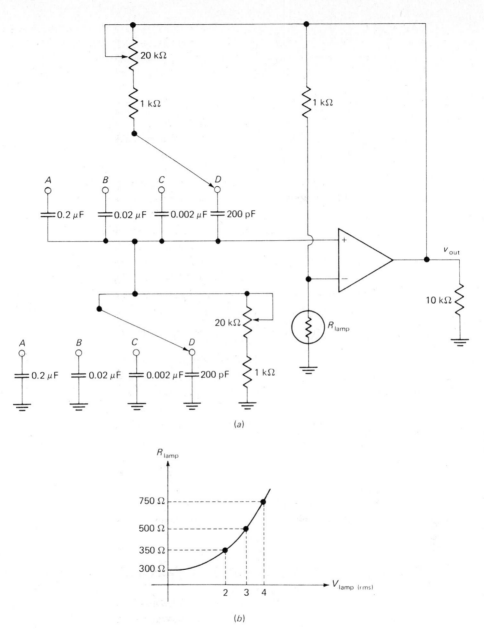

(a)

(b)

Figure 19-27.

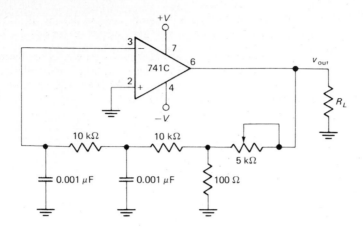

Figure 19-28.

19-11. A crystal has a thickness of t. If you reduce t by 1 percent, what happens to the frequency?

19-12. The ac equivalent circuit of a crystal has these values: $L = 1$ H, $C_s = 0.01$ pF, $R = 1000$ Ω, and $C_m = 20$ pF.
(a) What is the series resonant frequency?
(b) What is the Q at this frequency?

19-13. The output voltage of an oscillator is not easy to calculate because it depends on the f_T of the transistor and many other factors. You can get a rough estimate

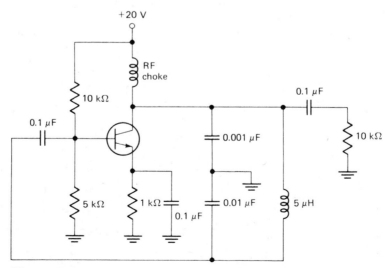

Figure 19-29.

as follows: the peak output voltage is V_{CEQ} or $I_{CQ}r_c$, whichever is smaller. If the r_c in Fig. 19-29 equals 1.5 kΩ, what is the peak output voltage?

19-14. To *pull* the crystal frequency means to change the frequency of oscillation slightly. Fig. 19-30 shows how it is done. The actual frequency of oscillation must lie between f_s and f_p. If the crystal has an f_s of 0.999 MHz and an f_p of 1.02 MHz, describe what you think the tuning capacitor is for.

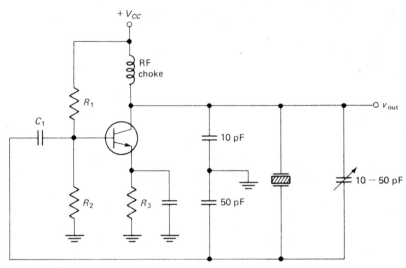

Figure 19-30.

19-15. The constant K of an X-cut crystal is 112.6 kHz in. What is the resonant frequency when t equals 0.1 in? When t equals 0.005 in?

19-16. If the 1N5160 of Fig. 19-31a is conducting, for what value of V will it stop? (Allow 0.7 V across the diode at the dropout point.)

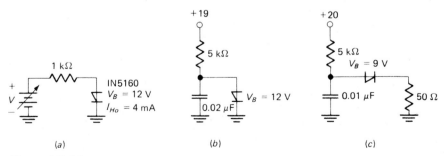

Figure 19-31.

19-17. With a supply of +19 V, it takes the capacitor of Fig. 19-31*b* exactly one time constant to charge to +12 V, the breakover voltage of the diode. What is the approximate frequency of the sawtooth output?

19-18. The current through the 50-Ω resistor of Fig. 19-31*c* is maximum just after the diode latches. If we allow 1 V across the latched diode, what is the maximum current?

19-19. The 2N4216 of Fig. 19-32*a* has a trigger current of 0.1 mA. If we allow 0.8 V across the gate-ground input, what value of V turns on the SCR?

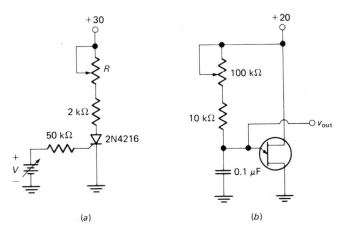

(a)   (b)

Figure 19-32.

19-20. The holding current of the 2N4216 of Fig. 19-32*a* is 3 mA. If the SCR is latched, for what value of R does the SCR stop conducting?

19-21. The intrinsic standoff ratio of the UJT shown in Fig. 19-32*b* is 0.63. What are the approximate values of the minimum and maximum output frequencies? (Assume flyback is down to 0 V.)

20 *voltage regulation*

Chapter 5 introduced the zener diode, a device used for voltage regulation. For voltage regulators capable of handling larger amounts of current, we need to combine the zener diode with negative-feedback amplifiers.

20-1 *simple regulators*

Figure 20-1 illustrates the idea of voltage regulation. Line voltage, nominally 120 V rms, drives an unregulated power supply like a full-wave rectifier with a capacitor-input filter. The voltage from this unregulated supply varies directly with line voltage. For instance, a 10 percent increase in line voltage results in approximately a 10 percent increase in the unregulated voltage. Furthermore, any changes in load current produce changes in unregulated voltage because of supply impedance, filter-capacitor discharge, etc.

This is where the *voltage regulator* comes in. As shown in Fig. 20-1, the voltage of the unregulated supply drives a voltage regulator, which produces the final output voltage. This output is ideally constant, varying neither with changes in line voltage nor in load current. In a practical regulator, the final output voltage is almost constant and the ripple is greatly reduced.

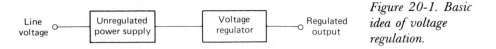

Figure 20-1. Basic idea of voltage regulation.

zener-diode regulator

A zener diode can be used for a voltage regulator. Figure 20-2 shows a bridge peak rectifier driving a zener-diode regulator. As you recall, the average voltage and ripple across the filter capacitor depend on the source resistance, the filter capacitance, and the load resistance. But as long as V_{IN} is greater than V_Z, the zener diode operates in the breakdown region. Ideally, this means the final output voltage is constant. To a second approximation, however, the zener impedance causes the final output to change slightly with changes in line voltage and load current.

The limitation on the zener-diode regulator is this. Changes in load current produce equal and opposite changes in zener current:

$$\Delta I_Z = -\Delta I_L \tag{20-1}$$

The changes in zener current flowing through the zener impedance produce changes in the final output voltage:

$$\Delta V_L = Z_Z \Delta I_Z \tag{20-2}***$$

The larger the changes in zener current, the larger the changes in the output voltage. If the changes in zener current are only a few milliamperes, the changes in load voltage may be acceptable. But when the changes are tens of milliamperes or more, the changes in load voltage become too large for most applications.

zener diode and emitter follower

What do we do if the changes in zener current are too large? The simplest approach is to add an *emitter follower,* as shown in Fig. 20-3. The load voltage

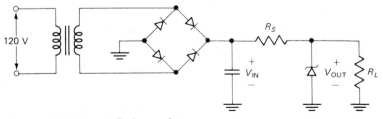

Figure 20-2. Zener-diode regulator.

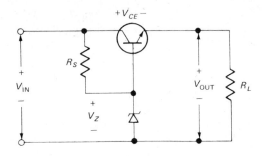

Figure 20-3. Zener diode and emitter follower.

still equals the zener voltage (less the V_{BE} drop of the transistor), but the changes in zener current are reduced by a factor of β:

$$\Delta I_Z = \frac{-\Delta I_L}{\beta} \tag{20-3}$$

When necessary, we can use a Darlington pair to get a larger β.

This circuit is an example of a *series* voltage regulator. The collector-emitter terminals are in series with the load. Because of this, the load current must *pass* through the transistor, and this is the reason the transistor is often called a *pass transistor*. The voltage across the pass transistor equals

$$V_{CE} = V_{IN} - V_{OUT} \tag{20-4}\star\star\star$$

and its power dissipation is

$$P_D = (V_{IN} - V_{OUT})\, I_L \tag{20-5}\star\star\star$$

Series regulators are the most widely used type.

EXAMPLE 20-1
In Fig. 20-3, $V_Z = 7.5$ V, $\beta = 100$, and $R_L = 100\ \Omega$. What is the load voltage? The load current? *(V_{IN} is greater than V_Z.)*

SOLUTION
The load voltage equals the zener voltage minus the V_{BE} drop:

$$V_{OUT} \cong 7.5\ \text{V} - 0.7\ \text{V} = 6.8\ \text{V}$$

The load current is

$$I_L = \frac{V_{OUT}}{R_L} = \frac{6.8\ \text{V}}{100\ \Omega} = 68\ \text{mA}$$

EXAMPLE 20-2

In the preceding example, $Z_{ZT} = 6 \ \Omega$. If R_L changes from 100 to 50 Ω, what is the change in load current? The change in zener current? The approximate change in load voltage?

SOLUTION

When $R_L = 50 \ \Omega$, the load current is

$$I_L = \frac{V_{OUT}}{R_L} = \frac{6.8 \ \text{V}}{50 \ \Omega} = 136 \ \text{mA}$$

Therefore, the change in load current is

$$\Delta I_L = 136 \ \text{mA} - 68 \ \text{mA} = 68 \ \text{mA}$$

The change in zener current is

$$\Delta I_Z = \frac{-\Delta I_L}{\beta} = \frac{-68 \ \text{mA}}{100} = -0.68 \ \text{mA}$$

And the approximate change in load voltage is

$$V_{OUT} = Z_Z \Delta I_Z = 6 \ \Omega \times (-0.68 \ \text{mA})$$
$$= -4.08 \ \text{mV}$$

20-2 *sp regulation*

In critical applications, zener voltages near 6 V are used because the temperature coefficient approaches zero. The highly stable zener voltage, sometimes called the *reference voltage,* can be amplified with an SP negative-feedback amplifier to get higher voltages with essentially the same temperature stability as the reference voltage.

a discrete sp regulator

Figure 20-4*a* shows a discrete SP regulator. Transistor Q_2 acts like an emitter follower as before. Transistor Q_1 provides voltage gain in a negative-feedback loop. Here is the basic idea behind circuit operation. Suppose the load voltage tries to increase. The feedback voltage V_F will increase. Since the emitter voltage of Q_1 is held constant by the zener diode, more collector current flows through Q_1; most of it flows through R_3 and causes the base voltage of Q_2 to decrease. In response, the emitter voltage of Q_2 decreases, offsetting almost all the original increase in load voltage.

Figure 20-4*b* shows the circuit redrawn so we can recognize the amplifier

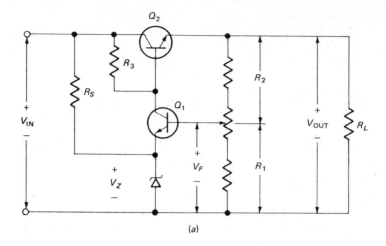

(a)

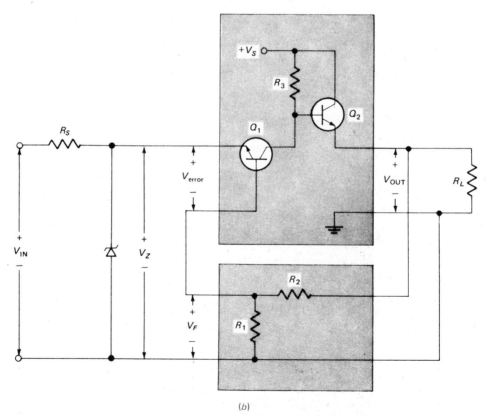

(b)

Figure 20-4. SP voltage regulator. (a) *Actual circuit.* (b) *Redrawn to emphasize* **A** *and* **B** *blocks.*

and feedback sections. The output voltage is fed back to the input side. Because of the SP negative feedback, the closed-loop gain is

$$A_{SP} \cong \frac{R_2}{R_1} + 1$$

Including the V_{BE} drop of Q_1, the final regulated load voltage equals

$$V_{OUT} = A_{SP}(V_Z + V_{BE}) \qquad (20\text{-}6)***$$

This means we can use a low zener voltage where the temperature coefficient approaches zero and still have a higher output voltage with an equally good temperature coefficient.

The potentiometer of Fig. 20-4a allows us to adjust the output voltage to the exact value required in a particular application. In this way, we can adjust for the tolerance in zener voltages, V_{BE} drops, and feedback resistors.

Here's an example. A 1N3157 has a zener voltage of 8.4 V $\pm$ 5 percent and a temperature coefficient of 0.001 percent per°C. If R_2/R_1 can be adjusted from 2 to 3 in Fig. 20-4a, what is the approximate range in regulated output voltage? The closed-loop voltage gain equals $R_2/R_1 + 1$; therefore, A_{SP} can be adjusted between 3 and 4. Using the nominal value of zener voltage and allowing 0.7 V for the V_{BE} drop of Q_1, the lower limit is

$$V_{OUT} = 3 \times (8.4 + 0.7) = 27.3 \text{ V}$$

and the upper limit is

$$V_{OUT} = 4 \times (8.4 + 0.7) = 36.4 \text{ V}$$

current limiting

The SP voltage regulator of Fig. 20-4a is a series regulator. As it now stands, it has no *short-circuit* protection. If we accidentally place a short across the load terminals, we get an enormous current through Q_2. Either Q_2 will be destroyed or a diode in the unregulated supply will burn out, or both. To avoid this possibility, regulated supplies often include *current limiting*.

Figure 20-5 shows one way to limit load current to safe values even though the output terminals are shorted. For normal load currents, the voltage drop across R_4 is small and Q_3 is off; under this condition, the regulator works as previously described. If excessive load current flows, however, the voltage across R_4 becomes large enough to turn on Q_3. The collector current of Q_3 flows through R_3; this decreases the base voltage of Q_2 and reduces the output voltage to prevent damage.

In Fig. 20-5 current limiting starts when the voltage across R_4 is around 0.6 to 0.7 V. At this point, Q_3 turns on and decreases the base drive for Q_2. Since R_4 is 1 Ω, current limiting begins when load current is in the vicinity of 600

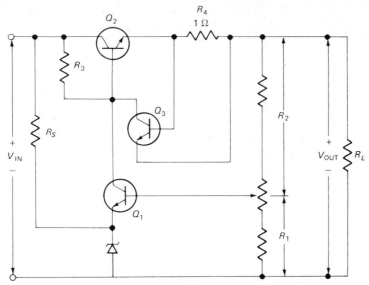

Figure 20-5. SP regulator including current limiting.

to 700 mA. By selecting other values of R_4, we can change the level of current limiting.

The current limiting of Fig. 20-5 is a simple example of how it is done. In more advanced approaches, Q_3 may be replaced by an op amp to increase the sharpness of the current limiting.

an op-amp regulator

In Fig. 20-5, Q_1 provides the open-loop gain. As discussed earlier, the greater the open-loop gain, the more precise the closed-loop gain. Because of this, we can improve performance by using an op amp in the place of Q_1.

Figure 20-6 shows an op-amp regulator. Zener voltage V_Z drives the noninverting input, and feedback voltage V_F drives the inverting input. Because we have eliminated the V_{BE} drop of Q_1, the regulated output voltage is

$$V_{OUT} = A_{SP}V_Z \qquad (20\text{-}7)\text{***}$$

where $A_{SP} = R_2/R_1 + 1$. This means the regulated output is as constant as the zener voltage. With temperature-compensated zener diodes, therefore, we can build extremely good voltage regulators. As before, we can use a Darlington pair in the place of Q_2 to increase current-handling capability.

A final point. To reduce the changes in zener voltage caused by changes in unregulated voltage, we can use the regulated output to drive the zener series-

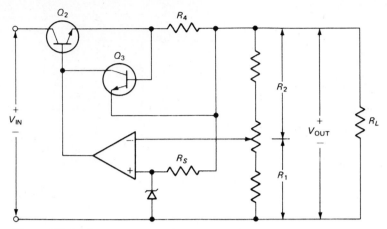

Figure 20-6. An op-amp regulator.

limiting resistor R_S as shown in Fig. 20-6. In this way, we get an almost rock-solid regulated output.

20-3 *early ic regulators*

In the late 1960s, IC manufacturers began producing a voltage regulator on a chip. These first-generation devices like the μA723 and LM300 include a zener diode, a high-gain amplifier, current-limiting, and other useful features. You supply the unregulated input voltage; the IC regulator then produces an almost constant output voltage.

Figure 20-7a shows the pin diagram of the LM300 (metal-can package). By connecting an external resistor to pin 1, we can set the level where current-limiting occurs. Pin 2, the booster output, is where an external pass transistor is connected to increase the load-current capability. The unregulated input goes to pin 3, while pin 4 is grounded. Pin 5 is for an external capacitor to bypass the zener diode; this reduces the noise it generates. Feedback voltage V_F goes to pin 6. We need to connect a compensating capacitor to pin 7; this rolls off the voltage gain at a rate of 20 dB per decade and prevents oscillations. Finally, we get the regulated output voltage from pin 8.

Figure 20-7b is a basic regulator that can handle load currents in tens of milliamperes. The 10-Ω current-limiting resistor sets the current limit at approximately 25 mA. The 0.1-μF capacitor bypasses the zener diode and reduces its noise. And the 47-pF compensating capacitor rolls off the voltage gain, preventing unwanted oscillations.

To increase the load current to hundreds of milliamperes, we need to add

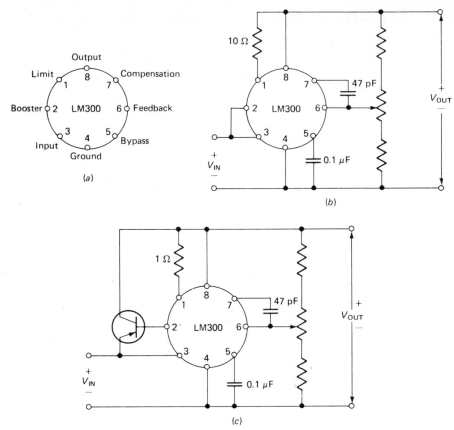

Figure 20-7. (a) *LM300 pin diagram.* (b) *Simple voltage regulator.* (c) *Boosting current capacity.*

an external pass transistor as shown in Fig. 20-7c. Here, the 1-Ω resistor sets the current limit around 250 mA. A circuit like this has better than 1 percent load and line regulation.

By adding a Darlington pair or other compound connection of transistors, we can increase the load-current capability to more than 5 A. Additional advantages of these early-generation regulators are adjustable output voltage and fast response to load and line transients.

The disadvantage of these early regulators is the need for external components, plus the eight terminals or pins that have to be connected in various ways to get what we want. Ideally, we should not have to connect many external components. Furthermore, an IC regulator ought to be a three-terminal device: one pin for the unregulated input voltage, another for the regulated output voltage, and the third for common (ground).

20-4 *three-terminal voltage regulators*

The newest IC regulators are three-terminal devices that can supply load currents from 100 mA to more than 3 A. These new IC regulators, available in plastic or metal packages, are easy to use and virtually blow-out proof.

the 340 series

The LM340 series is typical of the new breed of three-terminal positive voltage regulators. Figure 20-8 shows the block diagram. The built-in reference voltage drives the noninverting input of an amplifier. The feedback voltage comes from an internal voltage divider, preset to give output voltages from 5 to 24 V. Specifically, the following output voltages are available in the 340 series: 5, 6, 8, 10, 12, 15, 18, and 24 V. The chip includes a pass transistor that can handle more than 1.5 A of load current, provided adequate heat sinking is used. Also included are current limiting and *thermal shutdown.*

Thermal shutdown occurs when the internal temperature reaches 175°. At this point, the regulator turns off and prevents any further increase in chip temperature. Thermal shutdown is a precaution against excessive power dissipation, which depends on the ambient temperature, the heat sinking used, and other variables. Because of thermal shutdown and current limiting, the 340 series is almost indestructible.

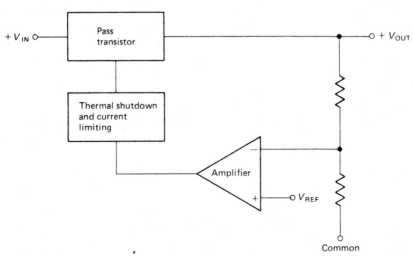

Figure 20-8. Equivalent circuit of 340 series.

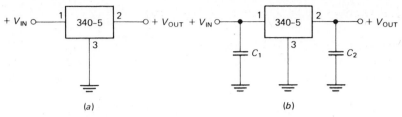

Figure 20-9. 340 voltage regulator.

a simple regulator

Figure 20-9a shows the 340 in its simplest configuration, a fixed voltage regulator. For concreteness, we have shown a 340-5, the device with a 5-V output (tolerance is typically ±2 percent). Pin 1 is the input, pin 2 is the output, and pin 3 is ground. A circuit like this not only regulates output voltage, it attenuates ripple. The 340-5 has a typical ripple rejection of 80 dB, equivalent to an attenuation of 10,000.

When the IC is more than a few inches from the supply filter capacitor, the lead inductance may produce oscillations within the IC because of power-supply feedback. This is why you often see a bypass capacitor C_1 on pin 1 (see Fig. 20-9b). Typical values for this bypass capacitor are from 0.1 to 1 μF. To improve the transient response of the 340, an output bypass capacitor C_2 is used; it's typically around 1 μF.

A 340-5 will regulate over an input range of 7 to 20 V. On the other hand, the 340-24 holds it output at 24 V for an input range of 27 to 38 V. In general, any device in the 340 series needs an input voltage at least 2 to 3 V greater than the regulated output; otherwise, it stops regulating.

two more applications

Figure 20-10a shows external components added to a 340 to get an adjustable output voltage. The common terminal of the 340 is not grounded, but rather is connected to the top of R_2. This means the regulated output V_{REG} is across R_1. A quiescent current I_Q comes out of pin 3 and flows down through R_2. Therefore, the output voltage from pin 2 to ground is

$$V_{OUT} = V_{REG} + \left(I_Q + \frac{V_{REG}}{R_1} \right) R_2 \qquad (20\text{-}8)\text{***}$$

Since I_Q shows little variation with line and load changes, the foregoing equation says V_{OUT} is regulated and adjustable. (For the 340 series, I_Q has a maximum value of 8 mA and varies only 1 mA over all line and load changes.)

Figure 20-10b is another application, this time a current source (or current

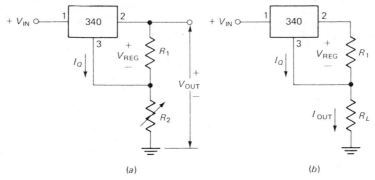

Figure 20-10. 340 applications. (a) Adjustable output voltage. (b) Current regulation.

regulator). A load resistor R_L takes the place of R_2. As before, the quiescent current I_Q and the R_1 current flow through R_L to ground. Therefore, the load current is

$$I_{OUT} = I_Q + \frac{V_{REG}}{R_1} \qquad (20\text{-}9)\text{***}$$

As an example, suppose $V_{REG} = 5$ V and $R_1 = 10 \; \Omega$. Then, I_{OUT} is approximately 500 mA, large enough to swamp out the small variations in I_Q. In other words, I_{OUT} is essentially constant and independent of R_L. This means we can change R_L and still have a fixed output current.

the 320 series

The 320 series is a group of negative voltage regulators with preset voltages from −5 to −24 V. Like the 340 series, load-current capability is approximately 1.5 A with adequate heat sinking. The 320 series is similar to the 340 series, and includes features like current limiting, thermal shutdown, and excellent ripple rejection.

By combining a 320 and a 340, we can regulate the output of a split supply (see Fig. 20-11). The 340 regulates the positive output, and the 320 handles

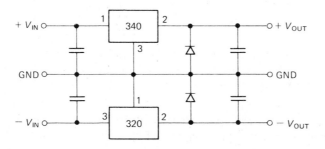

Figure 20-11. A split-supply voltage regulator.

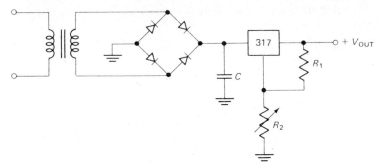

Figure 20-12. Bridge rectifier driving 317 regulator.

the negative output. The input capacitors prevent oscillations, and the output capacitors improve transient response. The diodes ensure that both regulators can turn on under all operating conditions.

the 317 adjustable regulator

The LM317 is a three-terminal positive voltage regulator that can supply approximately 1.5 A of load current over an adjustable output range of 1.25 to 37 V. The load regulation is 0.1 percent. The line regulation is 0.01 percent per volt; this means the output voltage changes only 0.01 percent for each volt of input change. The ripple rejection is 80 dB, equivalent to 10,000.

Figure 20-12 shows an unregulated supply driving a typical 317 circuit. The output voltage is given by

$$V_{OUT} = 1.25 \left(\frac{R_2}{R_1} + 1 \right) \qquad (20\text{-}10)***$$

Notice the unregulated supply has only a single filter capacitor, typical with the IC regulators now available. Since the regulator has about 80 dB of ripple rejection, it acts like a filter as well as a voltage regulator.

self-testing review

Read each of the following and provide the missing words. Answers appear at the beginning of the next question.

1. As long as V_{IN} is greater than V_Z, a zener diode operates in the _____ region. Ideally, this means the output voltage is constant. To a second approximation, the zener _____ causes the final output to change slightly with changes in line voltage and load current.

2. *(breakdown, impedance)* If changes in zener current are only a few milliamperes, the changes in load voltage may be acceptable. But when the changes in zener current

are tens of milliamperes, the changes in load _____ are too _____ for most applications.

3. *(voltage, large)* To reduce the changes in zener current, we can add an emitter _____. This reduces the zener current by a factor of beta. A circuit like this is an example of a _____ regulator, and the transistor is called a _____ transistor.

4. *(follower, series, pass)* In an SP regulator, the reference voltage drives the noninverting input and the _____ voltage drives the inverting input. To provide short-circuit protection, we can add another transistor to get _____ limiting.

5. *(feedback, current)* In an op-amp SP regulator, the open-loop gain is very _____, which means the output voltage is as constant as the reference voltage. To increase current-handling ability, we can use a _____ pair for the pass transistor.

6. *(high, Darlington)* First-generation IC regulators like the μA723 and LM300 include a zener diode, a high-gain _____, and current limiting. The main disadvantage of these early regulators is the number of external _____.

7. *(amplifier, components)* The latest IC regulators are three-terminal devices that can supply load currents from 100 mA to more than 3 A. Although external components are not normally needed, an input bypass _____ may be needed to prevent _____ caused by the lead inductance between the unregulated supply and the IC regulator.

8. *(capacitor, oscillations)* The LM340 series is typical of the new breed of three-terminal voltage regulators. It has many features including _____ shutdown, which prevents excessive power dissipation. It also has a ripple rejection of 80 dB, equivalent to an attenuation of _____.

9. *(thermal, 10,000)* The simplest application of a 340 device is a positive voltage regulator of preset voltage. By adding a fixed and variable external resistor, we can get an _____ output voltage. The 340 can also be connected to act like a _____ source.

10. *(adjustable, current)* The 320 series is the negative analog of the 340 series. With a 320 and a 340, we can regulate the output of a _____ supply; the 340 regulates the positive voltage, and the 320 handles the negative voltage.

11. *(split)* The LM317 is a three-terminal positive voltage regulator that can supply 1.5 A over a large adjustable output voltage range. The load regulation is 0.1 percent, and the line regulation is 0.01 percent per volt. Since the 317 has 80 dB of ripple rejection, it acts like a filter as well as a voltage regulator.

problems

20-1. In Fig. 20-13, $V_{IN} = 10$ V and $V_{BE} = 0.7$ V. What is V_{OUT}?

20-2. V_{IN} changes from 10 to 15 V in Fig. 20-13. If $Z_{ZT} = 5$ Ω, what is the approximate change in output voltage?

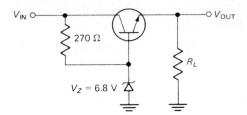

Figure 20-13.

20-3. If $\beta = 50$ and $R_L = 22$ Ω in Fig. 20-13, what is the base current? The zener current for $V_{IN} = 10$ V?

20-4. R_L changes from 22 to 500 Ω in Fig. 20-13. If $\beta = 50$ and $Z_{ZT} = 5$ Ω, what is the approximate change in output voltage? The approximate voltage regulation expressed in percent?

20-5. In Fig. 20-13, V_{IN} varies from 10 to 15 V. If R_L can vary from 22 to 500 Ω, what is the maximum power dissipation in the pass transistor?

20-6. In the SP regulator of Fig. 20-4a, $R_2 = 1000$ Ω and $R_1 = 250$ Ω. If $V_Z = 6.2$ V, what does the regulated output voltage equal?

20-7. In Fig. 20-4a, the feedback circuit consists of the following resistors (from top to bottom): 3.9 kΩ, 5-kΩ potentiometer, and 8.2 kΩ. If $V_Z = 7.5$ V, what is the adjustable range of V_{OUT}?

20-8. Q_3 just turns on when its $V_{BE} = 0.63$ V in Fig. 20-6. If current limiting is to start for a load current of 2 A, what is the required value of R_4?

20-9. In Fig. 20-6, $R_2 = 1.5$ kΩ, $R_1 = 700$ Ω, and $V_Z = 5.6$ V. What is the regulated output voltage?

20-10. In the SP regulator of Fig. 20-6, the feedback resistors have the following values: 4.7 kΩ, 5-kΩ potentiometer, and 6.8 kΩ (these are from top to bottom). If $V_Z = 5.6$ V, what is the adjustable range of output voltage?

20-11. In Fig. 20-6, $V_{IN} = 30$ V, $R_2 = 10$ kΩ, $R_1 = 5$ kΩ, $V_Z = 8.2$ V, and $R_L = 100$ Ω. What is the power dissipation in the pass transistor Q_2 if $R_4 = 1$ Ω?

20-12. The 340T-12 of Fig. 20-14a has a preset output voltage of 12 V. (The T stands for plastic tab package, similar to Fig. 10-18b.) If the tolerance on the regulated

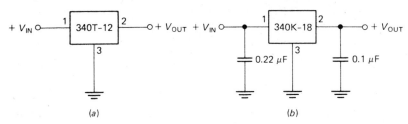

(a)   (b)

Figure 20-14.

output voltage is ±2 percent at $T_J = 25°$ and $V_{IN} = 20$ V, what are the minimum and maximum values of output voltage?

20-13. The data sheet of a 340T-12 indicates the output voltage is a nominal 12 V. The change in output voltage from no load to full load is 32 mV. What is the voltage regulation?

20-14. The 340T-12 of Fig. 20-14a has a ripple rejection of 72 dB. If the input ripple is 2 V rms, what is the output ripple? The ripple factor?

20-15. In Fig. 20-14b, the 340K-18 has a preset output voltage of 18 V. (The K stands for metal package, similar to Fig. 10-18a.) If the regulated output changes 38 mV from no load to full load, what is the voltage regulation?

20-16. The ripple rejection of the 340K-18 shown in Fig. 20-14b is 68 dB. If the input ripple is 3 V rms, what is the output ripple? The ripple factor?

20-17. If the line regulation of the 340K-18 of Fig. 20-14b is 0.01 percent per volt, how much does the output change when the input changes from 23 to 30 V?

20-18. In Fig. 20-15a, what is the adjustable range of output voltage if I_Q is 8 mA?

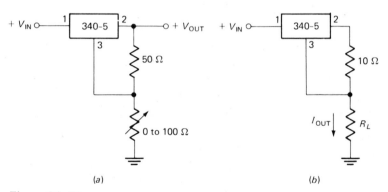

(a)   (b)

Figure 20-15.

20-19. The 50-Ω resistor of Fig. 20-15a is changed to 82 Ω. If $I_Q = 8$ mA, what is the adjustable range of output voltage?

20-20. In Fig. 20-15b, $I_Q = 8$ mA. What is the current through R_L?

20-21. If the 10-Ω resistor of Fig. 20-15b is changed to 15 Ω, what is the load current for an I_Q of 8 mA?

20-22. Figure 20-16 shows a 317 regulator with electronic shutdown. When the shutdown voltage is zero, the transistor is cut off and has no effect on the operation. But when the shutdown voltage is approximately 5 V, the transistor saturates.

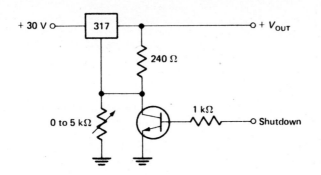

Figure 20-16. Regulator with electronic shutdown.

What is the adjustable range of output voltage when the shutdown voltage is zero? What does the output voltage equal when the shutdown voltage is 5 V?

20-23. The transistor of Fig. 20-16 is cut off. To get an output voltage of 15 V, what value should the adjustable resistor have?

21
op-amp applications

IC op amps are inexpensive, versatile, and easy to use. For this reason, they are used not only for negative-feedback amplifiers, but also for waveshaping, filtering, and mathematical operations. This chapter discusses some common uses of op amps.

21-1 *comparators*

The simplest way to use an op amp is open-loop (no feedback resistors), as shown in Fig. 21-1a. Because of the high gain of the op amp, the slightest error voltage (typically in microvolts) produces maximum output swing. For instance, when V_1 is greater than V_2, the error voltage is positive and the output voltage goes to its maximum positive value, typically 1 to 2 V less than the supply voltage. On the other hand, if V_1 is less than V_2, the output voltage swings to its maximum negative value.

Figure 21-1b summarizes the action. A positive error voltage drives the output to $+V_{SAT}$, the maximum positive value of output voltage. A negative error voltage produces an output of $-V_{SAT}$. When an op amp is used like this, it's called a *comparator* because all it can do is compare V_1 to V_2, producing a saturated positive or negative output, depending on whether V_1 is greater or less than V_2.

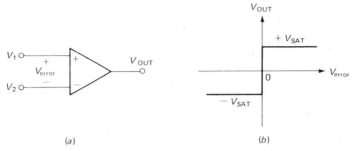

Figure 21-1. Comparator.

go/no-go detector

Figure 21-2*a* is one application of a comparator: a *go/no-go detector*. A reference voltage drives the inverting input. When V_{IN} is greater than V_{REF}, the output goes into positive saturation. If V_{IN} is less than V_{REF}, the output goes into negative saturation.

Figure 21-2*b* shows one way to set up the reference voltage and detect the output state. When V_{IN} exceeds V_{REF}, the output goes positive and turns on the green LED; this indicates the "go" condition. On the other hand, if V_{IN} is less than V_{REF}, the output goes negative and turns on the red LED, indicating "no-go." By selecting proper values for R_1 and R_2, we can set up any desired positive reference. If a negative reference is needed, we would connect the voltage divider to the negative supply.

input bias current

In Fig. 21-2*b* the reference voltage is ideally given by

$$V_{REF} = \frac{R_1}{R_1 + R_2} V \tag{21-1}$$

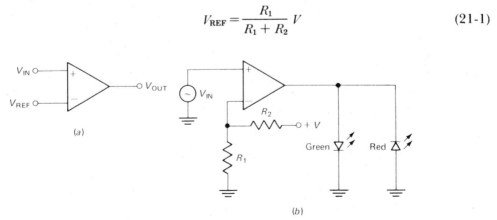

Figure 21-2. Go/no-go detector.

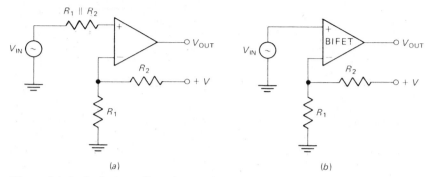

Figure 21-3. Reducing effect of input bias current. (a) *Resistive compensation.* (b) *BIFET.*

This is ideal because we are ignoring the input bias current from the inverting input. When this current flows through the equivalent Thevenin resistance of R_1 in parallel with R_2, it produces a small voltage which introduces some error into Eq. (21-1). In many applications, the error is negligible and Fig. 21-2b is practical as it stands.

In critical applications, however, we need to do something about the unbalance caused by input bias currents. One solution is *resistive compensation.* Figure 21-3a shows how it's done. We have added a resistance $R_1 \parallel R_2$ to the noninverting input. Now, the input bias currents flow through the same resistances, producing approximately the same common-mode voltage. As a result, the small voltages produced by input bias currents tend to cancel out.

Figure 21-3b shows an easier way to eliminate the effects of input bias currents: use a BIFET op amp. As you recall, these are op amps with a JFET input stage followed by bipolar stages. With a JFET, the input bias currents are so small that resistive compensation is not needed. This is one reason BIFET op amps are so popular; they eliminate most of the problems caused by input bias currents and input offset currents.

squaring circuit

Comparators are useful in *waveshaping,* producing a particular waveform from various input waveforms. Figure 21-4 is one example of a waveshaper. Since the inverting input is grounded, the reference voltage is zero. Therefore, when the input sine wave goes slightly positive, the output saturates positively. When the input goes slightly negative, the output saturates negatively. This is why the input sine wave produces an output square wave. Other symmetrical inputs like a triangular wave also result in a square-wave output.

An op amp like a 741 would be acceptable in Fig. 21-4, provided the slew rate is fast enough for the input frequency. But in most cases, a comparator

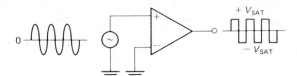

Figure 21-4. Squaring circuit.

should be able to change states as quickly as possible; this means the higher the slew rate, the better the comparator action.

In general, comparators are *uncompensated* op amps; they have no internal compensating capacitor to roll off the voltage gain. Furthermore, comparators are optimized for certain features not found in general-purpose op amps. Manufacturers' catalogs list comparators under a separate heading; examples of widely used IC comparators are the LM306, 311, and 710.

21-2 *amplifiers*

This section is about some common circuits used with noninverting and inverting amplifiers. Both configurations can act as voltage or current sources.

noninverting amplifier

The *noninverting* amplifier of Fig. 21-5a has high input impedance, low output impedance, and a stable voltage gain given by

$$\frac{V_{\text{OUT}}}{V_{\text{IN}}} = \frac{R_2}{R_1} + 1 \qquad (21\text{-}2)\text{***}$$

(Capital letters are used for V_{OUT} and V_{IN} because op amps can work directly with dc signals.) The noninverting amplifier of Fig. 21-5a is popular because it approaches the ideal voltage amplifier.

Figure 21-5b is the *voltage follower,* widely used because of its superb buffering qualities: extremely high input impedance, extremely low output impedance, and unity voltage gain. Because the negative feedback is maximum in a voltage follower, the bandwidth equals f_{unity}.

At times we want to source a fixed amount of current through a load. Figure 21-5c shows one way to do it. Since the error voltage is negligibly small, essentially all of V_{IN} appears across R, producing a current of

$$I_{\text{OUT}} = \frac{V_{\text{IN}}}{R}$$

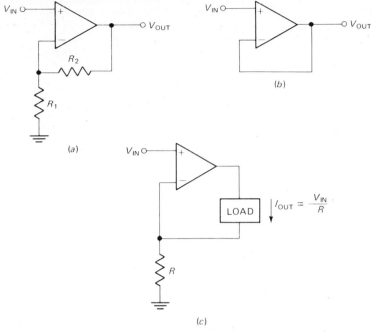

Figure 21-5. (a) Noninverting amplifier. (b) Voltage follower. (c) Current source.

All this current must flow through the load because negligible current flows into the op-amp inverting input. Depending on the application, the load may be a resistor, capacitor, inductor, or combination.

inverting amplifier

Figure 21-6a shows the *inverting* amplifier, a very popular op-amp circuit. The inverting terminal is at virtual ground, meaning its voltage with respect to ground is approximately zero. But since the virtual ground cannot sink current, all the input current is forced through R_2. As a result,

$$V_{IN} = I_{IN} R_1$$
$$V_{OUT} = -I_{IN} R_2$$

The minus sign occurs because of phase inversion. Taking the ratio of the preceding equations gives the voltage gain:

$$\frac{V_{OUT}}{V_{IN}} = -\frac{R_2}{R_1} \qquad\qquad (21\text{-}3)\text{***}$$

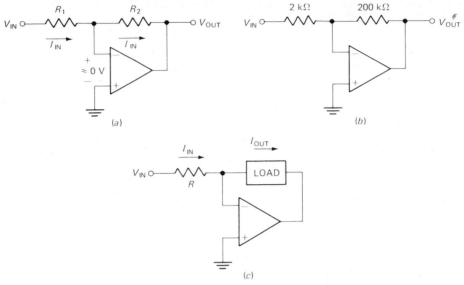

Figure 21-6. (a) *Inverting amplifier.* (b) *Example.* (c) *Current source.*

Also, the virtual ground means the input impedance is

$$Z_{IN} = R_1 \qquad (21\text{-}4)***$$

One reason for the popularity of the inverting amplifier is this: it allows us to set up a precise value of input impedance as well as voltage gain. There are many applications where we want to nail down the input impedance, along with the voltage gain. As an example, suppose we need an input impedance of 2 kΩ and a voltage gain of 100. Then, a circuit like Fig. 21-6*b* does the job.

Figure 21-6*c* shows the inverting amplifier used to source current through a load. Because of the virtual ground,

$$I_{OUT} = \frac{V_{IN}}{R}$$

This allows us to set up a precise value of I_{OUT}. The load may be a resistor, capacitor, inductor, or combination.

more than one input

Another advantage of the inverting amplifier is its ability to handle more than one input at a time, as shown in Fig. 21-7*a*. Because of the virtual ground,

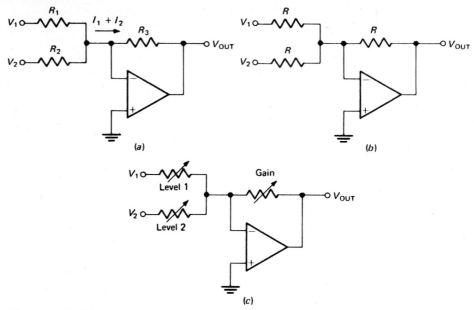

Figure 21-7. More than one input.

$$I_1 = \frac{V_1}{R_1}$$

$$I_2 = \frac{V_2}{R_2}$$

and
$$V_{OUT} = -(I_1 + I_2)R_3$$

or
$$V_{OUT} = -\frac{R_3}{R_1} V_1 - \frac{R_3}{R_2} V_2 \qquad (21\text{-}5)$$

This means we can have a different gain for each input; the output is the sum of the amplified inputs. The same idea applies to any number of inputs; add another input resistor for each new input signal.

Often, we need a circuit that adds two or more signals. In this case, we can use a *summer,* an inverting amplifier with several inputs all with unity gain. Figure 21-7*b* shows how it's done for two inputs. In this case, Eq. (21-5) reduces to

$$V_{OUT} = -(V_1 + V_2) \qquad (21\text{-}6)$$

Finally, Fig. 21-7*c* shows a convenient way to additively mix two signals. The adjustable input resistors allow us to set the level of each input, and the gain

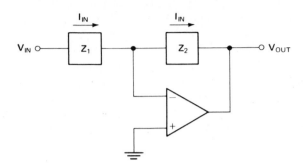

Figure 21-8. Inverting amplifier with complex impedances.

control allows us to control the output. A circuit like this is useful for combining audio signals.

using impedances with the inverting amplifier

Virtual ground, an unusually powerful feature, is the reason for the inverting amplifier's versatility. Figure 21-8 is another example of why virtual ground leads to important applications. In this circuit, the input and output voltages are phasors, and $\mathbf{Z}_1$ and $\mathbf{Z}_2$ are complex impedances. Again, the virtual ground is the key that allows us to write

$$\mathbf{V}_{IN} = \mathbf{I}_{IN}\mathbf{Z}_1$$

and

$$\mathbf{V}_{OUT} = -\mathbf{I}_{IN}\mathbf{Z}_2$$

The ratio is the complex voltage gain:

$$\frac{\mathbf{V}_{OUT}}{\mathbf{V}_{IN}} = -\frac{\mathbf{Z}_2}{\mathbf{Z}_1} \qquad (21\text{-}7)\text{***}$$

Why is this important? Because the complex voltage gain (also known as the transfer function) equals the ratio of two external impedances that we can select. This allows us to build *active filters*. More is said about this later.

21-3 *active diode circuits*

Op amps can enhance the performance of the diode circuits of Chap. 5. For one thing, an op amp eliminates the effect of diode offset voltage, allowing us to rectify, peak-detect, clip, and clamp *low-level* signals (those with amplitudes less than the diode offset voltage). And because of their buffering action, op amps can eliminate the effects of source and load on diode circuits. What follows is a discussion of *active* diode circuits, ones that use op amps to improve their performance.

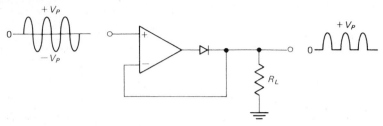

Figure 21-9. Active half-wave rectifier.

half-wave rectifier

Figure 21-9 is an active half-wave rectifier. When the input signal goes positive, the output goes positive and turns on the diode. The circuit then acts like a voltage follower, and the positive half cycle appears across the load resistor. On the other hand, when the input goes negative, the op-amp output goes negative and turns off the diode. Since the diode is open, no voltage appears across the load resistor. This is why the final output is almost a perfect half-wave signal.

The high gain of the op amp virtually eliminates the effect of diode offset voltage. For instance, if $\phi = 0.7$ V and $A = 100{,}000$, the input that just turns on the diode is

$$V_{IN} = \frac{0.7\ \text{V}}{100{,}000} = 7\ \mu\text{V}$$

When the input is greater than 7 μV, the diode turns on and the circuit acts like a voltage follower. The effect is equivalent to reducing the offset potential by a factor of A. In symbols,

$$\phi' = \frac{\phi}{A} \qquad\qquad (21\text{-}8)\text{***}$$

where ϕ' is the offset potential seen by the input signal.

Because ϕ' is so small, the active half-wave rectifier is useful with low-level signals. For instance, if we want to measure sinusoidal voltages in the millivolt region, we can add a milliammeter in series with the R_L of Fig. 21-9. With the proper value of R_L, we can calibrate the meter to indicate rms millivolts.

active peak detector

To peak-detect small signals, we can use an active peak detector like Fig. 21-10a. Again, the input offset potential ϕ' is in the microvolt region, which means we can peak-detect millivolt signals.

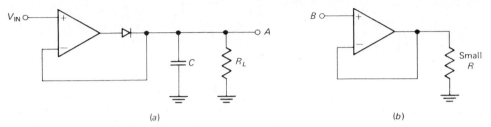

Figure 21-10. Active peak detector.

Here's the circuit action. When the diode is on, the heavy SP negative feedback produces a Thevenin output impedance approaching zero. For this reason, the charging time constant $R_{TH}C$ shrinks to a negligibly small value, eliminating source effects. Furthermore, the discharging time constant R_LC can be made much longer than the period of the input signal; this results in almost perfect peak detection.

If the peak-detected signal has to drive a small load, we can avoid loading effects by using an op-amp buffer. For instance, by connecting point A of Fig. 21-10a to point B of Fig. 21-10b, the final load is isolated from the peak detector.

One more idea. If possible, the R_LC time constant should be at least 100 times longer than the period of the lowest input frequency. If this condition is satisfied, the output voltage will be within 1 percent of the peak input.

active positive clipper

Figure 21-11 is an active positive clipper. With the wiper all the way to the left, V_{REF} is zero and the noninverting input is grounded. When V_{IN} goes positive, the error voltage drives the op-amp output negative and turns on the diode. This means the final output V_{OUT} is at virtual ground for any positive value of V_{IN}.

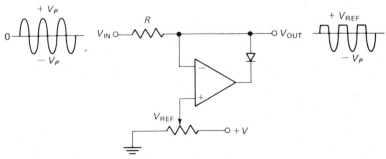

Figure 21-11. Positive clipper.

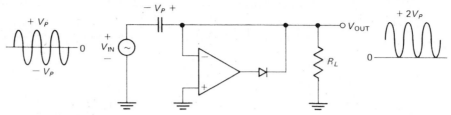

Figure 21-12. Positive clamper.

When V_{IN} goes negative, the op-amp output is positive, which turns off the diode and opens the loop. As this happens, the virtual ground is lost, and the final output V_{OUT} is free to follow the negative half cycle of input voltage. This is why the negative half cycle appears at the output.

To change the clipping level, all we do is adjust V_{REF} as needed. In this case, clipping occurs at V_{REF} as shown in Fig. 21-11.

As usual, the offset voltage is reduced to ϕ' at the input, which means the circuit is suitable for low-level inputs.

active positive clamper

Figure 21-12 is an active positive clamper. The first negative half cycle produces a positive op-amp output which turns on the diode. This allows the capacitor to charge to the peak value of the input with the polarity shown in Fig. 21-12.

Just beyond the negative peak, the diode turns off, the loop opens, and the virtual ground is lost. With Kirchhoff's voltage law,

$$V_{OUT} = V_{IN} + V_P$$

Since V_P is being added to a sinusoidal input voltage, the final output waveform is shifted positively through V_P volts. In other words, we get the positive clamped waveform of Fig. 21-12; it swings from 0 to $2V_P$. Again, the reduction of offset voltage allows excellent clamping with low-level inputs.

21-4 *special amplifiers*

Let's take a look at three special ways to amplify low-level signals: the differential amplifier, instrumentation amplifier, and the logarithmic amplifier.

differential amplifier

Figure 21-13 shows an op amp connected as a differential amplifier. The gain of the inverting input is

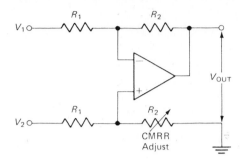

Figure 21-13. Differential amplifier.

$$\frac{V_{\text{OUT}}}{V_1} = -\frac{R_2}{R_1} \qquad (21\text{-}9)$$

The gain of the noninverting input is

$$\frac{V_{\text{OUT}}}{V_2} = \left(\frac{R_2}{R_1 + R_2}\right)\left(\frac{R_2}{R_1} + 1\right)$$

which reduces to

$$\frac{V_{\text{OUT}}}{V_2} = \frac{R_2}{R_1} \qquad (21\text{-}10)$$

Therefore, the gains are equal for each single-ended input.

If a differential input is used, the output voltage is given by

$$V_{\text{OUT}} = \frac{R_2}{R_1}\left(V_2 - V_1\right) \qquad (21\text{-}11)\text{***}$$

Note that the lower R_2 is adjustable in Fig. 21-13. This allows us to optimize the common-mode rejection ratio.

instrumentation amplifier

Figure 21-14 is an example of an *instrumentation amplifier,* a diff amp optimized for high input impedance, high CMRR, and other properties. In this simple version of an instrumentation amplifier, the input signals drive voltage followers, which then drive a diff amp. An amplifier like this is useful at the front end of measuring instruments because of its high input impedance.

Manufacturers can put the voltage followers and diff amps on a single chip to get an integrated instrumentation amplifier. The LF352 is a good example. This BIFET IC has JFETs at the input, followed by bipolar transistors. This results in an input impedance of approximately 2×10^{12} and input bias currents of only 3 pA. The JFETs also have extremely low noise, an essential characteristic of a good instrumentation amplifier. The LF352 has other outstanding features

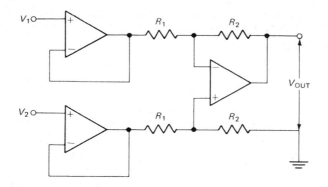

Figure 21-14. Instru-mentation amplifier.

like a CMRR of at least 110 dB, a supply current of only 1 mA, and a single external resistor to control gain.

logarithmic amplifier

Figure 21-15a is a *logarithmic* amplifier (log amp). Because of the virtual ground, I_{IN} is forced through the transistor, so that

$$I_C = I_{IN} = \frac{V_{IN}}{R}$$

The virtual ground also implies

$$V_{OUT} = -V_{BE}$$

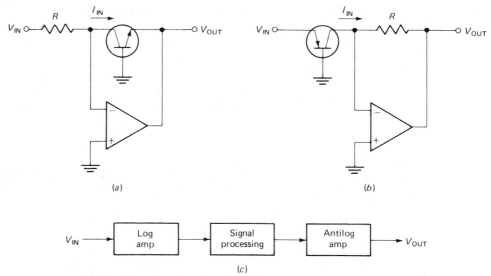

Figure 21-15. (a) *Log amp.* (b) *Antilog amp.* (c) *Signal processing.*

Because the transconductance curve of a transistor is exponential, the output voltage is logarithmically related to the input voltage. An advanced derivation shows that at room temperature,

$$V_{OUT} = -0.06 \log \frac{V_{IN}}{I_S R} \qquad (21\text{-}12)$$

where I_S is the reverse saturation current of the base-emitter diode, and the logarithm is to the base 10. The minus sign indicates phase inversion.

Equation (21-12) tells us the following. Each time the input voltage increases by a factor of 10 (a decade), the output voltage increases by 60 mV. One application of this is decibel conversion; each 60-mV increase in output voltage means the input voltage has increased 20 dB. Another application is range compression; signals with extremely wide dynamic range will saturate a linear amplifier but not a log amplifier.

After range compression, a signal can be processed as desired in a particular application. Then the original range can be restored with an *antilogarithmic* amplifier (antilog amp). Figure 21-15b is an example. In this case, changes in V_{IN} produce exponential changes in I_{IN}. And because of the current-to-voltage conversion,

$$V_{OUT} = -I_{IN} R$$

The exponential changes in input current have the effect of stretching the dynamic range of the input signal, equivalent to taking the antilogarithm. It can be shown that

$$V_{OUT} = -I_S R \text{ antilog} \frac{V_{IN}}{0.06} \qquad (21\text{-}13)$$

Figure 21-15c summarizes the idea of range compression, signal processing, and range decompression. An input signal whose amplitude can vary over five or more decades drives the log amp. (Such a signal would saturate a linear amp.) The log-amp output can then be processed (amplified, detected, wave-shaped, etc.). The antilog amp decompresses the signal range, restoring its linearity.

EXAMPLE 21-1
The log amp of Fig. 21-15a has an $I_S R$ of 1 mV. Calculate the output voltage for the following input voltages: 10 mV, 100 mV, 1 V, 10 V, and 100 V.

SOLUTION
For $V_{IN} = 10$ mV, Eq. (21-12) gives

$$V_{OUT} = -0.06 \log \frac{10 \text{ mV}}{1 \text{ mV}} = -0.06 \text{ V}$$

For $V_{IN} = 100$ mV,

$$V_{OUT} = -0.06 \log \frac{100 \text{ mV}}{1 \text{ mV}} = -0.12 \text{ V}$$

Each decade increase in input voltage makes the output 60 mV more negative; therefore, the remaining outputs are -0.18 V, -0.24 V, and -0.30 V.

Notice the range compression. An input range of 10 mV to 100 V produces an output range of only -0.06 to -0.30 V.

21-5 *the miller integrator*

Figure 21-16a is a Miller *integrator*. Because of the Miller effect, the effective input capacitance is

$$C_{IN} = C(A + 1)$$

and the time constant is

$$\tau = RC(A + 1)$$

Since A is typically more than 50,000, we can get incredibly long time constants.

The virtual ground gives

$$I_{IN} = \frac{V_{IN}}{R}$$

This input current flows to the capacitor and results in this rate of voltage change:

$$\frac{dV_{OUT}}{dt} = \frac{-I_{IN}}{C}$$

or

$$\frac{dV_{OUT}}{dt} = \frac{-V_{IN}}{RC} \qquad (21\text{-}14)\text{***}$$

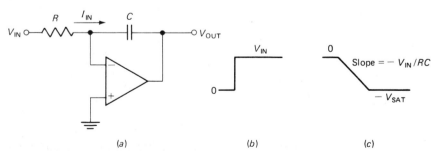

(a)   (b)   (c)

Figure 21-16. (a) *Miller integrator.* (b) *Input.* (c) *Output.*

The main use of a Miller integrator is to generate *ramps,* linearly changing output voltages. To get a ramp output, we need to use a step-voltage input as shown in Fig. 21-16b. This produces a constant input current, which forces the output to slew negatively as shown in Fig. 21-16c. Since the slope equals $-V_{IN}/RC$, we can control the rate of change by varying V_{IN} or by using different RC values. Linear ramps like this are used in digital voltmeters, oscilloscopes, and many other applications.

21-6 *the voltage-controlled oscillator*

A later chapter discusses the *phase-locked loop,* an important circuit in television, FM stereo, and space satellites. An essential part of a phase-locked loop is its *voltage-controlled oscillator* (VCO). Figure 21-17a shows one way to build a VCO. It's a Miller integrator with a four-layer diode across the capacitor. Since a

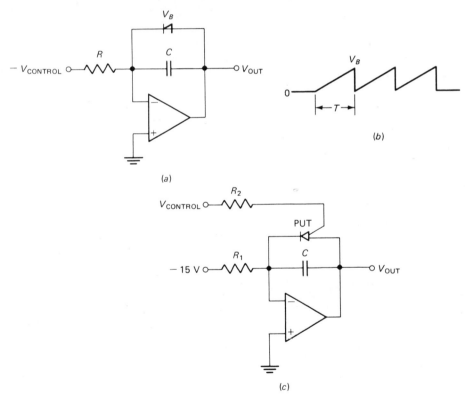

Figure 21-17. (a) *VCO with four-layer diode.* (b) *Sawtooth output.* (c) *VCO with PUT control.*

negative input voltage is used, the output ramp is positive-going, as shown in Fig. 21-17*b*.

The action is straightforward. When the ramp voltage reaches the breakover voltage, the diode latches and discharges the capacitor. The flyback ideally reaches 0 V, where the next cycle begins. This is why we get a train of sawtooth waves.

In Fig. 21-17*b*, the slope *m* of the ramp is

$$m = \frac{V_B}{T}$$

With Eq. (21-14) and $V_{IN} = -V_{CONTROL}$, the preceding equation becomes

$$\frac{V_{CONTROL}}{RC} = \frac{V_B}{T}$$

And since $T = 1/f$, we can rearrange the equation to get

$$f = \frac{V_{CONTROL}}{RCV_B} \tag{21-15}$$

This says frequency is directly proportional to control voltage. Therefore, we have a *voltage-to-frequency converter,* another name for a VCO.

Figure 21-17*c* shows another way to build a VCO. This time, a PUT is across the capacitor. The control voltage is applied to the PUT and determines the intrinsic standoff voltage; therefore, $V_{CONTROL}$ determines the frequency of the sawtooth output. With a circuit like Fig. 21-17*c*, we can control the output frequency over a large range.

21-7 *active filters*

By combining reactive elements and op amps, we can build *active filters.* These have advantages like voltage gain, negligible loading effects, and elimination of inductors.

low-pass filter

Figure 21-18*a* shows an active *low-pass filter.* Here's what it does. At low frequencies the capacitor appears open, and the circuit acts like an inverting amplifier with a voltage gain of $-R_2/R_1$. As the frequency increases, the capacitive reactance decreases, causing the voltage gain to drop off. As the frequency approaches infinity, the capacitor appears shorted and the voltage gain approaches zero.

Figure 21-18*b* illustrates the output response. The output signal is maximum

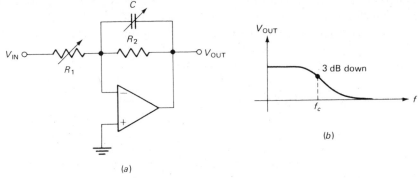

Figure 21-18. (a) *Active low-pass filter.* (b) *Output response.*

at low frequencies. When the frequency reaches the critical or cutoff frequency, the output is down 3 dB. Beyond this frequency, the gain rolls off at an ideal rate of 20 dB per decade or 6 dB per *octave* (a factor of two change in frequency).

Here's how to find the cutoff frequency. With Eq. (21-7),

$$\frac{V_{OUT}}{V_{IN}} = -\frac{Z_2}{Z_1}$$

where the minus sign stands for phase inversion. The complex admittance of the feedback network is

$$Y_2 = \frac{1}{R_2} + j\omega C = \frac{1 + j\omega R_2 C}{R_2}$$

The reciprocal of this is the impedance:

$$Z_2 = \frac{R_2}{1 + j\omega R_2 C}$$

The ratio of impedances is

$$\frac{Z_2}{Z_1} = \frac{R_2}{R_1} \frac{1}{1 + j\omega R_2 C}$$

and the magnitude of ratio is

$$\frac{Z_2}{Z_1} = \frac{R_2}{R_1} \frac{1}{\sqrt{1 + (\omega R_2 C)^2}}$$

The 3-dB corner frequency occurs when

$$\omega R_2 C = 1$$

or

$$f_c = \frac{1}{2\pi R_2 C} \qquad (21\text{-}16)\text{***}$$

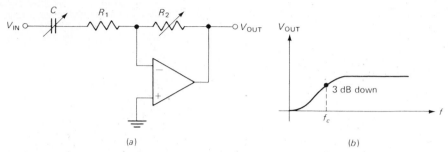

Figure 21-19. (a) *Active high-pass filter.* (b) *Output response.*

The adjustable C of Fig. 21-18a allows us to vary the corner frequency, and the adjustable R_1 lets us control the gain. If a fixed response is desired, we can eliminate the adjustments and use a fixed R_1 and C. Because of the voltage feedback, the Thevenin output impedance approaches zero, which means the active filter can drive low-impedance loads.

high-pass filter

Figure 21-19a is an active *high-pass filter.* At low frequencies the capacitor appears open, and the voltage gain approaches zero. At high frequencies the capacitor appears shorted, and the circuit becomes an inverting amplifier with a voltage gain of $-R_2/R_1$. Figure 21-19b is the response. By taking the ratio of impedances, we can derive this formula for the cutoff frequency:

$$f_c = \frac{1}{2\pi R_1 C} \qquad (21\text{-}17)\text{***}$$

faster rolloff filters

The preceding filters roll off at a rate of 20 dB per decade. By adding more components, we can increase the rolloff rate. Figure 21-20a shows a low-pass filter. At low frequencies both capacitors appear open, and the circuit becomes a voltage follower. As the frequency increases, the gain eventually breaks and is down 3 dB at the f_c. Because of the two capacitors, the rolloff rate beyond f_c is ideally 40 dB per decade. With an advanced derivation, it can be shown that

$$f_c = \frac{1}{2\pi RC\sqrt{2}} \qquad (21\text{-}18)$$

Figure 21-20b is a fast rolloff high-pass filter. At low frequencies the capacitors appear open and the voltage gain approaches zero. At high frequencies the

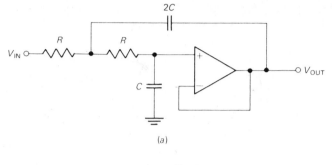

(a)

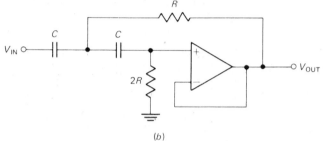

(b)

Figure 21-20. (a) *40-dB/decade low-pass filter.* (b) *40-dB/decade high-pass filter.*

capacitors appear shorted and the circuit becomes a voltage follower. Again, the presence of two capacitors causes the rolloff rate to become 40 dB per decade. The cutoff frequency is given by

$$f_c = \frac{1}{2\pi RC\sqrt{2}} \qquad (21\text{-}19)$$

Besides the filters described, it is possible to come up with bandpass and bandstop filters. Such filters are discussed in books specializing in op-amp theory and application.

self-testing review

Read each of the following and provide the missing words. Answers appear at the beginning of the next question.

1. An op amp used open-loop is called a _____. All it does is compare V_1 to V_2, producing a _____ positive or negative output, depending on whether V_1 is greater or less than V_2.

2. *(comparator, saturated)* In critical applications we can use resistive compensation to reduce the unbalance produced by input _____ currents. Resistive compensation can be used with any op-amp circuit discussed in this chapter. Another solution to the input-bias current problem is to use a _____ op amp.

3. *(bias, BIFET)* The inverting amplifier has a voltage gain of $-R_2/R_1$, and an input impedance of _____. An advantage of the inverting amplifier is its ability to handle more than _____ input at a time.

4. *(R_1, one)* Active diode circuits eliminate the effect of the diode _____ voltage, allowing us to rectify, peak-detect, clip, and clamp low-level signals.

5. *(offset)* An instrumentation amplifier is a diff amp optimized for high input impedance, high _____, and other properties. A _____ amplifier compresses the dynamic range of the input signal.

6. *(CMRR, logarithmic)* The main use of a Miller integrator is to generate _____, linearly changing output voltages. To get a ramp output, we need to use a step-voltage _____. Linear ramps are used in digital voltmeters, oscilloscopes, and other applications.

7. *(ramps, input)* An essential part of a phase-locked loop is its voltage-controlled _____. Another name for a VCO is a voltage-to-frequency _____.

8. *(oscillator, converter)* By combining reactive elements with op amps, we can build active filters. Twenty dB per decade is equivalent to 6 dB per octave. Active filters have advantages like voltage gain, negligible loading effects, and elimination of inductors.

problems

21-1. A sine wave drives the comparator of Fig. 21-21 a. If the open-loop gain is 200,000 and V_{SAT} is 12 V, what is the value of V_{IN} that just produces positive saturation?

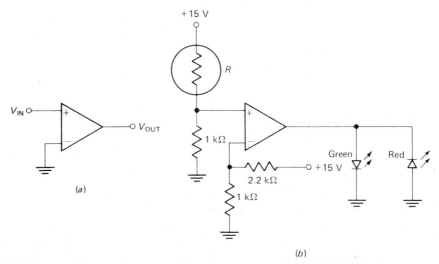

Figure 21-21.

21-2. In the preceding problem, the sine wave has a peak value of 5 V. If $V_{IN} = 5 \sin \omega t$, what does ωt equal in degrees when V_{IN} just produces positive saturation?

21-3. In Fig. 21-21a, the slew rate is 100 V per microsecond. If $V_{SAT} = 12$ V, how long will it take to slew from $-V_{SAT}$ to $+V_{SAT}$?

21-4. The go/no-go detector of Fig. 21-21b responds to temperature changes. The thermistor R has a value of 2 kΩ at 25°C. The temperature coefficient of the thermistor is -5 percent/°C. At what approximate temperature does the green LED just come on?

21-5. If the thermistor R of Fig. 21-21b has a value of 2 kΩ at 25°C and a temperature coefficient of -1 percent/°C, at what temperature does the red light just come on?

21-6. Thermistor R of Fig. 21-21b has a value of 2 kΩ. How much current is there through R? If R increases by 1 Ω, how much increase is there in the error voltage?

21-7. The input bias current is 100 nA in Fig. 21-21b. How much voltage does this bias current produce when it flows through the parallel combination of 1 kΩ and 2.2 kΩ?

21-8. If $R = 180$ kΩ in Fig. 21-22a, what is the voltage gain? To resistively compensate this circuit for input bias currents, how much resistance should we add between the noninverting input and ground?

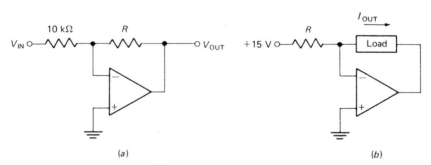

Figure 21-22.

21-9. $R = 68$ kΩ in Fig. 21-22a. If $V_{IN} = 15$ μV, what does V_{OUT} equal? What is the input impedance of the circuit? The input current?

21-10. What is the load current in Fig. 21-22b if $R = 6.8$ kΩ? What value of R do we need to get a load current of 2 mA?

21-11. A sine wave with a peak of 2 V drives the circuit of Fig. 21-23a. If the op amp has a gain of 100,000, what is the value of input voltage that just turns on the diode? What is the peak current through the load resistor? The average current over the cycle?

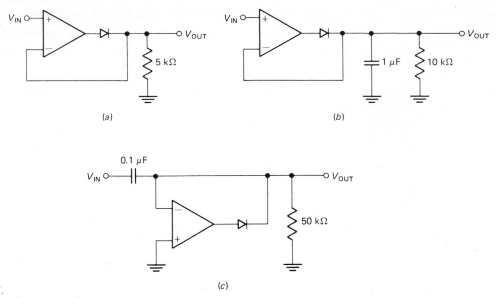

(a)

(b)

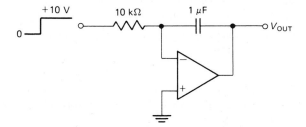

(c)

Figure 21-23.

21-12. If a sine wave with a peak of 500 mV drives the circuit of Fig. 21-23*b*, what is the approximate output voltage for normal operation? The discharging time constant is 100 times greater than the period of the input signal. What is the input frequency?

21-13. The input signal of Fig. 21-23*c* is a sine wave with a peak of 300 mV. What are the minimum and maximum values of output voltage? For good operation, the $R_L C$ time constant should be at least 100 times greater than the input period. What is the lowest frequency allowed?

21-14. The input signal to a log amp increases 60 dB. How much does the output voltage change? If the input signal changes by 120 dB, what is the change in output voltage?

21-15. If the op amp of Fig. 21-24 has a voltage gain of 250,000, what is the input Miller capacitance? The time constant?

Figure 21-24.

21-16. What is the rate of output voltage change for the circuit of Fig. 21-24? If the 10-kΩ resistor is changed to 100 kΩ, what is the rate of output voltage change?

21-17. If the capacitor of Fig. 21-24 is changed from 1 μF to 0.1 μF, what is the rate of output voltage change?

21-18. The output of a circuit like Fig. 21-24 can be used to provide the time base of an oscilloscope. By changing resistors and capacitors with range switches, we can produce different sweep speeds.
    (a) If $V_{IN} = 10$ V and $R = 1$ kΩ, what are the rates of output voltage change for capacitor values of 100 pF, 200 pF, 500 pF, and 1000 pF?
    (b) If $V_{IN} = 10$ V and $C = 1$ μF, what are the rates of output voltage change for resistor values of 1 kΩ, 2 kΩ, 5 kΩ, and 10 kΩ?

21-19. The step-input voltage of Fig. 21-24 is replaced by a dc input of -15 V. A four-layer diode with $V_B = 12$ V is connected across the capacitor. What is the frequency of the output sawtooth wave? If the dc input is changed to -10 V, what is the output frequency?

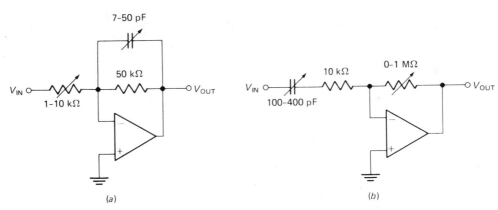

Figure 21-25.

21-20. What is the maximum voltage gain in Fig. 21-25a? The minimum voltage gain? Calculate the range of cutoff frequency.

21-21. What are the minimum and maximum voltage gains in Fig. 21-25b? The range of cutoff frequency?

$\mathit{22}$ *the frequency domain*

In our previous work, we have emphasized *time-domain analysis* by working out voltages from one instant to the next. But time-domain analysis is not the only method. This chapter discusses *frequency-domain analysis.*

22-1 *the fourier series*

Figure 22-1a shows a sine wave with a peak V_P and a period T. Ac-dc books concentrate on the sine wave because it is the most fundamental. Earlier chapters have likewise emphasized the sine wave. Now, we are ready to examine nonsinusoidal waves.

periodic waves

The *triangular* wave of Fig. 22-1b traces its basic pattern during period T; after this, each cycle is a repetition of the first cycle. It is the same with the sawtooth of Fig. 22-1c and the half-wave signal of Fig. 22-1d; all cycles are copies of the first cycle. Waveforms with repeating cycles are called *periodic;* they have a period T in which the size and shape of every cycle is determined.

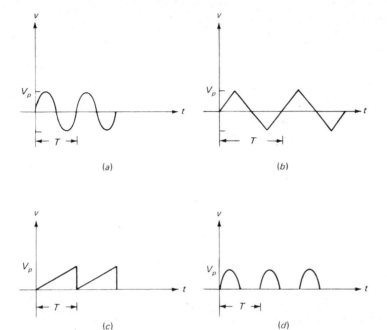

Figure 22–1. Periodic waves.

harmonics

The sine wave is extraordinary. By adding sine waves with the right amplitude and phase, we can produce the triangular wave of Fig. 22-1b. With a different combination of sine waves we can get the sawtooth wave of Fig. 22-1c. And with still another combination of sine waves we can produce the half-wave signal of Fig. 22-1d. In other words, *any periodic wave is a superposition of sine waves.*

These sine waves are harmonically related, meaning their frequencies are *harmonics* (multiples) of a *fundamental* (lowest frequency). Given a periodic wave, you can measure its period T on an oscilloscope. The reciprocal of T equals the fundamental frequency. Symbolically,

$$f_1 = \frac{1}{T} \qquad \text{(fundamental)} \qquad (22\text{-}1)***$$

The second harmonic has a frequency of

$$f_2 = 2f_1$$

The third harmonic has a frequency of

$$f_3 = 3f_1$$

In general, the nth harmonic has a frequency of

$$f_n = nf_1 \qquad\qquad (22\text{-}2)\text{***}$$

As an example, Fig. 22-2 shows a sawtooth wave on the left. This periodic wave is equivalent to the sum of the harmonically related sine waves on the right. Also included is a battery to account for the average or dc value of the sawtooth wave. Since period T equals 2 ms, the harmonic frequencies equal

$$f_1 = \frac{1}{0.002} = 500 \text{ Hz}$$
$$f_2 = 2(500 \text{ Hz}) = 1000 \text{ Hz}$$
$$f_3 = 3(500 \text{ Hz}) = 1500 \text{ Hz}$$

and so on.

formula for fourier series

As a word formula, here is what Fig. 22-2 says:

Periodic wave = dc component + first harmonic
+ second harmonic + third harmonic + . . . + nth harmonic

In precise mathematical terms,

$$v = V_0 + V_1 \sin(\omega t + \phi_1) + V_2 \sin(2\omega t + \phi_2)$$
$$+ V_3 \sin(3\omega t + \phi_3) + . . . + V_n \sin(n\omega t + \phi_n)$$

This famous equation is called the *Fourier series*. It says a periodic wave is a superposition of harmonically related sine waves. Voltage v is the value of the

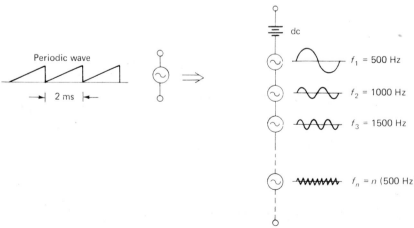

Figure 22–2. Periodic wave is a superposition of sine waves.

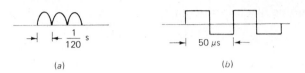

Figure 22–3.

periodic wave at any instant in time; we can calculate this value by adding the dc component and the instantaneous values of the harmonics.

The first term in the Fourier series is V_0; this is a constant and represents the dc component. The coefficients V_1, V_2, V_3, . . . , V_n are peak values of the harmonics. The angles ϕ_1, ϕ_2, ϕ_3, . . . , ϕ_n are phase angles of the harmonics. Radian frequency ω equals $2\pi f_1$; as we see, each succeeding term in the Fourier series represents the next higher harmonic.

Theoretically, the harmonics continue to infinity, that is, n has no upper limit. Often, five to ten harmonics are enough to synthesize a periodic wave to within 5 percent. With the right combination of amplitudes $(V_1, V_2, V_3, . . . , V_n)$ and angles $(\phi_1, \phi_2, \phi_3, . . . , \phi_n)$ we can produce any periodic waveform.

EXAMPLE 22-1
What frequencies do the harmonics of Fig. 22-3*a* have?

SOLUTION
The periodic wave is a full-wave-rectified signal with a period of 1/120 s. Therefore, the reciprocal of the period equals

$$f_1 = 120 \text{ Hz}$$

and the higher harmonics have frequencies of

$$f_2 = 240 \text{Hz}$$
$$f_3 = 360 \text{ Hz}$$

and so on.

EXAMPLE 22-2
What frequencies do the first three odd harmonics of Fig. 22-3*b* have?

SOLUTION
The period is 50 μs; therefore, the fundamental has a frequency of

$$f_1 = \frac{1}{50(10^{-6})} = 20 \text{ kHz}$$

The next odd harmonic has a frequency of

$$f_3 = 3(20 \text{ kHz}) = 60 \text{ kHz}$$

The third odd harmonic has a frequency of

$$f_5 = 5(20 \text{ kHz}) = 100 \text{ kHz}$$

EXAMPLE 22-3
Illustrate the superposition of harmonics to produce a square wave.

SOLUTION
The square wave has only odd harmonics (discussed in the next section). When we add the first and third harmonics shown by the dashed lines of Fig. 22-4a, we get a new wave. Already, this looks more like a square wave than a sine wave.

By adding more harmonics, we can approach the square wave. Figure 22-4b shows the addition of the fifth harmonic. If we continue like this, we eventually get a flat top and vertical sides.

The example brings out these ideas:

1. Each harmonic must have exactly the right amplitude and phase to produce a given periodic wave. For instance, if the peak value of the third harmonic in Fig. 22-4a is too large, the dip in the new wave will be too large; it will be impossible to correct for this excessive dip with higher harmonics.
2. The Fourier theorem (superposing sine waves to get a periodic wave) seems

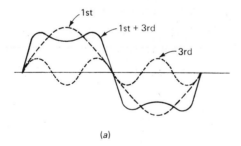

(a)

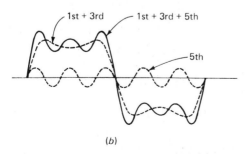

(b)

Figure 22-4. Adding sine waves to get a square wave.

plausible now. Limited as this example is, it does appear we can add sine waves to produce a periodic wave. The formal proof of the Fourier theorem is found in advanced books.

22-2 *the spectrum of a signal*

The Fourier theorem is the key to frequency-domain analysis. Since we already know a great deal about sine waves, we can reduce a periodic wave to its sine-wave components; then, by analyzing these sine waves, we are indirectly analyzing the periodic wave. In other words, there are two approaches in nonsinusoidal circuit analysis. We can figure out what a periodic wave does at each instant in time, or we can figure out what each harmonic does. Sometimes, the first approach (time-domain analysis) is faster. But often, the second approach (frequency-domain analysis) is superior.

spectral components

Suppose A represents the peak-to-peak value of a sawtooth wave. With advanced mathematics we can prove

$$V_n = \frac{A}{n\pi} \tag{22-3}$$

This says the peak value of the nth harmonic equals A divided by $n\pi$. For example, Fig. 22-5a shows a sawtooth wave with a peak-to-peak value of 100 V; the harmonics have these peak values:

$$V_1 = \frac{100}{\pi} = 31.8 \text{ V}$$

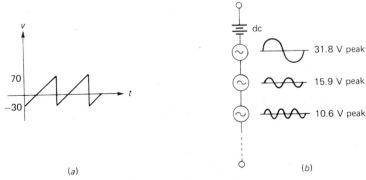

(a) (b)

Figure 22–5. Sawtooth wave and harmonics.

$$V_2 = \frac{100}{2\pi} = 15.9 \text{ V}$$

$$V_3 = \frac{100}{3\pi} = 10.6 \text{ V}$$

and so on. Figure 22-5b shows the first three harmonics for the sawtooth of Fig. 22-5a.

With an oscilloscope (a time-domain instrument) you see the periodic signal as a function of time (Fig. 22-5a). The vertical axis represents voltage and the horizontal axis stands for time. In effect, the oscilloscope displays the instantaneous value v of the periodic wave.

The *spectrum analyzer* differs from the oscilloscope. To begin with, a spectrum analyzer is a frequency-domain instrument; its horizontal axis represents frequency. With a spectrum analyzer we see the *harmonic peak values versus frequency*. For instance, if the sawtooth of Fig. 22-6a drives a spectrum analyzer, we will see the display of Fig. 22-6b. We call this kind of display a *spectrum;* the height of each line represents the harmonic peak value; the horizontal location gives the frequency.

Every periodic wave has a spectrum or set of vertical lines representing the harmonics. The spectrum normally differs from one periodic signal to the next. For instance, a square wave like Fig. 22-6c has a spectrum like Fig. 22-6d; this is different from the sawtooth spectrum of Fig. 22-6b.

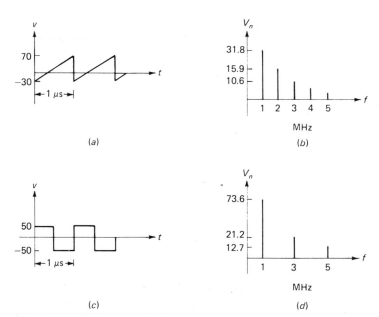

Figure 22–6. Waves and spectra.

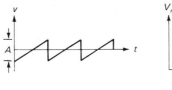
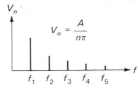

(a)

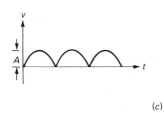
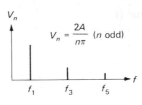

(b)

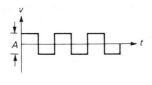
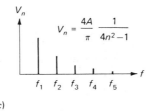

(c)

Figure 22–7. Some common waves and their spectra.

four basic spectra

For future reference, Fig. 22-7 shows four periodic waves and their spectra. In each of these, A is the peak-to-peak value of the periodic wave. For convenience, we have shown only the harmonics to $n = 5$. In each case, the formula for harmonic peak values is given. For example, with a full-wave signal like Fig. 22-7c, we can calculate the peak value of any harmonic by using

$$V_n = \frac{4A}{\pi} \frac{1}{4n^2 - 1} \qquad (22\text{-}4a)$$

Or if we want the peak values of harmonics for the square wave of Fig. 22-7b, we use

$$V_n = \frac{2A}{n\pi} \qquad (n \text{ odd only}) \qquad (22\text{-}4b)$$

half-wave symmetry

Of the waveforms in Fig. 22-7, two have *only odd harmonics*. Many other waveforms have this odd-harmonic property, and it helps to know a quick method for identifying such waveforms.

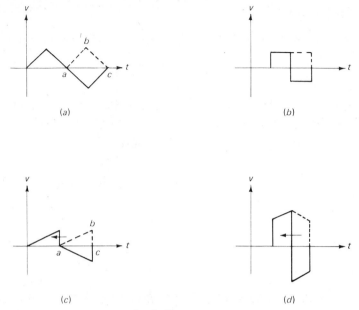

Figure 22–8. Testing for half-wave symmetry.

Any waveform with *half-wave symmetry* has the odd-harmonic property. Half-wave symmetry means you can invert the negative half cycle and get an exact duplicate of the positive half cycle. For example, Fig. 22-8*a* shows one period of a triangular wave. After inverting the negative half cycle, you have the *abc* half cycle (dashed line). This inverted half cycle is an exact duplicate of the positive half cycle. As a result, the triangular wave contains only odd harmonics. Similarly, the square wave of Fig. 22-8*b* has half-wave symmetry because the inverted negative cycle exactly duplicates the positive half; this means a square wave contains only odd harmonics.

In case of doubt, it helps to shift the inverted half cycle to the left; half-wave symmetry exists only if the shifted half cycle superimposes the positive half cycle. For instance, inverting the negative half cycle of a sawtooth wave produces the *abc* half cycle of Fig. 22-8*c*. When shifted left, the inverted half cycle exactly superimposes the positive half cycle; therefore, we are assured of half-wave symmetry and the odd-harmonic property.

The waveform of Fig. 22-8*d* does *not* have half-wave symmetry. When the negative half cycle is inverted and shifted left, it does not superimpose the positive half cycle. Therefore, Fig. 22-8*d* does not have the odd-harmonic property; it must contain at least one even harmonic.

the dc component

The dc component is the average value of the periodic wave, defined as

$$V_0 = \frac{\text{area under one cycle}}{\text{period}} \qquad (22\text{-}5)***$$

As an example, Fig. 22-9a shows a sawtooth wave with a peak of 10 V and a period of 2 s. The area under one cycle is shaded and equals

$$\text{Area} = \tfrac{1}{2}(\text{base} \times \text{height})$$
$$= \tfrac{1}{2}(2 \text{ s} \times 10 \text{ V}) = 10 \text{ V s}$$

Dividing by the period gives the average value of the sawtooth:

$$V_0 = \frac{10 \text{ V s}}{2 \text{ s}} = 5 \text{ V}$$

The area is positive when above the horizontal axis, but negative below. If part of a cycle is above and part below the horizontal axis, you algebraically add the positive and negative areas to get the net area under the cycle. Figure 22-9b shows a square wave that swings from $+70$ to -30 V. The first half cycle has an area above the horizontal axis; therefore, this area is positive. But the

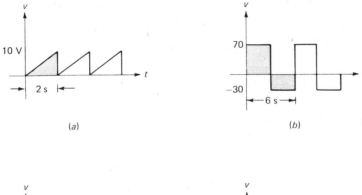

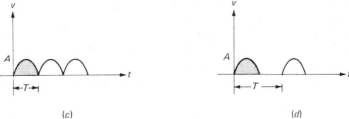

Figure 22–9. Areas and average values.

second half cycle is below the horizontal axis, so that the area is negative. Here is how to find the average value over the entire cycle:

Positive area = 3 s × 70 V = 210 V s
Negative area = 3 s × −30 V = −90 V s
Net area under one cycle = 210 − 90 = 120 V s

Dividing the net area by the period gives

$$V_0 = \frac{120 \text{ V s}}{6 \text{ s}} = 20 \text{ V}$$

This is the average value for the entire cycle of Fig. 22-9b; it is the value a dc voltmeter would read.

Using the area-over-period formula, you can calculate the average value of other periodic waves, provided the waves have linear segments. For instance, sawtooth, square, and triangular waves are made up of straight lines. Because of this, you can use well-known geometry formulas to calculate areas; after dividing by period, you have the average value.

But what do you do with half-wave and full-wave signals like Fig. 22-9c and d? The waveforms are nonlinear and no simple geometry formulas are available for areas. The only way to calculate the areas is with calculus; using calculus, we can derive these average values:

$$V_0 = 0.636A \qquad \text{(full wave)} \qquad (22\text{-}6)\text{***}$$

and
$$V_0 = 0.318A \qquad \text{(half wave)} \qquad (22\text{-}7)\text{***}$$

effect of dc component on spectrum

Figure 22-10a shows a triangular wave. Since it has half-wave symmetry, the spectrum contains only odd harmonics (Fig. 22-10b). If we add a dc component to the triangular wave, the wave moves up as shown in Fig. 22-10c. The only change in the spectrum is the appearance of a line at zero frequency (Fig. 22-10d). The height of this line represents the dc voltage. In general, adding a dc component to a waveform has no effect on the harmonics; the only spectral change is a new line at zero frequency.

EXAMPLE 22-4

A square wave has a peak-to-peak value of 25 V and a period of 5 μs. Calculate the peak value and frequency of the ninth harmonic.

SOLUTION

With Fig. 22-7b, the ninth-harmonic peak value is

$$V_n = \frac{2A}{n\pi} = \frac{2(25)}{9\pi} = 1.77 \text{ V}$$

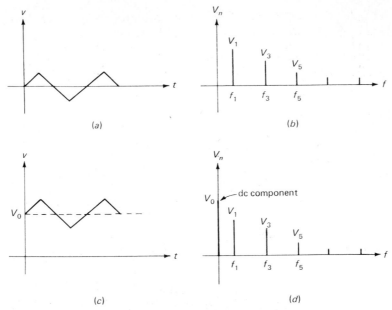

Figure 22–10. Effect of dc component on spectrum.

Since the period is 5 μs, the fundamental frequency equals

$$f_1 = \frac{1}{5(10^{-6})} = 200 \text{ kHz}$$

The frequency of the ninth harmonic is

$$f_9 = 9(200 \text{ kHz}) = 1.8 \text{ MHz}$$

EXAMPLE 22-5

A 100-V-peak sine wave drives a full-wave rectifier as shown in Fig. 22-11*a*. The ideal output is a full-wave signal with a 100-V peak. Calculate the dc component and peak values of the first three harmonics of rectified voltage.

SOLUTION

With Eq. (22-6),

$$V_0 = 0.636A = 0.636(100) = 63.6 \text{ V}$$

Referring to Fig. 22-7*c*,

$$V_n = \frac{4A}{\pi} \frac{1}{4n^2 - 1}$$

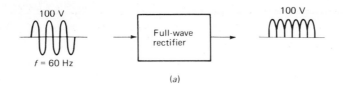

(a)

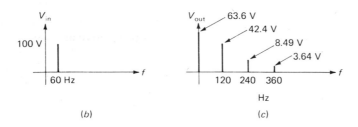

(b)

(c)

Figure 22–11. Full-wave rectification.

With this, we can calculate the peak values:

$$V_1 = 42.4 \text{ V}$$
$$V_2 = 8.49 \text{ V}$$
$$V_3 = 3.64 \text{ V}$$

EXAMPLE 22-6

The input sine wave of Fig. 22-11a has a frequency of 60 Hz. Calculate the first three harmonic frequencies of the rectified voltage and show the input and output spectra.

SOLUTION

The period of the input signal is 1/60 s. The full-wave signal has a period of half this value, or 1/120 s. Therefore, the fundamental frequency of the rectified output is

$$f_1 = 120 \text{ Hz}$$

and the next two harmonic frequencies are

$$f_2 = 240 \text{ Hz}$$

and

$$f_3 = 360 \text{ Hz}$$

The spectrum of a sine wave is a single line at the frequency of the sine wave. For this reason, the input spectrum contains only a single line with a peak value of 100 V and a frequency of 60 Hz (Fig. 22-11b).

Figure 22-11c shows the spectrum of rectified output voltage. The first line is the dc component with a value of 63.6 V (found in Example 22-5). The other lines are the harmonics of the rectified output signal; we have shown only the first three harmonics. Especially important, the fundamental frequency in the output spectrum is 120 Hz, exactly double the input frequency. This doubling effect happens in all full-wave rectifiers.

22-3 *filters*

Filters pass some sinusoidal frequencies but stop others. We first mentioned filters in our discussion of rectifiers; the idea was to remove all sinusoidal components, leaving only the dc component. Another example of a filter is the tuned class C amplifier. The collector current is a pulse waveform containing all harmonics of the fundamental frequency. The tuned tank removes all but one of the harmonics. This section describes filters and their effects on the spectrum of a signal.

low-pass filter

A *low-pass filter* passes low frequencies but stops high ones. Figure 22-12*a* shows the ideal response of a low-pass filter; this is a graph of voltage gain versus frequency. As we see, the gain is unity from zero up to the *cutoff frequency* f_c.

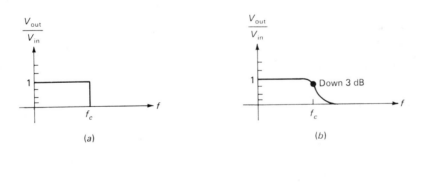

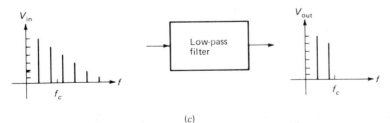

Figure 22–12. Low-pass filtering.

Beyond f_c, the voltage gain ideally drops to zero. The *passband* is the set of frequencies between zero and f_c; the *stopband*, the set of frequencies greater than f_c.

Any input sine wave with a frequency in the passband will appear at the output of the filter. But any sine wave whose frequency is greater than f_c cannot appear at the ouput of the low-pass filter. For instance, suppose a periodic signal has the spectrum shown in Fig. 22-12c. All sinusoidal frequencies up to f_c pass through the filter; all those beyond f_c are stopped. This is why the output spectrum contains only frequencies below f_c.

The ideal response of Fig. 22-12a is impossible to attain in practice, but we can get close. A real filter has a response like Fig. 22-12b. By definition, the gain is down 3 dB at the cutoff frequency. (The voltage gain in the passband may be slightly less than unity because of small losses in the filter.)

other responses

Figure 22-13b shows the ideal response of a *high-pass* filter; the passband is the set of frequencies greater than f_c.

The *bandpass filter* has the ideal response shown in Fig. 22-13c; here, the passband is the set of frequencies between f_{c1} and f_{c2}.

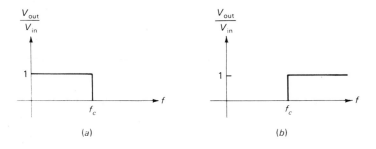

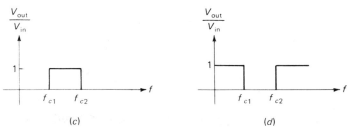

Figure 22–13. Ideal filter responses.

The fourth basic response is the ideal *bandstop* of Fig. 22-13*d;* it stops all frequencies between f_{c1} and f_{c2}.

In a real filter the rolloff is not vertical as shown in the ideal responses of Fig. 22-13; the cutoff frequency is defined as the frequency where the voltage gain is down 3 dB from the passband value. The rolloff outside the passband may be 20 dB per decade, 40 dB per decade, or any multiple of 20 dB, depending on the construction of the filter.

The bandpass filter (Fig. 22-13*c*) is often called a *window* because of its effect on a spectrum. As an example, Fig. 22-14*a* shows the spectrum of an input signal. Figure 22-14*b* is an ideal bandpass response. The easiest way to determine the output spectrum is to visualize the bandpass response superimposed on the input spectrum as shown in Fig. 22-14*c*. Since the gain is zero outside the passband and unity inside, only those components in the passband appear in the output spectrum of Fig. 22-14*d*. It's as though we are looking through a window at the spectrum of the signal.

In general, a filter is a frequency-domain device; it alters the spectrum of a signal. To determine the output spectrum, look at the input spectrum through the filter response, that is, with the filter response superimposed on the input spectrum; the output spectrum contains only those components in the filter passband.

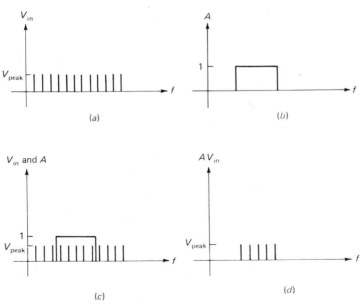

Figure 22–14. (a) *Input spectrum.* (b) *Ideal bandpass response.* (c) *Looking through the window.* (d) *Final output spectrum.*

class c amplifier

Frequency-domain analysis explains the action of a class C amplifier better than time-domain analysis. As discussed in Chap. 12, an input sine wave drives the transistor class C, producing narrow pulses of collector current. Because this nonsinusoidal current is periodic, it contains harmonics of the fundamental frequency. When the duty cycle is very small, the spectrum resembles Fig. 22-14a. The tuned tank is a bandpass filter; its response is wide enough to pass only a single harmonic. If we tune the tank to the fundamental frequency, we get an output spectrum containing only the fundamental. If we tune the tank to a higher harmonic, we have a frequency multiplier.

power-supply filter

Figure 22-15a shows the action of a full-wave rectifier and filter in the time domain. The input 60-Hz sine wave drives the rectifier to produce the 120-Hz full-wave signal. This full-wave signal is the input to a low-pass filter whose cutoff frequency is less than 120 Hz; therefore, the final output is a dc voltage because none of the sinusoidal components can get through the filter.

The filtering is easier to understand in the frequency domain. Figure 22-15b shows the input spectrum (a single line at 60 Hz) driving the rectifier. The output of the full-wave rectifier contains a dc component, a fundamental of 120 Hz, and higher harmonics. Because all sinusoidal frequencies are greater

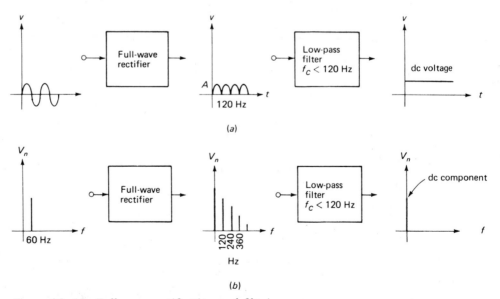

Figure 22–15. Full-wave rectification and filtering.

than the cutoff frequency of the low-pass filter, the output spectrum contains only the dc component. This dc component has an ideal value of

$$V_0 = 0.636A$$

where A is the peak value of the full-wave-rectified signal. The actual dc output voltage will be slightly less than this because of losses in the filter.

filter prototype circuits

Filters often use inductances and capacitances. Figure 22-16a shows the proto-type form a low-pass LC filter. At lower frequencies the inductors look almost shorted and the capacitors almost open; this allows lower frequencies to reach the output. At higher frequencies the inductors approach open circuits and the capacitors short circuits; for this reason, higher frequencies are stopped.

Figure 22-16b is the prototype of a high-pass LC filter; the action is complemen-tary to the low-pass. The capacitors look open and the inductors shorted at

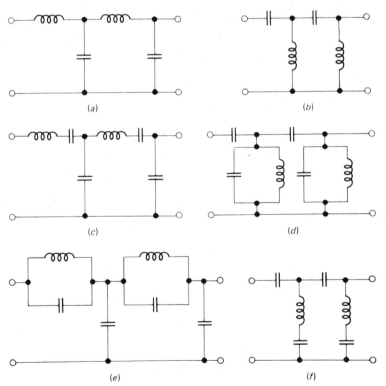

Figure 22–16. Filters. (a) *Low-pass.* (b) *High-pass.* (c) *Series resonant bandpass.* (d) *Parallel resonant bandpass.* (e) *Parallel resonant bandstop.* (f) *Series resonant bandstop.*

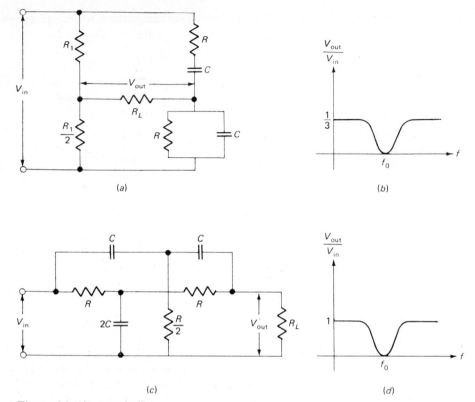

Figure 22-17. Notch filters.

lower frequencies; this prevents lower frequencies from reaching the output. On the other hand, higher frequencies pass through to the output because the capacitors look shorted and the inductors open.

Many designs are possible with bandpass filters. Figure 22-16c shows a series resonant type, while Fig. 22-16d illustrates a shunt resonant type. Both of these prototypes pass sine waves when the circuits are resonant.

Figure 22-16e and f shows two types of bandstop filters. When the circuits resonate, the filter stops the sine wave from reaching the output terminals.

EXAMPLE 22-7
Figure 22-17a shows a Wien bridge. How does the circuit affect the spectrum?

SOLUTION
As discussed in Chap. 19, the bridge balances at a frequency of

$$f_0 = \frac{1}{2\pi RC}$$

In other words, when the input sine wave has a frequency of f_0, the output voltage drops to zero.

Figure 22-17b shows the response of a Wien bridge. As mentioned earlier, the Wien bridge is sometimes called a *notch filter* because of its effect on the spectrum. If you visualize Fig. 22-17b superimposed on an input spectrum, you will see it notches frequency f_0 out of the spectrum.

Figure 22-17c shows another kind of notch filter known as the *twin T* (also called the parallel T). When the load resistance R_L is much greater than R, the voltage gain is approximately unity at low and high frequencies. As shown in Fig. 22-17d, the voltage gain drops to zero at frequency f_0. By using complex algebra, we can prove $f = 1/2\pi RC$.

22-4 *harmonic distortion*

As discussed in Sec. 8-5, when the amplified signal is small, only a small part of the transconductance curve is used. Because of this, the operation takes place over an almost linear arc of the curve. Operation like this is called *linear* because changes in output current are proportional to changes in input voltage. Linear operation means the shape of the amplified waveform is the same as the shape of the input waveform. That is, we get no distortion when the operation is linear or small-signal.

But when the signal is large, we can no longer treat the operation as linear; changes in output current no longer are proportional to changes in input voltage. Because of this, we get nonlinear distortion. This section examines distortion from the viewpoint of the frequency domain.

large-signal operation

When the signal swing is large, the operation becomes nonlinear. Figure 22-18 is an example of this. A sinusoidal V_{BE} voltage produces large swings along

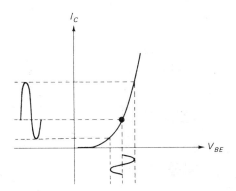

Figure 22-18. Nonlinear distortion.

the transconductance curve. Because of the nonlinearity of the curve, the resulting current is no longer sinusoidal. In other words, the shape of the output current no longer is a true duplication of the input shape. Since the output current flows through a load resistance, the output voltage will also have nonlinear distortion.

Figure 22-19a shows nonlinear distortion from the time-domain viewpoint. The input sine wave drives an amplifier. If the operation is large-signal, the amplified output voltage is no longer a pure sine wave. Arbitrarily, we have shown more gain on one half cycle than the other; this kind of distortion is sometimes called *amplitude distortion*.

The frequency domain gives us insight into amplitude distortion. Figure 22-19b shows how to visualize the same situation in the frequency domain. The input spectrum is a single line at f_1, the frequency of the input sine wave. The output signal is distorted but still periodic; therefore, it contains the dc component and harmonics shown. (Arbitrarily, we have stopped with the fourth harmonic.) The point is that a waveform with amplitude distortion contains a fundamental and harmonics; the strength of the higher harmonics is a clue to how bad the distortion is.

In fact, an alternative name for amplitude distortion is *harmonic distortion*. We use the term *amplitude distortion* when we visualize a signal in the time domain, and *harmonic distortion* when we are thinking of the signal in the frequency domain. When interested in the cause of the distortion, we use the term *nonlinear distortion*.

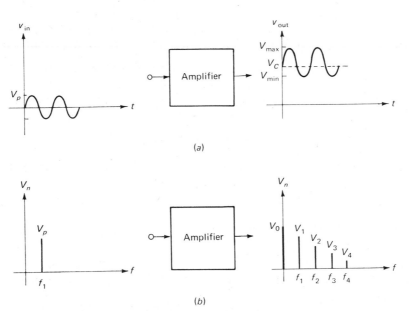

Figure 22-19. (a) *Amplitude distortion.* (b) *Harmonic distortion.*

All are synonyms for the kind of the distortion that occurs with *one input sine wave*. (The next chapter talks about distortion with two or more input sine waves.)

formula for individual harmonic distortion

How are we going to compare the harmonic distortion of one amplifier with another? The larger the peak values of the harmonics, the larger the harmonic distortion. The simplest way to compare different amplifiers is by taking the ratio of the harmonics to the fundamental. Specifically, we define the percent of second harmonic distortion as

$$\text{Percent second harmonic distortion} = \frac{V_2}{V_1} \times 100\% \qquad (22\text{-}8a)$$

The percent of third harmonic distortion is

$$\text{Percent third harmonic distortion} = \frac{V_3}{V_1} \times 100\% \qquad (22\text{-}8b)$$

and so on for any higher harmonic. In general, the percent of nth harmonic distortion is

$$\text{Percent } n\text{th harmonic distortion} = \frac{V_n}{V_1} \times 100\% \qquad (22\text{-}8c)\text{***}$$

As an example, suppose the output spectrum of Fig. 22-19b has $V_1 = 2$ V, $V_2 = 0.2$ V, $V_3 = 0.1$ V, and $V_4 = 0.05$ V. Then, the harmonic distortions are

$$\text{Percent second harmonic distortion} = 10\%$$
$$\text{Percent third harmonic distortion} = 5\%$$
$$\text{Percent fourth harmonic distortion} = 2.5\%$$

formula for total harmonic distortion

Data sheets usually give *total harmonic distortion,* all harmonics lumped together and compared to the fundamental. The formula for total harmonic distortion is

$$\text{Percent total harmonic distortion}$$
$$= \sqrt{(\text{percent second})^2 + (\text{percent third})^2 + \cdots} \qquad (20\text{-}9)\text{***}$$

As an example, for $V_1 = 2$ V, $V_2 = 0.2$ V, $V_3 = 0.1$ V, and $V_4 = 0.05$ V, the individual harmonic distortions are 10, 5, and 2.5 percent; the total harmonic distortion is

$$\text{Percent total harmonic distortion} = \sqrt{(10\%)^2 + (5\%)^2 + (2.5\%)^2} = 11.5\%$$

derivation of eq. (22-9)

When a spectrum like Fig. 22-20 appears across a resistor, each sinusoidal component produces its part of the total power *independent of the other harmonics.* In symbols,

$$p = p_1 + p_2 + p_3 + \cdots$$

In terms of voltage,

$$\frac{V^2}{R} = \frac{V_1^2}{R} + \frac{V_2^2}{R} + \frac{V_3^2}{R} + \cdots$$

where V, V_1, V_2, V_3, etc., are *rms voltages.* Dropping the R,

$$V^2 = V_1^2 + V_2^2 + V_3^2 + \cdots$$

Taking the square root of both sides gives

$$V = \sqrt{V_1^2 + V_2^2 + V_3^2 + \cdots} \qquad (22\text{-}10)$$

A formula like this is called a *quadratic sum* because we take the square root of the sum of squares.

Equation (22-10) gives the total rms voltage. If we want the rms voltage of the higher harmonics only, we can use

$$V_{\text{higher harmonics}} = \sqrt{V_2^2 + V_3^2 + V_4^2 + \cdots} \qquad (22\text{-}11)$$

This result follows directly from Eq. (22-10); it says the rms value of all harmonics above the fundamental equals the quadratic sum of the harmonic rms voltages. For instance, if $V_2 = 0.2$ V rms, $V_3 = 0.1$ V rms, and $V_4 = 0.05$ V rms, the rms value of these harmonics lumped together is

$$V_{\text{higher harmonics}} = \sqrt{0.2^2 + 0.1^2 + 0.05^2} = 0.229 \text{ V rms}$$

Total harmonic distortion is defined as the rms value of higher harmonics divided by the rms value of the fundamental. In symbols,

$$\text{Percent total harmonic distortion} = \frac{V_{\text{higher harmonics}}}{V_1} \times 100\% \qquad (22\text{-}12)$$

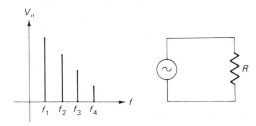

Figure 22-20. Total power of harmonic components is the superposition of individual powers.

When we substitute the right-hand member of Eq. (22-11) into this equation, we get

$$\text{Percent total harmonic distortion} = \frac{\sqrt{V_2^2 + V_3^2 + V_4^2 + \cdots}}{V_1} \times 100\%$$

which reduces to

Percent total harmonic distortion
$$= \sqrt{(\text{percent second})^2 + (\text{percent third})^2 + \cdots}$$

Therefore, we have proved Eq. (22-9).

harmonic distortion sometimes useful

If you are trying to amplify speech or music, the less harmonic distortion the better. In the case of speech, total harmonic distortion must be less than 10 percent or thereabouts to preserve intelligibility. High-fidelity music needs less than 1 percent total harmonic distortion for good quality.

Harmonic distortion is not always undesirable; sometimes we want as much as possible. For instance, in building a frequency multiplier (Sec. 12-6), we want as much nth harmonic as possible. Therefore, we deliberately optimize the circuit to produce maximum distortion at the nth harmonic. The resonant tank circuit can then separate the nth harmonic from the others, giving a pure output sine wave of frequency nf_1.

EXAMPLE 22-8
A spectrum analyzer shows the display of Fig. 22-21; the vertical scale is calibrated to read harmonic rms voltages rather than peak values (normal practice in a commercial analyzer).

1. Calculate the rms voltage for the periodic wave.
2. Calculate the rms voltage for the higher harmonics.
3. Calculate the total harmonic distortion.

SOLUTION

1. With Eq. (22-10),

$$V = \sqrt{8^2 + 4^2 + 3^2} = 9.43 \text{ V rms}$$

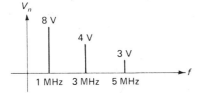

Figure 22-21.

If you measured the periodic wave with a true-rms voltmeter, you would read 9.43 V.

2. The rms value of the higher harmonics is found with Eq. (22-11):

$$V_{\text{higher harmonics}} = \sqrt{4^2 + 3^2} = 5 \text{ V rms}$$

If you notch the fundamental frequency out of the spectrum of Fig. 22-21, only the higher harmonics remain; measure these with a true-rms voltmeter and you will read 5 V.

3. Either Eq. (22-9) or (22-12) applies. Since we already have the rms value of the higher harmonics, Eq. (22-12) is more convenient:

$$\text{Percent total harmonic distortion} = \frac{5}{8} \times 100\% = 62.5\%$$

Alternatively, you calculate each individual distortion to get

$$\text{Percent third harmonic distortion} = 50\%$$
$$\text{Percent fifth harmonic distortion} = 37.5\%$$

The quadratic sum of these distortions is

$$\text{Percent total harmonic distortion} = \sqrt{(50\%)^2 + (37.5\%)^2} = 62.5\%$$

Incidentally, a *distortion analyzer* is an instrument that measures the rms value of all higher harmonics lumped together. By taking the ratio of this value to the fundamental, the distortion analyzer reads the value of total harmonic distortion.

22-5 *three-point distortion analysis*

Often, we encounter the special case of no harmonics above the second. For instance, the square-law curve of a FET causes amplitude distortion where the highest harmonic is the second (proved in the next chapter). We may also encounter this special case in a bipolar amplifier; when the input signal is increased beyond small-signal operation, the first higher harmonic to appear is the second harmonic. In other words, there is a level of input signal between small-signal and large-signal operation where the only significant distortion component is the second harmonic. We will refer to this special case as *square-law distortion*.

The transconductance curve of a bipolar or FET will force the fundamental and second harmonic to line up as shown in Fig. 22-22. When we add the fundamental and second harmonic (dashed waves), we get the total distorted waveform (solid). The equation for the distorted waveform is

$$v = V_0 + V_1 \sin \omega t + V_2 \sin (2\omega t - 90°) \tag{22-13}$$

At $\omega t = 0°$, this equation reduces to

$$V_C = V_0 - V_2 \qquad (22\text{-}13a)$$

At $\omega t = 90°$, Eq. (22-13) gives

$$V_{max} = V_0 + V_1 + V_2 \qquad (22\text{-}13b)$$

And when $\omega t = 270°$, Eq. (22-13) reduces to

$$V_{min} = V_0 - V_1 + V_2 \qquad (22\text{-}13c)$$

Solving the last three equations simultaneously gives

$$V_1 = \frac{V_{max} - V_{min}}{2} \qquad (22\text{-}14)$$

$$V_2 = \frac{V_{max} + V_{min} - 2V_C}{4} \qquad (22\text{-}15)$$

$$V_0 = V_C + V_2 \qquad (22\text{-}16)$$

How do we use these results? If we see a distorted wave like Fig. 22-22 on an oscilloscope, we can measure V_{max}, V_{min}, and V_C. Then, we can use Eqs. (22-14) through (22-16) to calculate V_1, V_2, and V_0. In this way, we find the *dc component* and the *peak values* of the fundamental and second harmonics. This is known as a *three-point analysis* because we measure three voltages: V_{max}, V_{min},

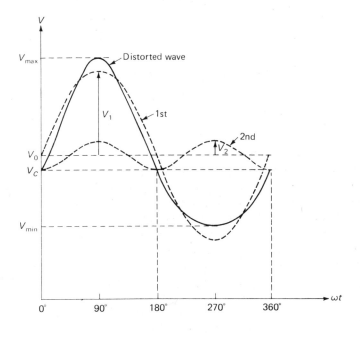

Figure 22-22. Adding fundamental and second harmonic to get distorted wave.

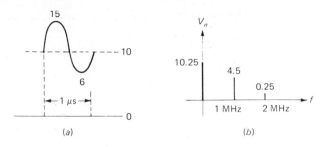

Figure 22-23.

and V_C. It gives exact values of V_1, V_2, and V_0 when there are no harmonics above the second; we can use it as a rough approximation even when there are smaller harmonics above the second.

Note that V_C is the voltage level passing through the 0°, 180°, and 360° points on the distorted wave.

EXAMPLE 22-9

Figure 22-23*a* shows a distorted output voltage. Assuming square-law distortion, calculate V_0, V_1, and V_2. (The 10-V level passes through the 0°, 180°, and 360° points on the distorted wave.)

SOLUTION

Figure 22-23*a* is the kind of picture that can appear on an oscilloscope. We read these three values:

$$V_{max} = 15 \text{ V}$$
$$V_{min} = 6 \text{ V}$$
$$V_C = 10 \text{ V}$$

The rest is straightforward calculation. With Eqs. (22-14) through (22-16), we get

$$V_1 = \frac{15 - 6}{2} = 4.5 \text{ V}$$

$$V_2 = \frac{15 + 6 - 2(10)}{4} = 0.25 \text{ V}$$

and
$$V_0 = 10 + 0.25 = 10.25 \text{ V}$$

Figure 22-23*b* shows these spectral components.

22-6 *other kinds of distortion*

For nonlinear operation of an amplifier, an input sine wave produces a distorted output signal in the time domain; in the frequency domain, nonlinear operation is equivalent to a single-line spectrum producing an output spectrum with many

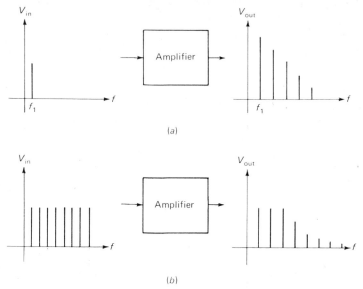

Figure 22-24. (a) *Harmonic distortion.* (b) *Frequency distortion.*

lines (see Fig. 22-24*a*). This is what we mean by harmonic distortion; a pure input sine wave produces a fundamental and harmonics.

frequency distortion

Frequency distortion is different. It has nothing to do with nonlinear distortion; frequency distortion can occur even with small-signal operation. The cause of frequency distortion is a change in amplifier gain with frequency. Figure 22-24*b* illustrates this kind of distortion. Arbitrarily, the input spectrum contains many equal-amplitude sinusoidal components. If the *break frequency* of the amplifier is *less than the highest sinusoidal frequency,* the higher frequencies in the output spectrum are attenuated as shown.

Frequency distortion, therefore, is nothing more than a change in the spectrum of the signal caused by amplifier filtering. This can affect the quality of speech or music signals. In other words, speech or music is a complex signal with many components in the spectrum. Unless all these components are in the bandwidth of the amplifier, the components in the output spectrum will not have the correct amplitudes. Because of this, the speech or music will sound different from the original input signal.

phase distortion

Phase distortion takes place when the phase of a harmonic is shifted with respect to the fundamental. As an example, the input signal of Fig. 22-25 shows the

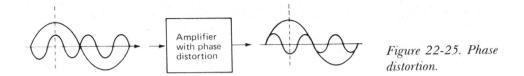

Figure 22-25. Phase distortion.

third harmonic peak in phase with the peak of the fundamental. If there is phase distortion, the third harmonic changes its phase with respect to the fundamental; arbitrarily, we have shown the third harmonic peak out of phase with the fundamental at the output.

Frequency and phase distortion almost always occur together. In the midband of an amplifier the voltage gain and phase shift are constant (either 0° or 180°). Because of this, no frequency or phase distortion can occur if all sinusoidal components are in the midband of the amplifier. Outside the midband, the voltage gain drops off and the phase angle changes; therefore, we get frequency and phase distortion if the spectrum contains components outside the midband.

EXAMPLE 22-10

An amplifier has a flat response from 20 Hz to 20 kHz. Harmonic distortion is negligible for input signals less than 0.1 V peak-to-peak. What kind of distortion occurs for each of these input signals:

1. The complex signal produced by middle C of a piano.
2. A wave whose spectrum has a fundamental of 12 kHz, plus *ultrasonic* components of 24 kHz and 36 kHz.

SOLUTION

If either input signal has a peak-to-peak value greater than 0.1 V, we will get amplitude or harmonic distortion; this means harmonics will appear in the output not present in the input.

If both input signals have a peak-to-peak value less than 0.1 V, the only possible distortions are frequency and phase distortion. In this case,

1. Middle C has a fundamental frequency of 256 Hz. A piano produces harmonics up to 10 kHz or thereabouts. Because of this, the spectrum of middle C contains a 256-Hz fundamental plus all harmonics up to 10 kHz or so. Since all these sinusoidal components are in the midband of the amplifier, no frequency or phase distortion takes place.
2. The ultrasonic components (those with frequencies greater than 20 kHz) lie above the midband of the amplifier; therefore, they get less gain and a different phase shift than the fundamental. As a result, frequency and phase distortion occur.

22-7 *distortion with negative feedback*

This section proves negative feedback reduces harmonic distortion.

open loop

Figure 22-26*a* shows an amplifier before we connect the feedback circuit. An input sine wave produces an amplified fundamental plus a second harmonic. For the moment, assume no harmonic distortion above the second. To account for the second harmonic in the output, we put a voltage source V_2 in series with the voltage source Av_{error}.

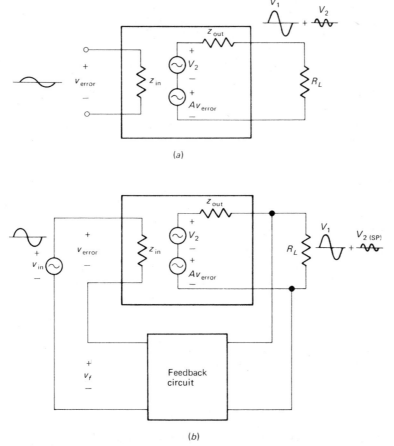

(a)

(b)

Figure 22-26. (a) *Distortion without feedback.* (b) *Distortion with feedback.*

closed loop

Figure 22-26*b* shows the same amplifier as part of an SP negative-feedback system. The feedback affects the second harmonic appearing across the output terminals; this is why we have labeled the second harmonic peak $V_{2(SP)}$. As we are about to prove, $V_{2(SP)}$ is much smaller than V_2.

In Fig. 22-26*b*, z_{out} is ideally small enough to neglect, so that

$$v_{out} = Av_{error} + V_2 = A(v_{in} - v_f) + V_2$$

or

$$v_{out} = A(v_{in} - Bv_{out}) + V_2 \qquad (22\text{-}17)$$

Expanding and solving for v_{out} gives

$$v_{out} = \frac{A}{1 + AB}v_{in} + \frac{V_2}{1 + AB}$$

Or since the sacrifice factor S equals $1 + AB$,

$$v_{out} = A_{SP}v_{in} + \frac{V_2}{S} \qquad (22\text{-}18)$$

The final result says the second harmonic is *reduced by the sacrifice factor.* For example, if the second harmonic has a peak value of 1 V with no feedback, then an SP feedback amplifier with a sacrifice factor of 1000 will have a second harmonic peak value of 1 mV.

Physically, here is why negative feedback reduces distortion. In Fig. 22-26*a*, the peak value of the second harmonic is V_2. When the amplifier is used in a negative-feedback system like Fig. 22-26*b*, the second harmonic is fed back to the input of the amplifier. Because of this, the error voltage contains both the input sine wave and a returning second-harmonic component. If you check the phase relations, you will find the amplified second-harmonic component arrives at the output 180° out of phase with the original distortion. Because of this, the resulting component $V_{2(SP)}$ is much smaller than V_2. Equation (22-18) exactly describes the amount of cancellation that takes place.

In Fig. 22-26*b*,

$$V_{2(SP)} = \frac{V_2}{S}$$

We can divide both sides by V_1 to get

$$\frac{V_{2(SP)}}{V_1} = \frac{1}{S}\frac{V_2}{V_1}$$

or

Percent second harmonic distortion $= \dfrac{1}{S}\dfrac{V_2}{V_1} \times 100\%$ $\qquad (22\text{-}19)$

This says the second-harmonic distortion with feedback equals the second-harmonic distortion without feedback reduced by the S factor. So, if an amplifier without feedback has a second-harmonic distortion of 5 percent, the same amplifier with feedback will have a second-harmonic distortion of 5 percent/S; if S is 1000, the second-harmonic distortion is 0.005 percent.

total harmonic distortion

The derivation we have gone through applies to all higher harmonics. In Fig. 22-26a, we can replace V_2 by $V_{\text{higher harmonics}}$. This substitution applies to all derived equations so that

$$\text{Percent total harmonic distortion} = \frac{1}{S}\frac{V_{\text{higher harmonics}}}{V_1} \times 100\% \quad (22\text{-}20)$$

This says the total harmonic distortion with feedback equals the total harmonic distortion without feedback reduced by the S factor.

As an example, if a data sheet gives a total harmonic distortion of 5 percent for a 10-V output signal, the same amplifier used in a negative-feedback system has a total harmonic distortion of 5 percent/S when the output signal is 10 V; if S is 100, the total harmonic distortion is only 0.05 percent.

self-testing review

Read each of the following and provide the missing words. Answers appear at the beginning of the next question.

1. Any periodic wave is a superposition of _____ waves. These harmonics are multiples of a lowest frequency called the _____.

2. *(sine, fundamental)* Any waveform with half-wave symmetry has only _____ harmonics. The dc component equals the area under one cycle divided by the _____.

3. *(odd, period)* The basic filter responses are low-pass, _____, bandpass, and _____. A bandpass filter is often called a _____.

4. *(high-pass, bandstop, window)* Another name for amplitude distortion is _____ distortion. It occurs because of nonlinear distortion of a sine wave.

5. *(harmonic)* Frequency distortion is a change in _____ of the signal caused by amplifier filtering. Phase distortion takes place when the _____ of a harmonic is shifted with respect to the fundamental.

6. *(spectrum, phase)* Negative feedback reduces nonlinear distortion. The harmonics above the fundamental are attenuated by the sacrifice factor.

problems

22-1. What are the frequencies of the first three harmonics in the spectrum of Fig. 22-27*a*?

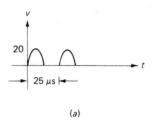

(a)

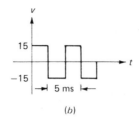

(b)

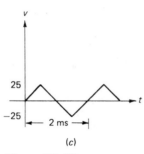

(c)

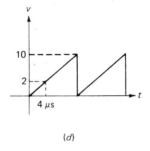

(d)

Figure 22-27.

22-2. Work out the first four odd harmonic frequencies for the square wave of Fig. 22-27*b*.

22-3. Figure 22-27*c* shows a triangular wave. What is the fundamental frequency? The frequency of the twenty-fifth harmonic?

22-4. The first cycle of a sawtooth wave is shown in Fig. 22-27*d*. What frequencies do the first three harmonics have?

22-5. Calculate the peak values of the first three odd harmonics of Fig. 22-27*b*.

22-6. What is the peak value of the fundamental in Fig. 22-27*c*? The peak value of the tenth harmonic?

22-7. Draw the spectrum for the sawtooth wave of Fig. 22-27*d*. Include the dc component and the first four harmonics.

22-8. What is the average voltage of the square wave in Fig. 22-28*a*?

22-9. Calculate the dc component of the pulse waveform shown in Fig. 22-28*b*.

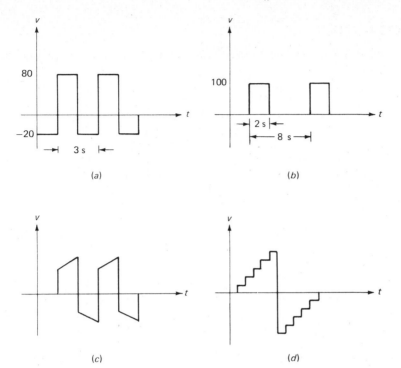

Figure 22-28.

22-10. Does the waveform of Fig. 22-28*c* have odd harmonics only? How about the staircase waveform of Fig. 22-28*d*?

22-11. If you shift the waveform of Fig. 22-28*a* vertically, can you find an average level that gives the waveform half-wave symmetry? If so, what is the new average value?

22-12. A bandpass filter has a lower cutoff frequency of 99 kHz and an upper cutoff frequency of 101 kHz; the gain of this filter is unity in the passband, but zero outside the passband. If a square wave drives this bandpass filter as shown in Fig. 22-29*a*, what is the peak value and frequency of the output signal?

22-13. The low-pass filter of Fig. 22-29*b* has a voltage gain of unity at zero frequency, a voltage gain of 0.005 at the fundamental frequency, and a voltage gain of zero for all other frequencies. What is the output dc voltage? How much ripple is there in the output?

22-14. What is the notch frequency of the twin-T filter shown in Fig. 22-30?

22-15. Suppose an amplifier has an input-output curve like Fig. 22-31*a*. What is the voltage gain when v_{in} equals 1 mV? When v_{in} equals 0.5 mV? Does this amplifier produce harmonic distortion?

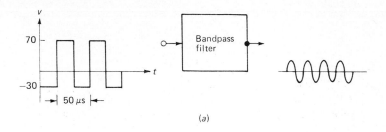

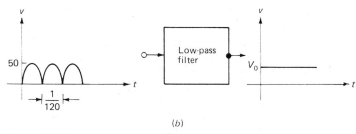

Figure 22-29.

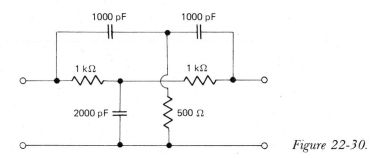

Figure 22-30.

22-16. Figure 22-31*b* shows the input-output curve of an amplifier. What is the voltage gain when v_{in} equals 1 mV? When v_{in} equals 0.5 mV? Does this amplifier produce nonlinear distortion?

22-17. A sine wave drives an amplifier. The output signal has a spectrum like Fig. 22-31*c*. Calculate the second harmonic distortion, the third harmonic distortion, and the total harmonic distortion.

22-18. An amplifier with a voltage gain of 500 has a total harmonic distortion of 0.2 percent for an input sine wave of 10 mV rms. If only the fundamental and third harmonics are present in the output, what is the rms value of the third harmonic?

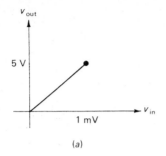

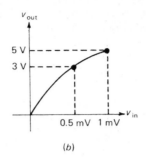

(a)

(b)

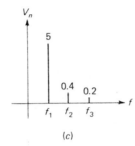

(c)

Figure 22-31.

22-19. The input voltage to an amplifier is changed in decade steps as follows: $v_{in} = 0.1$ mV, 1 mV, 10 mV, 100 mV, and 1 V. The corresponding output voltages are $v_{out} = 10$ mV, 100 mV, 1 V, 9 V, and 12 V. If you want to avoid harmonic distortion, what is the largest input voltage you should use?

22-20. An amplifier produces only square-law distortion. Figure 22-32a shows the distorted output signal produced by a single input sine wave; 9.3 V is the voltage level through the 0°, 180°, and 360° points on the waveform. Calculate V_0, V_1, and V_2. Also work out the total harmonic distortion.

22-21. If you measure a V_{max} of 10 V, a V_{min} of 2 V, and a V_C of 5 V, what is the peak value of the second harmonic voltage?

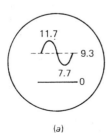

(a)

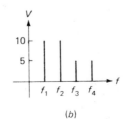

(b)

Figure 22-32.

22-22. The first two components of Fig. 22-32b have peak values of 10 V; the next two components have peak values of 5 V. Calculate the rms voltage of the periodic wave with this spectrum.

22-23. Calculate the total harmonic distortion for the signal whose spectrum is given by Fig. 22-32b.

22-24. An amplifier without feedback has output peaks of $V_1 = 10$ V and $V_2 = 0.5$ V. This amplifier is used in an SP negative-feedback system where the sacrifice factor S equals 500. If the fundamental output signal with feedback has a peak of 10 V, what is the peak value of the second harmonic?

22-25. A data sheet for an op amp says the total harmonic distortion equals 3 percent when the output signal is 20 V peak-to-peak. If you use this op amp in an SP negative-feedback system with a sacrifice factor of 5000, what will the total harmonic distortion equal when the output signal has a peak-to-peak value of 20 V?

22-26. An input signal with a single spectral line at f_1 drives an amplifier without feedback; the output spectrum is shown in Fig. 22-32b.
   (a) Calculate the total harmonic distortion.
   (b) If the amplifier is used in an SP negative-feedback system where the sacrifice factor equals 5000, what does the total harmonic distortion equal when the desired output component has a peak value of 10 V?

23 *intermodulation and mixing*

When a sine wave drives a nonlinear circuit, harmonics of this sine wave appear in the output. If two sine waves drive a nonlinear circuit, we get harmonics of each sine wave and we get *new frequencies*. These new frequencies are *not* harmonics of either input sine wave. This chapter describes the theory and application of these new output frequencies.

23-1 *nonlinearity*

Figure 23-1 shows a graph of output voltage versus input voltage. Arbitrarily, the curve is concave up. In a practical amplifier, it may concave up or down, depending on the number of stages and other factors.

small-signal operation

If the signal is small, the instantaneous operating point swings over a small part of the curve. In a case like this, the arc being used is almost linear. To a first approximation, the operation is linear. This allows us to write

$$v_{\text{out}} = Av_{\text{in}} \qquad (23\text{-}1a)$$

where A is a constant and v_{in} is the ac input signal. In Fig. 23-1, A represents the slope of the curve at point Q. In circuit terms, A is the small-signal voltage gain.

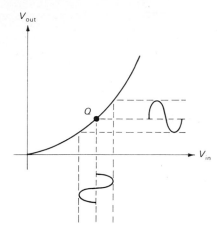

Figure 23-1. Non-linear input-output curve.

large-signal operation

When the signal swing is large, we can no longer use the linear approximation. As we saw in the preceding chapter, large-signal operation produces nonlinear distortion; the output signal no longer is sinusoidal. Equation (23-1*a*) cannot be used for large-signal operation. Instead, we must use a *power series:*

$$v_{\text{out}} = Av_{\text{in}} + Bv_{\text{in}}^2 + Cv_{\text{in}}^3 + Dv_{\text{in}}^4 + \cdot \ \cdot \ \cdot \qquad (23\text{-}1b)$$

Note that each succeeding term has v_{in} raised to the next higher power. A power series like this applies to any nonlinear curve like Fig. 23-1. For each combination of *A, B, C, D,* etc., we get a different nonlinear curve. In other words, if we have the values of *A, B, C, D,* and so on, we will know the exact relation between the input and output voltage.

The terms in Eq. (23-1*b*) have the following meanings:

Av_{in}—the linear term, the input signal amplified by a factor of *A*.
Bv_{in}^2—the quadratic term; it leads to second harmonic distortion.
Cv_{in}^3—the cubic term; it results in third harmonic distortion.

The higher terms have similar meanings; the fourth-power term produces fourth harmonic distortion, the fifth-power term fifth harmonic distortion, and so forth.

23-2 *linear operation*

Figure 23-2 illustrates small-signal operation. Because the arc being used is so small, it closely approximates a straight line. If the input signal is sinusoidal, so too is the output signal.

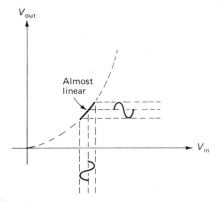

Figure 23-2. Small-signal operation.

higher-order terms drop out

Equation (23-1b) applies to any nonlinear curve and includes small-signal operation as a special case. When the signal is small, all higher-power terms in Eq. (23-1b) drop out, leaving only the Av_{in} term. For instance, suppose an amplifier has this equation describing its ac output voltage:

$$v_{out} = 50v_{in} + 4v_{in}^2 + 3v_{in}^3$$

with all higher terms negligible. Then, an input peak of 0.1 V gives an output peak of

$$v_{out} = 50(0.1) + 4(0.1)^2 + 3(0.1)^3$$
$$= 5 + 0.04 + 0.003$$
$$\cong 5 \text{ V peak}$$

We have neglected the last two terms because they are small compared to the linear term; as a guide, we will neglect a higher-power term when it is less than 1 percent of the linear term.

small-signal sinusoidal case

If the ac input signal is sine wave, we can express it by

$$v_{in} = V_x \sin \omega_x t$$

where V_x is the peak voltage and ω_x equals $2\pi f_x$. Figure 23-3a and b shows the input signal and its spectrum. If the signal is small, we get linear operation and the ac output voltage equals

$$v_{out} = Av_{in}$$
$$= AV_x \sin \omega_x t$$

This says the peak output voltage equals AV_x and the signal is sinusoidal with a radian frequency of ω_x. Figure 23-3c and d shows the output signal and its

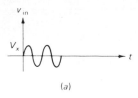

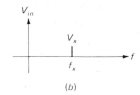

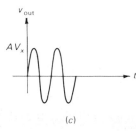

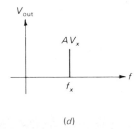

Figure 23-3. (a) *Input signal.* (b) *Input spectrum.* (c) *Output signal.* (d) *Output spectrum.*

spectrum. The output spectrum contains no harmonics; all that has happened is the spectral line at f_x has increased its height.

small-signal operation with two input sine waves

What happens if two sine waves drive an amplifier? We can express the first sine wave by

$$v_x = V_x \sin \omega_x t$$

and the second by

$$v_y = V_y \sin \omega_y t$$

Figure 23-4a shows these sine waves; arbitrarily, v_x has the higher frequency. If the two signal sources are in series as shown in Fig. 23-4b,

$$
\begin{aligned}
v_{\text{in}} &= v_x + v_y \\
&= V_x \sin \omega_x t + V_y \sin \omega_y t
\end{aligned}
\tag{23-1c}
$$

By adding the ordinates of v_x and v_y (Fig. 23-4a) at each instant in time, we get the waveform of Fig. 23-4c. Experimentally, we can get this waveform by driving an oscilloscope with two sine-wave sources in series. Each sinusoidal term in Eq. (23-1c) produces a spectral line at the corresponding frequency. In other words, if we put a signal like Fig. 23-4c into a spectrum analyzer, we will see a spectrum like Fig. 23-4d. The height of each spectral line equals the

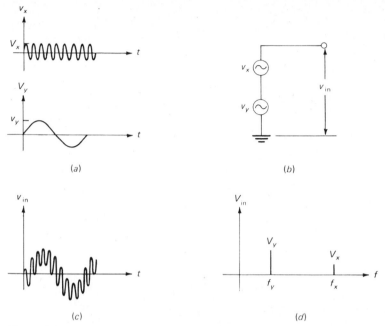

Figure 23-4. (a) *Two input signals.* (b) *Series connection of two signal sources.* (c) *Additive waveform.* (d) *Spectrum of additive waveform.*

peak value of the corresponding sine wave; the horizontal location gives the frequency. Since a spectrum analyzer is calibrated in cycle frequency f rather than radian frequency ω, the spectral lines appear at f_x and f_y.

If both signals are small, we get linear operation, and the ac output voltage from an amplifier equals

$$
\begin{aligned}
v_{\text{out}} &= A v_{\text{in}} \\
&= A(v_x + v_y) \\
&= A V_x \sin \omega_x t + A V_y \sin \omega_y t
\end{aligned}
$$

What does this equation say? The ac output voltage is the sum of each input sine wave amplified by A. Figure 23-5a shows how this output signal looks in the time domain; it is nothing more than the signal of Fig. 23-4c amplified by A. Figure 23-5b shows the spectrum of the amplified signal; it is the original input spectrum with each line amplified by A. Especially important, no harmonics or other lines appear in the output spectrum for small-signal operation.

additive waveforms

A waveform like Fig. 23-5a is called an *additive waveform* because it is the sum or superposition of two sine waves. You often see waveforms like this. As an

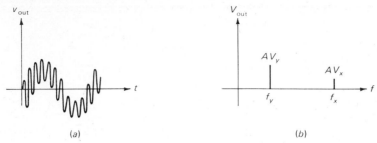

Figure 23-5. (a) *Amplified additive signal.* (b) *Spectrum.*

example, if you are amplifying a sine wave with frequency f_y, an interference signal with frequency f_x may somehow get into the amplifier and be added to the desired signal; in this case, you will see the additive waveform of Fig. 23-5*a* at the output of the amplifier.

The interference signal can get into the amplifier in a number of ways. It may be excessive power-supply ripple being added to the desired signal. Or it may be a signal induced by nearby electrical apparatus, or possibly a transmitted radio signal. In any event, whenever you see an additive waveform like Fig. 23-5*a*, remember it represents the sum of two sine waves.

linear operation with many sinusoidal components

What we have derived for two input sine waves applies to any number of input sine waves. In other words, if we have ten input sine waves, linear amplification results in ten output sine waves. The output spectrum will contain ten spectral lines with the same frequencies as the input spectrum; each component is amplified by A.

A good example of a multi-sine-wave input is the signal produced by middle C of a piano. This signal contains a fundamental with a frequency of 256 Hz plus harmonics to around 10,000 Hz (approximately 40 harmonics); the strength of the harmonics compared to the fundamental makes this note sound different from all others. If we *linearly* amplify this middle-C signal, the output spectrum contains only the original harmonics amplified by a factor A. Because of this, the output is louder but still has the distinct sound of middle C. A good high-fidelity amplifier therefore is a small-signal or linear amplifier; it does not add or subtract any spectral components; furthermore, it amplifies each component by the same amount.

23-3 *medium-signal operation with one sine wave*

For typical values of A, B, C, etc., in Eq. (23-1*b*), the first higher-power term to become important is the quadratic term. That is, there is a level of input signal between the small-signal and large-signal cases where only the first two terms of Eq. (23-1b) are important:

$$v_{\text{out}} = Av_{\text{in}} + Bv_{\text{in}}^2 \qquad (23\text{-}2)$$

Because this special case lies between small-signal and large-signal operation, we call it the *medium-signal* case. Any transistor amplifier can operate medium-signal, with only a linear and quadratic term in the equation for ac output voltage. With further increase in signal, a bipolar amplifier goes into the large-signal case where the cubic and higher-power terms become important.

the FET amplifier

Unlike a bipolar amplifier, a FET amplifier operates medium-signal all the way to saturation and cutoff. In other words, when we increase the input signal, the FET amplifier continues to operate medium-signal; only the linear and quadratic terms appear in the expression for output voltage. Equation (23-2) applies to a FET amplifier for any signal level, provided the output signal is not clipped. This is a direct consequence of the parabolic or square-law transconductance curve. We do not have the time to prove it, but it is possible to derive Eq. (23-2) from the transconductance equation of a FET. Because of this, as long as the instantaneous operating point remains along the transconductance curve, Eq. (23-2) applies to a FET amplifier.

quadratic term produces second harmonic

The quadratic term Bv_{in}^2 causes second harmonic distortion. Suppose the input voltage is a sine wave expressed by

$$v_{\text{in}} = V_x \sin \omega_x t$$

In the medium-signal case, the output voltage equals

$$\begin{aligned} v_{\text{out}} &= Av_{\text{in}} + Bv_{\text{in}}^2 \\ &= AV_x \sin \omega_x t + BV_x^2 \sin^2 \omega_x t \end{aligned} \qquad (23\text{-}3a)$$

A useful expansion formula proved in trigonometry is

$$\sin^2 A = \tfrac{1}{2} - \tfrac{1}{2} \cos 2A \qquad (23\text{-}3b)$$

where A represents angle. If we let $A = \omega_x t$, we can use Eq. (23-3b) to rearrange Eq. (23-3a); the result is

$$v_{\text{out}} = \tfrac{1}{2} BV_x^2 + AV_x \sin \omega_x t - \tfrac{1}{2} BV_x^2 \cos 2\omega_x t \qquad (23\text{-}4)$$

Each term in this expression is important. Briefly, here is what each means:

$\tfrac{1}{2} BV_x^2$—this has a constant value and represents a dc shift.

$AV_x \sin \omega_x t$—the linear output term; this is the amplified input sine wave.

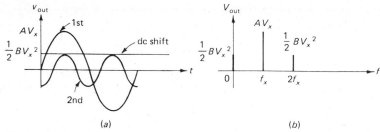

Figure 23-6. (a) x *output in time domain.* (b) *Spectrum of* x *output.*

$\frac{1}{2} BV_x^2 \cos 2\omega_x t$—this term has a radian frequency of $2\omega_x$, equivalent to a cycle frequency of $2f_x$; therefore, this term represents the second harmonic of the input sine wave.

Figure 23-6*a* shows each of these components in the time domain; if we add these components, we get the total waveform as it would appear on an oscilloscope. Figure 23-6*b* shows the spectrum; note the dc component, the fundamental with a frequency f_x, and the second harmonic with a frequency $2f_x$.

summary

For the medium-signal case, an input sine wave v_x produces a dc shift, an amplified fundamental, and a second harmonic. In later discussion, we will refer to the right-hand member of Eq. (23-4) as the *x output*. We can visualize the *x* output either in the time domain (Fig. 23-6*a*) or in the frequency domain (Fig. 23-6*b*).

23-4 *medium-signal operation with two sine waves*

When two input sine waves drive a medium-signal amplifier, something remarkable happens; in addition to harmonics produced, *new frequencies* appear in the output. To be specific, here is what we will prove in this section. When two input sine waves with frequencies f_x and f_y drive an amplifier in the medium-signal case, the output spectrum contains sinusoidal components with these frequencies:

f_x and f_y: the two input frequencies
$2f_x$ and $2f_y$: the second harmonics of the input frequencies
$f_x + f_y$: a new frequency equal to the sum of the input frequencies
$f_x - f_y$: a new frequency equal to the difference of the input frequencies

For instance, if the two input frequencies are 1 kHz and 20 kHz, the output spectrum contains sinusoidal frequencies of 1 kHz and 20 kHz (the two input frequencies), 2 kHz and 40 kHz (second harmonics), 21 kHz (the sum) and 19 kHz (the difference).

the cross product

When two input sine waves drive an amplifier, we can express the input voltage as

$$v_{\text{in}} = v_x + v_y$$

where v_x and v_y are sine waves. For the medium-signal case, the output voltage equals

$$
\begin{aligned}
v_{\text{out}} &= Av_{\text{in}} + Bv_{\text{in}}^2 \\
&= A(v_x + v_y) + B(v_x + v_y)^2 \\
&= Av_x + Av_y + Bv_x^2 + 2Bv_xv_y + Bv_y^2
\end{aligned}
$$

By rearranging, we get

$$v_{\text{out}} = \underbrace{Av_x + Bv_x^2}_{x\text{ output}} + \underbrace{Av_y + Bv_y^2}_{y\text{ output}} + \underbrace{2Bv_xv_y}_{\text{cross product}} \tag{23-5}$$

The first two terms in this equation are the x output discussed in the preceding section, that is, the output from a medium-signal amplifier when only v_x drives the amplifier; therefore, we already know these two terms result in a dc shift, an amplified fundamental f_x, and a second harmonic $2f_x$.

The next pair of terms $Av_y + Bv_y^2$ is called the y output; because they are identical to the x output except for the y subscript, the y-output terms result in a dc shift, an amplified fundamental f_y, and a second harmonic $2f_y$. In other words, the y output is what we get out of a medium-signal amplifier when a sine wave of frequency f_y is the only input to the amplifier.

The unusual thing about Eq. (23-5) is the *cross product* $2Bv_xv_y$. If the cross product were *not* present in this equation, the output would be the superposition of the x output and the y output. This would allow us to take each input separately, find how the amplifier responds to it, and sum the x and y output to get the total output. But the cross product is there, and because of it the superposition theorem gives us only part of the total output.

In other words, if only the v_x input were present, we would get only the first two terms in Eq. (23-5). Likewise, if only the v_y input were present, we would get only the next two terms in Eq. (23-5). But when both inputs are present simultaneously, we get a cross product in addition to the x and y outputs.

sum and difference frequencies

In Eq. (23-5), we already know the x-output terms represent a dc shift, an amplified fundamental f_x, and a second harmonic $2f_x$. Likewise, the y-output terms

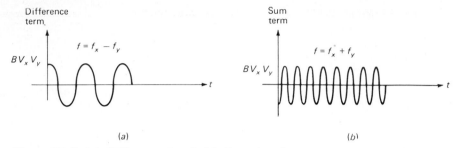

Figure 23-7. (a) *Difference signal.* (b) *Sum signal.*

represent another dc shift, an amplified fundamental f_y, and a second harmonic $2f_y$. All that remains now is the meaning of the cross product.

When we substitute

$$v_x = V_x \sin \omega_x t$$

and

$$v_y = V_y \sin \omega_y t$$

into the cross product, we get

$$2Bv_x v_y = 2B(V_x \sin \omega_x t)(V_y \sin \omega_y t)$$
$$= 2BV_x V_y(\sin \omega_x t)(\sin \omega_y t) \qquad (23\text{-}6)$$

The product of two sine waves can be expanded by the trigonometric identity

$$\sin A \sin B = \tfrac{1}{2} \cos (A - B) - \tfrac{1}{2} \cos (A + B)$$

By letting $A = \omega_x t$, and $B = \omega_y t$, we can expand and rearrange Eq. (23-6) to get

$$2Bv_x v_y = BV_x V_y \cos (\omega_x - \omega_y)t - BV_x V_y \cos (\omega_x + \omega_y)t \qquad (23\text{-}7)$$

The first term on the right side is a sinusoidal component with a radian frequency of $\omega_x - \omega_y$, equivalent to a cycle frequency of $f_x - f_y$; therefore, this first term represents the difference frequency. The second term is also a sinusoidal component but has a cycle frequency of $f_x + f_y$; so, this second term represents the sum frequency. Figure 23-7a and b shows how these two terms look in the time domain. As we see, they are sinusoids with a peak value $BV_x V_y$, but one is the difference term while the other is the sum term.

output spectrum

Equation (23-5) gives the output voltage from a medium-signal amplifier driven by two input sine waves. The first two terms are the x output; these terms represent a dc component, an amplified fundamental f_x, and a second harmonic $2f_x$; Fig. 23-8a shows the spectrum from this x output. In Eq. (23-5) the y-output

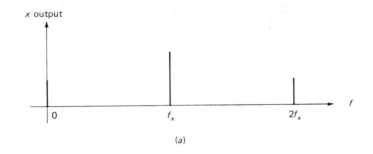

(a)

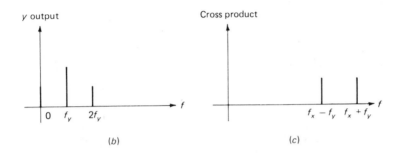

(b)

(c)

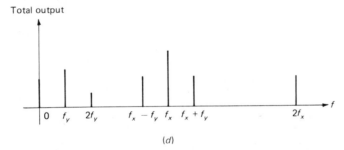

(d)

Figure 23-8. Spectra. (a) x *output.* (b) y *output.* (c) *Sum and difference components.*
(d) *Total output.*

terms represent another dc component, an amplified fundamental f_y, and a second harmonic $2f_y$; Fig. 23-8*b* illustrates the spectrum for the *y*-output terms.
In Eq. (23-5) the cross product represents sum and difference terms; Fig.
23-8*c* shows the spectrum of this cross product.

   The total output signal is the sum of the *x* output, the *y* output, and the
cross product. The signal is too complicated to draw in the time domain but
is easy to show in the frequency domain. Figure 23-8*d* is the output spectrum
of an amplifier driven by two sine waves in the medium-signal case. As we
see, the zero-frequency line is the sum of dc components. More important,

the remaining lines represent sinusoidal signals; we have each input frequency and its harmonic, plus sum and difference frequencies.

summary

We have found something new and useful. Put two sinusoidal signals into a medium-signal amplifier and out come six sinusoidal signals with frequencies f_x, $2f_x$, f_y, $2f_y$, a sum frequency of

$$\text{Sum} = f_x + f_y$$

and a difference frequency of

$$\text{Difference} = f_x - f_y$$

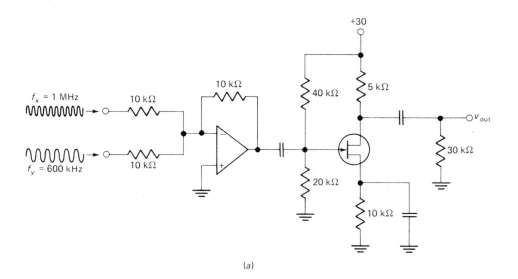

(a)

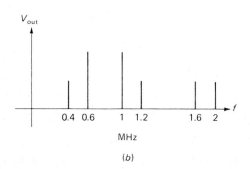

(b)

Figure 23-9.

Sometimes, we write these new frequencies as

$$\text{New } f\text{'s} = f_x \pm f_y \qquad (23\text{-}7a)\text{***}$$

EXAMPLE 23-1

Figure 23-9a shows an op-amp summing circuit driving a FET amplifier. What is the output spectrum of the FET amplifier?

SOLUTION

If the operation of the FET is small-signal, the only significant output will be the amplified input sine waves. That is, the output spectrum will contain two spectral lines, one at 600 kHz and the other at 1 MHz.

If either input signal is large enough to produce medium-signal operation, the output spectrum will contain six spectral lines at frequencies of 400 kHz (difference), 600 kHz (one input), 1MHz (another input), 1.2 MHz (second harmonic), 1.6 MHz (sum), and 2 MHz (second harmonic). Figure 23-9b shows these spectral lines. (The peak values are not important in this example.)

23-5 *large-signal operation with two sine waves*

What happens when two sine waves drive a bipolar amplifier in the large-signal mode? We get an output spectrum containing each input frequency, all harmonics of these frequencies, and sum and difference frequencies produced by every combination of harmonics.

derivation

For large-signal operation we need Eq. (23-1b) which says

$$v_{\text{out}} = Av_{\text{in}} + Bv_{\text{in}}^2 + Cv_{\text{in}}^3 + \cdot \cdot \cdot$$

This is an infinite series; there is no limit to the number of terms; each new term we add to the expression is the next higher power of v_{in}. With two input sine waves,

$$v_{\text{in}} = v_x + v_y$$

where v_x and v_y are sine waves. After substitution into the infinite series,

$$v_{\text{out}} = A(v_x + v_y) + V(v_x + v_y)^2 + C(v_x + v_y)^3 + \cdot \cdot \cdot$$

By applying the binomial theorem to each higher-power term, we can rearrange the expression to get

$$v_{\text{out}} = (Av_x + Bv_x^2 + Cv_x^3 + \cdot \cdot \cdot) + (Av_y + Bv_y^2 + Cv_y^3 + \cdot \cdot \cdot)$$
$$+ (2Bv_xv_y + 3Cv_x^2v_y + 3Cv_xv_y^2 + \cdot \cdot \cdot) \quad (23\text{-}8)$$

The first parenthesis in this equation is the *x output:*

$$x \text{ output} = Av_x + Bv_x^2 + Cv_x^3 + \cdots$$

This is the output we would get if v_x alone were driving the amplifier. With trigonometry we can prove v_x^n produces the nth harmonic of f_x. Therefore, the x output contains an amplified fundamental of frequency f_x, a second harmonic of frequency $2f_x$, a third harmonic of frequency $3f_x$, and so forth. Figure 23-10a shows the spectrum of the x output.

The second parenthesis in Eq. (23-8) is the *y output:*

$$y \text{ output} = Av_y + Bv_y^2 + Cv_y^3 + \cdots$$

This is the output we would get if only v_y were driving the amplifier. Again, the nth-power term introduces the nth harmonic; therefore, the spectrum of the y output contains spectral lines at f_y, $2f_y$, $3f_y$, and so forth as shown in Fig. 23-10b.

The third parenthesis in Eq. (23-8) is a collection of cross products from the binomial expansion of each power of $v_x + v_y$:

$$\text{Cross product} = 2Bv_xv_y + 3Cv_x^2v_y + 3Cv_xv_y^2 + \cdots$$

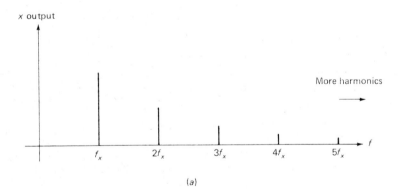

(a)

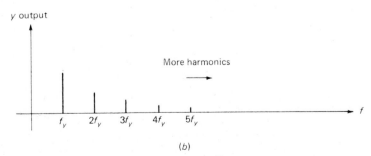

(b)

Figure 23-10. Spectra. (a) x *output.* (b) y *output.*

Following the last term are more terms of the form

$$Kv_x^m v_y^n$$

By advanced mathematics, we can prove each cross product produces a sum frequency of $mf_x + nf_y$ and a difference frequency of $mf_x - nf_y$. Or,

$$\text{New } f\text{'s} = mf_x \pm nf_y \qquad (23\text{-}9)***$$

where m and n can be any positive integers.

orderly calculations

Equation (23-9) is the formula for every possible sum and difference frequency generated in a large-signal amplifier driven by two sine waves. Here is an orderly way to use this important equation:

Group 1. First harmonic of v_x combining with each harmonic of v_y gives frequencies of
$$f_x \pm f_y$$
$$f_x \pm 2f_y$$
$$f_x \pm 3f_y$$
etc.

Group 2. Second harmonic of v_x combining with each harmonic of v_y gives frequencies of
$$2f_x \pm f_y$$
$$2f_x \pm 2f_y$$
$$2f_x \pm 3f_y$$
etc.

Group 3. Third harmonic of v_x combining with each harmonic of v_y gives frequencies of
$$3f_x \pm f_y$$
$$3f_x \pm 2f_y$$
$$3f_x \pm 3f_y$$
etc.

and so on. In this way, we can calculate any higher sum and difference frequencies of interest.

A concrete example will help. Suppose the two input sine waves to a large-signal amplifier have frequencies of $f_x = 100$ kHz and $f_y = 1$ kHz. Then we can calculate sum and difference frequencies as follows:

Group 1.
$$100 \text{ kHz} \pm 1 \text{ kHz} = 99 \text{ and } 101 \text{ kHz}$$
$$100 \text{ kHz} \pm 2 \text{ kHz} = 98 \text{ and } 102 \text{ kHz}$$
$$100 \text{ kHz} \pm 3 \text{ kHz} = 97 \text{ and } 103 \text{ kHz}$$
etc.

Group 2.

     200 kHz $\pm$ 1 kHz = 199 and 201 kHz
     200 kHz $\pm$ 2 kHz = 198 and 202 kHz
     200 kHz $\pm$ 3 kHz = 197 and 203 kHz
     etc.

Group 3.

     300 kHz $\pm$ 1 kHz = 299 and 301 kHz
     300 kHz $\pm$ 2 kHz = 298 and 302 kHz
     300 kHz $\pm$ 3 kHz = 297 and 303 kHz
     etc.

Continuing like this, we can find any sum and difference frequencies of interest. Figure 23-11a shows the spectrum for the first two groups of sum and difference frequencies.

In general, the spectrum of sum and difference frequencies looks like Fig. 23-11b. The total output spectrum contains these spectral lines plus the spectral lines of Fig. 23-10a and b.

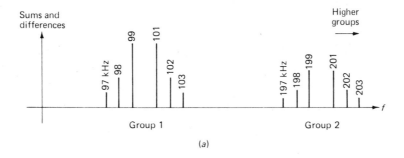

(a)

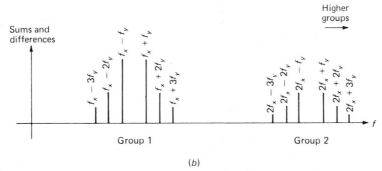

(b)

Figure 23-11. Sum and difference spectra for large-signal operation.

summary

Unquestionably, the situation is complex in a large-signal amplifier. Put two sine waves into an amplifier and out come two amplified sine waves, their harmonics, and every conceivable sum and difference frequency produced by the input frequencies and their harmonics. Theoretically, an infinite number of spectral components exist. As a practical matter, the size of these components decreases for higher values of m and n.

EXAMPLE 23-2

The two input sine waves to a large-signal amplifier have frequencies of $f_x = 10$ kHz and $f_y = 3$ kHz. Work out the first four sums and differences in group 1. Explain what to do with the negative frequency.

SOLUTION

Group 1 has these frequencies:

    10 kHz $\pm$ 3 kHz = 7 and 13 kHz

    10 kHz $\pm$ 6 kHz = 4 and 16 kHz

    10 kHz $\pm$ 9 kHz = 1 and 19 kHz

    10 kHz $\pm$ 12 kHz = -2 and 22 kHz

    etc.

Whenever a negative frequency turns up, you can take the absolute value. In this example, -2 kHz becomes 2 kHz. The reason you can disregard the negative sign lies in the identity used to expand cross products:

$$\sin A \sin B = \tfrac{1}{2} \cos (A - B) - \tfrac{1}{2} \cos (A + B)$$

The first term on the right is the difference component. If A is less than B, you get a negative angle. But the cosine of a negative angle is equal to the cosine of a positive angle with the same magnitude. For instance,

$$\cos (-60°) = \cos 60°$$

Because of this, a negative difference frequency is equivalent to a positive frequency of the same magnitude.

23-6 *intermodulation distortion*

If we amplify speech or music in a nonlinear amplifier, the sum and difference frequencies in the output make the speech or music sound radically different. *Intermodulation distortion* is the change in the spectrum caused by the sum and difference frequencies.

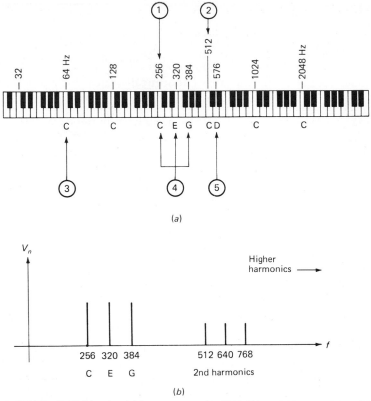

Figure 23-12. (a) *Piano keyboard.* (b) *C-E-G spectrum.*

chord C-E-G

For reasons not yet understood, music sounds good when the notes have a mathematical relation. In Fig. 23-12*a*, middle C (point 1) has a fundamental frequency of 256 Hz. The next C on the right is an *octave* higher (point 2); this C has a fundamental frequency of 512 Hz, exactly double that of middle C. Play two notes that are an octave apart and the combination sounds pleasant.

Chords may sound even better, Play chord C-E-G (point 4) and it sounds very pleasant. Here is something interesting. Chord C-E-G has these mathematical properties:

C	E	G	Notes in Chord
256	320	384	Fundamental frequency, Hz
1	1¼	1½	Ratio to middle C

The ratios suggest a chord sounds good when the fundamentals are quarter multiples of the lowest frequency.

Besides the fundamental frequencies shown, chord C-E-G contains harmonics to around 10 kHz. The amplitudes of these harmonics give the chord its distinct sound.

Figure 23-12b shows part of the spectrum for chord C-E-G. There are fundamental frequencies of 256-320-384, second harmonics of 512-640-768, third harmonics, fourth harmonics, and so on. The relative amplitudes of these harmonics must remain the same with respect to the fundamentals if we want to retain the exact sound. If this spectrum drives a *linear amplifier,* each spectral component receives the same gain, assuming all components are in the midband of the amplifier. In terms of the superposition theorem, the output is the sum of each amplified spectral component.

sum and difference frequencies

If the spectrum of chord C-E-G drives a *nonlinear amplifier,* the output spectrum contains all the input spectral lines plus the sums and differences of every possible combination of these components. How will this sound? Terrible! The ear immediately detects sum and difference frequencies because these frequencies sound like mistakes.

For instance, the first two spectral lines of Fig. 23-12b have frequencies of 256 and 320 Hz; these components produce a difference frequency of 64 Hz and a sum frequency of 576 Hz. The 64-Hz component corresponds to the C note two octaves below middle C (point 3 in Fig. 23-12a). This extra C note is not unpleasant because it is harmonically related to middle C; however, it does represent a new component not in the original sound. Much worse is the sum frequency 576 Hz; this corresponds to the D note one octave above middle C (point 5 in Fig. 23-12a). This D note does not belong in chord C-E-G and sounds *discordant;* the effect is the same as if the pianist made a mistake.

Besides the sum and difference frequencies of the first two fundamentals, we get many other sum and difference frequencies produced by other combinations. As a result, nonlinear amplification produces many discordant notes.

relation to nonlinear distortion

Nonlinear distortion causes harmonic and intermodulation distortion. Any device or circuit with a nonlinear input-output relation results in nonlinear distortion of the signal. In the time domain this means the shape of the periodic signal changes as it passes through the nonlinear circuit. In the frequency domain the result is a change in the spectrum of the signal. If only one input sine wave is present, only harmonic distortion occurs; if two or more input sine waves are involved, both harmonic and intermodulation distortion occur.

23-7 *frequency mixers*

A *frequency mixer* is used in almost every radio and television receiver; it is also used in many other electronic systems.

the basic idea

Figure 23-13 shows all the key ideas behind a frequency mixer. Two input sine waves drive a nonlinear circuit. As before, this results in all harmonics and intermodulation components. The bandpass filter then passes one of the intermodulation components, usually the difference frequency $f_x - f_y$. Therefore, the final output of a typical mixer is a sine wave with frequency $f_x - f_y$. In terms of spectra, a frequency mixer is a circuit that produces an output spectrum with a single line at $f_x - f_y$ when the input spectrum is a pair of lines at f_x and f_y.

A low-pass filter may be used in the place of a bandpass filter provided $f_x - f_y$ is less than f_x or f_y. For instance, if f_x is 2 MHz and f_y is 1.8 MHz, then

$$f_x - f_y = 2 \text{ MHz} - 1.8 \text{ MHz} = 0.2 \text{ MHz}$$

In this case, the difference frequency is lower than either input frequency; so we can use a low-pass filter if we wish. (Low-pass filters are usually easier to build than bandpass filters.)

But in applications where $f_x - f_y$ is between f_x and f_y, we must use a bandpass filter. As an example, if $f_x = 2$ MHz and $f_y = 0.5$ MHz, then

$$f_x - f_y = 2 \text{ MHz} - 0.5 \text{ MHz} = 1.5 \text{ MHz}$$

To pass only the difference frequency, we are forced to use a bandpass filter.

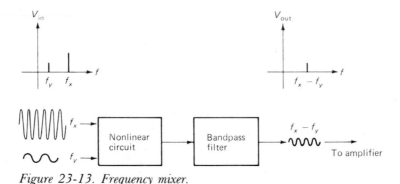

Figure 23-13. Frequency mixer.

usual size of input signals

In most applications, one of the input signals to the mixer will be large. This is necessary to ensure nonlinear operation; unless one of the signals is large, we cannot get intermodulation components. This large input signal is often supplied by an oscillator or signal generator.

The other input signal is usually small. By itself, this signal produces only small-signal operation of the mixer. One of the reasons this signal is small is because it often is a weak signal coming from an antenna (see Example 23-3).

The normal inputs to a mixer therefore are

1. A large signal adequate to produce medium- or large-signal operation of the mixer
2. A small signal that by itself can produce only small-signal operation

transistor mixer

Figure 23-14a shows a *transistor mixer*. The two input signals produce an additive waveform at the junction of R and C_B. This additive waveform drives the base-emitter diode over a significant part of the transconductance curve. The resulting collector current contains harmonics and intermodulation components. With the LC tank tuned to the difference frequency, the output signal has a frequency of $f_x - f_y$.

Figure 23-14b shows another way to couple the input signals. One signal drives the base, the other the emitter. The advantage is isolation of sources and sometimes a better impedance match.

diode mixers

Instead of using a transistor for the nonlinear device, we can use a diode. We drive the diode with an additive waveform containing frequencies f_x and f_y. The diode current then contains harmonics and intermodulation components. With a filter, we can remove the difference frequency.

Incidentally, *heterodyne* is another word for mix, and *beat frequency* is synonymous with difference frequency. In Fig. 23-14a, we are heterodyning two input signals to get a beat frequency of $f_x - f_y$.

conversion gain

Conversion gain refers to the power gain of the mixer. Specifically,

$$\text{Conversion gain} = \frac{p_{\text{out}}}{p_{\text{in}}} \qquad (23\text{-}10)\text{***}$$

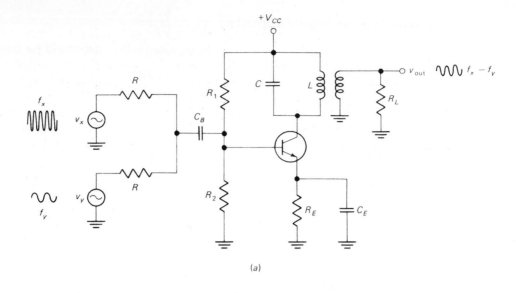

(a)

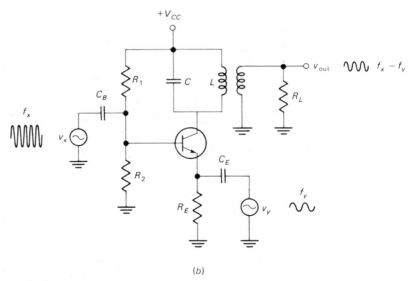

(b)

Figure 23-14. Transistor mixers.

where p_{out} is the output power of the difference signal and p_{in} is the input power of the smaller input signal. As an example, suppose the smaller signal in Fig. 23-14b delivers 10 μW to the emitter; if the output power of the difference signal is 40 μW, we have a conversion gain of

$$\text{Conversion gain} = \frac{40\ \mu\text{W}}{10\ \mu\text{W}} = 4$$

This is equivalent to a bel conversion gain of 6 dB.

Also useful is the *conversion voltage gain* defined as

$$\text{Conversion voltage gain} = \frac{v_{\text{out}}}{v_{\text{in}}} \qquad (23\text{-}11)$$

mixers in am receivers

Figure 23-15a shows the front end of a typical AM broadcast receiver. The antenna delivers a weak signal to the radio-frequency (RF) amplifier. Although this increases signal strength, the f_y signal driving the mixer is still small. The other mixer input comes from a *local oscillator* (LO); this signal is large enough to produce nonlinear distortion. As a result, the output of the mixer is a signal with a frequency $f_x - f_y$.

The difference signal now drives several stages called the *intermediate-frequency* (IF) amplifiers. These IF amplifiers provide most of the gain in the receiver. The large signal coming out of the IF amplifiers then drives other circuits.

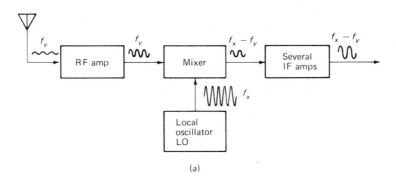

(a)

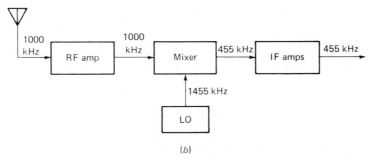

(b)

Figure 23-15. Front end of typical AM receiver.

Here is a numerical example. When you tune to a station with a frequency of 1000 kHz, you are doing the following:

1. Adjusting a capacitor to tune the RF amplifier to 1000 kHz
2. Adjusting another capacitor to set the LO frequency to 1455 kHz

As shown in Fig. 23-15*b*, the output of the mixer is 455 kHz. The IF amplifiers are fixed-tuned to this frequency; therefore, the 455-kHz signal receives maximum gain from the *IF strip* (group of amplifiers).

If you tune to a station with a frequency of 1200 kHz, the RF stage is tuned to 1200 kHz and the LO stage to 1655 kHz. The difference frequency is *still 455 kHz*. In other words, by deliberate design the frequency out of the mixer always equals 455 kHz no matter what station you tune to. We will discuss the reason for this in the next chapter. Briefly, it boils down to this: it is easier to design a group of amplifier stages tuned to a constant frequency than try to gang-tune these stages to each station frequency.

Another point. We used an IF frequency of 455 kHz because it is one of the common IF frequencies in commercial receivers. Other typical values are 456 kHz and 465 kHz. Many automobile radios use IF frequencies of 175 kHz or 262 kHz. Regardless of the exact value, any modern receiver has a group of IF amplifiers fix-tuned to the same frequency. The incoming signal frequency is converted down to this IF frequency.

EXAMPLE 23-3

Figure 23-16 shows a mixer. The large input signal v_x has a frequency of 100 kHz. The small input signal v_y has a frequency of 100.5 kHz. If the output

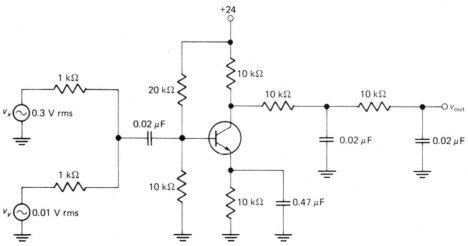

Figure 23-16. Bipolar mixer.

voltage has an rms value of 0.0125 V, what does the conversion voltage gain equal? What does the output lag network do?

SOLUTION
With Eq. (23-11),

$$\text{Conversion voltage gain} = \frac{0.0125}{0.01} = 1.25$$

This is equivalent to approximately 2 dB. (If you build the circuit of Fig. 23-16, you should get similar results.)

The lag networks in the collector circuit are low-pass filters; they stop all components above a cutoff frequency of approximately 800 Hz. Since the difference frequency is 500 Hz, it passes through to the output.

23-8 *spurious signals*

A serious problem may arise in mixer applications. This section describes the problem and its cure.

unwanted signals

The output filter of a mixer passes any signal whose frequency is in the passband. It is possible for small unwanted signals to get through the filter along with the desired difference frequency. As we saw in Sec. 23-5, many difference frequencies are generated in a mixer; these are given by

$$\text{Difference frequencies} = mf_x - nf_y$$

where m and n take on all integer values. It is possible to find values of m and n that produce difference frequencies close to $f_x - f_y$. Such difference frequencies are called *spurious signals* because they can pass through the filter along with the desired difference frequency.

Here is an example. Suppose $f_x = 10.5$ MHz and $f_y = 2.5$ MHz. The desired difference is

$$f_x - f_y = 10.5 \text{ MHz} - 2.5 \text{ MHz} = 8 \text{ MHz}$$

There are many values of m and n that produce difference frequencies near this. For instance, if $m = 2$ and $n = 5$

$$mf_x - nf_y = 2(10.5 \text{ MHz}) - 5(2.5 \text{ MHz}) = 8.5 \text{ MHz}$$

This frequency is close to the desired frequency; if enough of this 8.5-MHz signal reaches the output, it may interfere with the desired signal. We can find

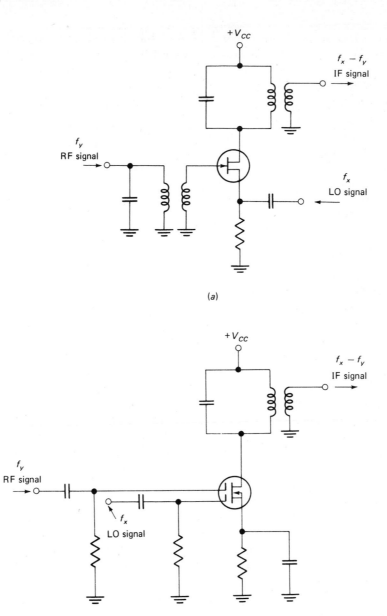

Figure 23-17. FET mixer. (a) JFET. (b) Dual-gate MOSFET.

other spurious frequencies close to the desired frequency. Because of this, the mixer output contains the desired signal plus many spurious signals.

Spurious signals occur for values of m and n greater than unity. For this reason, they are weak compared to the desired signal. Therefore, in some applications the bipolar mixer is acceptable even though it produces small spurious output signals.

the FET mixer

But there are many applications where spurious signals out of a bipolar mixer cause problems. The simplest way to eliminate spurious signals is to operate the mixer in the medium-signal case; the output voltage for this type of operation is

$$v_{out} = Av_{in} + Bv_{in}^2$$

As proved in Sec. 23-4, medium-signal operation produces only the input frequencies, their second harmonics, the sum frequency, and the difference frequency; there are no other difference frequencies; therefore, there can be no spurious signals with medium-signal operation.

Because the FET operates medium-signal all the way to cutoff and saturation, it is the ideal device to use in a mixer. With a FET mixer, spurious signals are almost eliminated. For this reason, the FET is superior to a bipolar transistor when it comes to mixer applications.

Figure 23-17a shows a JFET mixer. The RF signal drives the gate, and the stronger LO signal drives the source. For the FET to operate medium-signal, the LO signal must not drive the FET into saturation or cutoff; if this should happen, cubic and higher-power terms creep into the expression for output voltage. In other words, if the LO signal is large enough to cause clipping, the spurious signal content will increase. Ideally, the LO signal should cause the FET to swing over as much of the transconductance curve as possible without clipping; this results in the highest conversion gain with minimum spurious signals.

Figure 23-17b shows a dual-gate MOSFET mixer. The RF signal drives one of the gates, the LO signal the other gate. In a circuit like this, each input signal works into a high input impedance.

EXAMPLE 23-4
Explain the action of Fig. 23-18a and b.

SOLUTION
In Fig. 23-18a, each time we close the switch, the antenna radiates a signal. Because of this, we can transmit short bursts of energy (dots) or long bursts (dashes).

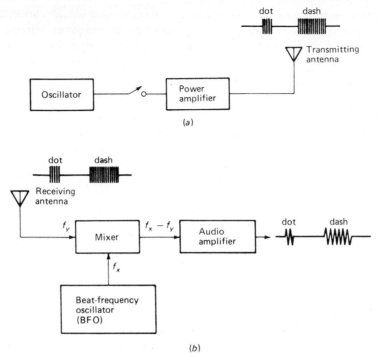

Figure 23-18. (a) *ICW transmitter.* (b) *ICW receiver.*

For antennas of practical size, the radiated frequencies are far above the audio range (greater than 20 kHz). To hear the dots and dashes, therefore, we must first mix the incoming signal with a *beat-frequency-oscillator* (BFO) signal (see Fig. 23-18b). The BFO signal has a frequency close to the received frequency, so close in fact that $f_x - f_y$ is an audio frequency. Because of this, the bursts of received energy are translated into bursts of audio frequencies, dots and dashes we can hear.

The system just described is called *interrupted-continuous-wave* (ICW) telegraphy.

23-9 *phase-locked loops*

A *phase-locked loop* (PLL) is a feedback loop with a *phase detector* (mixer used in special way), a low-pass filter, an amplifier, and a voltage-controlled oscillator. Rather than feeding back voltage and comparing it to the input, the PLL feeds back frequency and compares it to the incoming frequency. This allows the VCO to *lock* on (equal) the incoming frequency.

PLLs have many applications. TV receivers use PLLs to synchronize horizontal and vertical sweeps. FM stereo tuners use PLLs to improve performance. And because of its superior noise immunity, the PLL has been widely used to track signals from satellites. Other applications include frequency synthesizers, FM generators, and touch-tone telephone.

basic idea

Figure 23-19*a* shows a PLL. The incoming signal is one input to a phase detector; the returning VCO signal is the other input. The mixer output drives a low-pass filter, whose output is amplified and applied to the VCO. Initially, the VCO frequency is close to the incoming frequency. Because of this, the mixer output is a beat note (low-frequency signal). This causes the VCO frequency to vary until it equals the incoming frequency. At this point, the mixer output is a dc voltage, proportional to the phase difference between the VCO and incoming signal. It is this amplified dc voltage that controls the VCO frequency, keeping it locked on to the incoming frequency.

mathematical analysis

Section 23-4 discussed the mixing of two sinusoidal signals. And Eq. (23-5) came up with a cross product of $2Bv_xv_y$. This cross product is the key to the mixer action of a PLL. To begin with, when a PLL is locked on to the incoming

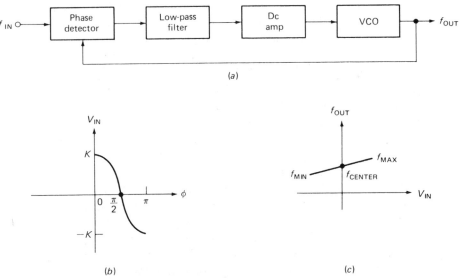

(a)

(b)   (c)

Figure 23-19. (a) *PLL* (b) *Phase-detector output.* (c) *VCO output.*

frequency, only a phase difference exists between the incoming signal and the VCO signal. In symbols,

$$v_x = V_x \sin \omega t$$
$$v_y = V_y \sin (\omega t + \phi)$$

where v_x is the incoming signal and v_y is the VCO signal. When these expressions are substituted into Eq. (23-6), the trigonometric expansion gives

$$2Bv_xv_y = BV_xV_y \cos \phi - BV_xV_y \cos (2\omega t + \phi)$$

The first term on the right-hand side is a dc voltage, proportional to the phase angle ϕ. The second term is the second harmonic. The low-pass filter of Fig. 23-19a blocks this second harmonic; only the dc component gets through to the amplifier. Because of the voltage gain, the VCO input is

$$V_{IN} = ABV_xV_y \cos \phi$$

or simply

$$V_{IN} = K \cos \phi$$

where K is a constant.

Figure 23-19b shows the graph of this control voltage. It's a cosine function with a maximum positive value when $\phi = 0°$ and a maximum negative value when $\phi = 180°$. Notice that the voltage is zero when $\phi = 90°$. Therefore, the control voltage can be positive, zero, or negative, which means the VCO frequency can vary above or below its center frequency, as shown in Fig. 23-19c.

For instance, suppose the VCO has a center frequency of 10 MHz, a minimum frequency of 9.5 MHz, and a maximum frequency of 10.5 MHz. When the incoming frequency is 10 MHz, no control input is needed for the VCO because it free runs at 10 MHz; in this case, the phase angle automatically goes to 90°, which means zero control voltage (see Fig. 23-19b and c). If the incoming frequency is 9.5 MHz and the VCO is locked on, the phase difference is 180°, which produces maximum negative voltage. If the incoming frequency is 10.5 MHz and the VCO is locked, the phase angle is 0° and the control voltage is maximum.

capture and lock range

When the VCO is locked on the incoming frequency, it can follow slow changes and remain locked. The *lock range* is band of frequencies over which the VCO can track the incoming signal after lock has occurred.

The *capture range* is different. If lock has not occurred yet, a beat note comes out of the phase detector. This beat note causes the VCO to sweep. Depending on the low-pass cutoff frequency and the closed-loop system gain, the VCO will be able to lock on a band of frequencies called capture range.

The capture range is smaller than the lock range. For instance, a PLL may have a lock range of ±5 percent of the VCO center frequency; the capture range may be only ±2 percent. If the VCO center frequency is 10 MHz, this means a lock range of ±0.5 MHz and a capture range of ±0.2 MHz.

pll outputs

In Fig. 23-19a, the input signal may be weak, almost buried in noise. Nevertheless, a PLL may be able to lock on such a signal and produce a strong output signal of the same frequency. The signal-to-noise ratio is greatly enhanced because the PLL filters noise outside the capture range.

Another important application is with *frequency-modulated* (FM) signals. Briefly, the idea is this. The voltage out of the amplifier is proportional to the phase difference. If the incoming signal is frequency-modulated (discussed in the next chapter), the voltage out of the amplifier will be an audio signal representing the voice or music being transmitted by the FM signal.

In other words, a PLL has two usable outputs: the VCO output and the amplifier output. Which of these is used depends on the particular application.

Manufacturers can produce integrated PLLs that contain a phase detector, amplifier, and VCO. By adding an external capacitor or RC network, we can control the low-pass cutoff frequency. The 560 family is the best known of the IC PLLs; it offers VCO frequencies from 0.1 to 15 MHz and lock ranges up to 60 percent.

23-10 noise

As mentioned earlier, noise contains sinusoidal components at all frequencies. Some of these noise components will mix with the LO signal and produce difference frequencies in the output of the mixer. Because of this, the remainder of the system amplifies the desired signal and unwanted noise out of the mixer.

In general, noise is any kind of unwanted signal not derived from or related to the input signal. This section describes noise.

some types of noise

Where does noise come from? Electric motors, neon signs, power lines, ignition systems, lightning, etc., set up electric and magnetic fields. These fields can induce noise voltages in electronic circuits. To reduce noise of this type, we can shield the circuit and its connecting cables.

Power-supply ripple is also classified as noise because it is independent of the desired signal. As we know, ripple can get into signal paths through biasing resistors (also by induction). With regulated power supplies and shielding, we can reduce ripple to an acceptably low level.

If you bump or jar a circuit, the vibrations may move capacitor plates, inductor windings, and so on. This results in a noise called *microphonics*. In this respect, the transistor is far superior to the vacuum tube; because of its solid-state construction, a transistor has negligible microphonics.

thermal noise

We can eliminate or at least minimize the effects of ripple, microphonics, and external field noise. But there is little we can do about *thermal noise*. Figure 23-20*a* shows the idea behind this kind of noise. Inside any resistor are conduction-band electrons. Since these electrons are loosely held by the atoms, they tend to move randomly in different directions as shown. The energy for this motion comes from the thermal energy of surrounding air; the higher the ambient temperature, the more active the electrons.

The motion of the billions of electrons is pure chaos. At some instants in time, more move up than down, producing a small negative voltage across the resistor. At other instants, more move down than up, producing a positive voltage. If amplified and viewed on an oscilloscope, this noise voltage would resemble Fig. 23-20*b*. Like any voltage, noise has an rms or heating value; as an approximation, the highest noise peaks are about four times the rms value.

The changing size and shape of the noise voltage implies components of

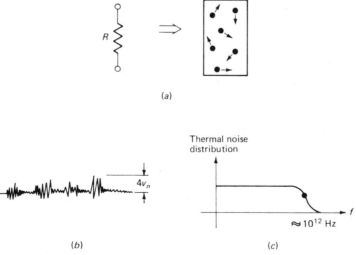

Figure 23-20. Thermal noise. (a) *Random electron motion.* (b) *Appearance on oscilloscope.* (c) *Spectral distribution.*

many different frequencies. An advanced derivation shows the noise spectrum is like Fig. 23-20c. As we see, noise is uniformly distributed throughout the practical frequency range. The break frequency of approximately 10^{12} Hz is far beyond the capability of electronic circuits. For this reason, most people say that noise contains sinusoidal components at all frequencies.

How much noise voltage does a resistor produce? This depends on the temperature, the bandwidth of the system, and the size of the resistance. Specifically,

$$v_n = \sqrt{4kTBR} \qquad (23\text{-}12)$$

where v_n = rms noise voltage

k = Boltzmann's constant (1.37×10^{-23})

T = absolute temperature, Celsius + 273°

B = noise bandwidth, Hz

R = resistance, Ω

In Eq. (23-12), the noise voltage increases with temperature, bandwidth, and resistance.

At room temperature (25°C), the formula reduces to

$$v_n = 1.28(10^{-10})\sqrt{BR} \qquad (23\text{-}12a)$$

The noise bandwidth B is approximately equal to the 3-dB bandwidth of the amplifier, mixer, or system being analyzed. As an example, a transistor radio has an overall bandwidth of about 10 kHz. This is the approximate value of B in Eq. (23-21a). A television receiver, on the other hand, has a bandwidth of about 4 MHz; this is the value of B to use for noise calculations in a television receiver.

With Eq. (23-12) or (23-12a), we can get an estimate for the amount of noise produced by any resistor in an amplifier. Those resistors near the input of the amplifier are most important because their noise will be amplified most and will dominate the final output noise. We can get a rough estimate of the input noise level by calculating the noise generated by the Thevenin resistance driving the amplifier or system.

EXAMPLE 23-5

The Thevenin source resistance driving an amplifier is 5 kΩ (Fig. 23-21). Estimate the amount of noise this resistance delivers to the amplifier in the bandwidth shown.

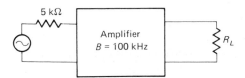

Figure 23-21.

SOLUTION

The bandwidth is 100 kHz. We will assume the ambient temperature is 25°C. With Eq. (23-12a),

$$v_n = 1.28(10^{-10}) \sqrt{10^5(5000)}$$
$$= 2.86 \ \mu V$$

This gives us an estimate of the input noise level. If the desired signal is smaller than this, it will be masked or covered by the noise. Therefore, the desired signal has to be much greater than 2.86 μV to stand out from source noise and noises generated inside the amplifier.

self-testing review

Read each of the following and provide the missing words. Answers appear at the beginning of the next question.

1. If both signals are small in an amplifier, we get _____ operation. In this case, two input sine waves produce _____ sine waves at the output. No harmonics or other spectral components appear at the output of a linear amplifier.

2. *(linear, two)* For the medium-signal case with one sinusoidal input signal, the output contains a dc component, an amplified _____, and a second _____.

3. *(fundamental, harmonic)* When two input sine waves drive a FET mixer, the output contains a dc component, the original frequencies, second harmonics of these original frequencies, a _____ frequency, and a _____ frequency.

4. *(sum, difference)* Two sinusoidal signals driving a large-signal amplifier produce new frequencies of $mf_x \pm nf_y$. Intermodulation _____ is the change in the spectrum caused by the sum and difference frequencies.

5. *(distortion)* In a typical frequency mixer, one input signal is large enough to produce nonlinear operation; the other input signal is small. Bipolar transistors, diodes, and FETs can be used for _____.

6. *(mixers)* AM receivers have a local _____; its signal is large enough to produce nonlinear operation of the mixer. The mixer output goes to the intermediate _____ (IF) stages. In the typical AM receiver, the IF frequency is approximately 455 kHz.

7. *(oscillator, frequency)* Spurious signals occur for large values of m and n. It is possible to find values of m and n that produce _____ frequencies close to $f_x - f_y$. These spurious signals pass through the filter along with the desired difference frequency.

8. *(difference)* A phase-locked loop has a phase detector, a low-pass _____, an amplifier, and a voltage-controlled _____. The VCO can lock on any frequency within the _____ range. Once locked, the VCO can track slow changes within its lock range.

9. *(filter, oscillator, capture)* Noise is any kind of signal not derived from or related to the input signal. Electric motors, neon signs, power lines, ignition systems, etc., can induce noise voltages into circuits. Thermal noise is produced by the random motion of charges inside a conductor. The higher the temperature, the greater the thermal noise.

problems

23-1. In Eq. (23-1*b*), $A = 50$, $B = 10$, and $C = 5$; all other coefficients are zero. If v_{in} is a sine wave with a peak of 1 V, what values do the positive and negative output peaks have?

23-2. Same coefficients as given in Prob. 23-1. Calculate the positive and negative output peaks if v_{in} is a sine wave with a peak of 1 mV.

23-3. A linear dc amplifier has a voltage gain of 50 and a bandwidth of 30 kHz. The input spectrum has a single line with a peak of 15 mV and a frequency of 1 kHz. What are the peak values and frequencies of the output spectral lines?

23-4. The input spectrum to a linear amplifier looks like Fig. 23-22*a*. If all components are within the midband of the amplifier and the voltage gain is 75, what does the output spectrum contain?

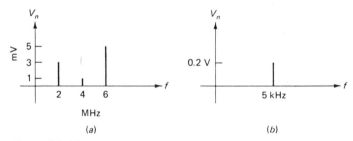

Figure 23-22.

23-5. An amplifier operates medium-signal with $A = 50$ and $B = 10$. The input spectrum is shown in Fig. 23-22*b*. Work out the value of

  (a) The dc shift
  (b) The peak output voltage of the fundamental
  (c) The peak output voltage of the second harmonic

23-6. Two sine waves drive a medium-signal amplifier. If the input frequencies are 56 kHz and 84 kHz, what frequencies does the output contain?

23-7. Three sine waves drive a medium-signal amplifier. If the input frequencies are 1 kHz, 4 kHz, and 9 kHz, what output frequencies are there? (Take two input frequencies at a time.)

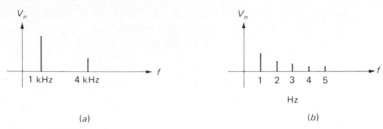

(a)   *(b)*

Figure 23-23.

23-8. Figure 23-23*a* shows the input spectrum to a large-signal amplifier. What harmonic frequencies does the output contain? What are the sum and difference frequencies of the first group?

23-9. The difference frequency $f_x - f_y$ equals 2 MHz and the sum equals 16 MHz. What are the input frequencies?

23-10. Figure 23-23*b* shows the output spectrum of an amplifier. If the amplifier is operating linearly, how many input sine waves are there? If the amplifier has harmonic distortion only, how many input sine waves are there?

23-11. On the piano keyboard of Fig. 23-12*a*, what is the fundamental frequency of the lowest C note? The highest C note? The highest E note?

23-12. Figure 23-24*a* shows the input spectrum to a medium-signal amplifier. What frequencies does the output spectrum contain?

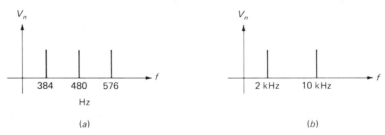

(a)   *(b)*

Figure 23-24.

23-13. Figure 23-24*b* shows the sum and difference frequencies out of a medium-signal amplifier. What are the input frequencies?

23-14. When tuned to channel 3, the mixer in a television set has a small input signal with a frequency of 63 MHz and LO signal with a frequency of 107 MHz. What is the difference frequency out of the mixer?

23-15. AM station frequencies are from 540 to 1600 kHz. In an AM radio, the received signal is one input to a mixer, and the LO signal is the other input. The frequency of the LO signal is 455 kHz greater than the received signal. What are the minimum and maximum LO frequencies? If a Hartley oscillator is used to generate the LO signal, what is the ratio of maximum to minimum tuning capacitance?

23-16. Suppose the transmitted frequency in Fig. 23-18 is 100 kHz. If the output of the audio amplifier in Fig. 23-19 sounds like a 2-kHz signal, what frequency does the BFO have?

23-17. A bipolar mixer has an input signal power of 3 μW and an output signal power of 1.5 μW. Express the conversion gain as an ordinary number and in decibels.

23-18. A FET mixer has an input signal voltage of 10 mV and an output signal voltage of 40 mV. Calculate the conversion voltage gain in decibels.

23-19. The data sheet of a diode mixer gives a conversion gain of −9 dB. If the input signal power to this mixer is 10 μW, how much output signal power is there?

23-20. Calculate the thermal noise voltage at room temperature for a bandwidth of 1 MHz and a resistance of (a) 1 MΩ, (b) 10 kΩ, and (c) 100 Ω.

23-21. How much thermal noise voltage is there at room temperature for a resistance of 10 kΩ and a bandwidth of 100 kHz? For a bandwidth of 1 kHz? 10 Hz?

23-22. The Thevenin resistance driving an amplifier is 1000 Ω. If the amplifier has a bandwidth of 1 MHz, approximately what is the input noise level at room temperature? If the amplifier has a voltage gain of 10,000, how large will the amplifier source noise be at the output?

24 *modulation*

Radio, television, and many other electronic systems would be impossible without *modulation;* it refers to a low-frequency signal controlling the amplitude, frequency, or phase of a high-frequency signal.

24-1 *amplitude modulation*

When the low-frequency signal controls the amplitude of the high-frequency signal, we get *amplitude modulation* (AM).

a simple modulator

Figure 24-1*a* shows a simple modulator. A high-frequency signal v_x is the input to a potentiometer; therefore, the amplitude of the output signal depends on the position of wiper. If we move the wiper up and down sinusoidally, we get the AM waveform of Fig. 24-1*b;* the amplitude or peak value of the high-frequency signal is varying at a low-frequency rate.

The high-frequency signal is called the *carrier,* and the low-frequency signal the *modulating signal.* Hundreds of carrier cycles normally occur during one cycle of the modulating signal. For this reason, an AM waveform on an oscilloscope looks like the signal of Fig. 24-1*c;* the positive peaks of the carrier are so closely

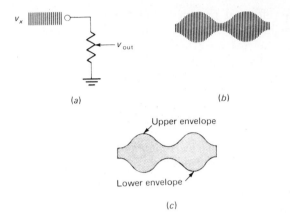

(a)   (b)

Upper envelope

Lower envelope

(c)

Figure 24-1. Amplitude modulation.

spaced they form a solid upper boundary known as the *upper envelope;* similarly, the negative peaks form the *lower envelope.*

a transistor modulator

Figure 24-2 is an example of a transistor modulator. Here is how it works. The carrier signal v_x is the input to a CE amplifier. The circuit amplifies the carrier by a factor of A, so that the output is Av_x. The modulating signal is

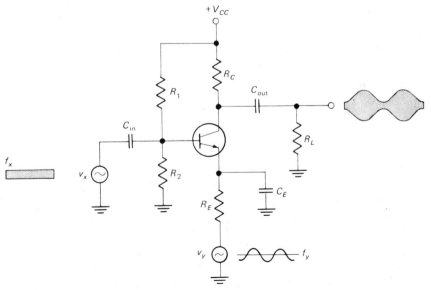

Figure 24-2. AM modulator.

part of the biasing; therefore, it produces low-frequency variations in emitter current; in turn, this means variations in r_e' and A. For this reason, the amplified carrier signal looks like the AM waveform shown; the peaks of the output vary sinusoidally with the modulating signal. Stated another way, the upper and lower envelopes have the shape of the modulating signal.

input voltages

For normal operation, Fig. 24-2 should have a small carrier. We do not want the carrier to influence voltage gain; only the modulating signal should do this. Therefore, the operation should be small-signal with respect to the carrier. On the other hand, the modulating signal is part of the biasing network. To produce noticeable changes in voltage gain, the modulating signal has to be large. For this reason, the operation is large-signal with respect to the modulating signal.

input frequencies

Usually, the carrier frequency f_x is much greater than the modulating frequency f_y. In Fig. 24-2, we need f_x at least 100 times greater than f_y. Here is the reason. The capacitors should look like low impedances to the carrier and like high impedances to the modulating signal; in this way, the carrier is coupled into and out of the circuit, but the modulating signal is blocked from the output.

EXAMPLE 24-1

In Fig. 24-3, the input peak of the carrier is 10 mV. The input peak of the modulating signal is 8 V. Calculate the minimum, quiescent, and maximum voltage gain.

SOLUTION

When the modulating voltage is zero, the voltage across the emitter resistor is 10 V. So, the emitter current is 1 mA and r_e' is approximately 25 Ω. The quiescent voltage gain therefore equals

$$A = \frac{r_C}{r_e'} = \frac{10,000 \parallel 1500}{25} = 52 \qquad \text{(quiescent)}$$

At the instant the modulating voltage reaches a positive peak of 8 V, only 2 V is across the 10-kΩ emitter resistor. At this instant, the emitter current is 0.2 mA and r_e' is 125 Ω. The corresponding voltage gain is

$$A = \frac{1300}{125} \cong 10 \qquad \text{(minimum)}$$

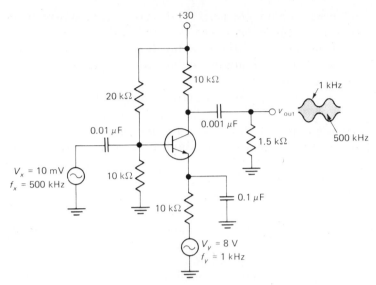

Figure 24-3.

At the negative peak of the modulating signal, the voltage across the emitter resistor is 18 V. The emitter current becomes 1.8 mA, r'_e decreases to 14 Ω, and

$$A = \frac{1300}{14} \cong 93 \qquad \text{(maximum)}$$

As far as the carrier is concerned, the CE amplifier changes its voltage gain from a low of 10 to a high of 93. Therefore, the output signal is an AM waveform as shown.

24-2 *need for modulation*

Without modulation, we would have no radio or television; we would also have to do without our modern telephone system. This section tells why.

practical antenna length

Books on antenna theory describe why the length of an antenna should be at least a quarter wavelength. In terms of frequency, a quarter wavelength equals

$$L = \frac{7.5(10^7)}{f} \qquad (24\text{-}1)$$

where L is in meters and f in hertz. If the antenna is shorter than this, it will not radiate signals efficiently.

Antennas would be immense at audio frequencies. For instance, to radiate 1000 Hz efficiently, we would need an antenna length of

$$L = \frac{7.5(10^7)}{1000} = 7.5(10^4) \text{ meters} \cong 47 \text{ miles}$$

This is much too long. For this reason, it is impractical to radiate audio frequencies directly into space. Instead, communication systems transmit radio frequencies (greater than 20 kHz).

am radio

AM broadcast signals use carrier frequencies between 540 and 1600 kHz. In the studio, the audio signal modulates the carrier to produce an AM signal like Fig. 24-4a. (Actually, the envelope is more complicated than this because voice and music contain many sinusoidal components.) The transmitting antenna radiates this AM signal into space.

When received, the AM signal is processed by different stages, and somewhere near the end of the receiver the original audio is retrieved (Fig. 24-4b). This audio signal is essentially the same as the original modulating signal. In this way, whatever goes into the microphone at the transmitting studio comes out the loudspeaker of the radio.

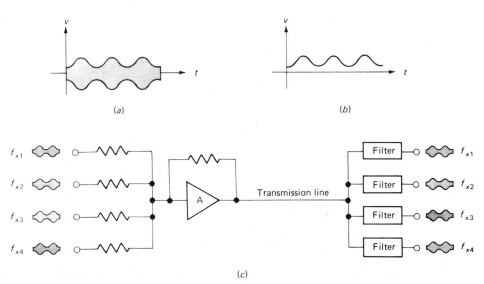

Figure 24-4. (a) *AM waveform.* (b) *Modulating signal.* (c) *Multiplexing four AM signals.*

telephone system

AM signals make modern telephone possible. If telephone systems had to use a single pair of wires for each conversation, the amount of the wire would be impractical. In today's telephone systems, a single pair of wires carries hundreds of conversations. Each conversation has a different carrier frequency, and this is the key to how we can separate the conversations.

Figure 24-4c shows the basic idea behind *multiplexing*, using the same transmission line to carry more than one signal. Four different AM signals are combined into a single-ended output by the op-amp summing circuit. These signals now travel down the transmission line (it may be miles long). At the receiving end, filters separate the different carrier frequencies, so that once again we have four separate AM signals. This approach can be extended to hundreds of AM signals and is used in modern telephone systems to reduce the amount of wire.

24-3 *percent modulation*

A sinusoidal modulating signal produces a sinusoidal variation in voltage gain expressed by

$$A = A_0(1 + m \sin \omega_y t) \tag{24-2}$$

where A_0 is the average gain and m is the *modulation coefficient*. As an example, if $A_0 = 100$ and $m = 0.5$, the graph of Eq. (24-2) looks like Fig. 24-5a. Here

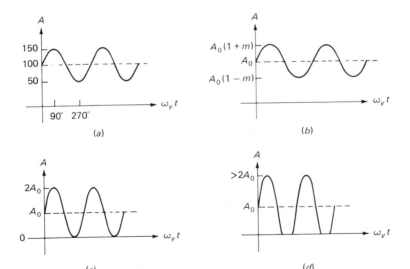

Figure 24-5. Modulator voltage gain. (a) m = 0.5. (b) *General case.* (c) m = 1. (d) m > 1.

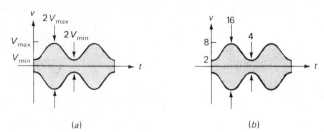

Figure 24-6.

we see the voltage gain swinging from a maximum of 150 to a minimum of 50.

In general, the graph of voltage gain looks like Fig. 24-5b. The maximum value is $A_0(1 + m)$, and the minimum is $A_0(1 - m)$. For the special case of m equal to 1, we get the variation shown in Fig. 24-5c. When m is greater than 1, the maximum voltage gain is greater than $2A_0$ (see Fig. 24-5d).

Percent modulation is used to describe the amount of modulation that has occurred. Percent modulation equals

$$\text{Percent modulation} = m \times 100\% \tag{24-3}$$

In Fig. 24-5a, m is 0.5; therefore, the percent modulation is 50 percent. In Fig. 24-5c, m is unity and percent modulation is 100 percent.

You can measure m as follows: Given a waveform like Fig. 24-6a, the maximum peak-to-peak voltage is $2V_{max}$ and the minimum peak-to-peak voltage is $2V_{min}$. It can be shown that

$$m = \frac{2V_{max} - 2V_{min}}{2V_{max} + 2V_{min}} \tag{24-4}***$$

As an example, if we see a waveform like Fig. 24-6b, we calculate

$$m = \frac{16 - 4}{16 + 4} = 0.6$$

which is equivalent to 60 percent modulation.

24-4 *am spectrum*

The output voltage of an AM modulator looks like Fig. 24-7a and equals

$$v_{out} = Av_x$$

If the carrier is sinusoidal, we may write

$$v_{out} = AV_x \sin \omega_x t$$

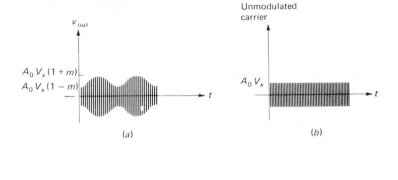

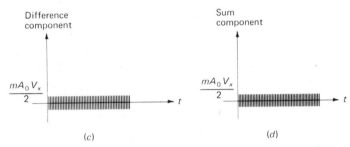

Figure 24-7. AM signal and its components (a) *AM wave.* (b) *Unmodulated carrier.* (c) *Difference component.* (d) *Sum component.*

where V_x is the peak value of the input carrier. With Eq. (24-2), the output voltage becomes

$$v_{out} = A_0(1 + m \sin \omega_y t) V_x \sin \omega_x t$$

or
$$v_{out} = A_0 V_x \sin \omega_x t + m A_0 V_x \sin \omega_y t \sin \omega_x t \qquad (24\text{-}5)$$

unmodulated carrier

The first term in Eq. (24-5) represents a sinusoidal component with a peak of $A_0 V_x$ and a frequency of f_x. Figure 24-7b is the graph of the first term. We call this the unmodulated carrier because it is the output voltage when m equals zero.

cross product

The second term in Eq. (24-5) is a cross product of two sine waves, similar to the cross products that occur in a mixer. As derived in the previous chapter, the product of two sine waves results in two new frequencies: a sum and a difference. Specifically, the second term of Eq. (24-5) equals

$$mA_0 V_x \sin \omega_y t \sin \omega_x t = \frac{mA_0 V_x}{2} \cos (\omega_x - \omega_y) t - \frac{mA_0 V_x}{2} \cos (\omega_x + \omega_y) t \quad (24\text{-}6)$$

The first term on the right is a sinusoid with a peak value of $mA_0 V_x/2$ and a difference frequency of $f_x - f_y$. The second term is also a sinusoid with a peak $mA_0 V_x/2$ but a sum frequency $f_x + f_y$. Figure 24-7c and d shows these sinusoidal components.

spectral components

In the time domain, an AM signal like Fig. 24-7a is the superposition of three sine waves (Fig. 24-7b through d). One sine wave has the same frequency as the carrier, another has the difference frequency, and the third has the sum frequency.

In terms of spectra, here is what AM means. Figure 24-8a is the input spectrum

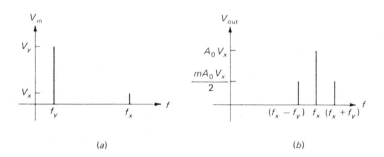

(a)   (b)

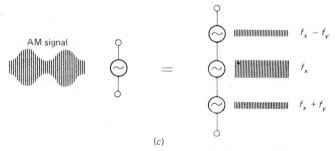

(c)

Figure 24-8. (a) *Input spectrum to modulator.* (b) *AM spectrum.* (c) *AM is superposition of three sine waves.*

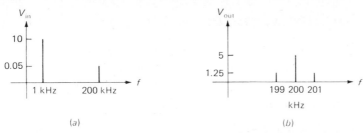

Figure 24-9. (a) *Input to AM modulator.* (b) *Output of AM modulator.*

to a modulator; the first line represents the large modulating signal with frequency f_y; the second line is for the small carrier with frequency f_x. Figure 24-8b is the output spectrum; here we see the amplified carrier between the difference and sum components. The difference component is sometimes called the *lower side frequency* and the sum is known as the *upper side frequency.*

The circuit implication of an AM signal is this. An AM signal is equivalent to three sine-wave sources in series as shown in Fig. 24-8c. This equivalence is not a mathematical fiction; the side frequencies really exist. In fact, with narrow-band filters we can separate the side frequencies from the carrier.

EXAMPLE 24-2

Figure 24-9a shows the inputs to a modulator. The modulating signal has a peak value of 10 V and a frequency of 1 kHz. The carrier has a peak of 0.05 V and a frequency of 200 kHz.

1. Find the frequencies in the output spectrum.
2. If A_0 is 100 and m is 0.5, calculate the peak values of the output spectral components.

SOLUTION

1. In the output spectrum, the center component has the same frequency as the carrier; in this case, 200 kHz. The lower side has a difference frequency of 199 kHz, and the upper side a sum of 201 kHz.
2. Referring to Fig. 24-8b, the unmodulated carrier has a peak value of $A_0 V_x$ and the side components have peaks of $m A_0 V_x/2$. So,

$$A_0 V_x = 100(0.05) = 5 \text{ V}$$

and

$$\frac{m A_0 V_x}{2} = \frac{0.5(100)0.05}{2} = 1.25 \text{ V}$$

Figure 24-9b shows the output spectrum.

24-5 *filtering am signals*

In a multiplexer like Fig. 24-4c, each filter must pass only the desired spectral components: the carrier and two side frequencies of an AM signal.

lower and upper sidebands

When the modulating signal is voice or music, the spectrum contains many sinusoidal components symbolized by the shaded region of Fig. 24-10a; the highest frequency is $f_{y(max)}$. As an example, if the spectrum is for piano music, $f_{y(max)}$ is approximately 10 kHz.

When voice or music modulates a carrier, the output spectrum looks like Fig. 24-10b. The shaded region to the left of the carrier is called the *lower sideband* because it contains all difference components. The region on the right of the carrier is the *upper sideband;* it contains all the sum components.

The spectrum of an AM signal is the set of frequencies in the lower and upper sidebands. Furthermore, the bandwidth of an AM signal is

$$B = [f_x + f_{y(max)}] - [f_x - f_{y(max)}]$$

or
$$B = 2f_{y(max)} \qquad\qquad (24\text{-}7)\text{***}$$

This says the bandwidth of an AM signal is twice the highest modulating frequency.

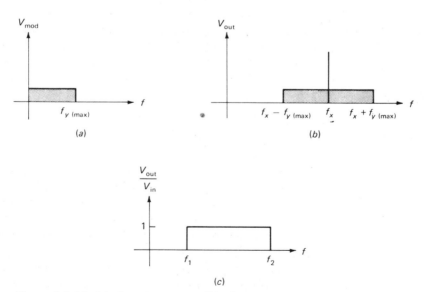

Figure 24-10. (a) *Input spectrum.* (b) *Output spectrum.* (c) *Window.*

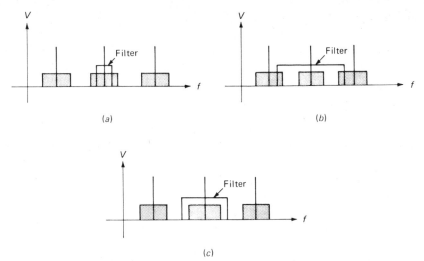

Figure 24-11. (a) *High-frequency distortion caused by narrow window.* (b) *Cross-talk caused by broad window.* (c) *Ideal window.*

frequency distortion

We have to be careful when we filter AM signals. If the filter has a passband as shown in Fig. 24-10c, we must be sure to pass the desired spectral components while stopping undesired ones. That is, the window must pass only the components belonging to the desired AM signal.

If the filter passband is too narrow, we get frequency distortion. For instance, Fig. 24-11a shows three AM signals. We are trying to pass the center signal while stopping the adjacent signals. In this case, the window is too narrow and we lose the sinusoidal components at the edges of the AM spectrum. This is equivalent to losing the higher modulating frequencies.

cross-talk

On the other hand, the filter passband may be too large as shown in Fig. 24-11b. We get not only the desired components but some components from adjacent signals. This results in *cross-talk,* an interference between channels of information. In a telephone system, this is equivalent to higher modulating frequencies from adjacent signals interfering with the desired conversation. In AM radio it amounts to higher modulating frequencies from adjacent stations interfering with the modulating signal of the desired station.

Figure 24-11c illustrates ideal filtering; we get only the components of one channel of information. There is no frequency distortion, nor is there any cross-talk.

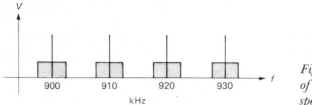

Figure 24-12. Part of AM broadcast spectrum.

am radio

AM radio transmission is an example of multiplexing because the same transmission medium (space) simultaneously transmits many channels of information (AM signals). The carrier frequencies are from 540 to 1600 kHz, spaced 10 kHz apart.

Figure 24-12 shows part of the AM broadcast spectrum. Especially notice that the spectra do not overlap. To avoid overlapping spectra, each AM station must keep all spectral components within 5 kHz of the carrier. This implies the highest modulating frequency $f_{y(\text{max})}$ is less than 5 kHz.

EXAMPLE 24-3
What is the bandwidth of the output signal in Fig. 24-13a?

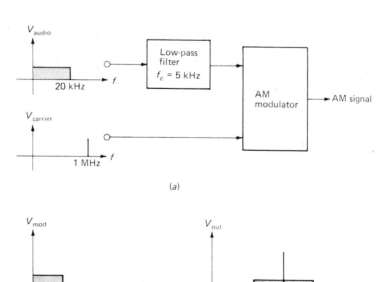

Figure 24-13.

SOLUTION

The input audio spectrum contains all components up to 20 kHz. This spectrum drives a low-pass filter with a cutoff frequency of 5 kHz. Ideally, the output of the filter contains only spectral components up to 5 kHz. For this reason, the highest modulating frequency is 5 kHz (Fig. 24-13*b*).

Since the carrier frequency is 1 MHz, the output spectrum has a lower sideband from 995 to 1000 kHz and an upper sideband from 1000 to 1005 kHz (Fig. 24-13*c*). So, the spectrum of the AM signal is from 995 to 1005 kHz; the bandwidth is 10 kHz, twice the highest modulating frequency.

Incidentally, this example indicates why AM radio has low fidelity. Since the highest modulating frequency is 5 kHz, many of the higher harmonics in voice and music are not transmitted.

24-6 *suppressed components*

In some communication systems, part of the AM spectrum is suppressed.

dsb-tc

Figure 24-14*a* shows the spectrum of an AM signal; a carrier component is between two sidebands. We classify this as *double sideband-transmitted carrier* (DSB-TC). In conventional AM radio, the transmitting antenna radiates a DSB-TC signal. The receiver picks up this signal and recovers the modulating signal.

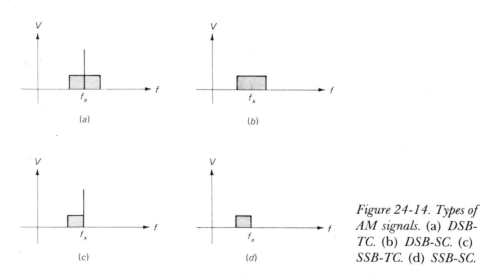

Figure 24-14. Types of AM signals. (a) DSB-TC. (b) DSB-SC. (c) SSB-TC. (d) SSB-SC.

dsb-sc

To transmit signals over great distances, the last stage in the transmitter must deliver large amounts of power to the antenna; even with a class C amplifier this implies a high power dissipation in the amplifying device. In some communication systems, this high dissipation is reduced by *suppressing the carrier*, that is, removing it as shown in Fig. 24-14*b*. This type of signal is classified as *double sideband-suppressed carrier* (DSB-SC). When the last stage of a transmitter amplifies a DSB-SC signal, the power dissipation is much less because there is no carrier power.

ssb-tc

Another way to reduce power dissipation is to use a *single sideband-transmitted carrier* (SSB-TC) signal illustrated by the spectrum of Fig. 24-14*c*. Either sideband may be suppressed. Since the last stage has to amplify only one sideband and the carrier, less power dissipation occurs in this stage.

ssb-sc

Single sideband-suppressed carrier (SSB-SC) is the most drastic suppression of all. We take out one sideband and the carrier (see Fig. 24-14*d*). Besides greatly reducing the power dissipation of the last stage, a SSB-SC signal has half the bandwidth of a DSB signal. Because of this, twice as many channels can be multiplexed in a given frequency range.

EXAMPLE 24-4

Figure 24-15*a* shows the spectrum of DSB-TC signal. If this signal drives a 1-kΩ resistor as shown in Fig. 24-15*b*, how much power does the resistor dissipate? If components are suppressed, how much power is there for each type of suppression?

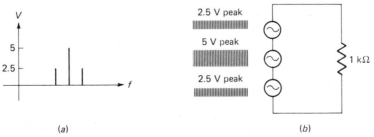

(a) (b)

Figure 24-15. (a) *AM Spectrum.* (b) *Equivalent circuit.*

SOLUTION

1. The carrier component delivers a power of

$$p_{\text{carrier}} = \frac{(0.707 \times 5)^2}{1000} = 12.5 \text{ mW}$$

The lower-side component produces

$$p_{\text{lower}} = \frac{(0.707 \times 2.5)^2}{1000} = 3.125 \text{ mW}$$

and the upper-side component,

$$p_{\text{upper}} = \frac{(0.707 \times 2.5)^2}{1000} = 3.125 \text{ mW}$$

Therefore, the total power produced by the DSB-TC signal is

$$p = (12.5 + 3.125 + 3.125) \text{ mW} = 18.75 \text{ mW}$$

2. In Fig. 24-15b, assume we remove the carrier. Then, we have a DSB-SC signal. In this case, the two side components deliver a power of

$$p = (3.125 + 3.125) \text{ mW} = 6.25 \text{ mW}$$

3. If we suppress one of the side components, the resulting SSB-TC signal produces a total power of

$$p = (12.5 + 3.125) \text{ mW} = 15.625 \text{ mW}$$

4. If the carrier and one side component are suppressed, the SSB-SC signal delivers a total power of

$$p = 3.125 \text{ mW}$$

This example gives you some idea of why suppressed systems are sometimes used. We could just as easily be dealing in watts or kilowatts of power instead of milliwatts. Suppressing the carrier cuts the power dissipation by a factor of three; suppressing the carrier and one side component reduces the power by a factor of six. Aside from lowering the required $P_{D(\text{max})}$ rating of the last stage of a transmitter, suppressing components also reduces the size of the power supply. In some transmitters, the reduced size and weight can be vitally important (like the transmitter in a spacecraft).

24-7 *am modulator circuits*

The simple transistor modulator of Fig. 24-2 works but is not as useful as other modulators. This section describes some modulators used in discrete and integrated circuits.

dsb-tc modulator

Figure 24-16 shows a diff-amp modulator. The carrier drives one input and is amplified by a factor of R_C/r'_e. The modulating signal drives Q_3, which delivers a common-mode signal to the diff amp. Because of this, the modulating signal is suppressed across the output. All that appears is

$$v_{out} = Av_x$$

Since the modulating signal controls the emitter current in each half of the diff amp, the voltage gain varies as the modulating signal. Because of this, the output is a DSB-TC signal as shown.

The diff-amp modulator of Fig. 24-16 has this advantage over the one-transistor modulator of Fig. 24-2: no restriction exists on the carrier and modulating frequencies. In other words, the simple modulator of Fig. 24-2 will not work properly unless the carrier frequency is at least 100 times greater than the modulating frequency. The diff-amp modulator, however, works properly even when f_x/f_y is less than 100.

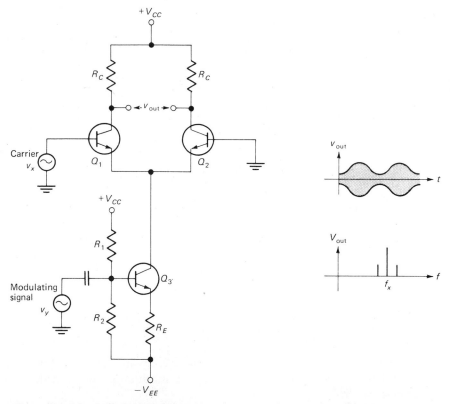

Figure 24-16. Diff-amp modulator.

For very high power requirements (kilowatts), DSB-TC modulators use a push-pull arrangement of vacuum tubes. This kind of modulation, called *plate modulation,* is discussed elsewhere.

dsb-sc modulator

Figure 24-17a is a DSB-SC modulator. The carrier drives Q_3 and produces common-mode operation. For this reason, the carrier is suppressed across the

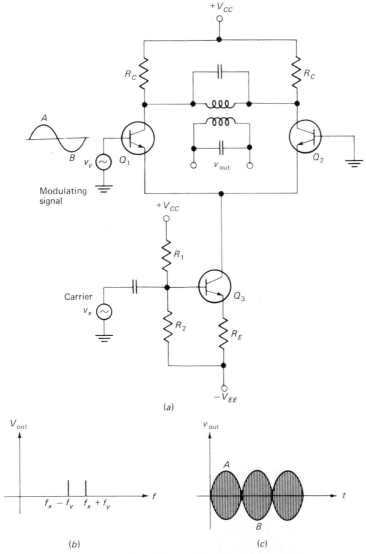

(a)

(b)   *(c)*

Figure 24-17. Balanced modulator. (a) *Circuit.* (b) *Output spectrum.* (c) *Output waveform.*

output. The modulating voltage produces large-signal operation and out-of-phase AM collector currents. The difference of these currents is a DSB-SC signal.

The proof is too lengthy to show here. Briefly, here is what happens. The tuned output blocks the low-frequency modulating signal because the resonant circuit is tuned to the carrier frequency. Since the carrier is suppressed by common-mode operation, all that reaches the final output are the lower and upper side frequencies. Because of this, the final output spectrum looks like Fig. 24-17b.

Figure 24-17c shows how a DSB-SC signal looks in the time domain. At first, this may appear like an ordinary AM signal. But look closely. The upper envelope resembles full-wave-rectified signal.

The modulator of Fig. 24-17a is often called a *balanced modulator* because each half of the diff amp is balanced (same gain) when no modulating signal is present. A balanced modulator suppresses the carrier.

ssb modulators

Given a DSB-TC or DSB-SC signal, we can produce SSB signals in either of two ways. First, we can use a bandpass filter to pass only one of the sidebands. Because the sidebands are so close together, we need an extremely sharp filter response. Usually, crystal filters are needed with this approach. Second, with the proper connection of modulators and phase shifters, we can make one of the sidebands cancel.

ic modulators

Some IC modulators are available. These use diff-amp modulators similar to Fig. 24-16. Examples are Motorola's MC1595 and 1596, which can act like DSB-TC or DSB-SC modulators. By cascading crystal filters, we can get SSB signals with these IC modulators.

24-8 *the envelope detector*

Once the AM signal is received, the carrier's work is over. Somewhere in the receiver is a special circuit that separates the modulating signal from the carrier. We call this circuit a *demodulator* or *detector*.

diode detector

Figure 24-18a shows one type of demodulator. Basically, it is a peak detector (discussed earlier in Sec. 5-5). Ideally, the peaks of the input signal are detected,

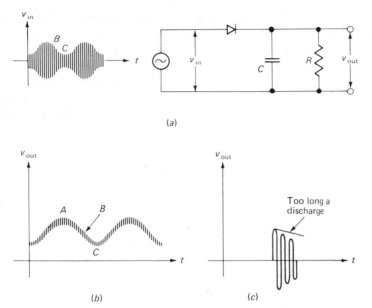

Figure 24-18. Envelope detector.

so the output is the upper envelope. For this reason, the circuit is called an *envelope detector.*

During each carrier cycle, the diode turns on briefly and charges the capacitor to the peak voltage of the particular carrier cycle. Between peaks, the capacitor discharges through the resistor. By making the *RC* time constant much greater than the period of the carrier, we get only a slight discharge between cycles. In this way, most of the carrier signal is removed. The output then looks like the upper envelope with a small ripple as shown in Fig. 24-18*b*.

required rc *time constant*

But here is a crucial idea. Between points *A* and *C* in Fig. 24-18*b*, each carrier peak is smaller than the preceding one. If the *RC* time constant is too long, the circuit cannot detect the next carrier peak (see Fig. 24-18*c*). The hardest part of the envelope to follow occurs at *B* in Fig. 24-18*b;* at this point, the envelope is decreasing at its fastest rate. With calculus, we can equate the rate of change of the envelope and the capacitor discharge to prove

$$f_{y(\text{max})} = \frac{1}{2\pi RCm} \qquad (24\text{-}8)\;\text{***}$$

where *m* is the modulation coefficient. With this equation, we can calculate the highest envelope frequency the detector can follow without attenuation. If the envelope frequency is greater than $f_{y(\text{max})}$, the detected output drops 20 dB per decade.

The stages following the envelope detector are usually audio amplifiers with an upper break frequency less than the carrier frequency. For this reason, the small carrier ripple in Fig. 24-18*b* is reduced by these audio stages. Occasionally, we may add a low-pass filter on the output of the detector to remove the small carrier ripple.

suppressed-carrier detection

If the received signal has a suppressed carrier, the envelope detector cannot recover the modulating signal. For instance, if we envelope-detect the DSB-SC signal of Fig. 24-17*c*, we get the upper envelope; this envelope has twice the frequency of the modulating signal.

What has to be done is this: the carrier frequency must be reinserted into the spectrum before detection. In other words, to receive a suppressed-carrier signal, we need an oscillator inside the receiver producing the same frequency f_x that was suppressed in the transmitter. This carrier component is added to the suppressed-carrier signal and the new signal is envelope-detected; the resulting output is the original modulating signal.

EXAMPLE 24-5
In Fig. 24-18*a*, suppose the incoming signal has an *m* of 0.5. If *R* equals 6.2 kΩ and *C* is 0.01 μF, what is the highest envelope frequency we can detect without attenuation?

SOLUTION
With Eq. (24-8),

$$f_{y(\text{max})} = \frac{1}{2\pi(6200)10^{-8}(0.5)} = 5.14 \text{ kHz}$$

This means the detector can follow any envelope frequency up to 5.14 kHz. Beyond this, we still get an output, but the amplitude falls off at 20 dB per decade.

24-9 *the tuned-radio-frequency receiver*

During the evolution of radio, the *tuned-radio-frequency* (TRF) receiver was used to receive AM signals. Today, a few special applications still use TRF receivers.

the basic idea

Figure 24-19*a* shows the block diagram of a TRF receiver. The idea is simple enough. The antenna picks up many AM signals, but the first four stages are tuned to only one of these signals. Therefore, the signal driving the detector is the AM signal from one station. The detected output then drives the audio amplifiers and the loudspeaker.

The TRF receiver relies on those first four tuned stages to separate one AM signal from the others. As mentioned, AM broadcast signals have bandwidths of 10 kHz; for this reason, the tuned RF stages in Fig. 24-19*a* have an overall bandwidth of approximately 10 kHz. Outside this bandwidth, the overall response should drop off as rapidly as possible.

selectivity

The *selectivity* of a receiver is its ability to separate the desired AM signal from all others received by the antenna. To have a high selectivity, a receiver must

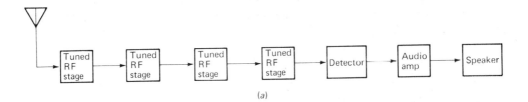

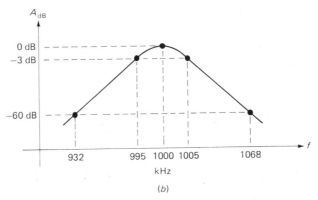

Figure 24-19. TRF receiver.

have a gain response that rolls off rapidly outside the 3-dB bandwidth. For a numerical definition, we will use

$$\text{Percent selectivity} = \frac{B_{\text{3-dB}}}{B_{\text{60-dB}}} \times 100\% \qquad (24\text{-}9)***$$

This formula compares the 3-dB bandwidth to the 60-dB bandwidth to give us a measure for how fast the gain rolls off. Maximum selectivity occurs when the 60-dB is almost as small as the 3-dB bandwidth. In this case, selectivity approaches 100 percent.

As an example, suppose Fig. 24-19*b* is the response of the four tuned RF stages of Fig. 24-19*a*. The receiver is tuned to a carrier frequency of 1000 kHz. The gain is down 3 dB at 995 and 1005 kHz, and down 60 dB at 932 and 1068 kHz. With Eq. (24-9),

$$\text{Percent selectivity} = \frac{1005 - 995}{1068 - 932} \times 100\% = 7.35\%$$

This is poor selectivity, but typical of a TRF receiver. Even a TRF receiver with six tuned stages has a selectivity less than 12 percent.

As mentioned, the best possible selectivity is 100 percent. A good communications receiver (not TRF) has a selectivity around 50 percent; this means the 60-dB bandwidth is only twice the 3-dB bandwidth. Small transistor radios have selectivities from 15 to 30 percent.

disadvantages

What is wrong with TRF receiver? Besides the poor selectivity (usually less than 10 percent), the problem of tuning is serious. When we want to change stations in Fig. 24-19*a,* we have to retune the first four stages. Gang-tuning these stages is difficult. Even worse, the bandwidth of the tuned stages changes with retuning because the Q of inductors changes with frequency. For these and other reasons, the TRF receiver is seldom used.

24-10 *the superheterodyne receiver*

The *superheterodyne* receiver (superhet) has a constant selectivity and is easier to tune over the frequency range.

the block diagram

Figure 24-20 is the block diagram of a superhet. Section 23-7 explained the mixing action near the front end of the superhet. As we recall, the received

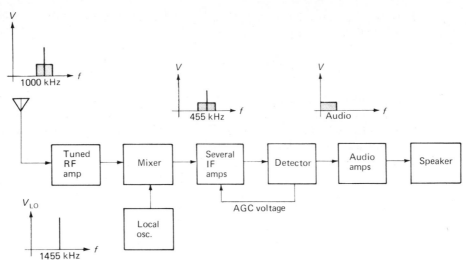

Figure 24-20. Superheterodyne receiver.

signal mixes with the LO signal to produce the IF signal. Because the LO of Fig. 24-20 operates 455 kHz above the RF signal, the IF spectrum has a center frequency of 455 kHz. The bandwidth of the IF signal is the same as the received signal. In AM radio, this means the RF and IF signals have bandwidths of 10 kHz.

The IF signal is amplified by *several* IF stages (see Fig. 24-20). The output of the last IF stage is envelope-detected to retrieve the modulating signal. This signal then drives the audio amplifiers and the loudspeaker. The output of the detector also feeds back a dc voltage to the IF amplifiers. This dc voltage is called the *AGC voltage* and is described in Sec. 24-11.

The major advantage of a superhet is the fixed-tuned IF stages. Because these stages are center-tuned to 455 kHz, the design is greatly simplified; furthermore, these IF amplifiers may be factory-tuned and left undisturbed; during normal use, only the RF and LO stages are ganged-tuned to different stations.

double-tuned if stages

To improve selectivity, the IF stages use double-tuned resonant circuits as shown in Fig. 24-21a. In a double-tuned stage like this, each resonant circuit is tuned to 455 kHz. *Critical coupling* occurs when the coefficient of coupling equals

$$k = \frac{1}{\sqrt{Q_1 Q_2}} \qquad (24\text{-}10)$$

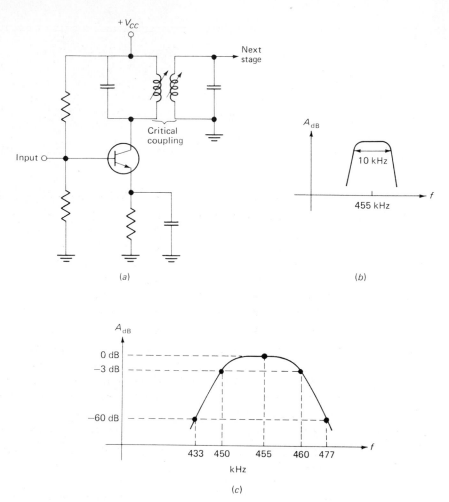

Figure 24-21. (a) *Double-tuned stage.* (b) *Maximally flat response.* (c) *Response of three double-tuned stages.*

where Q_1 is the Q of the primary resonant circuit and Q_2 the Q of the secondary resonant circuit.

When two resonant tanks are critically coupled, the response becomes *maximally flat* (see Fig. 24-21*b*). This means the response is flat over a broader range of frequencies than is possible with the single-tuned tank (compare Fig. 24-21*b* to 24-19*b*). Furthermore, the rolloff of a double-tuned stage is steeper than that of a single-tuned stage. For this reason, the double-tuned stage is more selective than the single-tuned stage.

As an example, Fig. 24-21*c* shows the overall response of three double-tuned IF stages. With a response like this, the selectivity is

$$\text{Percent selectivity} = \frac{460 - 450}{477 - 433} \times 100\% = 23\%$$

which is much better than the four-stage TRF receiver with a selectivity of only 7.35 percent (Fig. 24-19b). Furthermore, since the IF stages are fixed-tuned to 455 kHz, we get the same selectivity no matter what station we tune to.

superhet ics

As mentioned earlier, it's impractical to integrate inductors and large capacitors on a chip. This is why radio and television ICs contain only transistors and resistors; you have to connect external L's and C's to get tank circuits.

The LM1820 is an example of an integrated AM radio receiver. It contains an RF amplifier, oscillator, mixer, IF amplifiers, and AGC detector. By connecting external tank circuits, an audio amplifier like the LM386, and a loudspeaker, we get a complete AM radio.

summary

The superhet is the standard for most communication systems. The central idea is to use a mixer to shift the received spectrum down to the IF frequency; at this constant and lower frequency, the IF stages can efficiently amplify the signal with a maximally flat response. The maximally flat IF response minimizes frequency and phase distortion of the recovered modulating signal. For these reasons, the superhet is widely used in radio, television, radar, and many other electronic systems.

EXAMPLE 24-6
Some inexpensive receivers omit the RF stage as shown in Fig. 24-22a. What happens in such a receiver when the two given input AM signals are picked up by the antenna?

SOLUTION
The AM signal centered on 645 kHz will mix with the LO signal to produce an IF signal with a center frequency of

$$f_{out} = 1100 \text{ kHz} - 645 \text{ kHz} = 455 \text{ kHz}$$

Therefore, the lower of the two received AM signals produces a spectrum that will pass through the IF amplifiers.

The upper AM signal also mixes with the LO signal and produces an IF spectrum centered on

$$f_{out} = 1555 \text{ kHz} - 1100 \text{ kHz} = 455 \text{ kHz}$$

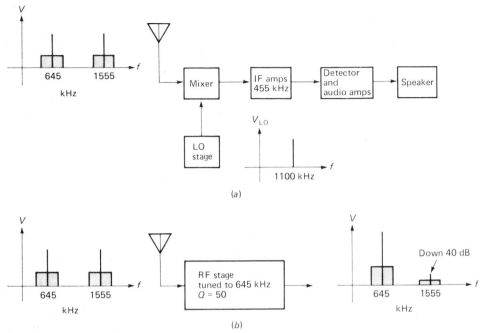

Figure 24-22. Image-frequency problem. (a) *Cross-talk produced by image frequency.* (b) *Tuned RF stage reduces effect of image signal.*

Therefore, the upper AM signal also produces a spectrum that passes through the IF amplifiers.

When both spectra are envelope-detected, we get audio information from both channels. Because of this, we hear both stations simultaneously (cross-talk).

This example brings out the need for a tuned RF stage. In the typical superhet, the desired RF spectrum is 455 kHz below the LO frequency, whereas the undesired RF spectrum is 455 kHz above the LO frequency; we call this undesired RF signal the *image signal*. The tuned RF stage in a superhet attenuates the image signal; by a straightforward calculation, we can show an RF stage with a Q of 50 provides the desired signal with 40 dB more gain than the image signal (Fig. 24-22b). In critical receiver applications, several tuned RF stages may be used.

24-11 *automatic gain control*

When a receiver is tuned from a weak to a strong station, the loudspeaker will blare unless the volume is immediately decreased. Or the volume may change

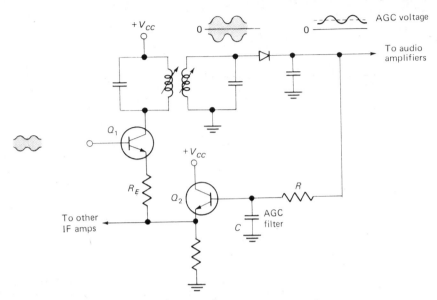

Figure 24-23. Automatic gain control.

because of fading. To counteract unwanted volume changes, most receivers use *automatic gain control* (AGC), as discussed in Sec. 14-8.

how it works

In Fig. 24-23, an AM signal drives the last IF stage Q_1. The amplified AM signal is envelope-detected to retrieve the audio signal. The average value of the detected output is labeled *AGC voltage;* this dc voltage is proportional to the unmodulated carrier. The stronger a received signal, the greater the AGC voltage.

The detected output drives audio amplifiers and also goes to a low-pass RC filter. The time constant of this AGC filter is typically 0.1 to 0.2 s. Because of this, the cutoff frequency is quite low, in the vicinity of 1 Hz.

The dc voltage out of the AGC filter drives the base of emitter follower Q_2. This positive AGC voltage causes the following to happen: the dc emitter current in each IF stage decreases, the value of r_e' increases, and the voltage gain decreases. Especially important, the larger the AGC voltage, the lower the voltage gain of the IF amplifiers. Therefore, when we tune from a weak to a strong station, the AGC voltage increases; this automatically reduces the IF gain and prevents the loudspeaker from blaring.

manual gain control

AGC action represents a coarse control of volume; we also need a manual control for fine adjustments. To avoid interacting with the AGC, the manual gain control

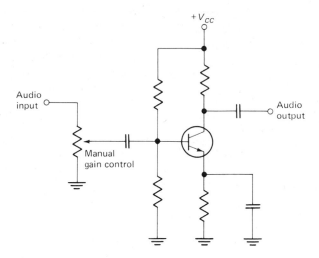

Figure 24-24. Manual gain control.

must be in a stage following the detector. Figure 24-24 shows a typical manual control. Changing the position of the wiper changes the size of the audio signal into the amplifier. Since the audio stage is after the detector, the manual gain control has no effect on the AM signal driving the detector. This is why AGC control does not interact with manual control.

other uses of agc

The idea of automatically controlling gain has spread to television receivers, signal generators, and other electronic systems. AGC is a form of dc modulation because a dc voltage controls the gain.

Commercial signal generators often use AGC to flatten or level variations

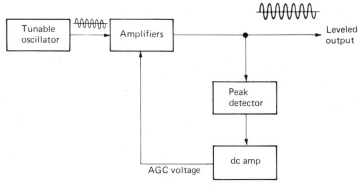

Figure 24-25. Signal generator with leveled output.

in signal strength. For instance, in Fig. 24-25 an oscillator drives amplifiers to produce an output signal. When we change the frequency of the oscillator, the signal strength changes because of variations in coil Q, transistor parameters, etc. Without AGC, the final output signal may easily vary 20 dB over the tuning range of the oscillator. But with AGC, we can usually keep the variation under 1 dB.

The basic idea of Fig. 24-25 is similar to the AGC of a receiver; the output signal is peak-detected to get a dc voltage, which is amplified and used as the AGC voltage to the amplifiers. When the output signal tries to increase, more AGC voltage is fed back to reduce the gain of the amplifiers. In this way, the *leveled output* remains almost constant over the entire frequency range.

EXAMPLE 24-7
Figure 24-26 shows a two-stage AGC amplifier, designed for the 50-kHz to 1-MHz range. For what output voltage does the AGC action become effective?

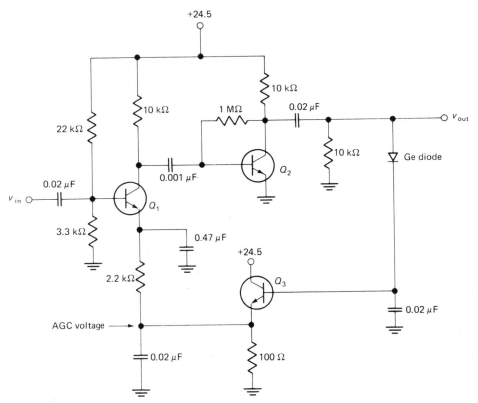

Figure 24-26.

SOLUTION

The first stage is voltage-divider biased with a base voltage of 3.2 V. Allowing 0.7 V across the base-emitter diode, we have 2.5 V at the top of the 2.2-kΩ emitter resistor. The AGC voltage at the bottom of this resistor is positive; therefore, the dc emitter current equals

$$I_E = \frac{2.5 - \text{AGC voltage}}{2200}$$

When the AGC voltage is small, the dc emitter current is large and so too is the voltage gain of the first stage. But for large AGC voltage, the emitter current and the gain both go down.

The AGC action becomes very effective when the AGC voltage approaches 2.5 V. Near this value, the dc emitter current I_E approaches zero. For this condition, further increases in the input signal are almost completely offset by gain reduction.

What is the approximate output voltage corresponding to 2.5 V of AGC voltage? Allowing 0.7 V for the V_{BE} of emitter follower Q_3, and 0.3 V for the peak-detector diode, we get

$$v_{\text{out(peak)}} = 2.5 + 0.7 + 0.3$$
$$= 3.5 \text{ V}$$

which is equivalent to 7 V peak-to-peak. This represents the upper limit (approximately) on the output voltage.

If you build the AGC amplifier of Fig. 24-26, here is what you will find. When the output voltage is much less than 7 V pp, the AGC voltage has no effect; for this weak-signal case, the input and output voltages are linearly related; a 100 percent increase in input voltage produces a 100 percent increase in output voltage.

But when the output voltage approaches 7 V pp, you will notice AGC action becomes quite effective. That is, when you increase the input signal, the gain is automatically reduced by almost the same factor; this results in an almost constant output. For instance, you may find that a 100 percent increase in input voltage produces less than a 1 percent change in output.

A final point about AGC amplifiers. Do not confuse an AGC amplifier with an overdriven amplifier. In an overdriven amplifier, an input sine wave produces an output square wave; when you increase the signal into the overdriven amplifier, you get no change in output amplitude. Similarly, in an AGC amplifier an increase in the input signal produces almost no increase in output voltage. But the crucial difference is this; the output of an AGC amplifier has the same shape as the input. In other words, an overdriven amplifier distorts the signal, but an AGC amplifier does not. For this reason, we use AGC amplifiers in radios, television receivers, signal generators, etc., where we want to limit the output signal with minimum distortion.

24-12 *frequency modulation*

When the modulation signal controls the frequency of the carrier, we get frequency modulation (FM). If the phase of the carrier is controlled, we have phase modulation (PM). FM is important in some communication systems; PM is rarely used. This section briefly describes the idea behind FM.

the basic idea

The key idea in an FM modulator is to vary the frequency of a sine wave, usually by changing the capacitance in an *LC* oscillator. For instance, Fig. 24-27*a* shows the tuning capacitor of an oscillator. If the capacitance has a fixed value, the oscillator produces a *quiescent frequency* f_{xQ}. Now, imagine that we *rock* the capacitor back and forth, causing it to vary from a minimum to a maximum value. The number of round trips per second between minimum and maximum capacitance is called the *rocking frequency*. If we are fast enough, we might manually reach a rocking frequency of 2 Hz.

Figure 24-27*b* shows the output of an oscillator whose capacitance is rocked back and forth. When the capacitance is maximum, we get a minimum frequency $f_{x(min)}$; for minimum capacitance, we get a maximum frequency $f_{x(max)}$. The number of round trips per second between $f_{x(min)}$ and $f_{x(max)}$ equals the rocking fre-

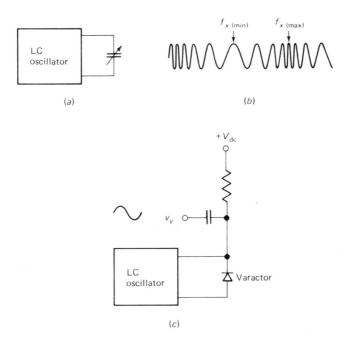

Figure 24-27. Frequency modulation.

quency. Figure 24-27*b* shows the FM signal in the time domain. All that happens is the frequency of oscillation rocks back and forth between $f_{x(\min)}$ and $f_{x(\max)}$.

frequency deviation

At any instant in time, the difference between the oscillator frequency f_x and the quiescent frequency f_{xQ} is called the *frequency deviation f_d*. Symbolically,

$$f_d = f_x - f_{xQ}$$

The maximum frequency deviation equals

$$f_{d(\max)} = f_{x(\max)} - f_{xQ} \qquad (24\text{-}11)$$

As an example, suppose $f_{x(\max)}$ equals 1075 kHz and f_{xQ} equals 1000 kHz. Then, the maximum frequency deviation is

$$f_{d(\max)} = 75 \text{ kHz}$$

In a practical FM modulator, we use a *varactor* in the place of a mechanically tuned capacitor. As discussed in Sec. 4-8, a reverse-biased diode acts like a capacitor; by varying the reverse bias, we can vary the capacitance. Figure 24-27*c* shows an FM modulator with a varactor determining the frequency of oscillation. The dc voltage sets up the quiescent value of capacitance; the modulating voltage v_y produces variations in this capacitance. Therefore, the modulating frequency equals the rocking frequency. For this reason, the modulating frequency equals the number of round trips per second between $f_{x(\min)}$ and $f_{x(\max)}$ in the FM wave of Fig. 24-27*b*.

If the modulating voltage is sinusoidal as shown in Fig. 24-27*c*, we may write

$$v_y = V_y \sin \omega_y t$$

where V_y is the peak voltage and ω_y the radian frequency. In a practical modulator, the frequency deviation is proportional to the modulating voltage. That is,

$$f_d = K V_y \sin \omega_y t$$

where K is a constant of proportionality that depends on the varactor and other factors. The maximum frequency deviation therefore equals

$$f_{d(\max)} = K V_y \qquad (24\text{-}12)$$

For instance, if K is 75 kHz/V, a peak modulating voltage V_y of 2 V produces

$$f_{d(\max)} = 150 \text{ kHz}$$

fm spectrum

It takes advanced mathematics to derive the spectrum of an FM signal. We will give the results without derivation. The spectrum of an FM signal contains the same frequencies as group 1 of a large-signal AM modulator. Specifically, an FM spectrum has these frequencies:

f_x
$f_x \pm f_y$
$f_x \pm 2f_y$
$f_x \pm 3f_y$
etc.

The carrier frequency f_x is in the center of the spectrum as shown by Fig. 24-28a; the distance between each component is f_y.

As an example, suppose f_x equals 1000 kHz and f_y is 15 kHz. Then the spectral components of an FM signal have frequencies of

1000 kHz
985 and 1015 kHz
970 and 1030 kHz
955 and 1045 kHz
etc.

The modulation coefficient of an FM signal equals

$$m = \frac{f_{d(\text{max})}}{f_y} \qquad (24\text{-}13)***$$

If $f_{d(\text{max})}$ is 75 kHz and f_y is 15 kHz, then m equals 5. As shown in Fig. 24-28a, the peak values are given by $J_0(m)$, $J_1(m)$, $J_2(m)$, etc. These particular J values

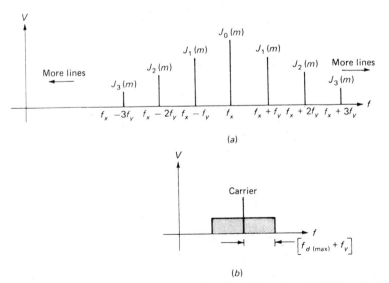

Figure 24-28. (a) FM spectrum. (b) Approximation of FM bandwidth.

are called *Bessel functions*. If you need the peak values of the spectral components, you can look up the *J* values in reference books.

As an example, if *m* equals 1, we want the values of $J_0(1)$, $J_1(1)$, $J_2(1)$, $J_3(1)$, etc. In any table of Bessel functions, we can read these values:

$J_0(1) = 0.765$ (carrier peak)
$J_1(1) = 0.440$ (two nearest side components)
$J_2(1) = 0.115$ (next nearest pair)
$J_3(1) = 0.0196$ (third nearest pair)

These *J* values represent the peak values of the spectral components relative to an unmodulated carrier peak of 1 V. To get the actual peak values, you multiply each *J* value by the peak value of the unmodulated carrier.

bandwidth of an fm signal

In practice, you will seldom bother looking up Bessel functions for an FM signal unless you are designing FM equipment or testing FM transmitters. But you should know the approximate *bandwidth of an FM signal* in case you ever have to filter FM signals.

In dealing with FM signals, many people use this convenient approximation: all significant side frequencies are within $[f_{d(\mathrm{max})} + f_y]$ of the carrier frequency as shown in Fig. 24-28b; when a component is further displaced than this, we can neglect it. In terms of bandwidth,

$$B = 2[f_{d(\mathrm{max})} + f_y] \qquad (24\text{-}14)***$$

As an example, suppose an FM broadcast signal has a carrier frequency of 100 MHz. By law, $f_{d(\mathrm{max})}$ must be less than 75 kHz and f_y less than 15 kHz. Therefore, in the worst case,

$$f_{d(\mathrm{max})} + f_y = 75 \text{ kHz} + 15 \text{ kHz} = 90 \text{ kHz}$$

The FM spectrum consists of a carrier frequency of 100 MHz and frequencies extending to approximately 90 kHz on each side of the carrier; this means a bandwidth of about 180 kHz.

advantage of fm over am

FM has some advantages over AM. We will only mention the *noise advantage*. External noise from electric motors, lightning, ignition systems, etc., tends to amplitude-modulate transmitted signals. In an AM receiver, you will hear this interfering noise. But in an FM receiver, the amplitude of the signal is unimportant; only the changing frequency is detected. Because of this, FM receivers are very quiet as far as external noise is concerned.

Commercially, frequency modulation is used in FM radio; it is also used for the sound portion of a television signal.

self-testing review

Read each of the following and provide the missing words. Answers appear at the beginning of the next question.

1. When a low-frequency signal controls the amplitude of a high-frequency signal, we get _____ modulation. The high-frequency signal is called the _____, and the low-frequency signal is the modulating signal.

2. *(amplitude, carrier)* AM radio uses carrier frequencies between 540 and 1600 kHz. In the studio an audio signal _____ the carrier to produce an _____ signal. In a multiplexed telephone system, each conversation has a different carrier frequency.

3. *(modulates, AM)* When voice or music amplitude-modulates a carrier, we get a _____ sideband that contains all the difference frequencies and an _____ sideband that contains all the sum frequencies. The spectrum of an AM signal is the set of frequencies in the lower and upper sidebands.

4. *(lower, upper)* In some communication systems, part of the AM spectrum is suppressed. In DSB-SC, the carrier is _____. But in SSB-TC, one of the _____ is suppressed. SSB-SC is the most drastic; here both the carrier and one of the sidebands are suppressed.

5. *(suppressed, sidebands)* To recover the audio signal from an AM signal, we need to demodulate or _____. If a peak detector is used, the circuit is called an _____ detector.

6. *(detect, envelope)* In a TRF receiver all stages before the detector are tuned to the carrier frequency. The disadvantage of a TRF receiver is its poor _____ and the problem of tuning the stages to the carrier frequency.

7. *(selectivity)* The superheterodyne receiver has a constant selectivity and is easier to tune over a large frequency range. To improve _____, the IF stages are double-tuned. This gives them a maximally flat response over a broader range of frequencies.

8. *(selectivity)* When a receiver is tuned from a weak to a strong station, the loudspeaker will blare unless the volume is immediately decreased. To counteract unwanted volume changes, most receivers use automatic _____ control (AGC).

9. *(gain)* When the modulating signal controls the frequency of the carrier, we get frequency modulation (FM). The advantage of FM over AM is less noise.

problems

24-1. In the transistor modulator of Fig. 24-3, suppose R_E is changed to 20 kΩ and V_y to 4 V. Calculate the quiescent, minimum, and maximum voltage gains. What are the corresponding peak carrier output voltages?

24-2. The input carrier to an AM modulator has a peak value of 20 mV. If the quiescent gain is 100, what is the quiescent peak output? If A changes $\pm$ 20 percent, what are the minimum and maximum output peak voltages?

24-3. How long is a quarter wavelength at each of these frequencies: 1 MHz (middle of AM broadcast band), 100 MHz (middle of FM broadcast band), and 200 MHz (channel 11 on TV)?

24-4. The lowest and highest carrier frequencies in AM radio are 540 kHz and 1600 kHz. Calculate the length of antennas for these carrier frequencies. (Use quarter wavelength.)

24-5. If an AM modulator has the gain variation of Fig. 24-29a, what is the value of m?

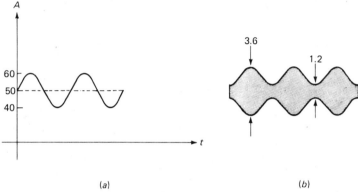

Figure 24-29.

24-6. If you see an AM signal like Fig. 24-29b on an oscilloscope, what is the modulation coefficient?

24-7. A modulating signal with spectral components from 20 Hz to 20 kHz amplitude-modulates a carrier with a frequency of 1 MHz. Calculate the side frequencies for f_y equal to 20 Hz. What are the side frequencies when f_y is 20 kHz?

24-8. An AM modulator has a quiescent voltage gain of 200 and a modulation coefficient of 0.3. If the peak carrier input V_x equals 10 mV, what are the peak values of the output spectral components?

24-9. Given an AM waveform like Fig. 24-29b, what are the peak values of the carrier and two side components?

24-10. Suppose the highest modulating frequency in an AM communication system is 20 kHz. What is the bandwidth value of the resulting AM signal?

24-11. The input frequencies to a DSB-SC modulator are 20 kHz and 1 MHz. What frequencies are in the output spectrum?

24-12. If the AM voltage of Fig. 24-29b appears across a 100-Ω resistor, how much carrier power is there? How much power is there in each sideband?

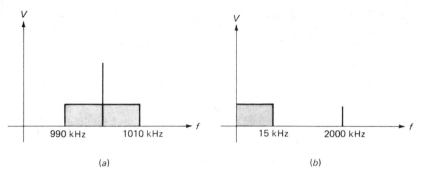

Figure 24-30.

24-13. An AM signal has the spectrum shown in Fig. 24-30a. What is the bandwidth of the AM signal? The bandwidth value?

24-14. If the signal whose spectrum is shown in Fig. 24-30a is changed to a SSB-SC signal, what does the new spectrum look like?

24-15. Figure 24-30b shows the input spectrum to a SSB-TC modulator. What does the output spectrum look like?

24-16. If the spectrum of Fig. 24-30b drives a DSB-SC modulator, what is the bandwidth value of the output spectrum?

24-17. An AM signal drives an envelope detector whose R is 10 kΩ and C is 1000 pF. If the modulation coefficient is 30 percent, what is the highest detectable modulating frequency without attenuation?

24-18. A receiver has a 3-dB bandwidth of 20 kHz and a 60-dB bandwidth of 80 kHz. What is the percent selectivity?

24-19. A superhet has an IF frequency of 455 kHz. What is the LO frequency when the received frequency is 540 kHz? When the received frequency is 1600 kHz?

24-20. An IF stage is critically coupled with a Q_1 of 50 and a Q_2 of 75. What is the value of k? In basic circuit books, you find this relation:

$$k = \frac{M}{\sqrt{L_1 L_2}}$$

If $L_1 = L_2 = 100$ μH, what is the value of M for critical coupling?

24-21. Figure 24-31 shows the response for the IF amplifiers in a receiver; calculate the percent selectivity.

24-22. The IF frequency equals 455 kHz in a receiver and the LO frequency is greater than the desired carrier frequency. What are the image frequencies for each of these cases:

(a) A desired carrier frequency of 550 kHz
(b) A LO frequency of 1355 kHz

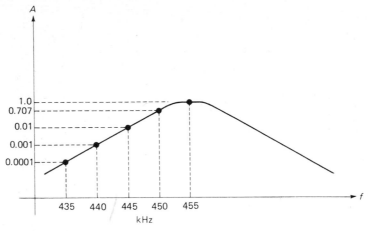

Figure 24-31.

24-23. In the AGC amplifier of Fig. 24-23, assume three identical IF stages receive the AGC voltage. The r_C seen by each collector is 1 kΩ and r_e' is 25 mV/I_E. With no AGC voltage, I_E equals 1 mA in each stage.

(a) What is the overall IF voltage gain when the AGC voltage is zero?
(b) When the AGC voltage reduces the dc emitter current to 0.5 mA, what is the total IF voltage gain?
(c) When I_E is 0.1 mA, what is the overall IF gain?

24-24. In an FM modulator, the highest oscillation frequency is 105.5 MHz. If the quiescent oscillation frequency is 105.3 MHz, what is the maximum frequency deviation?

24-25. In Eq. (24-12), K has a value of 10 kHz/V.

(a) If V_y equals 3 V, what does $f_{d(\text{max})}$ equal?
(b) If the maximum frequency deviation is 75 kHz, what is the peak modulating voltage?

24-26. The carrier frequency in an FM modulator is 20 MHz. If the modulating frequency is 100 kHz, what are the first three upper side frequencies in the FM spectrum? The first three lower side frequencies?

24-27. Calculate the modulation coefficient for an FM signal where the maximum frequency deviation is 50 kHz and the modulating frequency is 5 kHz.

24-28. In an FM modulator, the modulation coefficient equals 5 and the highest modulation frequency is 20 kHz. What is the approximate bandwidth value of the resulting FM signal?

25 *vacuum tubes*

Most historians feel electronics began when Edison discovered that a hot filament emits electrons (1883). Realizing the commercial value of Edison's discovery, Fleming developed the vacuum diode (1904). DeForest added a third electrode to get a vacuum *triode* (1906). Up to 1950, vacuum tubes dominated electronics; they were used in rectifiers, amplifiers, oscillators, modulators, etc.

Since 1950, solid-state devices like the bipolar transistor, FET, and IC have eliminated the vacuum tube from the mainstream of electronics. You will find vacuum tubes in older equipment, but new designs are solid-state; the exceptions are very high-power amplifiers, some microwave amplifiers, TV picture tubes, and a few other specialties. There is doubt about how long the tube can survive in its few applications; research may give us practical kilowatt transistors, solid-state TV picture tubes, etc.

The point is that the tube is now a specialty item with a shrinking future.

25-1 *vacuum diodes*

Figure 25-1*a* shows the schematic symbol for a vacuum diode. The *heater* glows during operation because it reaches incandescent temperatures. This warms the nearby *cathode,* a specially coated electrode that releases electrons by *thermionic emission.* In other words, thermal energy can free outer-orbit electrons near the

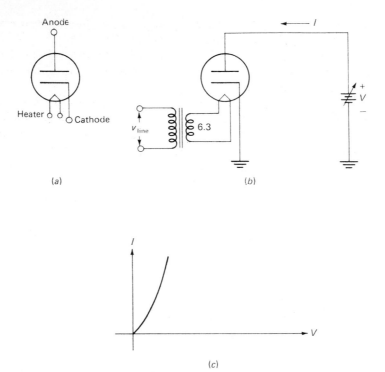

Figure 25-1. (a) *Vacuum diode.* (b) *Getting diode curve.* (c) *Diode curve.*

cathode surface. These emitted electrons are then in the vacuum between the cathode and the *anode.* If the anode of Fig. 25-1*a* is positive with respect to the cathode, it will attract and capture the emitted electrons.

Figure 25-1*b* is a circuit for measuring diode current. (The 6.3 V on the heater is a common value.) When anode voltage is positive, conventional current

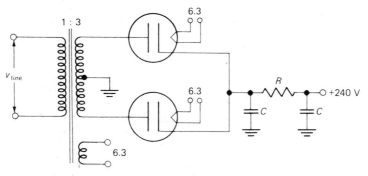

Figure 25-2. Full-wave peak rectifier.

I flows. The greater *V* is, the larger *I* is. On the other hand, when *V* is negative, *I* is zero because the anode no longer captures electrons.

Figure 25-1*c* is a typical graph of *I* versus *V*. Unlike the semiconductor diode, the vacuum diode has no barrier potential. Because of this, current *I* appears as soon as *V* goes positive.

Ideally, we can approximate the vacuum diode as a switch; the switch is closed when the diode is forward-biased, and open when reverse-biased. For this reason, we can use vacuum diodes in any of the rectifier circuits discussed in Chap. 5.

As an example, Fig. 25-2 shows a full-wave peak rectifier using vacuum diodes instead of semiconductor diodes. If peak line voltage is 160 V, then each half of the secondary has a peak voltage of 240 V. Ideally, the first capacitor charges to 240 V, which is the approximate value of output dc voltage.

25-2 *triodes*

Figure 25-3*a* shows the schematic symbol for a *triode,* a vacuum tube with three main electrodes. The *control grid* is between the cathode and anode. Because of this, it can influence the current. When the grid is negative with respect to

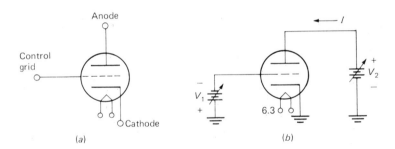

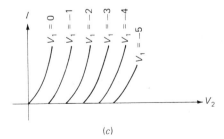

Figure 25-3. (a) *Vacuum triode.* (b) *Getting triode curves.* (c) *Triode curves.*

the cathode, it repels some of the electrons trying to flow from cathode to anode. The more negative the grid, the smaller the current.

triode curves

Figure 25-3*b* is a circuit for getting *triode curves* (similar to collector curves). By holding the grid voltage fixed, we can vary the anode voltage and measure the resulting anode current. For instance, if grid voltage V_1 is zero, we get the first curve of Fig. 25-3*c*. If we make V_1 equal to -1 V, it will take more anode voltage to overcome the effects of the negative grid. For this reason, the $V_1 = -1$ V curve has a knee as shown. Making the grid more negative increases the knee voltage and results in the other curves of Fig. 25-3*c*.

self-bias

For linear operation the triode has to be biased at a quiescent point that allows ac fluctuations without clipping. As far as biasing is concerned, the triode is similar to an *n*-channel JFET. Because of this, we can use any of the biasing circuits discussed in Chap. 14 for *n*-channel JFETs.

Figure 25-4 shows *self-bias*, the common way to bias a triode. Anode current flows through R_K, producing a plus-minus voltage across it. This makes the cathode positive with respect to the grid, equivalent to the grid being negative with respect to the cathode. Using the triode curves, we can select appropriate values for R_K and R to set up a desired Q point, similar to setting up the Q point of a transistor circuit. Once we have a Q point, the amplifier is ready for class A operation.

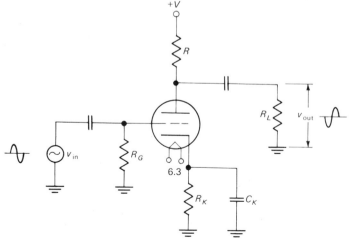

Figure 25-4. A triode amplifier.

voltage gain

C_K is nothing more than a bypass capacitor; it ac grounds the cathode to prevent loss of voltage gain. If C_K is omitted or accidentally opens during operation, we get a swamped amplifier whose voltage gain ideally equals

$$\frac{v_{\text{out}}}{v_{\text{in}}} \cong \frac{r_L}{R_K} \qquad \text{(ideal)} \qquad (25\text{-}1)$$

where r_L equals $R \parallel R_L$ (the ac load seen by the anode).

When C_K is present, the cathode is at ac ground. By an analysis similar to transistor analysis, we can prove the voltage gain of a triode amplifier equals

$$\frac{v_{\text{out}}}{v_{\text{in}}} = \frac{r_L}{r_p + r_L} \mu \qquad (25\text{-}2)$$

where μ and r_p are given on the data sheet of the tube.

As an example, suppose $R = 30 \text{ k}\Omega$ and $R_L = 60 \text{ k}\Omega$. Then,

$$r_L = 30,000 \parallel 60,000 = 20 \text{ k}\Omega$$

If the triode of Fig. 25-4 has a $\mu = 15$ and an $r_p = 5 \text{ k}\Omega$, the voltage gain equals

$$\frac{v_{\text{out}}}{v_{\text{in}}} = \frac{20,000}{5000 + 20,000} 15 = 12$$

In Eq. (25-2), $r_L/(r_p + r_L)$ is less than unity for all values of r_L and r_p. As a result, the maximum possible voltage gain of a triode amplifier is

$$A_{\text{max}} = \mu \qquad (25\text{-}3)$$

This is useful when examining data sheets for triodes; a glance at the value of μ immediately tells us the maximum possible gain of the particular triode.

A study shows that two-thirds of available triode types have μ's less than 50; about 90 percent have μ's less than 75. As a guide, therefore, two-thirds of triode amplifiers have voltage gains less than 50; 90 percent of them have voltage gains less than 75.

25-3 tetrodes and pentodes

Like transistors and FETs, the triode has internal capacitances which become part of lag networks in an amplifier. In particular, the capacitance between the grid and the anode results in a large input Miller capacitance that limits the usefulness of a triode at high frequencies.

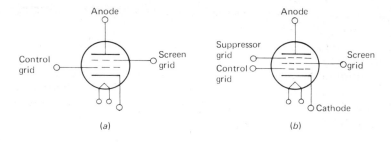

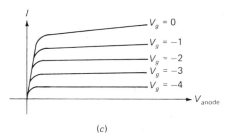

Figure 25-5. (a) *Tetrode.* (b) *Pentode.* (c) *Pentode curves.*

The *tetrode* of Fig. 25-5*a* uses a *screen grid* between the control grid and the anode. By applying a constant positive voltage to the screen grid, we can greatly reduce the capacitance between the control grid and the anode. This permits the tetrode to operate at higher frequencies than the triode.

The tetrode has a problem called *secondary emission* which limits its usefulness as a linear amplifier. What happens is this. Electrons arrive at the anode with a high enough velocity to dislodge outer-orbit electrons from the anode surface. Some of these emitted electrons are attracted by the positive screen grid; this reduces the anode current and leads to a distortion of the amplified signal.

The *pentode* of Fig. 25-5*b* uses a *suppressor grid* between the screen grid and the anode. This new grid is either grounded or connected to the cathode. Because of its low voltage, the suppressor grid forces secondarily emitted electrons to return to the anode; in this way, the undesirable effects of secondary emission are eliminated.

Figure 25-5*c* shows a typical set of pentode curves. As we see, these resemble JFET curves. For this reason, pentodes act like current sources. In fact, the ac analysis of a pentode amplifier is similar to a JFET amplifier and results in a voltage gain of

$$\frac{v_{\text{out}}}{v_{\text{in}}} \cong g_m r_L \qquad \text{(ac-grounded cathode)} \tag{25-4}$$

where g_m is the transconductance of the pentode. Similar to FETs, almost all pentodes have g_m's under $10,000\ \mu S$.

EXAMPLE 25-1

In the pentode amplifier of Fig. 25-6, the dc anode current equals 8 mA and the screen-grid current equals 2 mA. Calculate the dc voltage from anode to ground and from cathode to ground. How much dc voltage is there from the control grid to ground?

SOLUTION

The anode current flows through the 10-kΩ resistor, producing a dc voltage from anode to ground of

$$V = 250 - 0.008(10,000) = 170\ V$$

In a pentode, the dc cathode current is the sum of the anode current and screen-grid current. For this reason, the dc voltage from cathode to ground equals

$$V = 0.010(200) = 2\ V$$

Similar to a JFET, the controlling element has a negligible dc current. In other words, the control grid draws negligible current because it is negative with respect to the cathode. For this reason, the dc voltage is zero from the control grid to ground.

EXAMPLE 25-2

Calculate the voltage gain from input to output in Fig. 25-6.

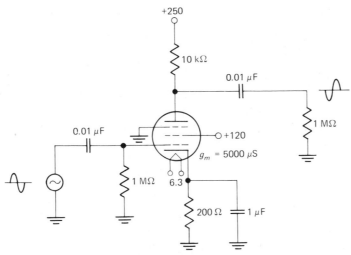

Figure 25-6. Pentode amplifier.

SOLUTION

The ac load resistance seen by the anode equals

$$r_L = 10,000 \parallel 1,000,000 \cong 10 \text{ k}\Omega$$

As shown in Fig. 25-6, the pentode has a g_m of 5000 μS; therefore, the voltage gain equals

$$\frac{v_{\text{out}}}{v_{\text{in}}} \cong g_m r_L = 0.005(10,000) = 50$$

In Fig. 25-6 the output sine wave is 180° out of phase with the input sine wave when the frequency is in the midband of the amplifier. As discussed in Chap. 16, coupling and bypass capacitors determine the lower break frequencies; internal and stray capacitances produce the upper break frequencies.

25-4 *transistors versus tubes*

This section briefly examines some of the differences between transistors and tubes.

input impedance

Vacuum tubes have very high input resistance because the control grid draws negligible dc current. If it weren't for FETs, we would need the vacuum tube for high-input-impedance applications. There are almost no applications, however, where the FET cannot replace the vacuum tube when input impedance is the main consideration. Because the FET is more convenient, it is replacing the vacuum tube in electronic voltmeters, oscilloscopes, and other instruments where high input impedance is needed.

voltage gain

You can get much more voltage gain with a bipolar transistor than with a vacuum tube. As mentioned, triode amplifiers normally have voltage gains less than 75. We can compare bipolars and pentodes as follows:

$$\frac{A_{\text{bipolar}}}{A_{\text{pentode}}} = \frac{g_{m(\text{bipolar})}}{g_{m(\text{pentode})}}$$

The typical pentode has a g_m under 10,000 μS, but a bipolar has a g_m of about 40,000 μS when I_E is 1 mA, 80,000 μS for 2 mA, 160,000 μS for 4 mA, and so on. Therefore, given the same ac load resistance, we can get much more gain with a bipolar than with a pentode.

voltages

Vacuum tubes run at much higher dc voltages than transistors. Typical power-supply voltages with receiver-type tubes are 100 to 300 V. More often than not, this is a disadvantage; lower voltages allow us to build portable, lightweight transistor equipment instead of heavier vacuum-tube equipment.

power dissipation

Most power transistors have dissipations under 300 W. Vacuum tubes, on the other hand, can easily have dissipations in the kilowatt region. Therefore, the areas where the tube is still needed are high-power amplifiers, transmitters, linear accelerators, industrial control systems, microwave systems, etc. In the microwave systems, the kinds of tubes we are talking about are *klystrons, magnetrons,* and *traveling-wave* tubes.

heater

The transistor needs no heater; the vacuum tube does. This is one of the most important advantages of a transistor over a vacuum tube. The typical heater requires more than a watt of power. In a system using many tubes, the required heater power leads to a bulky power supply. Furthermore, there is the problem of getting rid of the heat.

The heater limits the tube's useful life to a few thousand hours. Transistors, on the other hand, normally last for many years. This is the reason transistors are usually soldered permanently into a circuit whereas tubes are plugged into sockets.

miscellaneous

Low-power transistors are much smaller than vacuum tubes, roughly the size of a pea compared to a flashlight battery. This means transistor circuits can be more compact and lightweight. Furthermore, unlike the vacuum tube, a transistor is ready to operate as soon as power is applied.

Finally, transistors can be integrated along with resistors and diodes to produce incredibly small circuits.

In summary, solid-state devices have obsoleted the vacuum tube in most areas of electronics. Before long, the tube may be as anachronous as a zeppelin in the jet age.

self-testing review

Read each of the following and provide the missing words. Answers appear at the beginning of the next question.

1. In a vacuum diode the _____ glows during operation because it reaches incandescent temperatures. This warms the nearby _____, a specially coated electrode that releases electrons by thermionic emission.

2. *(heater, cathode)* A _____ is a vacuum tube with _____ electrodes. The control _____ is between the cathode and the anode. When the grid is negative with respect to the cathode, it repels some of the electrons trying to flow from cathode to anode.

3. *(triode, three, grid)* The tetrode uses a _____ grid between the control grid and the anode. The _____ has a suppressor grid between the screen grid and the anode.

4. *(screen, pentode)* Solid-state devices have obsoleted the vacuum tube in most areas of electronics. The exceptions are very-high power amplifiers, some microwave tubes, TV picture tubes, and a few other specialties.

problems

25-1. In a triode amplifier, μ equals 25, r_p is 6 kΩ, and r_L is 18 kΩ. Calculate the voltage gain.

25-2. The relation between μ, r_p, and g_m is $\mu = g_m r_p$. If a tube has a μ of 20 and an r_p of 5 kΩ, what does its g_m equal?

25-3. A pentode amplifier has a g_m of 6000 μS. If r_L equals 10 kΩ, what does the voltage gain equal?

25-4. If the turns ratio in Fig. 25-2 is changed to 1:5, what is the ideal value of output voltage?

25-5. In Fig. 25-6, suppose the dc anode current equals 10 mA and the screen-grid current equals 3 mA. What is the dc voltage from anode to ground? The dc voltage from cathode to ground?

25-6. If the 1-μF bypass capacitor is left out, the cathode no longer is at ac ground in Fig. 25-6. By a derivation similar to that given for the JFET, the voltage gain equals

$$\frac{v_{\text{out}}}{v_{\text{in}}} = \frac{r_L}{R_K + 1/g_m} \qquad \text{(no bypass)}$$

Calculate the voltage gain for this condition.

answers to selected odd-numbered problems

Chapter 1

1-1. 26 Ω   **1-3.** 1.26 Ω, 796 kHz   **1-5.** Between 13 Ω and 1.6 kΩ   **1-7.** $V_{TH} = 10$ V. $R_{TH} = 1.5$ kΩ   **1-9.** $V_{TH} = 0$, $R_{TH} = 4$ kΩ   **1-11.** $R_{TH} =$ infinity, both cases

Chapter 3

3-1. 0.6125 V, 0.5375 V, 0.3875 V   **3-3.** 64 nA, 2.048 µA   **3-5.** $2.5(10^6)$

Chapter 4

4-1. A is forward-biased; B is reverse-biased   **4-3.** A, both cases   **4-5.** Upper diode is forward-biased; lower diode is reverse-biased   **4-7.** 8 Ω   **4-9.** 10 Ω   **4-11.** 10 Ω   **4-13.** 2 mA   **4-15.** 2.5 mA; half-wave signal with a positive peak of 50 V   **4-17.** 1.5 mA   **4-19.** Half-wave signal with a positive peak of 8 V (ideal) or 7.44 V (second approximation)   **4-21.** 10 V   **4-23.** 0.34 mA   **4-25.** 24.8 mA, 7.44 V   **4-27.** $2.5(10^9)$ Ω, 2.5 MΩ   **4-29.** 300 mW   **4-31.** 7.5 Ω   **4-33.** 18.12 V   **4-35.** 12.5 mA, 0.1875 W   **4-37.** 25 V   **4-39.** 12.2 mA

Chapter 5

5-1. 148 V, 177 V   **5-3.** 170 mA, 54 mA   **5-5.** 84.9 V, 1N4002, 1N3070   **5-7.** 1N3070   **5-9.** 60 mA, 56.6 V, 30 mA   **5-11.** 135 mA, 67.5 mA, 85 V approximately   **5-13.** Yes   **5-15.** 14.9 V   **5-17.** 0.594 H, 4.7 H   **5-19.** 4 V rms   **5-21.** 0.5 V rms with Eq. (5-18)   **5-23.** Approximately 12 V, 0.3 V rms, 2.5%   **5-25.** 600 µF   **5-27.** 16.4 V, 5%, 0.82 V rms   **5-29.** (a) 0.895, (b) 0.865, (c) 0.845, (d) 0.825, (e) 0.79, (f) 0.765   **5-31.** 120 µF with Eq. (5-19); 95 µF with Fig. 5-13   **5-33.** 18.9 V, 80 µV   **5-35.** 10 V   **5-37.** 14.4 V, 0.8 mV   **5-39.** 66.7%, 2.5 kΩ   **5-41.** 19.8 V

5-43. 2036 V, 1357 V   **5-45.** 36.4%   **5-47.** 500 Ω   **5-49.** 12 V, 12.07 V   **5-51.** 0%, 0.278%

Chapter 6

6-1. 0.75 V, 0.65 V   **6-3.** 200,000; 9.8 million   **6-5.** 0.974   **6-7.** 4.7 V   **6-9.** 0.3 mA   **6-11.** 200 (approximate) or 199 (exact); 500 (approximate) or 499 (exact)   **6-13.** 600 mA   **6-15.** 10 mA; 0.111 mA   **6-17.** 0.125 mA   **6-19.** 0.1 mA, 20.5 mA   **6-21.** 100, 5 mA, 233 mA   **6-23.** 19.995 V, 50 V, no

Chapter 7

7-1. 12.9 V   **7-3.** 19.3 mA, 10.4 V   **7-5.** Saturation currents are 3 mA, 12 mA, and 60 mA; cutoff voltages equal 30 V   **7-7.** All stages have minimum of 9.28 V and maximum of 11.2 V   **7-9.** 26.9 mA, 7.8 V   **7-11.** Intercepts are 3.33 mA and 10 V; Q point at 0.364 mA and 8.9 V   **7-13.** Intercepts are 160 mA and 20 V; Q point at 79.6 mA and 10.1 V   **7-15.** Collector voltages are 28.2 V, 26.6 V, and 23.8 V; emitter voltages are 3.59 V, 6.8 V, and 9.3 V   **7-17.** 1.01 mA and 4.93 V, 2.33 mA and 5.35 V, 42.3 mA and 5.77 V   **7-19.** 131 kΩ (exact) or 130 kΩ (nearest practical value)   **7-21.** 2.11 mA and 9.47 V, 4.83 mA and 10.4 V, 87.7 mA and 11.2 V   **7-23.** 600 Ω   **7-25.** 150 Ω   **7-27.** 7.15 V, 3.4 V, 5.79 V   **7-29.** 1.22 mA and 12.9 V, 5.9 mA and 7.3 V, 1.45 mA and 15.2 V

Chapter 8

8-1. 2.5 µF   **8-3.** 100 µF (approximate) or 105 µF (exact)   **8-5.** 100 µF   **8-15.** Approximately 5 mA   **8-17.** 500 Ω   **8-19.** 25 Ω (ideal) or 26.9 Ω (second approximation)   **8-21.** 12.5 Ω (ideal) or 14.1 Ω (second approximation)   **8-23.** 20.9

735

Ω (ideal)   **8-25.** 200, 300   **8-27.** 0.1 mA, 0.5 mA

Chapter 9

9-1. 74.1 μA   **9-3.** 0.167 mA, 4.17 mV, 25
9-5. 1.29 V peak   **9-7.** 200   **9-9.** At the base, a sine wave with a peak value of 5 mV superimposed on 20 V dc; at the emitter, a horizontal line at the 20-V level (ideal) or at the 20.7-V level (second); at the collector, a sine wave with a peak of 1.5 V at the 10-V level (ideal) or at the 8.6-V level (second)   **9-11.** 14.1 including r_e'
9-13. 5 V and 2 mV; 4.3 V and 0; 14 V and 160 mV   **9-15.** 26   **9-17.** Transistor will saturate   **9-19.** Roughly 1.25 kΩ with the second approximation of I_E; 469 Ω **9-21.** (a) 10 kΩ, (b) 20 kΩ, (c) 25 kΩ, neglecting 10-kΩ R_E; or 24.4 kΩ including R_E, (d) 50 kΩ **9-23.** 0.526, 80, 42.1   **9-25.** Approximately 0.01 and 0.8
9-27. 1608   **9-29.** 1.73 V   **9-31.** 0.426, 0.421   **9-33.** 1.57 V assuming unity gain in first stage

Chapter 10

10-1. Ignoring V_{BE}, 10 mA and 10 V; including V_{BE}, 9.3 mA and 11.4 V   **10-3.** Including V_{BE}, 23.6 mA and 5.28 V   **10-5.** Ignoring V_{BE}, 10 mA and 10 V; including it, 9.3 mA and 10.7 V
10-7. 50 mA and 10 V   **10-9.** Load-line intercepts are 20 mA and 20 V   **10-11.** Load-line intercepts are 66.7 mA and 20 V   **10-13.** 7.05 V and 94 mA   **10-15.** 60 mA and 12 V neglecting V_{BE}; 62.8 mA and 12.6 V when V_{BE} is included
10-17. 20 V   **10-19.** Ideally, yes
10-21. Ignoring V_{BE}, 150 Ω; including V_{BE}, 657 Ω **10-23.** Ideally, 0.25 W
10-25. Approximately 0.5 W and 1 W
10-27. Approximately 1 W (both)
10-29. Ideally, 0.5 W   **10-31.** 4 W, 8 W
10-33. Ideally, 16.7%   **10-35.** Ideally, 33%
10-37. 0.167 Ω **10-39.** 0.0651 Ω
10-41. Ignoring V_{BE}, 20, 1000, 125 Ω
10-43. 3.31, 248, 450 Ω, and 34.6 Ω
10-45. 92.5°C   **10-47.** 189 mW
10-49. 139°C   **10-51.** 15 W

Chapter 11

11-1. 0.625 A   **11-3.** 0.938 A   **11-5.** Approximately 23.8 mA   **11-7.** 1.45 mA, 75 mA
11-9. 3.13 W   **11-11.** 0.625 W   **11-13.** 3.98 W
11-15. 10 W   **11-17.** 500 Ω, 0.909, 45.5
11-19. 74.3 Ω **11-21.** 1 W, 0.2 W
11-23. Ideally, 7.07 V rms

Chapter 12

12-1. 5 V, -5 V, -4.3 V, ideally 5 V   **12-3.** 200 mA, 20 V   **12-5.** 0.5%   **12-7.** 5 MHz
12-9. 5%, 50 kHz   **12-11.** 50, 5 kΩ **12-13.** 5 kΩ, 25   **12-15.** 200 mW   **12-17.** 4 mW, 96.2%   **12-19.** 16 W when the entire load line is used, 98.5%   **12-21.** 500 kHz, 1 MHz, 1.5 MHz   **12-23.** Approximately 4 MHz, or more closely 3.98 MHz   **12-25.** 100 kΩ

Chapter 13

13-1. 1.5(10^{11})   **13-3.** 2 V; 4.5 to 25 V
13-5. 0.009$(1 + V_{GS}/3)^2$; 2.25 mA   **13-7.** About 6.12 mA   **13-9.** 0.012$(1 + V_{GS}/5)^2$
13-11. Roughly -1.5 V for V_{GS}; about 6 V for V_P **13-13.** 6 V, 12 mA, 8.33 mA   **13-15.** 5(10^{11}) Ω **13-17.** 0.25 mA, 1 mA   **13-19.** 7.5(10^7) Ω **13-21.** 3000 μS   **13-23.** Approximately 4.88 mS

Chapter 14

14-1. 100 Ω **14-3.** 0.04 V   **14-5.** 1.9 mA, -1.56 V, 13.5 V   **14-7.** 0.26, 57.8 Ω
14-9. Approximately 2.13 mA   **14-11.** 3.46 mS, 8 mA, 1.14, 4 mA   **14-13.** 0.1 mV, 20 mV
14-15. (a) 10 V, (b) 10 V, (c) 1 V, (d) Because its gate voltage is only 1 V, which is less than the threshold voltage   **14-17.** 9.47, 8.87
14-19. 14.9   **14-21.** 3, 6
14-23. Approximately 500, exactly 488
14-25. 6.68 mA, 1.2 V, 0.383   **14-27.** 159 Ω, 13.5   **14-29.** 1500 μS, 750 μS, 150 μS
14-31. 5.92   **14-33.** (a) 2.04 V, (b) 2.04 V, (c) 0.5 mA, (d) 7500 MΩ

Chapter 15

15-1. 0, 10, 20, 30, 40, and 50 dB **15-3.** 12 dB, 42 dB, 62 dB **15-5.** 34.7712 dB, or three significant digits, 34.8 dB **15-7.** 26 dB, 400 **15-9.** 80 W **15-11.** 31 dB **15-13.** 6 dB, 26 dB, 46 dB, 66 dB **15-15.** 74 dB **15-17.** 40 dB, 34 dB, −6 dB; 68 dB; 2500 **15-19.** (a) 17 dB (approximate), (b) 46 dB, (c) 63 dB **15-21.** Neglecting effect of r_e' the total bel voltage gain is approximately 53 dB **15-23.** 1 Ω, 50 kΩ **15-25.** 50 μF, 500 pF **15-27.** 25, approximately 100 pF and 4 pF **15-29.** 338, 1.45 kΩ **15-31.** $1(10^{-4})$, $2(10^{-5})$ S **15-33.** 300 Ω **15-35.** With Table 15-3, 1, 175, 38.5 kΩ, and 34.3 Ω **15-37.** 69, 172, 4.98 kΩ, and 138 kΩ

Chapter 16

16-1. 15.9 MHz, 159 kHz, 1.59 kHz **16-3.** 79.6 kHz **16-5.** 16, 2.65 MHz, 2.65 MHz, −225° **16-7.** (a) 5, 3.18 MHz, (b) 10, 1.59 MHz, (c) 20, 796 kHz, (d) Inversely **16-9.** 7.96 MHz, drops to 3.98 MHz **16-11.** (a) 40, 15.9 MHz, (b) 80, 7.96 MHz, (c) 400, 1.59 MHz, (d) 10 kΩ, 1 kΩ **16-13.** 80 dB, 100 Hz, 40 dB, −90° **16-15.** −45°, −135°, −180° **16-17.** 2 V, 1.8 V, 70 kHz **16-19.** 35 μs, 3.5 μs **16-21.** 0.8 pF, 7.2 pF **16-23.** Gate, 194 kHz; drain, 5.89 MHz **16-25.** 1270 pF **16-27.** 11.5 MHz, 7.8 MHz **16-29.** 175 Hz

Chapter 17

17-1. 1.5 mA **17-3.** 1.46 mA **17-5.** 10 V (ideal); 10.3 V (second approximation) **17-7.** 200 mV peak **17-9.** 240 mV peak, 8 V **17-11.** 400; 400 mV **17-13.** 100,000; 100 dB **17-15.** 1 mA **17-17.** 10 V **17-19.** 10 V **17-21.** 45 nA, 10 nA **17-23.** 1.5 V/μs **17-25.** 3.33 V/μs **17-27.** 2.36 V/μs **17-29.** (a) 800 kHz, (b) 2 MHz, (c) 4 MHz

Chapter 18

18-1. 10^6, approximately 0.02, and 50 **18-3.** 100 mV, 2000 **18-5.** 150 **18-7.** 2000 MΩ **18-9.** 0.25 Ω, 0.75 Ω **18-11.** 100 mV, 1 V, 10 V **18-13.** 0.01 Ω, 0.003 Ω **18-15.** (a) 0.1 mA, (b) 100 mV, (c) 20 mV peak-to-peak **18-17.** 20,000 MΩ **18-19.** 0.5 mA, 0.667 mA, 1 mA **18-21.** 19.9 kΩ (exact), 20 kΩ (practical) **18-23.** 10 kHz, 2 MHz, 5 MHz **18-25.** 5 MHz, 2 MHz, 15 MHz **18-27.** 50 Hz, 2 MHz **18-29.** 40 kHz **18-31.** 333 kHz

Chapter 19

19-1. 2.6 kHz **19-3.** 37.9 kHz, 796 kHz **19-5.** Change the resistor in series with the lamp from 1 kΩ to 700 Ω **19-7.** 15.9 kHz **19-9.** 2.36 MHz, 0.1, 10 **19-11.** Increases 1% **19-13.** 8.95 V **19-15.** 1.126 MHz, 22.52 MHz **19-17.** 10 kHz **19-19.** 5.8 V **19-21.** Approximately 91 Hz and 1 kHz

Chapter 20

20-1. 6.1 V **20-3.** 5.55 mA, 6.3 mA **20-5.** 2.47 W **20-7.** 10.6 to 17.1 V **20-9.** 17.6 V **20-11.** 1.27 W **20-13.** 0.267% **20-15.** 0.211% **20-17.** 0.07% or 12.6 mV **20-19.** 5 to 11.9 V **20-21.** 341 mA **20-23.** 2.64 kΩ

Chapter 21

21-1. 60 μV **21-3.** 0.24 μs **21-5.** Approximately 15°C **21-7.** 68.8 μV **21-9.** −102 μV, 10 kΩ, 1.5 nA **21-11.** 7 μV, 0.4 mA, 0.127 mA **21-13.** 0 to 600 mV, 20 kHz **21-15.** 250,000 μF, 2500 s **21-17.** 10 V/ms **21-19.** 125 Hz, 83.3 Hz **21-21.** 0 to 100, 39.8 kHz to 159 kHz

Chapter 22

22-1. 40, 80, and 120 kHz **22-3.** 500 Hz, 12.5 kHz **22-5.** 19.1, 6.37, and 3.82 V **22-7.** The spectrum contains a dc component of 5 V, and harmonics with peak values of 3.18, 1.59, 1.06, and 0.796 V; the first four harmonic frequencies are 50, 100, 150, and 200 kHz **22-9.** 25 V **22-11.** Yes; 0 V **22-13.** 31.8 V, 0.106 V

22-15. 5000, 5000, no   **22-17.** 8%, 4%, and 8.94%   **22-19.** Of the given choices, 10 mV is the maximum input because the voltage gain still equals 100; somewhere beyond 10 mV but less than 100 mV, the voltage gain drops to less than 100 and nonlinear distortion appears   **22-21.** If no harmonics beyond the second are present, the peak value of the second harmonic voltage is 0.5 V; if harmonics beyond the second exist, 0.5 V is only rough estimate for the second harmonic peak voltage   **22-23.** 122% **22-25.** 0.0006%   **22-27.** 0.3 mV

Chapter 23

23-1. 65 V and −45 V   **23-3.** 0.75 V, 1 kHz **23-5.** 0.2 V, 10 V, 0.2 V   **23-7.** Inputs: 1, 4, 9 kHz; second harmonics: 2, 8, 18 kHz; sums: 5, 10, 13 kHz; differences: 3, 5, 8 kHz   **23-9.** $f_x =$ 9 kHz, $f_y = 7$ kHz   **23-11.** 32, 4096, 5120 Hz **23-13.** $f_x = 6$ kHz, $f_y = 4$ kHz   **23-15.** 995 to 2055 kHz; roughly 4 to 1 or exactly 4.27 to 1 **23-17.** 0.5, −3 dB   **23-19.** 1.25 μW **23-21.** 4.05 μV, 0.405 μV, 0.0405 μV

Chapter 24

24-1. 26, 15.7, 36.5; 260 mV, 157 mV, 365 mV   **24-3.** 75, 0.75, and 0.375 m **24-5.** 0.2   **24-7.** 1 MHz ± 20 Hz, 1 MHz ± 20 kHz   **24-9.** 2.4 V, 0.6 V   **24-11.** 1 MHz ± 20 kHz   **24-13.** 990 to 1010 kHz, 20 kHz **24-15.** 2000 kHz carrier plus either a lower sideband extending down to 1985 kHz or an upper sideband extending to 2015 kHz   **24-17.** 53.1 kHz   **24-19.** 995 kHz, 2055 kHz **24-21.** 33.3%   **24-23.** 64,000, 8000, 64 **24-25.** 30 kHz, 7.5 V   **24-27.** 10

Chapter 25

25-1. 18.75   **25-3.** 60   **25-5.** 150 V, 2.6 V

index